$118.25

2026 NATIONAL PLUMBING & HVAC ESTIMATOR

Edited by James A. Thomson

34th Edition

Craftsman Book Company
6058 Corte del Cedro, Carlsbad, CA 92011

Acknowledgments

The sample "Standard Form Subcontract" and "Subcontract Change Order" forms used in the final section of this book are reprinted with the permission of the publisher, the Associated General Contractors of America (National Office), 1957 E Street NW, Washington, District of Columbia 20006.

ISBN 978-1-57218-412-1
Published October 2025 for the year 2026.

Contents

How to Use This Book

This 2026 National Plumbing & HVAC Estimator is a guide to estimating labor and material costs for plumbing, heating, ventilating and air conditioning systems in residential, commercial and industrial buildings.

This manual is also available by subscription on the Web as part of *National Estimator Cloud*. For only a few dollars a month, you get all ten of Craftsman's 2026 construction cost estimating guides. Each has about 400 pages of current labor and material costs for construction – all neatly organized and indexed. Use these costs to build estimates, bids and invoices for nearly any type of project.

National Estimator Cloud:

- Prints estimates, bids and invoices as Word, Excel or PDF documents.
- Supports progress billing. National Estimator remembers what work has been billed and what hasn't.
- Runs as a secure app on the Web so you can write estimates anywhere you have a Web connection.
- Exports invoices to QuickBooks, either desktop or online.
- Bids and invoices can show as much or as little detail as you want.
- Material costs are updated regularly as prices change.
- Costs only a few dollars a month. Cancel any time you want.

Costs in This Manual will apply within a few percent on a wide variety of projects. Using the information given on the pages that follow will explain how to use these costs and suggest procedures to follow when compiling estimates. Reading the remainder of this section will help you produce more reliable estimates for plumbing and HVAC work.

Manhour Estimates in This Book will be accurate for some jobs and inaccurate for others. No manhour estimate fits all jobs because every construction project is unique. Expect installation times to vary widely from job to job, from crew to crew, and even for the same crew from day to day.

There's no way to eliminate all errors when making manhour estimates. But you can minimize the risk of a major error by:

1. Understanding what's included in the manhour estimates in this book, and
2. Adjusting the manhour estimates in this book for unusual job conditions.

The Craft@Hrs Column. Manhour estimates in this book are listed in the column headed *Craft@Hrs*. For example, on page 19 you'll see an estimate for installing a 6 gallon hot water heater. In the *Craft@Hrs* column opposite 6 gallon you'll see:

P1@.500

To the left of the @ symbol you see an abbreviation for the recommended work crew.

Page 7 shows the wage rates and craft codes used in this book.

To the right of the @ symbol you see a number. The number is the estimated manhours (not crew hours) required to install each unit of material listed. In the case of a 6 gallon hot water heater, P1@.500 means that .500 manhours are required to install 1 hot water heater.

Costs in the Labor $ Column are based on manhour estimates in the *Craft@Hrs* column. Multiply the manhour estimate by the assumed hourly labor cost to find the installation cost in the *Labor $* column. For example, .500 manhours times $44.45 (the average wage for crew P1) is $22.225 (rounded to the nearest dime, $22.23).

Manhour Estimates include all productive labor normally associated with installing the materials described. These estimates assume normal conditions: experienced craftsmen working on reasonably well planned and managed new construction with fair to good productivity. Labor estimates also assume that materials are standard grade, appropriate tools are on hand, work done by other crafts is adequate, layout and installation are relatively uncomplicated, and working conditions don't slow progress.

All manhour estimates include tasks such as:

- Unloading and storing construction materials, tools and equipment on site.
- Working no more than two floors above or below ground level.

- Working no more than 10 feet above an uncluttered floor.
- Normal time lost due to work breaks.
- Moving tools and equipment from a storage area or truck not more than 200 feet from the work area.
- Returning tools and equipment to the storage area or truck at the end of the day.
- Planning and discussing the work to be performed.
- Normal handling, measuring, cutting and fitting.
- Regular cleanup of construction debris.
- Infrequent correction or repairs required because of faulty installation.

If the work you're estimating won't be done under these conditions, you need to apply a correction factor to adjust the manhour estimates in this book to fit your job.

Applying Correction Factors. Analyze your job carefully to determine whether a labor correction factor is needed. Failure to consider job conditions is probably the most common reason for inaccurate estimates.

Use one or more of the recommended correction factors in Table 1 to adjust for unusual job conditions. To make the adjustment, multiply the manhour estimate by the appropriate conversion factor. On some jobs, several correction factors may be needed. A correction factor less than 1.00 means that favorable working conditions will reduce the manhours required.

Condition	Correction Factor
Work in large open areas, no partitions	.85
Prefabrication under ideal conditions, bench work	.90
Large quantities of repetitive work	.90
Very capable tradesmen	.95
Work 300' from storage area	1.03
Work 400' from storage area	1.05
Work 500' from storage area	1.07
Work on 3rd through 5th floors	1.05
Work on 6th through 9th floors	1.10
Work on 10th through 13th floors	1.15
Work on 14th through 17th floors	1.20
Work on 18th through 21st floors	1.25
Work over 21 floors	1.35
Work in cramped shafts	1.30
Work in commercial kitchens	1.10
Work above a sloped floor	1.25
Work in attic space	1.50
Work in crawl space	1.20
Work in a congested equipment room	1.20
Work 15' above floor level	1.10
Work 20' above floor level	1.20
Work 25' above floor level	1.30
Work 30' above floor level	1.40
Work 35' to 40' above floor level	1.50

Table 1 Recommended Correction Factors

Supervision Expense to the installing contractor is not included in the labor cost. The cost of supervision and non-productive labor varies widely from job to job. Calculate the cost of supervision and non-productive labor and add this to the estimate.

Hourly Labor Costs also vary from job to job. This book assumes an average manhour labor cost of $51.49 for plumbers and $49.64 for sheet metal workers. If these hourly labor costs are not accurate for your jobs, adjust the labor costs up or down by an appropriate percentage. Instructions on the next page explain how to make these adjustments. If you're using National Estimator Cloud, it's easy to set your own wage rates.

Hourly labor costs in this book include the basic wage, fringe benefits, the employer's contribution to welfare, pension, vacation and apprentice funds, and all tax and insurance charges based on wages. Table 2 at the top of the next page shows how hourly labor costs in this book were calculated. It's important that you understand what's included in the figures in each of the six columns in Table 2. Here's an explanation:

Column 1, the base wage per hour, is the craftsman's hourly wage. These figures are representative of what many contractors are paying plumbers, sheet metal workers and helpers in 2026.

Column 2, taxable fringe benefits, includes vacation pay, sick leave and other taxable benefits. These fringe benefits average about 6.37% of the base wage for many plumbing and HVAC contractors. This benefit is in addition to the base wage.

Column 3, insurance and employer-paid taxes in percent, shows the insurance and tax rate for the craft workers. The cost of insurance in this column includes workers' compensation and contractor's casualty and liability coverage. Insurance rates vary

Column Number	1	2	3	4	5	6
Craft	Base wage per hour	Taxable fringe benefits (at 6.37% of base wage)	Insurance and employer taxes (%)	Insurance and employer taxes ($)	Non-taxable fringe benefits (at 5.63% of base wage)	Total hourly cost used in this book
Laborer	25.94	1.65	30.27%	8.35	1.46	37.40
Plumber	37.64	2.40	23.29%	9.33	2.12	51.49
Sheet Metal Worker	35.94	2.29	24.57%	9.39	2.02	49.64
Operating Engineer	37.21	2.37	23.79%	9.42	2.09	51.09
Sprinkler Fitter	36.99	2.36	23.88%	9.40	2.08	50.83
Electrician	36.52	2.33	19.14%	7.44	2.06	48.35
Cement Mason	32.20	2.05	22.18%	7.60	1.81	43.66

Craft Code	Crew Composition	Average Hourly Cost per Manhour
ER	4 building plumbers, 2 building laborers, 1 operating engineer	47.41
SN	4 building sheet metal workers, 2 building laborers, 1 operating engineer	46.35
P1	1 building plumber and 1 building laborer	44.45
ST	1 sprinkler fitter	50.83
SK	4 sprinkler fitters, 2 building laborers, 1 operating engineer	47.03
SL	1 sprinkler fitter and 1 laborer	44.12
S2	1 building sheet metal worker, 1 building laborer	43.52
BE	1 electrician	48.35
CF	1 cement mason	43.66
SW	1 sheet metal worker	49.64

Table 2 Labor Costs Used in This Book

widely from state to state and depend on a contractor's loss experience. Note that taxes and insurance increase the hourly labor cost by approximately 30%. There is no legal way to avoid these costs.

Column 4, insurance and employer taxes in dollars, shows the hourly cost of taxes and insurance. Insurance and taxes are paid on the costs in both columns 1 and 2.

Column 5, non-taxable fringe benefits, includes employer paid non-taxable benefits such as medical coverage and tax-deferred pension and profit sharing plans. These fringe benefits average 5.63% of the base wage for many plumbing and HVAC contractors. The employer pays no taxes or insurance on these benefits.

Column 6, the total hourly cost in dollars, is the sum of columns 1, 2, 4, and 5. The labor costs in Column 6 were used to compute costs in the Labor $ column of this book.

Adjusting Costs in the Labor $ Column. The hourly labor costs used in this book may apply within a few percent on many of your jobs. But wage rates may be much higher or lower in some areas. If the hourly costs shown in Column 6 of Table 2 are not accurate for your work, adjust labor costs to fit your jobs.

For example, suppose your hourly labor costs are as follows:

Plumber	$29.00
Laborer	$19.00
Total hourly crew cost	$48.00

Your average cost per manhour would be $24.00 ($48.00 per crew hour divided by 2 because this is a crew of two).

A labor cost of $24.00 is about 54% of the $44.45 labor cost used for crew P1. Multiply costs in the Labor $ column by .539 to find your estimated cost.

Adjusting the labor costs in this book will make your estimates much more accurate. Making adjustments to labor costs is both quick and easy if you use National Estimator Cloud.

Equipment Cost will vary according to need and application. It is typically $33.20 per day for a 2-ton chain hoist.

Material Costs in this manual are intended to reflect what medium- to low-volume contractors will be paying in 2026 after applying normal discounts. These costs include charges for delivery to within 25 to 30 miles of the supplier.

Overhead and Profit for the installing contractor are not included in the costs in this manual unless specifically identified in the text. Markup can vary widely with local economic conditions, competition and the installing contractor's operating expenses. Add the markup that's appropriate for your company, the job and the competitive environment.

How Accurate Are These Figures? As accurate as possible considering that the editors don't know your material suppliers, haven't seen the plans or specifications, don't know what building code applies or where the job is, had to project material costs at least six months into the future, and had no record of how much work the crew that will be assigned to the job can handle.

You wouldn't bid a job under those conditions. And I don't claim that all plumbing and HVAC work is done at these prices.

Estimating Is an Art, not a science. There is no one price that applies on all jobs. On many jobs the range between high and low bid will be 10% or more. There's room for legitimate disagreement on what the correct costs are, even when complete plans and specifications are available, the date and site are established, and labor and material costs are identical for all bidders.

No estimate fits all jobs. Good estimates are custom made for a particular project and a single contractor through judgment, analysis and experience. This book is not intended as a substitute for judgment, analysis and sound estimating practice. It's an aid in developing an informed opinion of cost, not an answer book.

Additional Costs to Consider

Here's a checklist of additional costs to consider before submitting any bid.

1. Sales taxes
2. Mobilization costs
3. Payment and performance bond costs
4. Permits and fees
5. Storage container rental costs
6. Utility costs
7. Tool costs
8. Callback costs during warranty period
9. Demobilization costs

Exclusions and Clarifications

Neither the job specifications nor the contract may identify exactly what work should be included in the plumbing and HVAC bid. Obviously, you have to identify what work is included in the job.

The most efficient way to define the scope of the work is to prepare a list of tasks not normally performed by your company and attach that list to each bid submitted. Here's a good list of work that should be excluded from your bid.

Your Bid Should Exclude

Final cleaning of plumbing fixtures

Backings for plumbing fixtures

Toilet room accessories

Electrical work, including motor starters

Electrical wiring and conduit over 100 volts

Temporary utilities

Painting, priming and surface preparation

Structural cutting, patching or repairing

Fire protection and landscape sprinklers

Equipment supports

Surveying and layout of control lines

Removal or stockpiling of excess soil

Concrete work, including forming and rebar

Setting of equipment furnished by others

Equipment, unless shown, and personnel hoisting

Wall and floor blockouts

Pitch pockets

The costs of performance or payment bonds

Site utilities

Asbestos removal or disposal

Contaminated soil removal or disposal

Major increases in copper material prices

Fire dampers not shown on the plans

Your Bid Should Include

Trash sweep-up only. Others haul it away

Site utilities from building to property line only

Piping to 5 feet outside the building only

Plumbing & HVAC permits for your work only

Beware of Price Changes

There's no way to be sure what prices will be in three to six months. All labor, equipment, material and subcontract prices in a bid should be based on costs anticipated when the project is expected to be built, not when the estimate is compiled. That presents a problem. Except for the installation of underground utilities, most plumbing and HVAC work is done six months to a year after the bid is submitted. When possible, get price protection in writing from your suppliers and subcontractors. If your suppliers and subs won't guarantee prices, include an escalation allowance in your bid to cover anticipated price increases.

Material Pricing Conditions

All equipment and material prices quoted by your vendors will be conditional. They usually don't include sales tax and are subject to specific payment and shipping terms. Every estimator should understand the meaning of common shipping terms. They define who pays the freight and who has responsibility for processing freight-damage claims. Here's a summary of important conditions you should understand.

F.O.B. Factory (Free On Board at the Factory): Title passes to the buyer when the goods are delivered by the seller to the freight carrier. The buyer pays the freight and is responsible for freight-damage claims.

F.O.B. Factory F.F.A. (Free On Board at the Factory, Full Freight Allowed): The title passes to the buyer when the goods are delivered by the seller to the freight carrier. The seller pays the freight charges, but the buyer is responsible for freight-damage claims.

F.O.B. (city of destination) (Free On Board to your city): The title passes to the buyer when the goods are delivered by the seller to the freight terminal in the city, or nearest city, of destination. The seller pays the freight and is responsible for freight-damage claims to the terminal. The buyer pays the freight charge and is responsible for freight-damage claims from the terminal to the final destination.

F.O.B. Job Site (Free On Board at job site, or contractor's shop): The title passes to the buyer when the goods are delivered to the job site (or shop). The seller pays the freight and is responsible for freight-damage claims.

F.A.S. Port [of a specific city] (Free Alongside Ship at the nearest port): The title passes to the buyer when goods are delivered to the ship dock or port terminal. The seller pays the freight and is responsible for freight-damage claims to the ship dock or port terminal only. The buyer pays the freight and is responsible for freight-damage claims from the ship dock or port terminal to the designated delivery point.

Obviously, it's to your advantage to instruct all vendors to quote costs F.O.B. the job site or your shop.

Reducing Costs

Most construction specifications allow the use of alternative equipment and materials. It's the estimator's responsibility to select the most cost-effective products. Research and compare your costs before making any decisions. Avoid selecting any material or equipment simply because that's what you've always done.

Don't recommend plastic products such as ABS, PVC, or polypropylene pipe or corrugated flexible ducts until you've checked local code requirements. Most building codes prohibit use of these materials inside public buildings such as schools, care centers and hospitals.

It's wise to select 100% factory-packaged equipment. Beware of equipment labeled "Some assembly required." Field labor costs for mounting loose coils, motors and similar equipment are very high.

Value Engineering

Let's suppose you've submitted a combined plumbing and HVAC bid for $233,000. Your cutthroat competitor put in a bid at $4,000 less, $229,000. Obviously there's no way you're going to get the job. Right?

Not so fast! Maybe value engineering can help you win that contract — while fattening your profit margin.

Suppose the proposal you submitted had two parts. Part I is the bid for $233,000, based entirely on job plans and specs, just the way they were written. But appended to your proposal is Part II, a list of suggestions for saving money without sacrificing any of the capacity or quality designed into the system. Here's an example of what might be in Part II:

1. Deduct for providing pipe hanger spacings per UPC in lieu of specified spacings: $1,750.00
2. Deduct for reducing heating hot water pipe sizes by using 40 degrees F Delta T in lieu of specified 20 degrees F Delta T: $4,600.00
3. Deduct for providing pressure/temperature taps at air handling units, pumps and chillers in lieu of specified thermometers and pressure gauges: $875.00

4. Deduct for eliminating water treatment in closed piping systems: $1,800.00

5. Deduct for piping chilled and heating hot water pumps in parallel in lieu of providing 100% stand-by pumps: $2,900.00

Total deductions: $11,925.00

Adopting these suggestions would make you low bidder by nearly $8,000. A saving like that will be tempting to most owners, especially if the owner understands that your suggestions result in a system that is every bit as good and maybe better than the system as originally designed.

You're not offering to undercut the competition. Far from it. You're using knowledge and experience to create better value for the owner. That's called value engineering and it's likely to win the respect of nearly all cost-conscious owners.

Notice that reducing costs is only part of what value engineering is all about. You don't cut costs at the expense of system quality, integrity, capacity or performance.

Don't waste your time, and your client's, by offering to substitute cheaper or lower-quality fixtures or equipment. Any cutthroat contractor with a price list can do that. Recommend the use of inferior materials and you'll be associated with the inferior goods you promote. Some owners consider even the suggestion to be insulting.

The recommendations you make (like most of those in the example) will require design changes. You can expect to be examined (or even challenged) on these points. Be ready to explain and defend each of your suggestions. Convince the client (or the design engineer) that your ideas are based on sound engineering principles and you're well on the way to winning the owner's confidence and the contract.

Now, let's go back to the list and see how we might justify the five value engineering recommendations.

1. **Pipe Hanger Spacing.** The pipe hanger spacings recommended in the Uniform Plumbing Code (UPC) are calculated by experienced, professional structural engineers. The safety factors used in these calculations are very conservative. They've been widely used for many years and have proved to be more than adequate. There's no need for more hangers than the UPC requires.

2. **Changing HHW Delta T.** In hydronic heating systems, heat measured in Btus is pumped to terminal units. The proposed change of the Delta T, from 20 degrees F to 40 degrees F, has no effect whatsoever on how many Btus the system delivers. You're not changing anything but the volume of water being pumped. At lower volume levels, the size of the pump, the pipe and the pipe insulation can all be reduced. Not one of these changes will affect the system's ability to transmit heat. Furthermore, operating costs will also drop, since less pump horsepower will be needed to run the smaller pump.

3. **Thermometers/Pressure Gauges.** Thermometers and pressure gauges installed on or near vibrating machinery have a very short life expectancy. Gauges quickly lose accuracy under harsh conditions. Readings will become less and less reliable. That's potentially dangerous. You can avoid this problem by using insertion-type pressure/temperature taps instead. Store these sensitive gauges in a desk drawer or a tool crib when not in use. Safely stored, they're protected from damage. They'll give accurate readings longer and won't need to be replaced as often. And they're simple to use. Just insert a gauge in one of the conveniently located taps. Make the reading, then remove the gauge and put it away.

4. **Water Treatment.** ITT Bell & Gossett has done studies on corrosion in closed hydronic systems that have a make-up water rate of no more than 5% per year. These studies show that corrosion virtually stops when entrained air is either removed or depleted. No water treatment is needed in this closed system.

5. **100% Standby Pumps.** Two pumps piped and operated in parallel are more economical. Even if one pump fails, the other pump can maintain delivery at 75 to 80% of the designed flow rate. That's usually adequate for emergency operation.

These cost-saving ideas are small, but could tip the balance in your favor. I hope they demonstrate the potential that value engineering has when bidding jobs. Any time you're compiling an estimate, keep an eye out for ways to save money or reduce the owner's cost. Jot a note to yourself about each potential saving you identify. Before submitting the bid, make a list of your alternate suggestions. Maybe best of all, markup on your value engineering suggestions can be higher than your normal markup. If value engineering can cut costs by $10,000, maybe as much as $4,000 of that should end up in your pocket!

Value Engineering: Surplus Materials

Value engineering doesn't begin and end with job plans and specs. Value engineering means getting the most value at the least cost, no matter whether it's value to the owner or value to the contractor. Smart mechanical contractors learn to build extra value into their jobs by controlling shrinkage of materials. Nearly every significant plumbing and HVAC job ends with at least some surplus material on hand. Material left over when the job is done tends to be discarded as waste or hauled off the job in the back of a truck that doesn't have your company name on the door. And why not? It's surplus — not needed. The owner didn't need it. So now it's up for grabs.

Not quite. Let's consider who actually owns that surplus material. When your company has been paid, every piece of material your crew installed belongs to the building owner. But what about those fittings, hangers and valves delivered to the job site but never actually used? Almost certainly, those materials were included in your bid. So aren't they the property of the owner? Not in my opinion. The owner contracted for a mechanical system and (presumably) has one. Unless it's a cost-plus job or a labor-only job, the owner didn't buy materials delivered to the job site. The owner bought a mechanical system and has one — completely separate and apart from any surplus materials. In my mind, the property owner has no more claim to left-over materials than the same owner would have claim to labor hours not expended or equipment not used on the same job.

Unless there's some provision in your contract to the contrary, surplus material belongs to the installing contractor. But your right to that material and the chance of actually getting it back to your shop are two very different propositions. I see recovery of surplus material as a training issue. As a matter of company policy, make it clear to your crews that surplus material belongs to your company. The supervisor on every job should be accountable for recovery of excess material. Every significant job will have at least some surplus. Accounting for that surplus should be part of your routine close-out procedure. Fortunately, it's not difficult. I'll explain.

Control of surplus materials begins with a good checklist, or form. I recommend the Materials, Equipment and Tool form, "MET" for short. A blank MET form appears following this section. Your MET should show both what's delivered to the job site (material, equipment and tools) and surplus "drops" returned to your shop at project close-out. A MET ensures that the estimator, the shop inventory manager and your field supervisor are on the same page. Your MET establishes accountability. Nothing falls through the cracks. Job input equals job output plus returns. Everything delivered to the job and not expended should be returned to your shop.

Here's how it works:

1. Based on the estimate that won you the job, the items needed are purchased for the job and staged for delivery to the job site.
2. As materials, equipment and tools are delivered to the job site, your supervisor completes the first three columns of the MET form: Description, Quantity and Date.
3. As work is completed, the same supervisor completes the four columns under Returned to Inventory: Quantity Returned, Date, Status Code and Value. The status code will be either "RS" (Returned and Salvaged) or "RN" (Returned New).
4. Back at your shop, both RS and RN materials should be restored to inventory.
5. If your company has an inventory manager, have that manager assign the return value to each item returned. If you're using QuickBooks Pro, the "Adjust Inventory" feature can handle this task quite easily. Add two new categories under "Inventory Stock on Hand by Vendor." The first new category is Returned Salvage. The second is Returned New. Be sure the value of RS materials includes the cost of any reconditioning done to restore salvaged materials (such as pumps and boilers) to serviceable condition.
6. Comparing MET deployed to the job site with MET returned to inventory yields MET actually used on the job. That's a very important number to every plumbing and HVAC estimator. Be sure actual usage gets entered on the Project Summary form.
7. When the take-off on your next estimate is complete, compare that materials list with a summary of RS and RN materials on hand from prior jobs.
8. Evaluate which returned materials can be redeployed on the new job.
9. It's a management decision to either (1) charge the new job for the cost of RS and RN materials already on hand, or (2) consider materials on hand as "free" and a competitive advantage in winning the new bid. Either way, RN and RS materials are an asset to your company.

Plumbing and HVAC materials are expensive. Every mechanical contractor has an interest in MET tracking. Everyone in your company should be aware of the need for good materials management. Used correctly, the MET form in this book can help engineer more value into your jobs.

Maximizing the Value of Old Estimates

There should be two profits in every job. The first is money in the bank — a return on time and expenses. The second is what you learn from the job — primarily by comparing the estimate you made with what turns out to be your actual cost. On some jobs, the value of lessons learned may outweigh net revenue.

Every plumbing and HVAC contractor has marginal jobs. That's normal. What *shouldn't* be normal is repeating mistakes. The best way to avoid trouble in your future is to keep track of your past. Keeping old estimates available for reference can help prevent errors on new estimates.

As your file of completed estimates grows, organization becomes more important. You need an easy way to find similar projects with the same components and comparable scope of work. If your estimating file is in QuickBooks Pro, searching by keyword may be enough. Otherwise, I recommend creating a short summary for each completed job, and an index that references all summaries available for comparison. You'll find a blank Project Summary form at the end of this section. To make reference easier, create an index by type of job and equipment used. You may choose to use an alphabetical index based on client name or project ID.

How to complete the Project Summary form is obvious. The many ways to use this form may not be so obvious, so here are a few pointers.

1. Use your index of Project Summary forms to find completed jobs most similar to the job you're bidding. Believe it or not, Project Summary forms with the widest margin of error will be most useful. Ask yourself: Who worked on those projects? Who was the field superintendent? Who were the vendors? Did the errors result from poor estimating or the poor performance of vendors, supervisors or crews? The most common estimating errors occur when (a) inspecting the job site, (b) examining the plans or (c) reading the specifications. What did you miss and why? Look for pitfalls to avoid in the job now being estimated. Identify the biggest two or three mistakes made when bidding that job. Make a notation about each on the Project Summary form.
2. Now look at your bid for the current job. Which mistakes made on a prior job might you expect on this job? Concentrate on the big three oversights to avoid: Inspecting the job site; examining the plans; and reading the specifications.
3. Unless there's a major error in take-off, your estimate of material costs should be within about 5 percent of the actual costs of materials. However, it's common for labor cost estimates to vary 20 percent or more from actual labor costs. This is precisely where data from old jobs comes in handy. If your Project Summary files show that some project types are consistent money-losers, either shift your company's focus to another class of work, factor more contingency into your bids, or find some way to wring inefficiencies out of the labor component. Poor staging, delivery and retrieval procedures drag down labor productivity on any job.
4. Use your file of Project Summary forms to spot any common thread that runs through either money-making jobs or money-losing jobs. For example, if the names of certain subcontractors or vendors are prominent on low-margin jobs, maybe there's a relationship between your profit margin and choice of subs and suppliers. Even the best and most reliable vendors can become complacent if not challenged occasionally.
5. Project Summary forms should note changes and extras identified after the contract was signed — both for which your company was paid and changes done without additional compensation. Projects with changes and extras that exceed about 4 percent of the contract price deserve special scrutiny. Jobs with changes beyond about 4 percent aren't good for business, at least in my opinion. Nearly all changes have a negative impact on your job schedule and require a disproportionate investment of management resources. Too many changes can antagonize the owner and design staff, even if they were responsible for the altered plans. You may know of a mechanical contractor with a reputation for capitalizing on change orders. But I've rarely seen a job plagued with changes that turned into a money-maker for anyone — except the attorneys. Your file of Project Summary forms will show job types that carry change order risk. Before finalizing and submitting any bid, consider whether the job will get mired in disputes over changes and extras. If similar jobs have ended on the courthouse steps, factor that risk into your estimate.

Utility of a Project Summary forms file is limited only by your ingenuity. The important point is to keep and organize the source of your second profit available on every job. What you learn can be more valuable than what you earn.

The Estimating Procedure

Every plumbing and HVAC estimator works under deadline pressure. You'll seldom have the luxury of spending as much time as you would like on an estimate. Estimators who aren't organized waste valuable time and tend to make careless errors. Try to be well-organized and consistent in your approach to estimating. For most projects, I recommend that you follow the procedures listed below and in the order listed:

1. Get a second set of project drawings and specifications for use by your suppliers and subcontractors. Remember that your subs and suppliers need access to the plans and specs and time to prepare their quotes.
2. Study the plans and specs carefully. Highlight important items. Make a list of specific tasks that require labor unit correction factors. The estimate is never complete until you're totally familiar with the project and the applicable construction codes.
3. Get the general contractor or owner to identify the proposed construction schedule and subcontractor lay-down (storage) area. Work schedule and site conditions always affect your costs.
4. Contact all potential suppliers and subcontractors as early as possible. Set a time when each can come to your office to make their take-offs from the spare set of contract documents.

When this important preliminary work is done, or in progress, it's time to begin your detailed take-off.

Guidelines for Good Estimating

You can compile estimates on a legal pad, a printed estimating form or on a computer. Regardless of the method, these guidelines will apply:

List Each Cost Separately on your take-off sheet. Don't combine system estimates, even if the materials are the same type. A combined system estimate may have to be completely redone if materials for one system are changed at a later date. Use the Estimate Detail Sheet on page 16 if you don't already have a good material take-off form.

Use Engineer's Identification Numbers when listing equipment. The word pump without any other description is ambiguous when there are several pumps included in the project.

Don't Forget Labor Adjustment factors if your labor costs are significantly higher or lower than the costs used in this book. See instructions on page 7 for adjusting labor costs.

Use Colored Pencils or highlighters to mark the items you've taken off and listed. Use a different color for each piping or ducting system.

Log Telephone Quotes and other important phone conversations on a telephone quote form. See the sample on page 18.

Project Estimated Costs for labor, material and equipment to the time when the work is expected to be done, not when the job is being estimated.

The only good estimate is a complete estimate. You've probably heard this saying, "He who makes the most mistakes is likely to be low bidder, and live to regret it."

Preparing the Proposal

It's both common courtesy and good business practice to deliver an unpriced copy of your bid or proposal letter to the general contractor three or four days before the bid deadline date. This gives the contractor time to study your proposal and obtain alternate pricing for items you may have excluded. To avoid misunderstandings, make sure your proposals include, as a minimum, the following elements:

1. The complete name and address of the proposed project.
2. Specification title and issue date.
3. A complete listing of drawings and their issue or revision date.
4. A complete list of addenda and their dates of issue.
5. A list of specification section numbers covered by your proposal.
6. A list of exclusions, clarifications and assumptions.

Your final bid can be phoned in or sent by fax, but it should reach the general contractor or owner no more than five or ten minutes before the bid deadline. Prices submitted too early may have to be revised because of last-minute price changes by subcontractors or suppliers.

MET Worksheet

Material, Equipment and Tool Delivery and Surplus Return Record

Project ID ______________________________ Job Location ______________________________

Supervisor ______________________________ Start Date ______________________________

Description of Material, Equipment or Tool Delivered or Returned	Delivered to Job Site		Returned to Inventory			
	Quantity Delivered	Date Delivered	Quantity Returned	Date Returned	Status Code RN or RS	Value at Return

PROJECT SUMMARY

Project ID ______________________ Job Location ______________________

Short description ______________________

Supervisor ______________________

Index ID ______________________ Start Date ______________________

Estimator ______________________ Client ______________________

Major vendors ______________________ Subcontractors ______________________

Sources of cost deviation ______________________

Related Projects by ID Number ______________________

Thumbnail Summary	Labor	Material	Equipment	Subcontract	Deployed RN/RS	Total
Actual cost						
Estimate Over/(Under)						
Full Summary						
Bid amount						
Estimated cost						
Projected profit						
Cost overrun						
Bid profit						
Change orders						
Cost of changes						
Total profit						
Total profit with RN/RS						
Redeployment						

Estimate Detail Sheet

Data carried forward from Take-Off Quantity Survey Sheet(s)

Company/Department ______________________ Estimator ______________________ Date __________

Project ______________________ Checked by ______________________ Date __________

Address ______________________ Notes: ______________________

Job description __________ Estimate # __________ ______________________

CSI Division/Account __________ Estimate due __________ ______________________

Item Description	Quantity	Unit	Crew @ MH/Unit	Manhours Ext.	Materials		Labor		Equipment		Subcontract		Total $
					Unit $	Ext. $	Unit $	Ext. $	Unit $	Ext. $	Unit $	Ext. $	

Totals This Sheet	Manhours	Material $	Labor $	Equipment $	Subcontract $	Total $

Carry totals forward to Estimate Summary Sheet Estimate # __________ Estimate Detail Sheet ______ of ______

Quotation Sheet

Job: ____

Supplier: ____

Salesperson: ______________________ Phone No: ______________

Per Plans/Specs: ____________ Freight: ____________ Terms: ____________

Description	Delivery Time	Price

By: ________________________________

Record of Telephone Conversation

Date:__________ Time:__________ Project: __________

Telecon with:__________

Company: __________ Phone No:__________

Subject: __________

Details of Conversation: __________

By:__________

Electric domestic hot water heater (residential). Set in place only (floor models). Make additional allowances for pipe and electrical connections. (See below)

Description	Craft@Hrs	Unit	Material $	Labor $	Equipment $	Total $
6 gallon						
1.5 KW/110V	P1@.500	Ea	495.00	22.20	—	517.20
10 gallon						
1.5 KW/110V	P1@.500	Ea	554.00	22.20	—	576.20
15 gallon						
1.5 KW/110V	P1@.750	Ea	582.00	33.30	—	615.30
20 gallon						
1.5 KW/110V	P1@.750	Ea	547.00	33.30	—	580.30
30 gallon						
1.5 KW/110V	P1@1.00	Ea	561.00	44.50	—	605.50
40 gallon						
1.5 KW/110V	P1@1.20	Ea	587.00	53.30	—	640.30
50 gallon						
3 KW/110V	P1@1.30	Ea	634.00	57.80	—	691.80
12 gallon						
3 KW/220V	P1@.500	Ea	488.00	22.20	—	510.20
20 gallon						
3 KW/220V	P1@.750	Ea	534.00	33.30	—	567.30
30 gallon						
3 KW/220V	P1@1.00	Ea	609.00	44.50	—	653.50
40 gallon						
3 KW/220V	P1@1.20	Ea	663.00	53.30	—	716.30
50 gallon						
3 KW/220V	P1@1.30	Ea	710.00	57.80	—	767.80

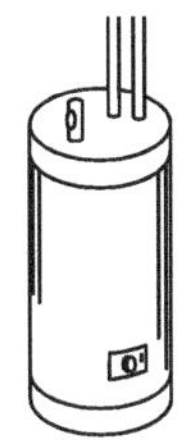

Electric domestic hot water heater (commercial), 208/240 volt. Set in place only. Make additional allowances for pipe and electrical connections. (See below)

Description	Craft@Hrs	Unit	Material $	Labor $	Equipment $	Total $
96 gallon, 12 kw	P1@1.50	Ea	2,750.00	66.70	—	2,816.70
96 gallon, 18 kw	P1@1.50	Ea	3,730.00	66.70	—	3,796.70
96 gallon, 36 kw	P1@1.50	Ea	3,860.00	66.70	—	3,926.70
120 gallon, 18 kw	P1@2.00	Ea	3,960.00	88.90	—	4,048.90
120 gallon, 36 kw	P1@2.00	Ea	4,080.00	88.90	—	4,168.90
120 gallon, 54 kw	P1@2.00	Ea	4,830.00	88.90	—	4,918.90
120 gallon, 63 kw	P1@2.00	Ea	5,210.00	88.90	—	5,298.90

Gas-fired domestic hot water heater (residential). Set in place only, Make additional allowances for pipe and combustion venting connections. (See below)

Description	Craft@Hrs	Unit	Material $	Labor $	Equipment $	Total $
30 gallon	P1@1.00	Ea	565.00	44.50	—	609.50
40 gallon	P1@1.00	Ea	913.00	44.50	—	957.50
50 gallon	P1@1.50	Ea	1,040.00	66.70	—	1,106.70

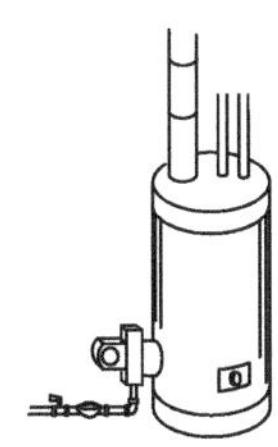

Description	Craft@Hrs	Unit	Material $	Labor $	Equipment $	Total $

Gas-fired domestic hot water heater (commercial), standard efficiency. Set in place only, Make additional allowances for pipe and combustion venting connections. (See below)

50 gal./ 95 gph	P1@2.00	Ea	2,740.00	88.90	—	2,828.90
67 gal./106 gph	P1@2.00	Ea	3,250.00	88.90	—	3,338.90
76 gal./175 gph	P1@2.00	Ea	4,340.00	88.90	—	4,428.90
91 gal./291 gph	P1@2.00	Ea	5,240.00	88.90	—	5,328.90

Gas-fired domestic hot water heater (commercial), energy miser. Set in place only, Make additional allowances for pipe and combustion venting connections. (See below)

50 gal./ 95 gph	P1@2.00	Ea	6,760.00	88.90	—	6,848.90
67 gal./106 gph	P1@2.00	Ea	7,060.00	88.90	—	7,148.90
76 gal./175 gph	P1@2.00	Ea	8,750.00	88.90	—	8,838.90
91 gal./291 gph	P1@2.00	Ea	10,400.00	88.90	—	10,488.90

Tankless natural gas water heaters. Ambient pressure. DOE and UL rated. For residential, multi-dwelling and light commercial potable water applications. Add the cost of piping, tempering valve, circulating pump, controls, and electrical connection, post-installation inspection by both the fire marshal and the mechanical inspector to validate federal, state and local energy tax credits or energy tax credit offsets. For larger arrays (laundries, institutional facilities, food processing plants), develop an estimate based on the required capacity and multiply these costs by the number of heaters required. Rated in Btus and gallons per minute capacity. (1 Mbh = 1,000 Btus)

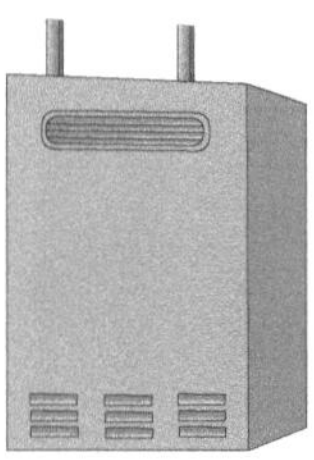

19.5-140 Mbh, .75-5.8 Gpm	P1@16.0	Ea	1,950.00	711.00	—	2,661.00
11-199 Mbh, .5-7 Gpm	P1@20.0	Ea	2,310.00	889.00	—	3,199.00
25-235 Mbh, .75-9.6 Gpm	P1@20.0	Ea	3,000.00	889.00	—	3,889.00

Tankless electric point-of-use water heaters. Ambient pressure, DOE and UL rated. For residential, multi-dwelling and light commercial potable water applications. Cost does not include piping, tempering valve, circulating pump, controls, storage tank, electrical connection. Add the cost of post-installation inspection by the mechanical inspector to validate federal, state and local energy tax credits or energy tax credit offsets. In rated gallons per minute capacity.

5.5 Kw/40 Amp, .75-2 Gpm	P1@4.00	Ea	492.00	178.00	—	670.00
9.5 Kw/50 Amp, .75-2.5 Gpm	P1@4.25	Ea	582.00	189.00	—	771.00
19 Kw/100 Amp, 1-3.5 Gpm	P1@4.50	Ea	969.00	200.00	—	1,169.00
28 Kw/120 Amp, 1.5-5 Gpm	P1@4.75	Ea	1,770.00	211.00	—	1,981.00

Description	Craft@Hrs	Unit	Material $	Labor $	Equipment $	Total $

Domestic hot water heater connection assembly. Includes supply, return, recirculation and relief piping and fittings (copper), relief and isolation valves. Make additional allowances for gas and venting connections where applicable.

Description	Craft@Hrs	Unit	Material $	Labor $	Equipment $	Total $
¾" residential	P1@1.75	Ea	342.00	77.80	—	419.80
¾" commercial	P1@2.25	Ea	459.00	100.00	—	559.00
1" commercial	P1@2.75	Ea	804.00	122.00	—	926.00
1¼" commercial	P1@3.50	Ea	985.00	156.00	—	1,141.00
1½" commercial	P1@3.75	Ea	1,020.00	167.00	—	1,187.00
2" commercial	P1@4.50	Ea	1,090.00	200.00	—	1,290.00
2½" commercial	P1@5.75	Ea	2,270.00	256.00	—	2,526.00
3" commercial	P1@6.50	Ea	3,480.00	289.00	—	3,769.00

Domestic water heater combustion vent connection. Make additional allowances for piping distances greater than 25'.

Description	Craft@Hrs	Unit	Material $	Labor $	Equipment $	Total $
2" B-vent	P1@.090	LF	7.18	4.00	—	11.18
3" B-vent	P1@.100	LF	8.88	4.45	—	13.33
4" B-vent	P1@.110	LF	11.80	4.89	—	16.69
6" B-vent	P1@.130	LF	14.40	5.78	—	20.18
Tankless heater vent kit	P1@2.50	Ea	681.00	111.00	—	792.00
Power vent kit	P1@2.00	Ea	1,670.00	88.90	—	1,758.90

Description	Craft@Hrs	Unit	Material $	Labor $	Equipment $	Total $

Water softener, time clock controller. Including brine tank, brine well and pick-up tube. Labor includes setting in place, connecting the unit to an existing domestic water distribution system, start up and testing.

Description	Craft@Hrs	Unit	Material $	Labor $	Equipment $	Total $
20,000 grain water softener, TCC	P1@4.50	Ea	634.00	200.00	—	834.00
30,000 grain water softener, TCC	P1@4.50	Ea	676.00	200.00	—	876.00
45,000 grain water softener, TCC	P1@4.50	Ea	752.00	200.00	—	952.00
50,000 grain water softener, TCC	P1@4.75	Ea	848.00	211.00	—	1,059.00
60,000 grain water softener, TCC	P1@4.75	Ea	1,000.00	211.00	—	1,211.00
75,000 grain water softener, TCC	P1@5.00	Ea	1,080.00	222.00	—	1,302.00
90,000 grain water softener, TCC	P1@5.50	Ea	1,460.00	244.00	—	1,704.00
120,000 grain water softener, TCC	P1@5.75	Ea	1,570.00	256.00	—	1,826.00

Water softener, mechanically-metered controller. Including brine tank, brine well and pick up tube. Labor includes setting in place, connecting the unit to an existing domestic water distribution system, start up and testing.

Description	Craft@Hrs	Unit	Material $	Labor $	Equipment $	Total $
20,000 grain water softener, MMC	P1@4.50	Ea	824.00	200.00	—	1,024.00
30,000 grain water softener, MMC	P1@4.50	Ea	860.00	200.00	—	1,060.00
45,000 grain water softener, MMC	P1@4.50	Ea	936.00	200.00	—	1,136.00
50,000 grain water softener, MMC	P1@4.75	Ea	1,030.00	211.00	—	1,241.00
60,000 grain water softener, MMC	P1@4.75	Ea	1,200.00	211.00	—	1,411.00
75,000 grain water softener, MMC	P1@5.00	Ea	1,290.00	222.00	—	1,512.00
90,000 grain water softener, MMC	P1@5.50	Ea	1,650.00	244.00	—	1,894.00
120,000 grain water softener, MMC	P1@5.75	Ea	1,760.00	256.00	—	2,016.00

Description	Craft@Hrs	Unit	Material $	Labor $	Equipment $	Total $

Water softener, electronically-metered controller. Including brine tank, brine well and pick up tube. Labor includes setting in place, connecting the unit to an existing domestic water distribution system, start up and testing.

Description	Craft@Hrs	Unit	Material $	Labor $	Equipment $	Total $
20,000 grain water softener, EMC	P1@4.50	Ea	874.00	200.00	—	1,074.00
30,000 grain water softener, EMC	P1@4.50	Ea	900.00	200.00	—	1,100.00
45,000 grain water softener, EMC	P1@4.50	Ea	987.00	200.00	—	1,187.00
50,000 grain water softener, EMC	P1@4.75	Ea	1,080.00	211.00	—	1,291.00
60,000 grain water softener, EMC	P1@4.75	Ea	1,270.00	211.00	—	1,481.00
75,000 grain water softener, EMC	P1@5.00	Ea	1,330.00	222.00	—	1,552.00
90,000 grain water softener, EMC	P1@5.50	Ea	1,700.00	244.00	—	1,944.00
120,000 grain water softener, EMC	P1@5.75	Ea	1,810.00	256.00	—	2,066.00

Water softener accessories

Description	Craft@Hrs	Unit	Material $	Labor $	Equipment $	Total $
By-pass valve	P1@.400	Ea	85.50	17.80	—	103.30
Manifold adapter kit	P1@.200	Ea	23.10	8.89	—	31.99
Turbulator	P1@.400	Ea	42.20	17.80	—	60.00

Iron filter, electronically-metered controller. Manganese green sand filter. Labor includes setting in place, connecting the unit to an existing domestic water distribution system, start-up and testing.

Description	Craft@Hrs	Unit	Material $	Labor $	Equipment $	Total $
42,000 iron filter (1.5 cf media), 5 gpm	P1@4.00	Ea	826.00	178.00	—	1,004.00
65,000 iron filter (2.0 cf media), 6 gpm	P1@4.50	Ea	978.00	200.00	—	1,178.00
84,000 iron filter (2.5 cf media), 8 gpm	P1@4.75	Ea	1,040.00	211.00	—	1,251.00
Replacement green sand media	P1@1.20	CF	49.80	53.30	—	103.10

Iron filter accessories

Description	Craft@Hrs	Unit	Material $	Labor $	Equipment $	Total $
By-pass valve	P1@.400	Ea	85.80	17.80	—	103.60
Air vent	P1@.200	Ea	68.10	8.89	—	76.99
Air controller	P1@.400	Ea	76.90	17.80	—	94.70

Description	Craft@Hrs	Unit	Material $	Labor $	Equipment $	Total $

Combination iron filter/water softener. Zeolite resins soften water and remove iron and manganese. Controller automatically controls PH level. Labor includes setting in place, connecting the unit to an existing domestic water distribution system, start-up and testing.

Description	Craft@Hrs	Unit	Material $	Labor $	Equipment $	Total $
40,000 iron filter, 1.3 cf media	P1@4.00	Ea	1,600.00	178.00	—	1,778.00
60,000 iron filter, 1.7 cf media	P1@4.50	Ea	1,730.00	200.00	—	1,930.00
80,000 iron filter, 2.5 cf media	P1@4.75	Ea	2,510.00	211.00	—	2,721.00

Hot water softener, time clock controller. Brass valve construction. Designed for 150 F. maximum operating temperature. Includes brine tank, brine well and pick-up tube. Labor includes setting in place, connecting the unit to an existing domestic water distribution system, start-up and testing.

Description	Craft@Hrs	Unit	Material $	Labor $	Equipment $	Total $
20,000 grain hot water softener	P1@4.50	Ea	2,000.00	200.00	—	2,200.00
30,000 grain hot water softener	P1@4.50	Ea	2,130.00	200.00	—	2,330.00
40,000 grain hot water softener	P1@4.50	Ea	2,230.00	200.00	—	2,430.00
60,000 grain hot water softener	P1@4.75	Ea	2,630.00	211.00	—	2,841.00

Pressure tank, fiberglass wound. Labor includes setting in place, connecting the tank to a domestic water distribution system and testing.

Description	Craft@Hrs	Unit	Material $	Labor $	Equipment $	Total $
Fiberglass pressure tank, 20 gallon	P1@2.00	Ea	325.00	88.90	—	413.90
Fiberglass pressure tank, 30 gallon	P1@2.00	Ea	366.00	88.90	—	454.90
Fiberglass pressure tank, 80 gallon	P1@2.75	Ea	593.00	122.00	—	715.00
Fiberglass pressure tank, 120 gallon	P1@3.50	Ea	782.00	156.00	—	938.00
Brass tank tee assembly, ¾"	P1@3.50	Ea	50.40	156.00	—	206.40
Brass tank tee assembly, 1"	P1@3.50	Ea	93.90	156.00	—	249.90
Brass tank tee assembly, 1¼"	P1@3.50	Ea	164.00	156.00	—	320.00

Description	Craft@Hrs	Unit	Material $	Labor $	Equipment $	Total $

Ultra-violet water disinfection unit. Stainless steel reactor, audible and visible alarm, lamp end-of-life indicator and 7-day override. Gpm rating at 30,000 mj/m2 output. Labor includes setting in place, connecting to the water system and testing.

Description	Craft@Hrs	Unit	Material $	Labor $	Equipment $	Total $
UV system, 1 gpm, ¼" in/out	P1@3.00	Ea	276.00	133.00	—	409.00
UV system, 6 gpm, ½" in/out	P1@3.00	Ea	537.00	133.00	—	670.00
UV system, 8 gpm, ¾" in/out	P1@4.00	Ea	621.00	178.00	—	799.00
UV system, 12 gpm, ¾" in/out	P1@4.00	Ea	796.00	178.00	—	974.00
UV replacement lamp, 20W, 1 gpm	P1@.750	Ea	61.80	33.30	—	95.10
UV replacement lamp, 32W, 6 gpm	P1@.750	Ea	70.10	33.30	—	103.40
UV replacement lamp, 39W, 8-12 gpm	P1@.750	Ea	89.70	33.30	—	123.00
UV replacement ballast, 420 Mv/110V	P1@1.00	Ea	270.00	44.50	—	314.50

Kitchen equipment booster heater

Description	Craft@Hrs	Unit	Material $	Labor $	Equipment $	Total $
1,000 watt	P1@4.00	Ea	876.00	178.00	—	1,054.00

Dishwasher

Description	Craft@Hrs	Unit	Material $	Labor $	Equipment $	Total $
Built-in	P1@5.00	Ea	949.00	222.00	—	1,171.00

Garbage disposal

Description	Craft@Hrs	Unit	Material $	Labor $	Equipment $	Total $
½ HP	P1@2.00	Ea	197.00	88.90	—	285.90
¾ HP	P1@2.00	Ea	329.00	88.90	—	417.90

Grease and oil interceptor

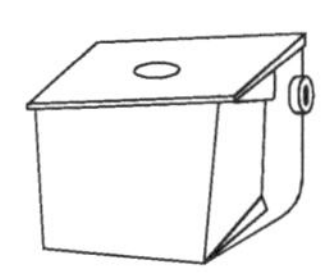

Description	Craft@Hrs	Unit	Material $	Labor $	Equipment $	Total $
4 GPM	P1@4.00	Ea	395.00	178.00	—	573.00
10 GPM	P1@5.00	Ea	643.00	222.00	—	865.00
15 GPM	P1@7.00	Ea	959.00	311.00	—	1,270.00
20 GPM	P1@8.00	Ea	1,160.00	356.00	—	1,516.00

Hair and lint interceptor

Description	Craft@Hrs	Unit	Material $	Labor $	Equipment $	Total $
1½"	P1@.650	Ea	225.00	28.90	—	253.90
2"	P1@.750	Ea	320.00	33.30	—	353.30

All bronze ¾" to 1½" in-line NPT pump

Description	Craft@Hrs	Unit	Material $	Labor $	Equipment $	Total $
1/12 HP	P1@1.50	Ea	843.00	66.70	—	909.70
1/6 HP	P1@1.50	Ea	1,260.00	66.70	—	1,326.70
1/4 HP	P1@1.50	Ea	1,470.00	66.70	—	1,536.70

Description	Craft@Hrs	Unit	Material $	Labor $	Equipment $	Total $
Kitchen appliance gas trim						
½"	P1@1.15	Ea	48.60	51.10	—	99.70
¾"	P1@1.30	Ea	88.80	57.80	—	146.60
1"	P1@1.60	Ea	103.00	71.10	—	174.10
1¼"	P1@2.10	Ea	170.00	93.30	—	263.30
1½"	P1@2.50	Ea	215.00	111.00	—	326.00
2"	P1@3.00	Ea	287.00	133.00	—	420.00
Hot and cold water supply						
½"	P1@1.10	Ea	58.90	48.90	—	107.80
¾"	P1@1.55	Ea	83.60	68.90	—	152.50
1"	P1@1.90	Ea	114.00	84.50	—	198.50
1¼"	P1@2.50	Ea	160.00	111.00	—	271.00
1½"	P1@3.00	Ea	200.00	133.00	—	333.00
Continuous waste						
2-part	P1@.250	Ea	71.60	11.10	—	82.70
3-part	P1@.350	Ea	122.00	15.60	—	137.60
4-part	P1@.450	Ea	156.00	20.00	—	176.00
Indirect waste						
½"	P1@1.05	Ea	18.70	46.70	—	65.40
¾"	P1@1.50	Ea	31.70	66.70	—	98.40
1"	P1@1.90	Ea	50.90	84.50	—	135.40
1¼"	P1@2.15	Ea	75.10	95.60	—	170.70
1½"	P1@2.60	Ea	98.90	116.00	—	214.90
2"	P1@3.00	Ea	151.00	133.00	—	284.00
Kitchen fixture waste tailpiece						
1½"	P1@.100	Ea	17.90	4.45	—	22.35
Kitchen fixture trap with solder bushing						
1½"	P1@.250	Ea	60.40	11.10	—	71.50
2"	P1@.300	Ea	83.70	13.30	—	97.00

Water closet, floor-mounted, flush tank, white vitreous china, lined tank. Complete with trim. Make additional allowances for rough-in. Based on American Standard Cadet series. ADA means American Disabilities Act compliant. (Wheelchair accessible)

Description	Craft@Hrs	Unit	Material $	Labor $	Equipment $	Total $
Round bowl	P1@2.10	Ea	312.00	93.30	—	405.30
Elongated bowl	P1@2.10	Ea	377.00	93.30	—	470.30
ADA, 18" high	P1@2.10	Ea	513.00	93.30	—	606.30

Water closet, floor-mounted, flush valve, white vitreous china. Complete with trim. Make additional allowances for rough-in. Based on American Standard. ADA means American Disabilities Act compliant. (Wheelchair accessible)

Description	Craft@Hrs	Unit	Material $	Labor $	Equipment $	Total $
Elongated bowl	P1@2.60	Ea	488.00	116.00	—	604.00
Elongated bowl, ADA 18" high	P1@2.60	Ea	581.00	116.00	—	697.00
Elongated bowl with a bedpan cleanser	P1@4.10	Ea	843.00	182.00	—	1,025.00
Elongated bowl, ADA 18" high with a bedpan cleanser	P1@4.10	Ea	913.00	182.00	—	1,095.00

Water closet, wall-hung, flush valve, white vitreous china. Complete with fixture carrier and all trim. Make additional allowances for rough-in. Based on American Standard Afwall series.

Description	Craft@Hrs	Unit	Material $	Labor $	Equipment $	Total $
Elongated bowl	P1@3.55	Ea	773.00	158.00	—	931.00
Elongated bowl with electronic flush valve	P1@3.80	Ea	1,380.00	169.00	—	1,549.00
Elongated bowl with bedpan cleanser	P1@5.05	Ea	1,120.00	224.00	—	1,344.00
Electronic flush valve, add	P1@.600	Ea	611.00	26.70	—	637.70

Urinal, wall-hung, flush valve, white vitreous china. Complete with trim. Make additional allowances for rough-in.

Description	Craft@Hrs	Unit	Material $	Labor $	Equipment $	Total $
Siphon-jet type	P1@3.15	Ea	773.00	140.00	—	913.00
Wash-out type	P1@3.10	Ea	631.00	138.00	—	769.00
Wash-down type	P1@3.00	Ea	446.00	133.00	—	579.00
Urinal carrier, add	P1@.600	Ea	135.00	26.70	—	161.70
Electronic flush valve, add	P1@.600	Ea	494.00	26.70	—	520.70

Urinal, stall-type, flush valve, white vitreous china. Complete with trim. Make additional allowances for rough-in.

Description	Craft@Hrs	Unit	Material $	Labor $	Equipment $	Total $
Stall urinal	P1@5.00	Ea	1,470.00	222.00	—	1,692.00

Description	Craft@Hrs	Unit	Material $	Labor $	Equipment $	Total $

Lavatory, wall-hung, white vitreous china. Complete with trim and fixture carrier. Make additional allowances for rough-in. ADA means American Disabilities Act compliant. (Wheelchair accessible)

Wall-hung lav	P1@2.70	Ea	598.00	120.00	—	718.00
Wall-hung, ADA	P1@2.70	Ea	871.00	120.00	—	991.00
Fixture carrier	P1@.500	Ea	134.00	22.20	—	156.20

Countertop lavatory, white. Complete with trim. Make additional allowances for rough-in.

Vitreous china	P1@2.00	Ea	435.00	88.90	—	523.90
Enameled steel	P1@2.00	Ea	360.00	88.90	—	448.90
Acrylic	P1@2.00	Ea	262.00	88.90	—	350.90

Bathtub, white, 60" x 32". Complete with trim, including shower head. Make additional allowances for rough-in.

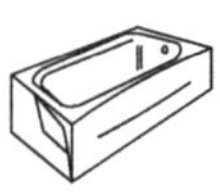

Enameled steel	P1@2.50	Ea	611.00	111.00	—	722.00
Cast iron	P1@3.50	Ea	1,200.00	156.00	—	1,356.00
Fiberglass	P1@2.50	Ea	622.00	111.00	—	733.00
Acrylic	P1@2.50	Ea	665.00	111.00	—	776.00

Tub and shower combination, fiberglass, white. Complete with trim. Make additional allowances for rough-in.

One-piece	P1@4.50	Ea	1,400.00	200.00	—	1,600.00
Two-piece (reno)	P1@5.50	Ea	1,790.00	244.00	—	2,034.00
Four-piece (reno)	P1@6.25	Ea	1,900.00	278.00	—	2,178.00

Shower stall, white, 36" x 36". Complete with trim. Make additional allowances for rough-in.

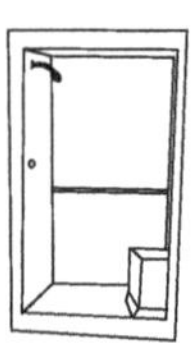

Fiberglass one-piece	P1@3.50	Ea	886.00	156.00	—	1,042.00
Fiberglass three-piece	P1@4.25	Ea	1,140.00	189.00	—	1,329.00
Acrylic one-piece	P1@3.50	Ea	1,330.00	156.00	—	1,486.00
Acrylic three-piece	P1@4.25	Ea	1,750.00	189.00	—	1,939.00

Shower basin, 36" x 36". Complete with trim (faucet, shower head and strainer). Make additional allowances for rough-in.

Fiberglass	P1@2.50	Ea	570.00	111.00	—	681.00
Acrylic	P1@2.50	Ea	613.00	111.00	—	724.00
Molded stone	P1@2.65	Ea	593.00	118.00	—	711.00

Description	Craft@Hrs	Unit	Material $	Labor $	Equipment $	Total $
Kitchen sink, double compartment. Complete with trim. Make additional allowances for rough-in.						
Stainless steel	P1@2.15	Ea	448.00	95.60	—	543.60
Cast iron	P1@2.50	Ea	1,400.00	111.00	—	1,511.00
Acrylic	P1@2.15	Ea	530.00	95.60	—	625.60
Kitchen sink, single compartment. Complete with trim. Make additional allowances for rough-in.						
Stainless steel	P1@2.00	Ea	378.00	88.90	—	466.90
Cast iron	P1@2.10	Ea	873.00	93.30	—	966.30
Acrylic	P1@2.00	Ea	394.00	88.90	—	482.90
Bar sink. Complete with trim. Make additional allowances for rough-in.						
Stainless steel	P1@2.00	Ea	321.00	88.90	—	409.90
Acrylic	P1@2.00	Ea	216.00	88.90	—	304.90
Exam room sink. Complete with trim. Make additional allowances for rough-in.						
Stainless steel	P1@2.10	Ea	491.00	93.30	—	584.30
Acrylic	P1@2.10	Ea	419.00	93.30	—	512.30
Laboratory sink. Complete with trim. Make additional allowances for rough-in.						
Stainless steel	P1@2.25	Ea	562.00	100.00	—	662.00
Acrylic	P1@2.25	Ea	491.00	100.00	—	591.00
Laundry sink, double compartment. Complete with trim. Make additional allowances for rough-in.						
Cast iron	P1@3.50	Ea	626.00	156.00	—	782.00
Acrylic	P1@2.25	Ea	275.00	100.00	—	375.00
Laundry sink, single compartment. Complete with trim. Make additional allowances for rough-in.						
Cast iron	P1@2.75	Ea	1,080.00	122.00	—	1,202.00
Acrylic	P1@2.00	Ea	190.00	88.90	—	278.90
Mop sink, floor-mounted, 36" x 24". Complete with trim. Make additional allowances for rough-in.						
Molded stone	P1@2.65	Ea	800.00	118.00	—	918.00
Terrazzo	P1@2.65	Ea	953.00	118.00	—	1,071.00
Acrylic	P1@2.35	Ea	617.00	104.00	—	721.00

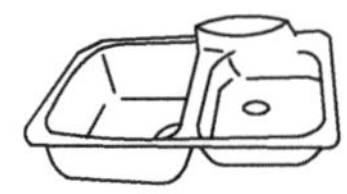

Description	Craft@Hrs	Unit	Material $	Labor $	Equipment $	Total $

Slop sink, enameled cast iron with P-trap, standard. Complete with trim. Make additional allowances for rough-in.

Slop sink with P-trap, std.	P1@3.50	Ea	1,520.00	156.00	—	1,676.00

Floor sink, recessed, enameled steel, white. Add 40% to material prices for acid-resisting finish. Complete with strainer. Make additional allowances for rough-in.

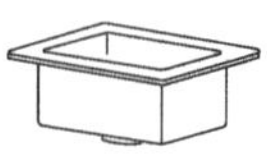

9" x 9"	P1@1.00	Ea	112.00	44.50	—	156.50
12" x 12"	P1@1.00	Ea	131.00	44.50	—	175.50
15" x 15"	P1@1.15	Ea	143.00	51.10	—	194.10
18" x 18"	P1@1.25	Ea	164.00	55.60	—	219.60
24" x 24"	P1@1.50	Ea	215.00	66.70	—	281.70

Drinking fountain, refrigerated, stainless steel. Complete with trim. Make additional allowances for rough-in. ADA means American Disabilities Act compliant. (Wheelchair accessible)

Free-standing	P1@2.00	Ea	1,800.00	88.90	—	1,888.90
Semi-recessed	P1@2.50	Ea	2,410.00	111.00	—	2,521.00
Fully-recessed	P1@2.50	Ea	4,160.00	111.00	—	4,271.00
Wall-hung	P1@2.00	Ea	1,690.00	88.90	—	1,778.90
Wall-hung, ADA	P1@2.50	Ea	4,160.00	111.00	—	4,271.00

Drinking fountain, non-refrigerated. Complete with trim. Make additional allowances for rough-in. ADA means American Disabilities Act compliant. (Wheelchair accessible) S.S. means stainless steel.

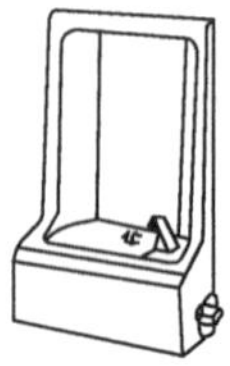

Recessed, china	P1@2.50	Ea	1,430.00	111.00	—	1,541.00
Wall-hung, china	P1@2.00	Ea	813.00	88.90	—	901.90
Recessed, S.S.	P1@2.50	Ea	1,600.00	111.00	—	1,711.00
Wall-hung, S.S.	P1@2.00	Ea	850.00	88.90	—	938.90
ADA, S.S.	P1@2.50	Ea	1,450.00	111.00	—	1,561.00

Commercial plumbing fixture rough-in.

Commercial plumbing fixture rough-in. Includes type L copper supply pipe and DWV copper (to 2½") or cast iron (MJ) DWV (over 2½") drain and vent piping. Make additional allowances for plumbing fixtures and trim. Use these costs for preliminary estimates.

Description	Craft@Hrs	Unit	Material $	Labor $	Equipment $	Total $
Water closet, wall-hung, flush valve, with carrier	P1@2.25	Ea	1,080.00	100.00	—	1,180.00
Water closet, wall-hung, flush valve, no carrier	P1@1.95	Ea	993.00	86.70	—	1,079.70
Water closet, floor-mounted, flush valve	P1@2.75	Ea	877.00	122.00	—	999.00
Water closet, floor-mounted, tank type	P1@2.25	Ea	672.00	100.00	—	772.00
Bidet	P1@2.00	Ea	468.00	88.90	—	556.90
Urinal, wall-hung, flush valve, with carrier	P1@3.10	Ea	1,180.00	138.00	—	1,318.00
Urinal, wall-hung, flush valve, without carrier	P1@2.35	Ea	672.00	104.00	—	776.00
Lavatory, wall-hung, with carrier	P1@2.40	Ea	974.00	107.00	—	1,081.00
Lavatory	P1@1.90	Ea	468.00	84.50	—	552.50
Sink	P1@1.90	Ea	504.00	84.50	—	588.50
Bath tub	P1@2.35	Ea	720.00	104.00	—	824.00
Shower	P1@2.60	Ea	844.00	116.00	—	960.00
Mop sink	P1@2.40	Ea	599.00	107.00	—	706.00
Slop sink	P1@2.60	Ea	428.00	116.00	—	544.00
Laundry tub	P1@1.95	Ea	509.00	86.70	—	595.70
Wash fountain	P1@2.10	Ea	548.00	93.30	—	641.30
Lab sink, glass drainage	P1@3.80	Ea	2,160.00	169.00	—	2,329.00
Lab sink, acid resistant plastic drainage	P1@2.65	Ea	344.00	118.00	—	462.00
Drinking fountain	P1@2.20	Ea	373.00	97.80	—	470.80
Emergency eyewash and shower	P1@1.75	Ea	141.00	77.80	—	218.80
Washing machine	P1@2.25	Ea	547.00	100.00	—	647.00

Description	Craft@Hrs	Unit	Material $	Labor $	Equipment $	Total $

Commercial plumbing fixture group rough-in. Includes Type L copper supply pipe and DWV copper (to 2½") or cast iron (MJ) DWV (over 2½") drain and vent piping. Make additional allowances for plumbing fixtures and trim. Use these costs for preliminary estimates.

Description	Craft@Hrs	Unit	Material $	Labor $	Equipment $	Total $
3-piece washroom group	P1@5.50	Ea	1,290.00	244.00	—	1,534.00
3-piece washroom group back to back	P1@9.75	Ea	2,390.00	433.00	—	2,823.00
Kitchen sink, back to back	P1@2.15	Ea	695.00	95.60	—	790.60
Battery of water closets, floor-mounted, tank type, per water closet	P1@1.75	Ea	541.00	77.80	—	618.80
Battery of water closets, floor-mounted, flush valve, per water closet	P1@2.20	Ea	707.00	97.80	—	804.80
Battery of water closets, wall-hung, flush valve, with carrier, per water closet	P1@1.80	Ea	930.00	80.00	—	1,010.00
Battery of water closets, wall-hung, flush valve, without carrier, per water closet	P1@1.50	Ea	272.00	66.70	—	338.70
Battery of urinals, wall-hung, flush valve with carrier, per urinal	P1@2.45	Ea	1,120.00	109.00	—	1,229.00
Battery of urinals, wall-hung, flush valve without carrier, per urinal	P1@1.90	Ea	576.00	84.50	—	660.50
Battery of lavatory basins, wall-hung, with carrier, per lavatory	P1@2.00	Ea	870.00	88.90	—	958.90
Battery of lavatory basins, without carrier, per lavatory	P1@1.50	Ea	393.00	66.70	—	459.70

Residential plumbing fixture rough-in. Includes polyethylene (PE) supply pipe and ABS DWV drain and vent piping. Make additional allowances for plumbing fixtures and trim. Use these costs for preliminary estimates.

Description	Craft@Hrs	Unit	Material $	Labor $	Equipment $	Total $
Water closet, floor-mounted, tank type	P1@2.00	Ea	143.00	88.90	—	231.90
Bidet	P1@1.85	Ea	107.00	82.20	—	189.20
Lavatory	P1@1.75	Ea	107.00	77.80	—	184.80
Counter sink	P1@1.75	Ea	118.00	77.80	—	195.80
Bathtub	P1@2.10	Ea	107.00	93.30	—	200.30
Shower	P1@2.45	Ea	157.00	109.00	—	266.00
Laundry tub	P1@1.75	Ea	97.90	77.80	—	175.70
Washing machine	P1@2.00	Ea	121.00	88.90	—	209.90

Copper, Type K with Brazed Joints

Type K hard-drawn copper pipe with wrought copper fittings and brazed joints is used in a wide variety of plumbing and HVAC systems such as potable water, heating hot water, chilled water, compressed air, refrigerant and A.C. condensate.

Brazed joints are those made with silver or other alloy filler metals having melting points at, or above, 1,000 degrees F. Maximum working pressure/temperature relationships for brazed joints are approximately as follows:

Maximum Working Pressure (PSIG)*			
Water Temperature	**Nominal Pipe Size (inches)**		
(degrees F.)	**Up to 1**	**1¼ to 2**	**2½ to 4**
Up to 350	270	190	150

**For copper pipe and solder-type fittings using brazing alloys melting at, or above 1,000 degrees F.*

This section has been arranged to save the estimator's time by including all normally-used system components such as pipe, fittings, valves, hanger assemblies, riser clamps and miscellaneous items under one heading. Additional items can be found under "Plumbing and Piping Specialties." The cost estimates in this section are based on the conditions, limitations and wage rates described in the section "How to Use This Book" beginning on page 5.

Description Craft@Hrs Unit Material $ Labor $ Equipment $ Total $

Type K copper pipe with brazed joints, installed horizontally. Complete installation including six to twelve wrought copper tees and six to twelve wrought copper elbows every 100', and hangers spaced to meet plumbing code *(a tee & elbow every 8.3' for ½" dia., a tee and elbow every 16.6' for 4" dia.).*

Description	Craft@Hrs	Unit	Material $	Labor $	Equipment $	Total $
½"	P1@.117	LF	4.50	5.20	—	9.70
¾"	P1@.127	LF	7.88	5.65	—	13.53
1"	P1@.147	LF	12.50	6.53	—	19.03
1¼"	P1@.167	LF	17.20	7.42	—	24.62
1½"	P1@.187	LF	21.20	8.31	—	29.51
2"	P1@.217	LF	35.80	9.65	—	45.45
2½"	P1@.247	LF	59.00	11.00	—	70.00
3"	P1@.277	LF	68.10	12.30	—	80.40
4"	P1@.317	LF	123.00	14.10	—	137.10

Type K copper pipe with brazed joints, installed risers. Complete installation including a wrought copper reducing tee every floor and a riser clamp every other floor.

Description	Craft@Hrs	Unit	Material $	Labor $	Equipment $	Total $
½"	P1@.064	LF	4.09	2.84	—	6.93
¾"	P1@.074	LF	7.38	3.29	—	10.67
1"	P1@.079	LF	11.20	3.51	—	14.71
1¼"	P1@.084	LF	15.50	3.73	—	19.23
1½"	P1@.104	LF	18.50	4.62	—	23.12
2"	P1@.134	LF	31.40	5.96	—	37.36
2½"	P1@.154	LF	46.80	6.85	—	53.65
3"	P1@.174	LF	65.60	7.73	—	73.33
4"	P1@.194	LF	103.00	8.62	—	111.62

Type K copper pipe with brazed joints, pipe only

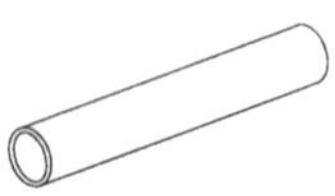

Description	Craft@Hrs	Unit	Material $	Labor $	Equipment $	Total $
½"	P1@.032	LF	3.59	1.42	—	5.01
¾"	P1@.035	LF	6.65	1.56	—	8.21
1"	P1@.038	LF	8.37	1.69	—	10.06
1¼"	P1@.042	LF	11.20	1.87	—	13.07
1½"	P1@.046	LF	12.20	2.04	—	14.24
2"	P1@.053	LF	21.80	2.36	—	24.16
2½"	P1@.060	LF	28.40	2.67	—	31.07
3"	P1@.066	LF	37.20	2.93	—	40.13
4"	P1@.080	LF	66.50	3.56	—	70.06

Type K copper 45-degree ell C x C with brazed joints

Description	Craft@Hrs	Unit	Material $	Labor $	Equipment $	Total $
½"	P1@.129	Ea	.97	5.73	—	6.70
¾"	P1@.181	Ea	1.86	8.05	—	9.91
1"	P1@.232	Ea	5.49	10.30	—	15.79
1¼"	P1@.284	Ea	7.43	12.60	—	20.03
1½"	P1@.335	Ea	11.10	14.90	—	26.00
2"	P1@.447	Ea	17.50	19.90	—	37.40
2½"	P1@.550	Ea	44.80	24.40	—	69.20
3"	P1@.654	Ea	59.20	29.10	—	88.30
4"	P1@.860	Ea	114.00	38.20	—	152.20

Type K copper 90-degree ell C x C with brazed joints

Description	Craft@Hrs	Unit	Material $	Labor $	Equipment $	Total $
½"	P1@.129	Ea	.57	5.73	—	6.30
¾"	P1@.181	Ea	1.20	8.05	—	9.25
1"	P1@.232	Ea	4.19	10.30	—	14.49
1¼"	P1@.284	Ea	6.76	12.60	—	19.36
1½"	P1@.335	Ea	10.20	14.90	—	25.10
2"	P1@.447	Ea	17.70	19.90	—	37.60
2½"	P1@.550	Ea	37.00	24.40	—	61.40
3"	P1@.654	Ea	46.90	29.10	—	76.00
4"	P1@.860	Ea	99.20	38.20	—	137.40

Description	Craft@Hrs	Unit	Material $	Labor $	Equipment $	Total $

Type K copper 90-degree ell ftg. x C with brazed joints

Description	Craft@Hrs	Unit	Material $	Labor $	Equipment $	Total $
½"	P1@.129	Ea	1.20	5.73	—	6.93
¾"	P1@.181	Ea	3.95	8.05	—	12.00
1"	P1@.232	Ea	5.93	10.30	—	16.23
1¼"	P1@.284	Ea	8.26	12.60	—	20.86
1½"	P1@.335	Ea	12.30	14.90	—	27.20
2"	P1@.447	Ea	24.20	19.90	—	44.10
2½"	P1@.550	Ea	50.40	24.40	—	74.80
3"	P1@.654	Ea	56.70	29.10	—	85.80
4"	P1@.860	Ea	151.00	38.20	—	189.20

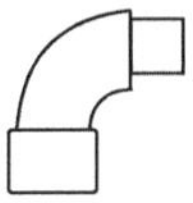

Type K copper tee C x C x C with brazed joints

Description	Craft@Hrs	Unit	Material $	Labor $	Equipment $	Total $
½"	P1@.156	Ea	.92	6.93	—	7.85
¾"	P1@.204	Ea	2.48	9.07	—	11.55
1"	P1@.263	Ea	9.34	11.70	—	21.04
1¼"	P1@.322	Ea	14.90	14.30	—	29.20
1½"	P1@.380	Ea	21.80	16.90	—	38.70
2"	P1@.507	Ea	33.60	22.50	—	56.10
2½"	P1@.624	Ea	68.10	27.70	—	95.80
3"	P1@.741	Ea	105.00	32.90	—	137.90
4"	P1@.975	Ea	176.00	43.30	—	219.30

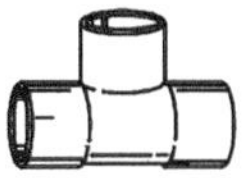

Description	Craft@Hrs	Unit	Material $	Labor $	Equipment $	Total $
Type K copper branch reducing tee C x C x C with brazed joints						
½" x 3/8"	P1@.129	Ea	1.87	5.73	—	7.60
¾" x ½"	P1@.182	Ea	4.34	8.09	—	12.43
1" x ½"	P1@.209	Ea	11.10	9.29	—	20.39
1" x ¾"	P1@.234	Ea	11.10	10.40	—	21.50
1¼" x ½"	P1@.241	Ea	16.80	10.70	—	27.50
1¼" x ¾"	P1@.264	Ea	16.80	11.70	—	28.50
1¼" x 1"	P1@.287	Ea	16.80	12.80	—	29.60
1½" x ½"	P1@.291	Ea	20.60	12.90	—	33.50
1½" x ¾"	P1@.307	Ea	20.60	13.60	—	34.20
1½" x 1"	P1@.323	Ea	20.60	14.40	—	35.00
1½" x 1¼"	P1@.339	Ea	20.60	15.10	—	35.70
2" x ½"	P1@.348	Ea	26.90	15.50	—	42.40
2" x ¾"	P1@.373	Ea	26.90	16.60	—	43.50
2" x 1"	P1@.399	Ea	26.90	17.70	—	44.60
2" x 1¼"	P1@.426	Ea	26.90	18.90	—	45.80
2" x 1½"	P1@.452	Ea	26.90	20.10	—	47.00
2½" x ½"	P1@.468	Ea	80.20	20.80	—	101.00
2½" x ¾"	P1@.485	Ea	80.20	21.60	—	101.80
2½" x 1"	P1@.502	Ea	80.20	22.30	—	102.50
2½" x 1½"	P1@.537	Ea	80.20	23.90	—	104.10
2½" x 2"	P1@.555	Ea	80.20	24.70	—	104.90
3" x 1¼"	P1@.562	Ea	91.00	25.00	—	116.00
3" x 2"	P1@.626	Ea	94.40	27.80	—	122.20
3" x 2½"	P1@.660	Ea	94.40	29.30	—	123.70
4" x 1¼"	P1@.669	Ea	164.00	29.70	—	193.70
4" x 1½"	P1@.718	Ea	164.00	31.90	—	195.90
4" x 2"	P1@.768	Ea	164.00	34.10	—	198.10
4" x 2½"	P1@.819	Ea	164.00	36.40	—	200.40
4" x 3"	P1@.868	Ea	164.00	38.60	—	202.60

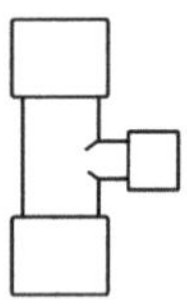

Description	Craft@Hrs	Unit	Material $	Labor $	Equipment $	Total $

Type K copper reducer with brazed joints

Description	Craft@Hrs	Unit	Material $	Labor $	Equipment $	Total $
½" x 3/8"	P1@.148	Ea	.67	6.58	—	7.25
¾" x ½"	P1@.155	Ea	1.39	6.89	—	8.28
1" x ¾"	P1@.206	Ea	2.28	9.16	—	11.44
1¼" x ½"	P1@.206	Ea	3.82	9.16	—	12.98
1¼" x ¾"	P1@.232	Ea	3.82	10.30	—	14.12
1¼" x 1"	P1@.258	Ea	3.82	11.50	—	15.32
1½" x ½"	P1@.232	Ea	5.83	10.30	—	16.13
1½" x ¾"	P1@.257	Ea	5.83	11.40	—	17.23
1½" x 1"	P1@.283	Ea	5.83	12.60	—	18.43
1½" x 1¼"	P1@.308	Ea	5.83	13.70	—	19.53
2" x ½"	P1@.287	Ea	8.79	12.80	—	21.59
2" x ¾"	P1@.277	Ea	8.79	12.30	—	21.09
2" x 1"	P1@.338	Ea	8.79	15.00	—	23.79
2" x 1¼"	P1@.365	Ea	8.79	16.20	—	24.99
2" x 1½"	P1@.390	Ea	8.79	17.30	—	26.09
2½" x 1"	P1@.390	Ea	23.90	17.30	—	41.20
2½" x 1¼"	P1@.416	Ea	23.90	18.50	—	42.40
2½" x 1½"	P1@.442	Ea	23.90	19.60	—	43.50
2½" x 2"	P1@.497	Ea	23.90	22.10	—	46.00
3" x 1¼"	P1@.468	Ea	28.00	20.80	—	48.80
3" x 1½"	P1@.493	Ea	28.00	21.90	—	49.90
3" x 2"	P1@.548	Ea	28.00	24.40	—	52.40
3" x 2½"	P1@.600	Ea	28.00	26.70	—	54.70
4" x 2"	P1@.652	Ea	53.90	29.00	—	82.90
4" x 2½"	P1@.704	Ea	53.90	31.30	—	85.20
4" x 3"	P1@.755	Ea	53.90	33.60	—	87.50

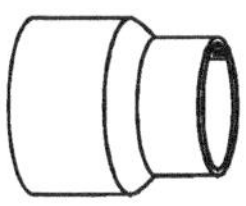

Type K copper adapter C x MPT

Description	Craft@Hrs	Unit	Material $	Labor $	Equipment $	Total $
½"	P1@.090	Ea	1.83	4.00	—	5.83
¾"	P1@.126	Ea	2.59	5.60	—	8.19
1"	P1@.162	Ea	6.44	7.20	—	13.64
1¼"	P1@.198	Ea	12.00	8.80	—	20.80
1½"	P1@.234	Ea	13.60	10.40	—	24.00
2"	P1@.312	Ea	16.90	13.90	—	30.80
2½"	P1@.384	Ea	54.90	17.10	—	72.00
3"	P1@.456	Ea	91.00	20.30	—	111.30

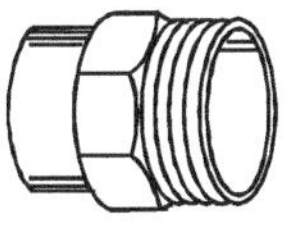

Type K copper adapter C x FPT

Description	Craft@Hrs	Unit	Material $	Labor $	Equipment $	Total $
½"	P1@.090	Ea	2.76	4.00	—	6.76
¾"	P1@.126	Ea	5.01	5.60	—	10.61
1"	P1@.162	Ea	7.56	7.20	—	14.76
1¼"	P1@.198	Ea	16.00	8.80	—	24.80
1½"	P1@.234	Ea	19.20	10.40	—	29.60
2"	P1@.312	Ea	28.60	13.90	—	42.50
2½"	P1@.384	Ea	72.00	17.10	—	89.10
3"	P1@.456	Ea	99.00	20.30	—	119.30

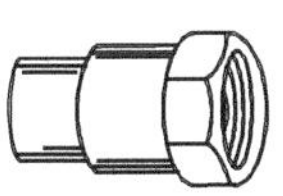

Type K copper flush bushing with brazed joints

Description	Craft@Hrs	Unit	Material $	Labor $	Equipment $	Total $
½" x 3/8"	P1@.148	Ea	1.67	6.58	—	8.25
¾" x ½"	P1@.155	Ea	2.87	6.89	—	9.76
1" x ¾"	P1@.206	Ea	7.43	9.16	—	16.59
1" x ½"	P1@.206	Ea	8.48	9.16	—	17.64
1¼" x 1"	P1@.258	Ea	8.81	11.50	—	20.31
1½" x 1¼"	P1@.308	Ea	11.20	13.70	—	24.90
2" x 1½"	P1@.390	Ea	20.50	17.30	—	37.80

Type K copper union with brazed joints

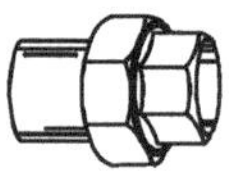

Description	Craft@Hrs	Unit	Material $	Labor $	Equipment $	Total $
½"	P1@.146	Ea	6.75	6.49	—	13.24
¾"	P1@.205	Ea	8.05	9.11	—	17.16
1"	P1@.263	Ea	18.10	11.70	—	29.80
1¼"	P1@.322	Ea	32.90	14.30	—	47.20
1½"	P1@.380	Ea	40.40	16.90	—	57.30
2"	P1@.507	Ea	58.20	22.50	—	80.70

Type K copper dielectric union with brazed joints

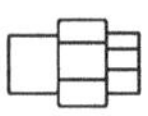

Description	Craft@Hrs	Unit	Material $	Labor $	Equipment $	Total $
½"	P1@.146	Ea	6.32	6.49	—	12.81
¾"	P1@.205	Ea	6.32	9.11	—	15.43
1"	P1@.263	Ea	13.10	11.70	—	24.80
1¼"	P1@.322	Ea	22.30	14.30	—	36.60
1½"	P1@.380	Ea	30.70	16.90	—	47.60
2"	P1@.507	Ea	41.00	22.50	—	63.50

Type K copper pressure cap with brazed joints

Description	Craft@Hrs	Unit	Material $	Labor $	Equipment $	Total $
½"	P1@.083	Ea	.56	3.69	—	4.25
¾"	P1@.116	Ea	1.12	5.16	—	6.28
1"	P1@.149	Ea	2.54	6.62	—	9.16
1¼"	P1@.182	Ea	3.64	8.09	—	11.73
1½"	P1@.215	Ea	5.31	9.56	—	14.87
2"	P1@.286	Ea	9.24	12.70	—	21.94
2½"	P1@.352	Ea	23.50	15.60	—	39.10
3"	P1@.418	Ea	36.70	18.60	—	55.30
4"	P1@.550	Ea	56.40	24.40	—	80.80

Type K copper coupling with brazed joints

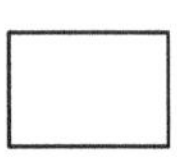

Description	Craft@Hrs	Unit	Material $	Labor $	Equipment $	Total $
½"	P1@.129	Ea	.26	5.73	—	5.99
¾"	P1@.181	Ea	.66	8.05	—	8.71
1"	P1@.232	Ea	2.41	10.30	—	12.71
1¼"	P1@.284	Ea	4.20	12.60	—	16.80
1½"	P1@.335	Ea	6.07	14.90	—	20.97
2"	P1@.447	Ea	8.93	19.90	—	28.83
2½"	P1@.550	Ea	21.10	24.40	—	45.50
3"	P1@.654	Ea	32.10	29.10	—	61.20
4"	P1@.860	Ea	63.00	38.20	—	101.20

Description	Craft@Hrs	Unit	Material $	Labor $	Equipment $	Total $

Class 125 bronze body gate valve, brazed joints

½"	P1@.240	Ea	22.80	10.70	—	33.50
¾"	P1@.300	Ea	28.50	13.30	—	41.80
1"	P1@.360	Ea	40.10	16.00	—	56.10
1¼"	P1@.480	Ea	50.40	21.30	—	71.70
1½"	P1@.540	Ea	68.00	24.00	—	92.00
2"	P1@.600	Ea	114.00	26.70	—	140.70
2½"	P1@1.00	Ea	188.00	44.50	—	232.50
3"	P1@1.50	Ea	270.00	66.70	—	336.70

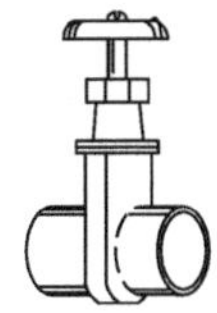

Class 125 iron body gate valve, flanged joints

2"	P1@.500	Ea	435.00	22.20	—	457.20
2½"	P1@.600	Ea	586.00	26.70	—	612.70
3"	P1@.750	Ea	636.00	33.30	—	669.30
4"	P1@1.35	Ea	934.00	60.00	—	994.00
5"	P1@2.00	Ea	1,780.00	88.90	—	1,868.90
6"	P1@2.50	Ea	1,780.00	111.00	—	1,891.00

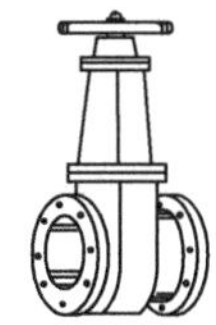

Class 125 bronze body globe valve, with brazed joints

½"	P1@.240	Ea	42.20	10.70	—	52.90
¾"	P1@.300	Ea	56.20	13.30	—	69.50
1"	P1@.360	Ea	96.90	16.00	—	112.90
1¼"	P1@.480	Ea	136.00	21.30	—	157.30
1½"	P1@.540	Ea	185.00	24.00	—	209.00
2"	P1@.600	Ea	299.00	26.70	—	325.70

Class 125 iron body globe valve, flanged joints

2½"	P1@.650	Ea	493.00	28.90	—	521.90
3"	P1@.750	Ea	591.00	33.30	—	624.30
4"	P1@1.35	Ea	798.00	60.00	—	858.00

200 PSIG iron body butterfly valve, lug-type, lever operated

2"	P1@.450	Ea	179.00	20.00	—	199.00
2½"	P1@.450	Ea	186.00	20.00	—	206.00
3"	P1@.550	Ea	196.00	24.40	—	220.40
4"	P1@.550	Ea	244.00	24.40	—	268.40

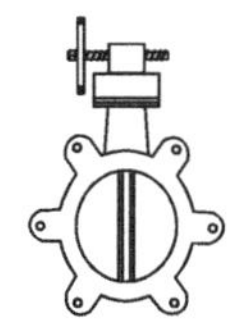

200 PSIG iron body butterfly valve, wafer-type, lever operated

2"	P1@.450	Ea	164.00	20.00	—	184.00
2½"	P1@.450	Ea	167.00	20.00	—	187.00
3"	P1@.550	Ea	179.00	24.40	—	203.40
4"	P1@.550	Ea	217.00	24.40	—	241.40

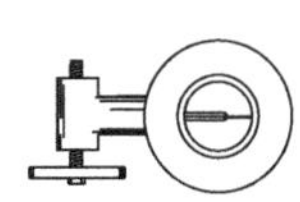

Description	Craft@Hrs	Unit	Material $	Labor $	Equipment $	Total $

Class 125 bronze body 2-piece ball valve, brazed joints

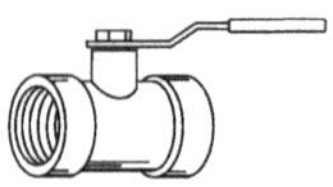

½"	P1@.240	Ea	11.60	10.70	—	22.30
¾"	P1@.300	Ea	15.40	13.30	—	28.70
1"	P1@.360	Ea	26.70	16.00	—	42.70
1¼"	P1@.480	Ea	44.60	21.30	—	65.90
1½"	P1@.540	Ea	60.30	24.00	—	84.30
2"	P1@.600	Ea	76.40	26.70	—	103.10
3"	P1@1.50	Ea	524.00	66.70	—	590.70
4"	P1@1.80	Ea	684.00	80.00	—	764.00

Class 125 bronze body swing check valve, brazed joints

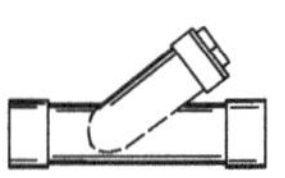

½"	P1@.240	Ea	23.90	10.70	—	34.60
¾"	P1@.300	Ea	34.10	13.30	—	47.40
1"	P1@.360	Ea	44.50	16.00	—	60.50
1¼"	P1@.480	Ea	63.60	21.30	—	84.90
1½"	P1@.540	Ea	89.60	24.00	—	113.60
2"	P1@.600	Ea	148.00	26.70	—	174.70
2½"	P1@1.00	Ea	299.00	44.50	—	343.50
3"	P1@1.50	Ea	430.00	66.70	—	496.70

Class 125 iron body swing check valve, flanged joints

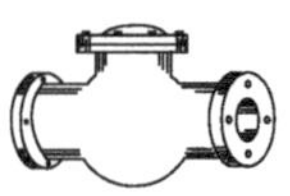

2"	P1@.500	Ea	209.00	22.20	—	231.20
2½"	P1@.600	Ea	264.00	26.70	—	290.70
3"	P1@.750	Ea	329.00	33.30	—	362.30
4"	P1@1.35	Ea	483.00	60.00	—	543.00

Class 125 iron body silent check valve, wafer-type

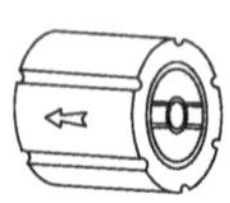

2"	P1@.500	Ea	153.00	22.20	—	175.20
2½"	P1@.600	Ea	170.00	26.70	—	196.70
3"	P1@.750	Ea	195.00	33.30	—	228.30
4"	P1@1.35	Ea	260.00	60.00	—	320.00

Class 125 bronze body strainer, threaded joints

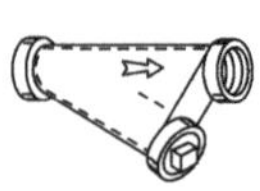

½"	P1@.230	Ea	37.40	10.20	—	47.60
¾"	P1@.260	Ea	49.10	11.60	—	60.70
1"	P1@.330	Ea	60.20	14.70	—	74.90
1¼"	P1@.440	Ea	84.20	19.60	—	103.80
1½"	P1@.495	Ea	126.00	22.00	—	148.00
2"	P1@.550	Ea	220.00	24.40	—	244.40

Class 125 iron body strainer, flanged joints

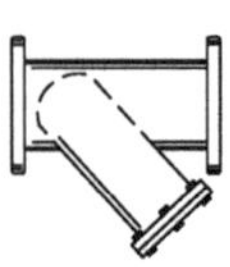

2"	P1@.500	Ea	168.00	22.20	—	190.20
2½"	P1@.600	Ea	187.00	26.70	—	213.70
3"	P1@.750	Ea	218.00	33.30	—	251.30
4"	P1@1.35	Ea	367.00	60.00	—	427.00

Installation of 2-way control valve, threaded joints

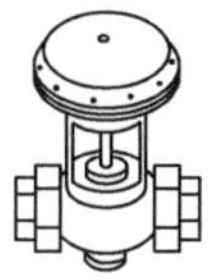

Description	Craft@Hrs	Unit	Material $	Labor $	Equipment $	Total $
½"	P1@.210	Ea	—	9.33	—	9.33
¾"	P1@.275	Ea	—	12.20	—	12.20
1"	P1@.350	Ea	—	15.60	—	15.60
1¼"	P1@.430	Ea	—	19.10	—	19.10
1½"	P1@.505	Ea	—	22.40	—	22.40
2"	P1@.675	Ea	—	30.00	—	30.00
2½"	P1@.830	Ea	—	36.90	—	36.90
3"	P1@.990	Ea	—	44.00	—	44.00

Installation of 3-way control valve, threaded joints

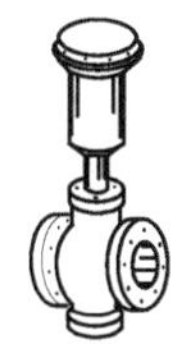

Description	Craft@Hrs	Unit	Material $	Labor $	Equipment $	Total $
½"	P1@.260	Ea	—	11.60	—	11.60
¾"	P1@.365	Ea	—	16.20	—	16.20
1"	P1@.475	Ea	—	21.10	—	21.10
1¼"	P1@.575	Ea	—	25.60	—	25.60
1½"	P1@.680	Ea	—	30.20	—	30.20
2"	P1@.910	Ea	—	40.40	—	40.40
2½"	P1@1.12	Ea	—	49.80	—	49.80
3"	P1@1.33	Ea	—	59.10	—	59.10

Companion flange, 150 pound cast brass

Description	Craft@Hrs	Unit	Material $	Labor $	Equipment $	Total $
2"	P1@.290	Ea	330.00	12.90	—	342.90
2½"	P1@.380	Ea	410.00	16.90	—	426.90
3"	P1@.460	Ea	410.00	20.40	—	430.40
4"	P1@.600	Ea	655.00	26.70	—	681.70

Bolt and gasket sets

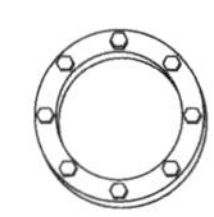

Description	Craft@Hrs	Unit	Material $	Labor $	Equipment $	Total $
2"	P1@.500	Ea	4.87	22.20	—	27.07
2½"	P1@.650	Ea	5.69	28.90	—	34.59
3"	P1@.750	Ea	9.65	33.30	—	42.95
4"	P1@1.00	Ea	16.10	44.50	—	60.60

Thermometer with well

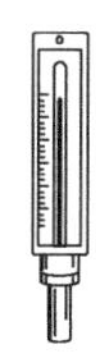

Description	Craft@Hrs	Unit	Material $	Labor $	Equipment $	Total $
7"	P1@.250	Ea	186.00	11.10	—	197.10
9"	P1@.250	Ea	192.00	11.10	—	203.10

Dial-type pressure gauge

Description	Craft@Hrs	Unit	Material $	Labor $	Equipment $	Total $
2½"	P1@.200	Ea	38.50	8.89	—	47.39
3½"	P1@.200	Ea	51.20	8.89	—	60.09

Description	Craft@Hrs	Unit	Material $	Labor $	Equipment $	Total $

Pressure/temperature tap

Description	Craft@Hrs	Unit	Material $	Labor $	Equipment $	Total $
Tap	P1@.150	Ea	18.60	6.67	—	25.27

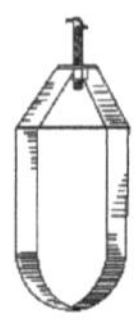

Hanger with swivel assembly

Description	Craft@Hrs	Unit	Material $	Labor $	Equipment $	Total $
½"	P1@.250	Ea	4.46	11.10	—	15.56
¾"	P1@.250	Ea	4.93	11.10	—	16.03
1"	P1@.250	Ea	5.15	11.10	—	16.25
1¼"	P1@.300	Ea	5.28	13.30	—	18.58
1½"	P1@.300	Ea	5.69	13.30	—	18.99
2"	P1@.300	Ea	5.93	13.30	—	19.23
2½"	P1@.350	Ea	8.08	15.60	—	23.68
3"	P1@.350	Ea	9.98	15.60	—	25.58
4"	P1@.350	Ea	11.00	15.60	—	26.60

Riser clamp

Description	Craft@Hrs	Unit	Material $	Labor $	Equipment $	Total $
½"	P1@.100	Ea	3.30	4.45	—	7.75
¾"	P1@.100	Ea	4.87	4.45	—	9.32
1"	P1@.100	Ea	4.91	4.45	—	9.36
1¼"	P1@.105	Ea	5.90	4.67	—	10.57
1½"	P1@.110	Ea	6.22	4.89	—	11.11
2"	P1@.115	Ea	6.62	5.11	—	11.73
2½"	P1@.120	Ea	6.93	5.33	—	12.26
3"	P1@.120	Ea	7.53	5.33	—	12.86
4"	P1@.125	Ea	9.61	5.56	—	15.17

Type K hard-drawn copper pipe with wrought copper fittings and soft-soldered joints is used in a wide variety of plumbing and HVAC systems such as potable water, heating hot water, chilled water and A.C. condensate.

Soft-soldered joints are those made with solders having melting points in the 350 degree F. to 500 degree F. range. Maximum working pressure/temperature relationships for soft-soldered joints are approximately as follows:

Maximum Working Pressures (PSIG)*				
Soft-solder Type	Water Temperature (degrees F.)	Nominal Pipe Size (inches)		
		Up to 1	1¼ to 2	2½ to 4
50-50 tin-lead**	100	200	175	150
	150	150	125	100
	200	100	90	75
	250	85	75	50
95-5 tin-antimony	100	500	400	300
	150	400	350	275
	200	300	250	200
	250	200	175	150

**For copper pipe and solder-type fittings using soft-solders melting at approximately 350 degrees F. to 500 degrees F.*

***The use of any solder containing lead is not allowed in potable water systems.*

This section has been arranged to save the estimator's time by including all normally-used system components such as pipe, fittings, valves, hanger assemblies, riser clamps and miscellaneous items under one heading. Additional items can be found under "Plumbing and Piping Specialties." The cost estimates in this section are based on the conditions, limitations and wage rates described in the section "How to Use This Book" beginning on page 5.

Type K copper pipe with soft-soldered joints installed horizontally. Complete installation including six to twelve wrought copper tees and six to twelve wrought copper elbows every 100', and hangers spaced to meet plumbing code (*a tee & elbow every 8.3' for ½" dia., a tee and elbow every 16.6' for 4" dia.*)

Description	Craft@Hrs	Unit	Material $	Labor $	Equipment $	Total $
½"	P1@.112	LF	4.50	4.98	—	9.48
¾"	P1@.122	LF	7.88	5.42	—	13.30
1"	P1@.142	LF	12.50	6.31	—	18.81
1¼"	P1@.162	LF	17.20	7.20	—	24.40
1½"	P1@.182	LF	21.20	8.09	—	29.29
2"	P1@.212	LF	35.80	9.42	—	45.22
2½"	P1@.242	LF	59.00	10.80	—	69.80
3"	P1@.272	LF	68.10	12.10	—	80.20
4"	P1@.312	LF	123.00	13.90	—	136.90

Copper, Type K with Soft-Soldered Joints

Type K copper pipe with soft-soldered joints, installed risers.

Complete installation including a wrought copper reducing tee every floor and a riser clamp every other floor.

Description	Craft@Hrs	Unit	Material $	Labor $	Equipment $	Total $
½"	P1@.062	LF	4.09	2.76	—	6.85
¾"	P1@.072	LF	7.38	3.20	—	10.58
1"	P1@.077	LF	11.20	3.42	—	14.62
1¼"	P1@.082	LF	15.50	3.64	—	19.14
1½"	P1@.102	LF	18.50	4.53	—	23.03
2"	P1@.132	LF	31.40	5.87	—	37.27
2½"	P1@.152	LF	46.80	6.76	—	53.56
3"	P1@.172	LF	65.60	7.65	—	73.25
4"	P1@.192	LF	103.00	8.53	—	111.53

Type K copper pipe with soft-soldered joints, pipe only

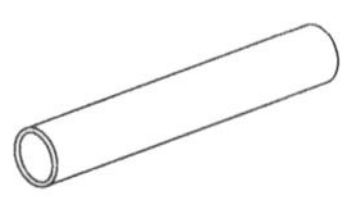

Description	Craft@Hrs	Unit	Material $	Labor $	Equipment $	Total $
½"	P1@.032	LF	3.59	1.42	—	5.01
¾"	P1@.035	LF	6.65	1.56	—	8.21
1"	P1@.038	LF	8.37	1.69	—	10.06
1¼"	P1@.042	LF	11.20	1.87	—	13.07
1½"	P1@.046	LF	12.20	2.04	—	14.24
2"	P1@.053	LF	21.80	2.36	—	24.16
2½"	P1@.060	LF	28.40	2.67	—	31.07
3"	P1@.066	LF	37.20	2.93	—	40.13
4"	P1@.080	LF	66.50	3.56	—	70.06

Type K copper 45-degree ell C x C with soft-soldered joints

Description	Craft@Hrs	Unit	Material $	Labor $	Equipment $	Total $
½"	P1@.107	Ea	.97	4.76	—	5.73
¾"	P1@.150	Ea	1.85	6.67	—	8.52
1"	P1@.193	Ea	5.44	8.58	—	14.02
1¼"	P1@.236	Ea	7.40	10.50	—	17.90
1½"	P1@.278	Ea	11.00	12.40	—	23.40
2"	P1@.371	Ea	17.40	16.50	—	33.90
2½"	P1@.457	Ea	44.80	20.30	—	65.10
3"	P1@.543	Ea	58.90	24.10	—	83.00
4"	P1@.714	Ea	114.00	31.70	—	145.70

Description	Craft@Hrs	Unit	Material $	Labor $	Equipment $	Total $

Type K copper 90-degree ell C x C with soft-soldered joints

Description	Craft@Hrs	Unit	Material $	Labor $	Equipment $	Total $
½"	P1@.107	Ea	.59	4.76	—	5.35
¾"	P1@.150	Ea	1.20	6.67	—	7.87
1"	P1@.193	Ea	4.16	8.58	—	12.74
1¼"	P1@.236	Ea	6.76	10.50	—	17.26
1½"	P1@.278	Ea	10.10	12.40	—	22.50
2"	P1@.371	Ea	17.70	16.50	—	34.20
2½"	P1@.457	Ea	36.80	20.30	—	57.10
3"	P1@.543	Ea	46.50	24.10	—	70.60
4"	P1@.714	Ea	99.20	31.70	—	130.90

Type K copper 90-degree ell ftg. x C with soft-soldered joint

Description	Craft@Hrs	Unit	Material $	Labor $	Equipment $	Total $
½"	P1@.107	Ea	.92	4.76	—	5.68
¾"	P1@.150	Ea	1.98	6.67	—	8.65
1"	P1@.193	Ea	5.89	8.58	—	14.47
1¼"	P1@.236	Ea	8.16	10.50	—	18.66
1½"	P1@.278	Ea	12.10	12.40	—	24.50
2"	P1@.371	Ea	24.20	16.50	—	40.70
2½"	P1@.457	Ea	50.20	20.30	—	70.50
3"	P1@.543	Ea	56.10	24.10	—	80.20
4"	P1@.714	Ea	114.00	31.70	—	145.70

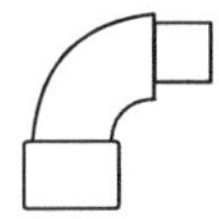

Type K copper tee C x C x C with soft-soldered joints

Description	Craft@Hrs	Unit	Material $	Labor $	Equipment $	Total $
½"	P1@.129	Ea	.92	5.73	—	6.65
¾"	P1@.181	Ea	2.48	8.05	—	10.53
1"	P1@.233	Ea	9.34	10.40	—	19.74
1¼"	P1@.285	Ea	14.90	12.70	—	27.60
1½"	P1@.337	Ea	21.40	15.00	—	36.40
2"	P1@.449	Ea	33.30	20.00	—	53.30
2½"	P1@.552	Ea	68.10	24.50	—	92.60
3"	P1@.656	Ea	105.00	29.20	—	134.20
4"	P1@.863	Ea	174.00	38.40	—	212.40

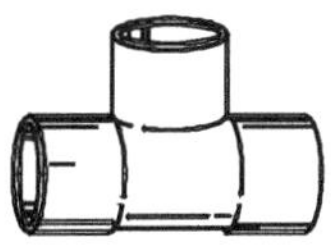

Description	Craft@Hrs	Unit	Material $	Labor $	Equipment $	Total $
Type K copper branch reducing tee C x C x C with soft-soldered joints						
½" x 3/8"	P1@.129	Ea	1.87	5.73	—	7.60
¾" x ½"	P1@.182	Ea	3.97	8.09	—	12.06
1" x ½"	P1@.195	Ea	7.78	8.67	—	16.45
1" x ¾"	P1@.219	Ea	7.78	9.73	—	17.51
1¼" x ½"	P1@.225	Ea	11.70	10.00	—	21.70
1¼" x ¾"	P1@.247	Ea	11.70	11.00	—	22.70
1¼" x 1"	P1@.268	Ea	11.70	11.90	—	23.60
1½" x ½"	P1@.272	Ea	14.40	12.10	—	26.50
1½" x ¾"	P1@.287	Ea	14.40	12.80	—	27.20
1½" x 1"	P1@.302	Ea	14.40	13.40	—	27.80
1½" x 1¼"	P1@.317	Ea	14.40	14.10	—	28.50
2" x ½"	P1@.325	Ea	18.80	14.40	—	33.20
2" x ¾"	P1@.348	Ea	18.80	15.50	—	34.30
2" x 1"	P1@.373	Ea	18.80	16.60	—	35.40
2" x 1¼"	P1@.398	Ea	18.80	17.70	—	36.50
2" x 1½"	P1@.422	Ea	18.80	18.80	—	37.60
2½" x ½"	P1@.437	Ea	56.80	19.40	—	76.20
2½" x ¾"	P1@.453	Ea	56.80	20.10	—	76.90
2½" x 1"	P1@.469	Ea	56.80	20.80	—	77.60
2½" x 1½"	P1@.502	Ea	56.80	22.30	—	79.10
2½" x 2"	P1@.519	Ea	56.80	23.10	—	79.90
3" x 1¼"	P1@.525	Ea	64.20	23.30	—	87.50
3" x 2"	P1@.585	Ea	66.40	26.00	—	92.40
3" x 2½"	P1@.617	Ea	66.40	27.40	—	93.80
4" x 1¼"	P1@.625	Ea	87.00	27.80	—	114.80
4" x 1½"	P1@.671	Ea	87.00	29.80	—	116.80
4" x 2"	P1@.718	Ea	87.00	31.90	—	118.90
4" x 2½"	P1@.765	Ea	87.00	34.00	—	121.00
4" x 3"	P1@.811	Ea	87.00	36.00	—	123.00

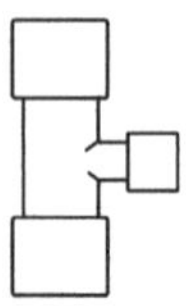

Description	Craft@Hrs	Unit	Material $	Labor $	Equipment $	Total $

Type K copper reducer with soft-soldered joints

½" x 3/8"	P1@.124	Ea	.70	5.51	—	6.21
¾" x ½"	P1@.129	Ea	1.30	5.73	—	7.03
1" x ¾"	P1@.172	Ea	2.28	7.65	—	9.93
1¼" x ½"	P1@.172	Ea	3.82	7.65	—	11.47
1¼" x ¾"	P1@.193	Ea	3.82	8.58	—	12.40
1¼" x 1"	P1@.215	Ea	3.82	9.56	—	13.38
1½" x ½"	P1@.193	Ea	5.83	8.58	—	14.41
1½" x ¾"	P1@.214	Ea	5.83	9.51	—	15.34
1½" x 1"	P1@.236	Ea	5.83	10.50	—	16.33
1½" x 1¼"	P1@.257	Ea	5.83	11.40	—	17.23
2" x ½"	P1@.239	Ea	8.79	10.60	—	19.39
2" x ¾"	P1@.261	Ea	8.79	11.60	—	20.39
2" x 1"	P1@.282	Ea	8.79	12.50	—	21.29
2" x 1¼"	P1@.304	Ea	8.79	13.50	—	22.29
2" x 1½"	P1@.325	Ea	8.79	14.40	—	23.19
2½" x 1"	P1@.325	Ea	23.90	14.40	—	38.30
2½" x 1¼"	P1@.347	Ea	23.90	15.40	—	39.30
2½" x 1½"	P1@.368	Ea	23.90	16.40	—	40.30
2½" x 2"	P1@.414	Ea	23.90	18.40	—	42.30
3" x 2"	P1@.457	Ea	28.00	20.30	—	48.30
3" x 2½"	P1@.500	Ea	28.00	22.20	—	50.20
4" x 2"	P1@.543	Ea	53.90	24.10	—	78.00
4" x 2½"	P1@.586	Ea	53.90	26.00	—	79.90
4" x 3"	P1@.629	Ea	53.90	28.00	—	81.90

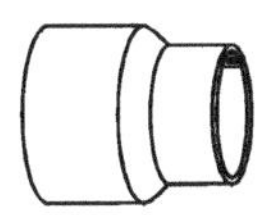

Type K copper adapter C x MPT with soft-soldered joint

½"	P1@.075	Ea	1.65	3.33	—	4.98
¾"	P1@.105	Ea	2.59	4.67	—	7.26
1"	P1@.134	Ea	6.44	5.96	—	12.40
1¼"	P1@.164	Ea	12.00	7.29	—	19.29
1½"	P1@.194	Ea	13.60	8.62	—	22.22
2"	P1@.259	Ea	16.90	11.50	—	28.40
2½"	P1@.319	Ea	54.90	14.20	—	69.10
3"	P1@.378	Ea	91.00	16.80	—	107.80

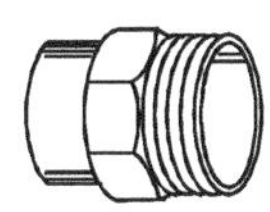

Copper, Type K with Soft-Soldered Joints

Description	Craft@Hrs	Unit	Material $	Labor $	Equipment $	Total $

Type K copper adapter C x FPT with soft-soldered joint

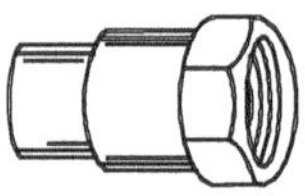

Description	Craft@Hrs	Unit	Material $	Labor $	Equipment $	Total $
½"	P1@.075	Ea	2.76	3.33	—	6.09
¾"	P1@.105	Ea	5.01	4.67	—	9.68
1"	P1@.134	Ea	7.56	5.96	—	13.52
1¼"	P1@.164	Ea	16.00	7.29	—	23.29
1½"	P1@.194	Ea	19.20	8.62	—	27.82
2"	P1@.259	Ea	28.60	11.50	—	40.10
2½"	P1@.319	Ea	72.00	14.20	—	86.20
3"	P1@.378	Ea	99.00	16.80	—	115.80

Type K copper flush bushing with soft-soldered joints

Description	Craft@Hrs	Unit	Material $	Labor $	Equipment $	Total $
½" x 3/8"	P1@.124	Ea	1.67	5.51	—	7.18
¾" x ½"	P1@.129	Ea	2.87	5.73	—	8.60
1" x ¾"	P1@.172	Ea	7.43	7.65	—	15.08
1" x ½"	P1@.172	Ea	8.48	7.65	—	16.13
1¼" x 1"	P1@.215	Ea	8.81	9.56	—	18.37
1½" x 1¼"	P1@.257	Ea	11.20	11.40	—	22.60
2" x 1½"	P1@.325	Ea	20.50	14.40	—	34.90

Type K copper union with soft-soldered joint

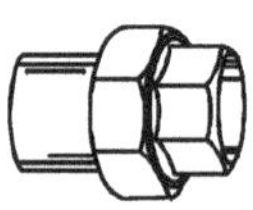

Description	Craft@Hrs	Unit	Material $	Labor $	Equipment $	Total $
½"	P1@.121	Ea	6.75	5.38	—	12.13
¾"	P1@.170	Ea	8.05	7.56	—	15.61
1"	P1@.218	Ea	18.10	9.69	—	27.79
1¼"	P1@.267	Ea	32.90	11.90	—	44.80
1½"	P1@.315	Ea	40.40	14.00	—	54.40
2"	P1@.421	Ea	58.20	18.70	—	76.90

Type K copper dielectric union with soft-soldered joint

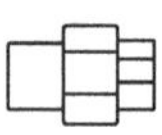

Description	Craft@Hrs	Unit	Material $	Labor $	Equipment $	Total $
½"	P1@.121	Ea	6.32	5.38	—	11.70
¾"	P1@.170	Ea	6.32	7.56	—	13.88
1"	P1@.218	Ea	13.10	9.69	—	22.79
1¼"	P1@.267	Ea	22.30	11.90	—	34.20
1½"	P1@.315	Ea	30.70	14.00	—	44.70
2"	P1@.421	Ea	41.00	18.70	—	59.70

Type K copper cap with soft-soldered joint

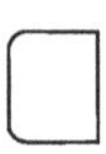

Description	Craft@Hrs	Unit	Material $	Labor $	Equipment $	Total $
½"	P1@.069	Ea	.56	3.07	—	3.63
¾"	P1@.096	Ea	1.12	4.27	—	5.39
1"	P1@.124	Ea	2.54	5.51	—	8.05
1¼"	P1@.151	Ea	3.64	6.71	—	10.35
1½"	P1@.178	Ea	5.31	7.91	—	13.22
2"	P1@.237	Ea	9.24	10.50	—	19.74
2½"	P1@.292	Ea	23.50	13.00	—	36.50
3"	P1@.347	Ea	36.70	15.40	—	52.10
4"	P1@.457	Ea	56.40	20.30	—	76.70

Description	Craft@Hrs	Unit	Material $	Labor $	Equipment $	Total $
Type K copper coupling with soft-soldered joints						
½"	P1@.107	Ea	.36	4.76	—	5.12
¾"	P1@.150	Ea	.91	6.67	—	7.58
1"	P1@.193	Ea	2.42	8.58	—	11.00
1¼"	P1@.236	Ea	4.21	10.50	—	14.71
1½"	P1@.278	Ea	6.09	12.40	—	18.49
2"	P1@.371	Ea	8.95	16.50	—	25.45
2½"	P1@.457	Ea	21.20	20.30	—	41.50
3"	P1@.543	Ea	32.30	24.10	—	56.40
4"	P1@.714	Ea	63.10	31.70	—	94.80

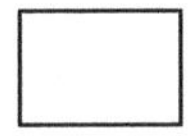

Class 125 bronze body gate valve, soft-soldered joints						
½"	P1@.200	Ea	22.60	8.89	—	31.49
¾"	P1@.249	Ea	28.50	11.10	—	39.60
1"	P1@.299	Ea	40.10	13.30	—	53.40
1¼"	P1@.398	Ea	50.40	17.70	—	68.10
1½"	P1@.448	Ea	67.60	19.90	—	87.50
2"	P1@.498	Ea	113.00	22.10	—	135.10
2½"	P1@.830	Ea	187.00	36.90	—	223.90
3"	P1@1.24	Ea	268.00	55.10	—	323.10

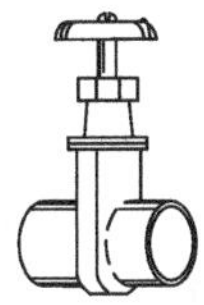

Class 125 iron body gate valve, flanged joints						
2"	P1@.500	Ea	435.00	22.20	—	457.20
2½"	P1@.600	Ea	586.00	26.70	—	612.70
3"	P1@.750	Ea	636.00	33.30	—	669.30
4"	P1@1.35	Ea	934.00	60.00	—	994.00
5"	P1@2.00	Ea	1,780.00	88.90	—	1,868.90
6"	P1@2.50	Ea	1,780.00	111.00	—	1,891.00

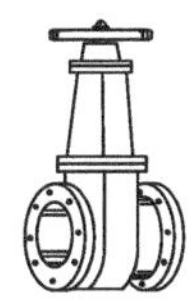

Class 125 bronze body globe valve, with soft-soldered joints						
½"	P1@.200	Ea	42.20	8.89	—	51.09
¾"	P1@.249	Ea	56.20	11.10	—	67.30
1"	P1@.299	Ea	96.90	13.30	—	110.20
1¼"	P1@.398	Ea	136.00	17.70	—	153.70
1½"	P1@.448	Ea	185.00	19.90	—	204.90
2"	P1@.498	Ea	299.00	22.10	—	321.10

Class 125 iron body globe valve, flanged ends						
2½"	P1@.600	Ea	493.00	26.70	—	519.70
3"	P1@.750	Ea	591.00	33.30	—	624.30
4"	P1@1.35	Ea	793.00	60.00	—	853.00

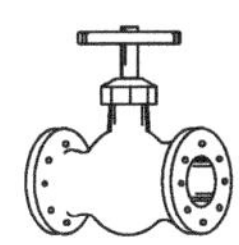

Description	Craft@Hrs	Unit	Material $	Labor $	Equipment $	Total $

200 PSIG iron body butterfly valve, lug-type, lever operated

2"	P1@.450	Ea	178.00	20.00	—	198.00
2½"	P1@.450	Ea	185.00	20.00	—	205.00
3"	P1@.550	Ea	195.00	24.40	—	219.40
4"	P1@.550	Ea	243.00	24.40	—	267.40

200 PSIG iron body butterfly valve, wafer-type, lever operated

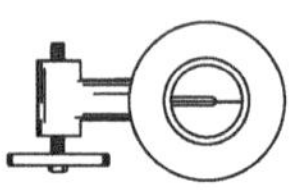

2"	P1@.450	Ea	164.00	20.00	—	184.00
2½"	P1@.450	Ea	167.00	20.00	—	187.00
3"	P1@.550	Ea	179.00	24.40	—	203.40
4"	P1@.550	Ea	217.00	24.40	—	241.40

Class 125 bronze body 2-piece ball valve, soft-soldered joints

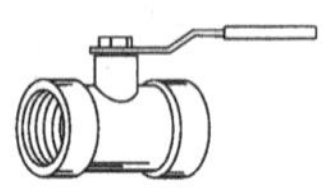

½"	P1@.200	Ea	11.60	8.89	—	20.49
¾"	P1@.249	Ea	13.90	11.10	—	25.00
1"	P1@.299	Ea	26.70	13.30	—	40.00
1¼"	P1@.398	Ea	44.60	17.70	—	62.30
1½"	P1@.448	Ea	60.30	19.90	—	80.20
2"	P1@.498	Ea	76.40	22.10	—	98.50
3"	P1@1.24	Ea	524.00	55.10	—	579.10
4"	P1@1.45	Ea	684.00	64.50	—	748.50

Class 125 bronze body swing check valve, soft-soldered joints

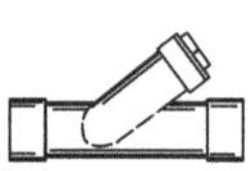

½"	P1@.200	Ea	23.90	8.89	—	32.79
¾"	P1@.249	Ea	34.10	11.10	—	45.20
1"	P1@.299	Ea	44.50	13.30	—	57.80
1¼"	P1@.398	Ea	63.60	17.70	—	81.30
1½"	P1@.448	Ea	89.60	19.90	—	109.50
2"	P1@.498	Ea	148.00	22.10	—	170.10
2½"	P1@.830	Ea	299.00	36.90	—	335.90
3"	P1@1.24	Ea	430.00	55.10	—	485.10

Class 125 iron body swing check valve, flanged ends

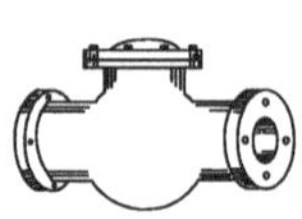

2"	P1@.500	Ea	209.00	22.20	—	231.20
2½"	P1@.600	Ea	264.00	26.70	—	290.70
3"	P1@.750	Ea	329.00	33.30	—	362.30
4"	P1@1.35	Ea	483.00	60.00	—	543.00

Class 125 iron body silent check valve, wafer-type

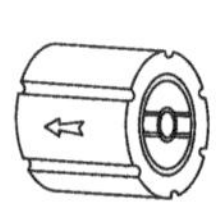

2"	P1@.500	Ea	153.00	22.20	—	175.20
2½"	P1@.600	Ea	170.00	26.70	—	196.70
3"	P1@.750	Ea	195.00	33.30	—	228.30
4"	P1@1.35	Ea	260.00	60.00	—	320.00

Description	Craft@Hrs	Unit	Material $	Labor $	Equipment $	Total $

Class 125 bronze body strainer, threaded ends

½"	P1@.230	Ea	37.40	10.20	—	47.60
¾"	P1@.260	Ea	49.10	11.60	—	60.70
1"	P1@.330	Ea	60.20	14.70	—	74.90
1¼"	P1@.440	Ea	84.20	19.60	—	103.80
1½"	P1@.495	Ea	126.00	22.00	—	148.00
2"	P1@.550	Ea	220.00	24.40	—	244.40

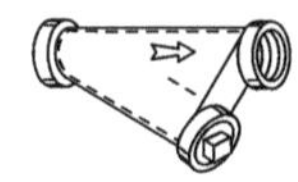

Class 125 iron body strainer, flanged ends

2"	P1@.500	Ea	168.00	22.20	—	190.20
2½"	P1@.600	Ea	187.00	26.70	—	213.70
3"	P1@.750	Ea	218.00	33.30	—	251.30
4"	P1@1.35	Ea	367.00	60.00	—	427.00

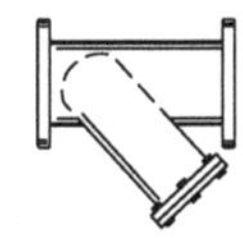

Installation of copper 2-way control valve, threaded joints

½"	P1@.210	Ea	—	9.33	—	9.33
¾"	P1@.275	Ea	—	12.20	—	12.20
1"	P1@.350	Ea	—	15.60	—	15.60
1¼"	P1@.430	Ea	—	19.10	—	19.10
1½"	P1@.505	Ea	—	22.40	—	22.40
2"	P1@.675	Ea	—	30.00	—	30.00
2½"	P1@.830	Ea	—	36.90	—	36.90
3"	P1@.990	Ea	—	44.00	—	44.00

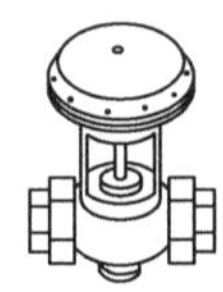

Installation of copper 3-way control valve, threaded joints

½"	P1@.260	Ea	—	11.60	—	11.60
¾"	P1@.365	Ea	—	16.20	—	16.20
1"	P1@.475	Ea	—	21.10	—	21.10
1¼"	P1@.575	Ea	—	25.60	—	25.60
1½"	P1@.680	Ea	—	30.20	—	30.20
2"	P1@.910	Ea	—	40.40	—	40.40
2½"	P1@1.12	Ea	—	49.80	—	49.80
3"	P1@1.33	Ea	—	59.10	—	59.10

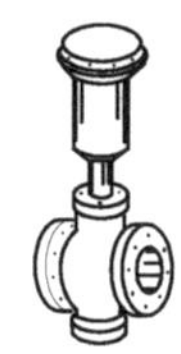

Companion flange, 150 pound cast brass

2"	P1@.290	Ea	330.00	12.90	—	342.90
2½"	P1@.380	Ea	410.00	16.90	—	426.90
3"	P1@.460	Ea	410.00	20.40	—	430.40
4"	P1@.600	Ea	655.00	26.70	—	681.70

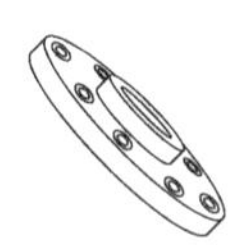

Bolt and gasket sets

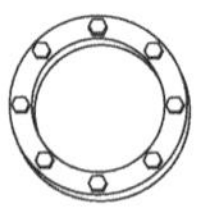

Description	Craft@Hrs	Unit	Material $	Labor $	Equipment $	Total $
2"	P1@.500	Ea	4.65	22.20	—	26.85
2½"	P1@.650	Ea	5.46	28.90	—	34.36
3"	P1@.750	Ea	9.30	33.30	—	42.60
4"	P1@1.00	Ea	15.60	44.50	—	60.10

Thermometer with well

Description	Craft@Hrs	Unit	Material $	Labor $	Equipment $	Total $
7"	P1@.250	Ea	183.00	11.10	—	194.10
9"	P1@.250	Ea	188.00	11.10	—	199.10

Dial-type pressure gauge

Description	Craft@Hrs	Unit	Material $	Labor $	Equipment $	Total $
2½"	P1@.200	Ea	37.80	8.89	—	46.69
3½"	P1@.200	Ea	50.20	8.89	—	59.09

Pressure/temperature tap

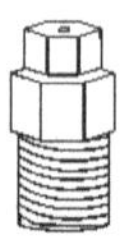

Description	Craft@Hrs	Unit	Material $	Labor $	Equipment $	Total $
Tap	P1@.150	Ea	18.30	6.67	—	24.97

Hanger with swivel assembly

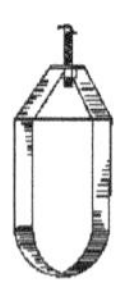

Description	Craft@Hrs	Unit	Material $	Labor $	Equipment $	Total $
½"	P1@.250	Ea	4.38	11.10	—	15.48
¾"	P1@.250	Ea	4.85	11.10	—	15.95
1"	P1@.250	Ea	5.07	11.10	—	16.17
1¼"	P1@.300	Ea	5.20	13.30	—	18.50
1½"	P1@.300	Ea	5.59	13.30	—	18.89
2"	P1@.300	Ea	5.83	13.30	—	19.13
2½"	P1@.350	Ea	7.93	15.60	—	23.53
3"	P1@.350	Ea	9.82	15.60	—	25.42
4"	P1@.350	Ea	10.80	15.60	—	26.40

Riser clamp

Description	Craft@Hrs	Unit	Material $	Labor $	Equipment $	Total $
½"	P1@.100	Ea	3.24	4.45	—	7.69
¾"	P1@.100	Ea	4.79	4.45	—	9.24
1"	P1@.100	Ea	4.83	4.45	—	9.28
1¼"	P1@.105	Ea	5.83	4.67	—	10.50
1½"	P1@.110	Ea	6.16	4.89	—	11.05
2"	P1@.115	Ea	6.52	5.11	—	11.63
2½"	P1@.120	Ea	6.86	5.33	—	12.19
3"	P1@.120	Ea	7.43	5.33	—	12.76
4"	P1@.125	Ea	9.19	5.56	—	14.75

Type L hard-drawn copper pipe with wrought copper fittings and brazed joints is used in a wide variety of plumbing and HVAC systems such as potable water, heating hot water, chilled water, compressed air, refrigerant and A.C. condensate.

Brazed joints are those made with silver or other alloy filler metals having melting points at, or above, 1,000 degrees F. Maximum working pressure/temperature relationships for brazed joints are approximately as follows:

Maximum Working Pressure (PSIG)*			
Water Temperature (degrees F.)	Nominal Pipe Size (inches)		
	Up to 1	1¼ to 2	2½ to 4
Up to 350	270	190	150

**For copper pipe and solder-type fittings using brazing alloys melting at, or above, 1,000 degrees F.*

This section has been arranged to save the estimator's time by including all normally-used system components such as pipe, fittings, valves, hanger assemblies, riser clamps and miscellaneous items under one heading. Additional items can be found under "Plumbing and Piping Specialties." The cost estimates in this section are based on the conditions, limitations and wage rates described in the section "How to Use This Book" beginning on page 5.

Type L copper pipe with brazed joints installed horizontally. Complete installation including six to twelve wrought copper tees and six to twelve wrought copper elbows every 100', and hangers spaced to meet plumbing code *(a tee & elbow every 8.3' for ½" dia., a tee and elbow every 16.6' for 4" dia.)*

Description	Craft@Hrs	Unit	Material $	Labor $	Equipment $	Total $
½"	P1@.116	LF	3.10	5.16	—	8.26
¾"	P1@.126	LF	4.77	5.60	—	10.37
1"	P1@.146	LF	8.57	6.49	—	15.06
1¼"	P1@.166	LF	12.20	7.38	—	19.58
1½"	P1@.186	LF	16.20	8.27	—	24.47
2"	P1@.216	LF	29.50	9.60	—	39.10
2½"	P1@.246	LF	52.10	10.90	—	63.00
3"	P1@.276	LF	72.10	12.30	—	84.40
4"	P1@.316	LF	104.00	14.00	—	118.00

Description	Craft@Hrs	Unit	Material $	Labor $	Equipment $	Total $

Type L copper pipe with brazed joints, installed risers. Complete installation including a wrought copper reducing tee every floor and a riser clamp every other floor.

½"	P1@.063	LF	3.06	2.80	—	5.86
¾"	P1@.073	LF	4.83	3.24	—	8.07
1"	P1@.073	LF	8.20	3.24	—	11.44
1¼"	P1@.083	LF	11.60	3.69	—	15.29
1½"	P1@.103	LF	14.40	4.58	—	18.98
2"	P1@.133	LF	37.90	5.91	—	43.81
2½"	P1@.153	LF	40.80	6.80	—	47.60
3"	P1@.173	LF	53.00	7.69	—	60.69
4"	P1@.193	LF	78.80	8.58	—	87.38

Type L copper pipe with brazed joints, pipe only

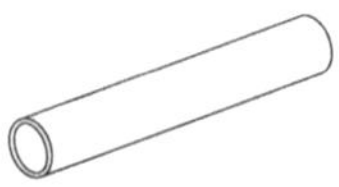

½"	P1@.032	LF	2.60	1.42	—	4.02
¾"	P1@.035	LF	4.10	1.56	—	5.66
1"	P1@.038	LF	5.67	1.69	—	7.36
1¼"	P1@.042	LF	8.14	1.87	—	10.01
1½"	P1@.046	LF	10.40	2.04	—	12.44
2"	P1@.053	LF	17.80	2.36	—	20.16
2½"	P1@.060	LF	26.10	2.67	—	28.77
3"	P1@.066	LF	35.10	2.93	—	38.03
4"	P1@.080	LF	55.10	3.56	—	58.66

Type L copper 45-degree ell C x C with brazed joints

½"	P1@.129	Ea	.97	5.73	—	6.70
¾"	P1@.181	Ea	1.86	8.05	—	9.91
1"	P1@.232	Ea	5.49	10.30	—	15.79
1¼"	P1@.284	Ea	7.43	12.60	—	20.03
1½"	P1@.335	Ea	11.10	14.90	—	26.00
2"	P1@.447	Ea	17.50	19.90	—	37.40
2½"	P1@.550	Ea	44.80	24.40	—	69.20
3"	P1@.654	Ea	59.20	29.10	—	88.30
4"	P1@.860	Ea	114.00	38.20	—	152.20

Type L copper 90-degree ell C x C with brazed joints

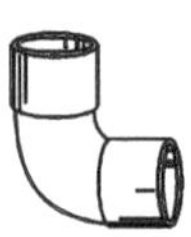

½"	P1@.129	Ea	.57	5.73	—	6.30
¾"	P1@.181	Ea	1.20	8.05	—	9.25
1"	P1@.232	Ea	4.19	10.30	—	14.49
1¼"	P1@.284	Ea	6.76	12.60	—	19.36
1½"	P1@.335	Ea	10.20	14.90	—	25.10
2"	P1@.447	Ea	17.70	19.90	—	37.60
2½"	P1@.550	Ea	37.00	24.40	—	61.40
3"	P1@.654	Ea	46.90	29.10	—	76.00
4"	P1@.860	Ea	99.20	38.20	—	137.40

Description	Craft@Hrs	Unit	Material $	Labor $	Equipment $	Total $

Type L copper 90-degree ell ftg. x C with brazed joint

½"	P1@.129	Ea	.92	5.73	—	6.65
¾"	P1@.181	Ea	1.98	8.05	—	10.03
1"	P1@.232	Ea	5.93	10.30	—	16.23
1¼"	P1@.284	Ea	8.26	12.60	—	20.86
1½"	P1@.335	Ea	12.30	14.90	—	27.20
2"	P1@.447	Ea	24.20	19.90	—	44.10
2½"	P1@.550	Ea	50.40	24.40	—	74.80
3"	P1@.654	Ea	56.70	29.10	—	85.80
4"	P1@.860	Ea	114.00	38.20	—	152.20

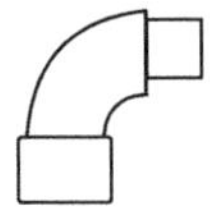

Type L copper tee C x C x C with brazed joints

½"	P1@.156	Ea	.96	6.93	—	7.89
¾"	P1@.204	Ea	2.60	9.07	—	11.67
1"	P1@.263	Ea	9.81	11.70	—	21.51
1¼"	P1@.322	Ea	15.60	14.30	—	29.90
1½"	P1@.380	Ea	22.90	16.90	—	39.80
2"	P1@.507	Ea	35.20	22.50	—	57.70
2½"	P1@.624	Ea	71.60	27.70	—	99.30
3"	P1@.741	Ea	111.00	32.90	—	143.90
4"	P1@.975	Ea	185.00	43.30	—	228.30

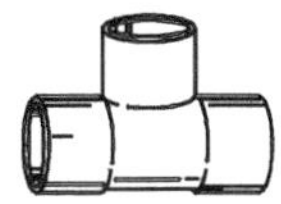

Type L copper branch reducing tee C x C x C with brazed joints

½" x 3/8"	P1@.129	Ea	1.87	5.73	—	7.60
¾" x ½"	P1@.182	Ea	3.97	8.09	—	12.06
1" x ½"	P1@.209	Ea	7.78	9.29	—	17.07
1" x ¾"	P1@.234	Ea	7.78	10.40	—	18.18
1¼" x ½"	P1@.241	Ea	11.70	10.70	—	22.40
1¼" x ¾"	P1@.264	Ea	11.70	11.70	—	23.40
1¼" x 1"	P1@.287	Ea	11.70	12.80	—	24.50
1½" x ½"	P1@.291	Ea	14.40	12.90	—	27.30
1½" x ¾"	P1@.307	Ea	14.40	13.60	—	28.00
1½" x 1"	P1@.323	Ea	14.40	14.40	—	28.80
1½" x 1¼"	P1@.339	Ea	14.40	15.10	—	29.50
2" x ½"	P1@.348	Ea	18.80	15.50	—	34.30
2" x ¾"	P1@.373	Ea	18.80	16.60	—	35.40
2" x 1"	P1@.399	Ea	18.80	17.70	—	36.50
2" x 1¼"	P1@.426	Ea	18.80	18.90	—	37.70
2" x 1½"	P1@.452	Ea	18.80	20.10	—	38.90
2½" x ½"	P1@.468	Ea	56.80	20.80	—	77.60
2½" x ¾"	P1@.485	Ea	56.80	21.60	—	78.40
2½" x 1"	P1@.502	Ea	56.80	22.30	—	79.10
2½" x 1½"	P1@.537	Ea	56.80	23.90	—	80.70
2½" x 2"	P1@.555	Ea	56.80	24.70	—	81.50
3" x 1¼"	P1@.562	Ea	64.20	25.00	—	89.20
3" x 2"	P1@.626	Ea	66.40	27.80	—	94.20
3" x 2½"	P1@.660	Ea	66.40	29.30	—	95.70
4" x 1¼"	P1@.669	Ea	116.00	29.70	—	145.70
4" x 1½"	P1@.718	Ea	116.00	31.90	—	147.90
4" x 2"	P1@.768	Ea	116.00	34.10	—	150.10
4" x 2½"	P1@.819	Ea	116.00	36.40	—	152.40
4" x 3"	P1@.868	Ea	116.00	38.60	—	154.60

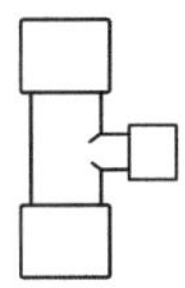

Description	Craft@Hrs	Unit	Material $	Labor $	Equipment $	Total $
Type L copper reducer with brazed joints						
½" x 3/8"	P1@.148	Ea	.67	6.58	—	7.25
¾" x ½"	P1@.155	Ea	1.26	6.89	—	8.15
1" x ¾"	P1@.206	Ea	2.28	9.16	—	11.44
1¼" x ½"	P1@.206	Ea	3.82	9.16	—	12.98
1¼" x ¾"	P1@.232	Ea	3.82	10.30	—	14.12
1¼" x 1"	P1@.258	Ea	3.82	11.50	—	15.32
1½" x ½"	P1@.232	Ea	5.83	10.30	—	16.13
1½" x ¾"	P1@.257	Ea	5.83	11.40	—	17.23
1½" x 1"	P1@.283	Ea	5.83	12.60	—	18.43
1½" x 1¼"	P1@.308	Ea	5.83	13.70	—	19.53
2" x ½"	P1@.287	Ea	8.79	12.80	—	21.59
2" x ¾"	P1@.277	Ea	8.79	12.30	—	21.09
2" x 1"	P1@.338	Ea	8.79	15.00	—	23.79
2" x 1¼"	P1@.365	Ea	8.79	16.20	—	24.99
2" x 1½"	P1@.390	Ea	8.79	17.30	—	26.09
2½" x 1"	P1@.390	Ea	23.90	17.30	—	41.20
2½" x 1¼"	P1@.416	Ea	23.90	18.50	—	42.40
2½" x 1½"	P1@.442	Ea	23.90	19.60	—	43.50
2½" x 2"	P1@.497	Ea	23.90	22.10	—	46.00
3" x 1¼"	P1@.468	Ea	28.00	20.80	—	48.80
3" x 1½"	P1@.493	Ea	28.00	21.90	—	49.90
3" x 2"	P1@.548	Ea	28.00	24.40	—	52.40
3" x 2½"	P1@.600	Ea	28.00	26.70	—	54.70
4" x 2"	P1@.652	Ea	53.90	29.00	—	82.90
4" x 2½"	P1@.704	Ea	53.90	31.30	—	85.20
4" x 3"	P1@.755	Ea	53.90	33.60	—	87.50

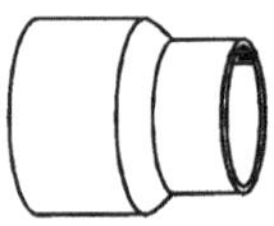

Description	Craft@Hrs	Unit	Material $	Labor $	Equipment $	Total $
Type L copper adapter C x MPT with brazed joint						
½"	P1@.090	Ea	1.65	4.00	—	5.65
¾"	P1@.126	Ea	2.59	5.60	—	8.19
1"	P1@.162	Ea	6.44	7.20	—	13.64
1¼"	P1@.198	Ea	12.00	8.80	—	20.80
1½"	P1@.234	Ea	13.60	10.40	—	24.00
2"	P1@.312	Ea	16.90	13.90	—	30.80
2½"	P1@.384	Ea	54.90	17.10	—	72.00
3"	P1@.456	Ea	91.00	20.30	—	111.30

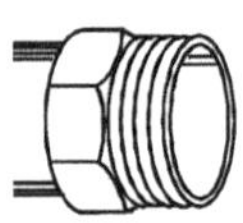

Description	Craft@Hrs	Unit	Material $	Labor $	Equipment $	Total $
Type L copper adapter C x FPT with brazed joint						
½"	P1@.090	Ea	2.76	4.00	—	6.76
¾"	P1@.126	Ea	5.01	5.60	—	10.61
1"	P1@.162	Ea	7.56	7.20	—	14.76
1¼"	P1@.198	Ea	16.00	8.80	—	24.80
1½"	P1@.234	Ea	19.20	10.40	—	29.60
2"	P1@.312	Ea	28.60	13.90	—	42.50
2½"	P1@.384	Ea	72.00	17.10	—	89.10
3"	P1@.456	Ea	99.00	20.30	—	119.30

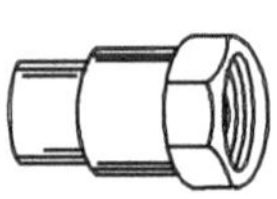

Type L copper flush bushing with brazed joints

Description	Craft@Hrs	Unit	Material $	Labor $	Equipment $	Total $
½" x 3/8"	P1@.148	Ea	1.67	6.58	—	8.25
¾" x ½"	P1@.155	Ea	2.87	6.89	—	9.76
1" x ¾"	P1@.206	Ea	7.43	9.16	—	16.59
1" x ½"	P1@.206	Ea	8.48	9.16	—	17.64
1¼" x 1"	P1@.258	Ea	8.81	11.50	—	20.31
1½" x 1¼"	P1@.308	Ea	11.20	13.70	—	24.90
2" x 1½"	P1@.390	Ea	20.50	17.30	—	37.80

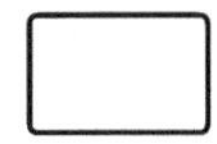

Type L copper union with brazed joints

Description	Craft@Hrs	Unit	Material $	Labor $	Equipment $	Total $
½"	P1@.146	Ea	6.75	6.49	—	13.24
¾"	P1@.205	Ea	8.05	9.11	—	17.16
1"	P1@.263	Ea	18.10	11.70	—	29.80
1¼"	P1@.322	Ea	32.90	14.30	—	47.20
1½"	P1@.380	Ea	40.40	16.90	—	57.30
2"	P1@.507	Ea	58.20	22.50	—	80.70

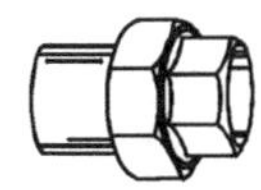

Type L copper dielectric union with brazed joints

Description	Craft@Hrs	Unit	Material $	Labor $	Equipment $	Total $
½"	P1@.146	Ea	6.32	6.49	—	12.81
¾"	P1@.205	Ea	6.32	9.11	—	15.43
1"	P1@.263	Ea	13.10	11.70	—	24.80
1¼"	P1@.322	Ea	22.30	14.30	—	36.60
1½"	P1@.380	Ea	30.70	16.90	—	47.60
2"	P1@.507	Ea	41.00	22.50	—	63.50

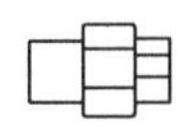

Type L copper cap with brazed joints

Description	Craft@Hrs	Unit	Material $	Labor $	Equipment $	Total $
½"	P1@.083	Ea	.56	3.69	—	4.25
¾"	P1@.116	Ea	1.12	5.16	—	6.28
1"	P1@.149	Ea	2.54	6.62	—	9.16
1¼"	P1@.182	Ea	3.64	8.09	—	11.73
1½"	P1@.215	Ea	5.31	9.56	—	14.87
2"	P1@.286	Ea	9.24	12.70	—	21.94
2½"	P1@.352	Ea	23.50	15.60	—	39.10
3"	P1@.418	Ea	36.70	18.60	—	55.30
4"	P1@.550	Ea	56.40	24.40	—	80.80

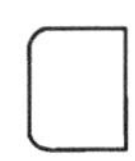

Type L copper coupling with brazed joints

Description	Craft@Hrs	Unit	Material $	Labor $	Equipment $	Total $
½"	P1@.129	Ea	.38	5.73	—	6.11
¾"	P1@.181	Ea	.90	8.05	—	8.95
1"	P1@.232	Ea	2.41	10.30	—	12.71
1¼"	P1@.284	Ea	4.20	12.60	—	16.80
1½"	P1@.335	Ea	6.07	14.90	—	20.97
2"	P1@.447	Ea	8.93	19.90	—	28.83
2½"	P1@.550	Ea	21.10	24.40	—	45.50
3"	P1@.654	Ea	32.10	29.10	—	61.20
4"	P1@.860	Ea	63.00	38.20	—	101.20

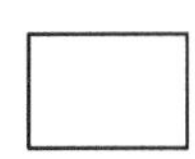

Description	Craft@Hrs	Unit	Material $	Labor $	Equipment $	Total $

Class 125 bronze body gate valve, brazed joints

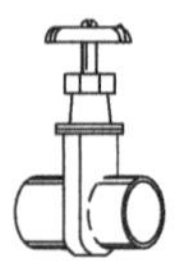

½"	P1@.240	Ea	22.80	10.70	—	33.50
¾"	P1@.300	Ea	28.50	13.30	—	41.80
1"	P1@.360	Ea	40.10	16.00	—	56.10
1¼"	P1@.480	Ea	50.40	21.30	—	71.70
1½"	P1@.540	Ea	68.00	24.00	—	92.00
2"	P1@.600	Ea	114.00	26.70	—	140.70
2½"	P1@1.00	Ea	188.00	44.50	—	232.50
3"	P1@1.50	Ea	270.00	66.70	—	336.70

Class 125 iron body gate valve, flanged joints

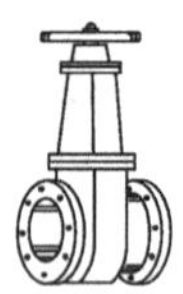

2"	P1@.500	Ea	435.00	22.20	—	457.20
2½"	P1@.600	Ea	586.00	26.70	—	612.70
3"	P1@.750	Ea	636.00	33.30	—	669.30
4"	P1@1.35	Ea	934.00	60.00	—	994.00
5"	P1@2.00	Ea	1,780.00	88.90	—	1,868.90
6"	P1@2.50	Ea	1,780.00	111.00	—	1,891.00

Class 125 bronze body globe valve, with brazed joints

½"	P1@.240	Ea	42.20	10.70	—	52.90
¾"	P1@.300	Ea	56.20	13.30	—	69.50
1"	P1@.360	Ea	96.90	16.00	—	112.90
1¼"	P1@.480	Ea	136.00	21.30	—	157.30
1½"	P1@.540	Ea	185.00	24.00	—	209.00
2"	P1@.600	Ea	299.00	26.70	—	325.70

Class 125 iron body globe valve, flanged joints

2½"	P1@.650	Ea	493.00	28.90	—	521.90
3"	P1@.750	Ea	591.00	33.30	—	624.30
4"	P1@1.35	Ea	798.00	60.00	—	858.00

200 PSIG iron body butterfly valve, lug-type, lever operated

2"	P1@.450	Ea	179.00	20.00	—	199.00
2½"	P1@.450	Ea	186.00	20.00	—	206.00
3"	P1@.550	Ea	196.00	24.40	—	220.40
4"	P1@.550	Ea	244.00	24.40	—	268.40

200 PSIG iron body butterfly valve, wafer-type, lever operated

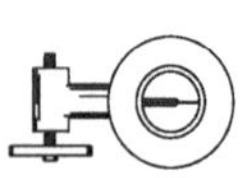

2"	P1@.450	Ea	164.00	20.00	—	184.00
2½"	P1@.450	Ea	167.00	20.00	—	187.00
3"	P1@.550	Ea	179.00	24.40	—	203.40
4"	P1@.550	Ea	217.00	24.40	—	241.40

Description	Craft@Hrs	Unit	Material $	Labor $	Equipment $	Total $
Class 125 bronze body 2-piece ball valve, brazed joints						
½"	P1@.240	Ea	11.60	10.70	—	22.30
¾"	P1@.300	Ea	15.40	13.30	—	28.70
1"	P1@.360	Ea	26.70	16.00	—	42.70
1¼"	P1@.480	Ea	44.60	21.30	—	65.90
1½"	P1@.540	Ea	60.30	24.00	—	84.30
2"	P1@.600	Ea	76.40	26.70	—	103.10
3"	P1@1.50	Ea	524.00	66.70	—	590.70
4"	P1@1.80	Ea	684.00	80.00	—	764.00
Class 125 bronze body swing check valve, brazed joints						
½"	P1@.240	Ea	23.90	10.70	—	34.60
¾"	P1@.300	Ea	34.10	13.30	—	47.40
1"	P1@.360	Ea	44.50	16.00	—	60.50
1¼"	P1@.480	Ea	63.60	21.30	—	84.90
1½"	P1@.540	Ea	89.60	24.00	—	113.60
2"	P1@.600	Ea	148.00	26.70	—	174.70
2½"	P1@1.00	Ea	299.00	44.50	—	343.50
3"	P1@1.50	Ea	430.00	66.70	—	496.70
Class 125 iron body swing check valve, flanged joints						
2"	P1@.500	Ea	209.00	22.20	—	231.20
2½"	P1@.600	Ea	264.00	26.70	—	290.70
3"	P1@.750	Ea	329.00	33.30	—	362.30
4"	P1@1.35	Ea	483.00	60.00	—	543.00
Class 125 iron body silent check valve, wafer-type						
2"	P1@.500	Ea	153.00	22.20	—	175.20
2½"	P1@.600	Ea	170.00	26.70	—	196.70
3"	P1@.750	Ea	195.00	33.30	—	228.30
4"	P1@1.35	Ea	260.00	60.00	—	320.00
Class 125 bronze body strainer, threaded joints						
½"	P1@.230	Ea	37.40	10.20	—	47.60
¾"	P1@.260	Ea	49.10	11.60	—	60.70
1"	P1@.330	Ea	60.20	14.70	—	74.90
1¼"	P1@.440	Ea	84.20	19.60	—	103.80
1½"	P1@.495	Ea	126.00	22.00	—	148.00
2"	P1@.550	Ea	220.00	24.40	—	244.40
Class 125 iron body strainer, flanged joints						
2"	P1@.500	Ea	168.00	22.20	—	190.20
2½"	P1@.600	Ea	187.00	26.70	—	213.70
3"	P1@.750	Ea	218.00	33.30	—	251.30
4"	P1@1.35	Ea	367.00	60.00	—	427.00

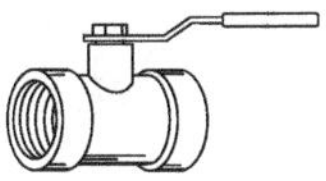
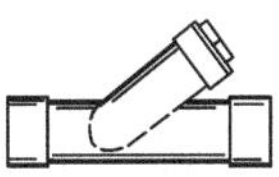
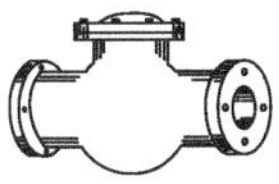
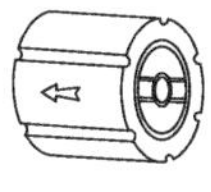
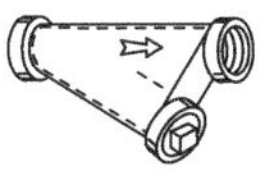
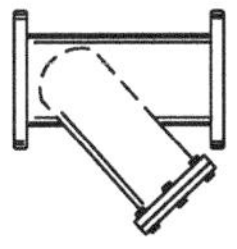

Description	Craft@Hrs	Unit	Material $	Labor $	Equipment $	Total $

Installation of 2-way control valve, threaded joints

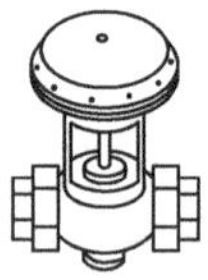

½"	P1@.210	Ea	—	9.33	—	9.33
¾"	P1@.275	Ea	—	12.20	—	12.20
1"	P1@.350	Ea	—	15.60	—	15.60
1¼"	P1@.430	Ea	—	19.10	—	19.10
1½"	P1@.505	Ea	—	22.40	—	22.40
2"	P1@.675	Ea	—	30.00	—	30.00
2½"	P1@.830	Ea	—	36.90	—	36.90
3"	P1@.990	Ea	—	44.00	—	44.00

Installation of 3-way control valve, threaded joints

½"	P1@.260	Ea	—	11.60	—	11.60
¾"	P1@.365	Ea	—	16.20	—	16.20
1"	P1@.475	Ea	—	21.10	—	21.10
1¼"	P1@.575	Ea	—	25.60	—	25.60
1½"	P1@.680	Ea	—	30.20	—	30.20
2"	P1@.910	Ea	—	40.40	—	40.40
2½"	P1@1.12	Ea	—	49.80	—	49.80
3"	P1@1.33	Ea	—	59.10	—	59.10

Companion flange, 150 pound cast brass

2"	P1@.290	Ea	327.00	12.90	—	339.90
2½"	P1@.380	Ea	403.00	16.90	—	419.90
3"	P1@.460	Ea	403.00	20.40	—	423.40
4"	P1@.600	Ea	641.00	26.70	—	667.70

Bolt and gasket sets

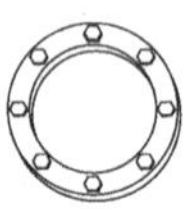

2"	P1@.500	Ea	4.60	22.20	—	26.80
2½"	P1@.650	Ea	5.38	28.90	—	34.28
3"	P1@.750	Ea	9.12	33.30	—	42.42
4"	P1@1.00	Ea	15.30	44.50	—	59.80

Thermometer with well

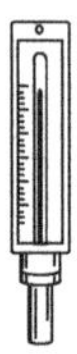

7"	P1@.250	Ea	183.00	11.10	—	194.10
9"	P1@.250	Ea	188.00	11.10	—	199.10

Dial-type pressure gauge

2½"	P1@.200	Ea	38.30	8.89	—	47.19
3½"	P1@.200	Ea	50.50	8.89	—	59.39

Type L hard-drawn copper pipe with wrought copper fittings and soft-soldered joints is used in a wide variety of plumbing and HVAC systems such as potable water, heating hot water, chilled water, and A.C. condensate.

Soft-soldered joints are those made with solders having melting points in the 350 degree F. to 500 degree F. range. Maximum working temperature/pressure relationships for soft-soldered joints are approximately as follows:

Maximum Working Pressures (PSIG)*				
Soft-solder Type	Water Temperature (degrees F.)	Nominal Pipe Size (inches)		
		Up to 1	1¼ to 2	2½ x 4
50-50 tin-lead**	100	200	175	150
	150	150	125	100
	200	100	90	75
	250	85	75	50
95-5 tin-antimony	100	500	400	300
	150	400	350	275
	200	300	250	200
	250	200	175	150

**For copper pipe and solder-type fittings using soft-solders melting at approximately 350 degrees F. to 500 degrees F.*

***The use of any solder containing lead is not allowed in potable water systems.*

This section has been arranged to save the estimator's time by including all normally-used system components such as pipe, fittings, valves, hanger assemblies, riser clamps and miscellaneous items under one heading. Additional items can be found under "Plumbing and Piping Specialties." The cost estimates in this section are based on the conditions, limitations and wage rates described in the section "How to Use This Book" beginning on page 5.

Type L copper pipe with soft-soldered joints installed horizontally. Complete installation including eight wrought copper tees and eight wrought copper elbows every 100', and hangers spaced to meet plumbing code. Use these figures for preliminary estimates.

Description	Craft@Hrs	Unit	Material $	Labor $	Equipment $	Total $
½"	P1@.111	LF	2.85	4.93	—	7.78
¾"	P1@.121	LF	4.42	5.38	—	9.80
1"	P1@.141	LF	7.99	6.27	—	14.26
1¼"	P1@.161	LF	11.40	7.16	—	18.56
1½"	P1@.181	LF	15.40	8.05	—	23.45
2"	P1@.211	LF	27.80	9.38	—	37.18
2½"	P1@.241	LF	48.00	10.70	—	58.70
3"	P1@.271	LF	67.20	12.00	—	79.20
4"	P1@.311	LF	96.90	13.80	—	110.70

Description	Craft@Hrs	Unit	Material $	Labor $	Equipment $	Total $

Type L copper pipe with soft-soldered joints, installed risers.

Complete installation including a wrought copper reducing tee every floor and a riser clamp every other floor.

Description	Craft@Hrs	Unit	Material $	Labor $	Equipment $	Total $
½"	P1@.061	LF	3.12	2.71	—	5.83
¾"	P1@.071	LF	4.83	3.16	—	7.99
1"	P1@.076	LF	8.20	3.38	—	11.58
1¼"	P1@.081	LF	11.60	3.60	—	15.20
1½"	P1@.101	LF	14.40	4.49	—	18.89
2"	P1@.131	LF	37.90	5.82	—	43.72
2½"	P1@.151	LF	40.80	6.71	—	47.51
3"	P1@.171	LF	53.00	7.60	—	60.60
4"	P1@.191	LF	72.00	8.49	—	80.49

Type L copper pipe with soft-soldered joints, pipe only

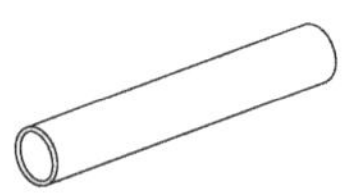

Description	Craft@Hrs	Unit	Material $	Labor $	Equipment $	Total $
½"	P1@.032	LF	2.60	1.42	—	4.02
¾"	P1@.035	LF	4.10	1.56	—	5.66
1"	P1@.038	LF	5.67	1.69	—	7.36
1¼"	P1@.042	LF	8.14	1.87	—	10.01
1½"	P1@.046	LF	10.40	2.04	—	12.44
2"	P1@.053	LF	17.80	2.36	—	20.16
2½"	P1@.060	LF	26.10	2.67	—	28.77
3"	P1@.066	LF	35.10	2.93	—	38.03
4"	P1@.080	LF	55.10	3.56	—	58.66

Type L copper 45-degree ell C x C with soft-soldered joints

Description	Craft@Hrs	Unit	Material $	Labor $	Equipment $	Total $
½"	P1@.107	Ea	.97	4.76	—	5.73
¾"	P1@.150	Ea	1.86	6.67	—	8.53
1"	P1@.193	Ea	6.71	8.58	—	15.29
1¼"	P1@.236	Ea	9.08	10.50	—	19.58
1½"	P1@.278	Ea	11.10	12.40	—	23.50
2"	P1@.371	Ea	17.50	16.50	—	34.00
2½"	P1@.457	Ea	44.80	20.30	—	65.10
3"	P1@.543	Ea	59.20	24.10	—	83.30
4"	P1@.714	Ea	114.00	31.70	—	145.70

Type L copper 90-degree ell C x C with soft-soldered joints

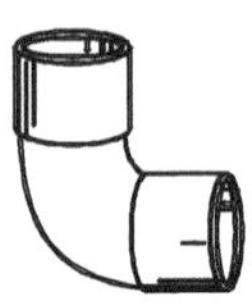

Description	Craft@Hrs	Unit	Material $	Labor $	Equipment $	Total $
½"	P1@.107	Ea	.57	4.76	—	5.33
¾"	P1@.150	Ea	1.20	6.67	—	7.87
1"	P1@.193	Ea	4.19	8.58	—	12.77
1¼"	P1@.236	Ea	6.76	10.50	—	17.26
1½"	P1@.278	Ea	10.20	12.40	—	22.60
2"	P1@.371	Ea	17.70	16.50	—	34.20
2½"	P1@.457	Ea	37.00	20.30	—	57.30
3"	P1@.543	Ea	46.90	24.10	—	71.00
4"	P1@.714	Ea	99.20	31.70	—	130.90

Description	Craft@Hrs	Unit	Material $	Labor $	Equipment $	Total $

Type L copper 90-degree ell ftg. x C with soft-soldered joint

½"	P1@.107	Ea	1.20	4.76	—	5.96
¾"	P1@.150	Ea	3.95	6.67	—	10.62
1"	P1@.193	Ea	5.93	8.58	—	14.51
1¼"	P1@.236	Ea	8.26	10.50	—	18.76
1½"	P1@.278	Ea	12.30	12.40	—	24.70
2"	P1@.371	Ea	24.20	16.50	—	40.70
2½"	P1@.457	Ea	50.40	20.30	—	70.70
3"	P1@.543	Ea	56.70	24.10	—	80.80
4"	P1@.714	Ea	151.00	31.70	—	182.70

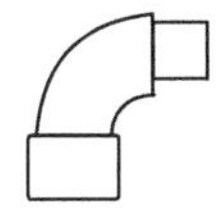

Type L copper tee C x C x C with soft-soldered joints

½"	P1@.129	Ea	.92	5.73	—	6.65
¾"	P1@.181	Ea	2.48	8.05	—	10.53
1"	P1@.233	Ea	9.34	10.40	—	19.74
1¼"	P1@.285	Ea	14.90	12.70	—	27.60
1½"	P1@.337	Ea	21.80	15.00	—	36.80
2"	P1@.449	Ea	33.60	20.00	—	53.60
2½"	P1@.552	Ea	68.10	24.50	—	92.60
3"	P1@.656	Ea	105.00	29.20	—	134.20
4"	P1@.863	Ea	176.00	38.40	—	214.40

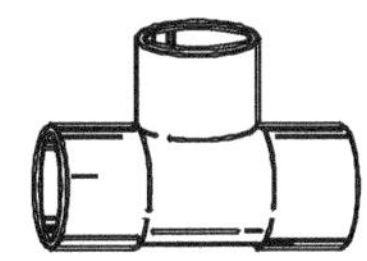

Type L copper branch reducing tee C x C x C with soft-soldered joints

½" x 3/8"	P1@.129	Ea	1.87	5.73	—	7.60
¾" x ½"	P1@.182	Ea	3.97	8.09	—	12.06
1" x ½"	P1@.195	Ea	7.78	8.67	—	16.45
1" x ¾"	P1@.219	Ea	7.78	9.73	—	17.51
1¼" x ½"	P1@.225	Ea	11.70	10.00	—	21.70
1¼" x ¾"	P1@.247	Ea	11.70	11.00	—	22.70
1¼" x 1"	P1@.268	Ea	11.70	11.90	—	23.60
1½" x ½"	P1@.272	Ea	14.40	12.10	—	26.50
1½" x ¾"	P1@.287	Ea	14.40	12.80	—	27.20
1½" x 1"	P1@.302	Ea	14.40	13.40	—	27.80
1½" x 1¼"	P1@.317	Ea	14.40	14.10	—	28.50
2" x ½"	P1@.325	Ea	18.80	14.40	—	33.20
2" x ¾"	P1@.348	Ea	18.80	15.50	—	34.30
2" x 1"	P1@.373	Ea	18.80	16.60	—	35.40
2" x 1¼"	P1@.398	Ea	18.80	17.70	—	36.50
2" x 1½"	P1@.422	Ea	18.80	18.80	—	37.60
2½" x ½"	P1@.437	Ea	56.80	19.40	—	76.20
2½" x ¾"	P1@.453	Ea	56.80	20.10	—	76.90
2½" x 1"	P1@.469	Ea	56.80	20.80	—	77.60
2½" x 1½"	P1@.502	Ea	56.80	22.30	—	79.10
2½" x 2"	P1@.519	Ea	56.80	23.10	—	79.90
3" x 1¼"	P1@.525	Ea	64.20	23.30	—	87.50
3" x 2"	P1@.585	Ea	66.40	26.00	—	92.40
3" x 2½"	P1@.617	Ea	66.40	27.40	—	93.80
4" x 1¼"	P1@.625	Ea	116.00	27.80	—	143.80
4" x 1½"	P1@.671	Ea	116.00	29.80	—	145.80
4" x 2"	P1@.718	Ea	116.00	31.90	—	147.90
4" x 2½"	P1@.765	Ea	116.00	34.00	—	150.00
4" x 3"	P1@.811	Ea	116.00	36.00	—	152.00

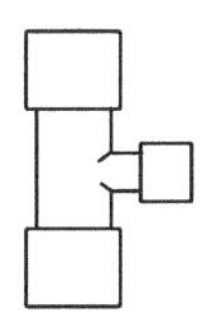

Description	Craft@Hrs	Unit	Material $	Labor $	Equipment $	Total $

Type L copper reducer with soft-soldered joints

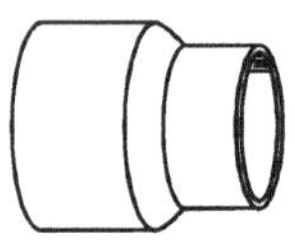

Description	Craft@Hrs	Unit	Material $	Labor $	Equipment $	Total $
½" x 3/8"	P1@.124	Ea	.67	5.51	—	6.18
¾" x ½"	P1@.129	Ea	1.26	5.73	—	6.99
1" x ¾"	P1@.172	Ea	2.28	7.65	—	9.93
1¼" x ½"	P1@.172	Ea	3.82	7.65	—	11.47
1¼" x ¾"	P1@.193	Ea	3.82	8.58	—	12.40
1¼" x 1"	P1@.215	Ea	3.82	9.56	—	13.38
1½" x ½"	P1@.193	Ea	5.83	8.58	—	14.41
1½" x ¾"	P1@.214	Ea	5.83	9.51	—	15.34
1½" x 1"	P1@.236	Ea	5.83	10.50	—	16.33
1½" x 1¼"	P1@.257	Ea	5.83	11.40	—	17.23
2" x ½"	P1@.239	Ea	8.79	10.60	—	19.39
2" x ¾"	P1@.261	Ea	8.79	11.60	—	20.39
2" x 1"	P1@.282	Ea	8.79	12.50	—	21.29
2" x 1¼"	P1@.304	Ea	8.79	13.50	—	22.29
2" x 1½"	P1@.325	Ea	8.79	14.40	—	23.19
2½" x 1"	P1@.325	Ea	23.90	14.40	—	38.30
2½" x 1¼"	P1@.347	Ea	23.90	15.40	—	39.30
2½" x 1½"	P1@.368	Ea	23.90	16.40	—	40.30
2½" x 2"	P1@.414	Ea	23.90	18.40	—	42.30
3" x 2"	P1@.457	Ea	28.00	20.30	—	48.30
3" x 2½"	P1@.500	Ea	28.00	22.20	—	50.20
4" x 2"	P1@.543	Ea	53.90	24.10	—	78.00
4" x 2½"	P1@.586	Ea	53.90	26.00	—	79.90
4" x 3"	P1@.629	Ea	53.90	28.00	—	81.90

Type L copper adapter C x MPT with soft-soldered joint

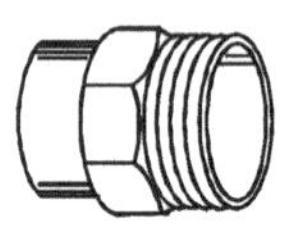

Description	Craft@Hrs	Unit	Material $	Labor $	Equipment $	Total $
½"	P1@.075	Ea	1.65	3.33	—	4.98
¾"	P1@.105	Ea	2.59	4.67	—	7.26
1"	P1@.134	Ea	6.44	5.96	—	12.40
1¼"	P1@.164	Ea	12.00	7.29	—	19.29
1½"	P1@.194	Ea	13.60	8.62	—	22.22
2"	P1@.259	Ea	16.90	11.50	—	28.40
2½"	P1@.319	Ea	54.90	14.20	—	69.10
3"	P1@.378	Ea	91.00	16.80	—	107.80

Type L copper adapter C x FPT with soft-soldered joint

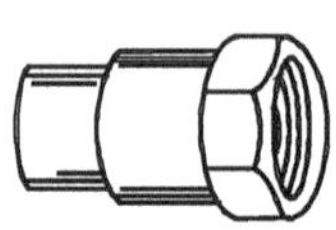

Description	Craft@Hrs	Unit	Material $	Labor $	Equipment $	Total $
½"	P1@.075	Ea	2.76	3.33	—	6.09
¾"	P1@.105	Ea	5.01	4.67	—	9.68
1"	P1@.134	Ea	7.56	5.96	—	13.52
1¼"	P1@.164	Ea	16.00	7.29	—	23.29
1½"	P1@.194	Ea	19.20	8.62	—	27.82
2"	P1@.259	Ea	28.60	11.50	—	40.10
2½"	P1@.319	Ea	72.00	14.20	—	86.20
3"	P1@.378	Ea	99.00	16.80	—	115.80

Type L copper flush bushing with soft-soldered joints

Description	Craft@Hrs	Unit	Material $	Labor $	Equipment $	Total $
½" x 3/8"	P1@.124	Ea	1.67	5.51	—	7.18
¾" x ½"	P1@.129	Ea	2.87	5.73	—	8.60
1" x ¾"	P1@.172	Ea	7.43	7.65	—	15.08
1" x ½"	P1@.172	Ea	8.48	7.65	—	16.13
1¼" x 1"	P1@.215	Ea	8.81	9.56	—	18.37
1½" x 1¼"	P1@.257	Ea	11.20	11.40	—	22.60
2" x 1½"	P1@.325	Ea	20.50	14.40	—	34.90

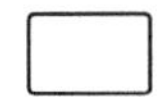

Type L copper union with soft-soldered joint

Description	Craft@Hrs	Unit	Material $	Labor $	Equipment $	Total $
½"	P1@.121	Ea	6.75	5.38	—	12.13
¾"	P1@.170	Ea	8.05	7.56	—	15.61
1"	P1@.218	Ea	18.10	9.69	—	27.79
1¼"	P1@.267	Ea	32.90	11.90	—	44.80
1½"	P1@.315	Ea	40.40	14.00	—	54.40
2"	P1@.421	Ea	58.20	18.70	—	76.90

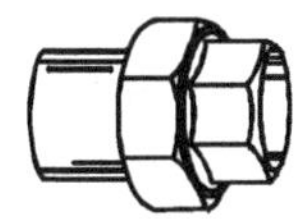

Type L copper dielectric union with soft-soldered joint

Description	Craft@Hrs	Unit	Material $	Labor $	Equipment $	Total $
½"	P1@.121	Ea	6.32	5.38	—	11.70
¾"	P1@.170	Ea	6.32	7.56	—	13.88
1"	P1@.218	Ea	13.10	9.69	—	22.79
1¼"	P1@.267	Ea	22.30	11.90	—	34.20
1½"	P1@.315	Ea	30.70	14.00	—	44.70
2"	P1@.421	Ea	41.00	18.70	—	59.70

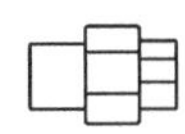

Type L copper cap with soft-soldered joint

Description	Craft@Hrs	Unit	Material $	Labor $	Equipment $	Total $
½"	P1@.069	Ea	.56	3.07	—	3.63
¾"	P1@.096	Ea	1.12	4.27	—	5.39
1"	P1@.124	Ea	2.54	5.51	—	8.05
1¼"	P1@.151	Ea	3.64	6.71	—	10.35
1½"	P1@.178	Ea	5.31	7.91	—	13.22
2"	P1@.237	Ea	9.24	10.50	—	19.74
2½"	P1@.292	Ea	23.50	13.00	—	36.50
3"	P1@.347	Ea	36.70	15.40	—	52.10
4"	P1@.457	Ea	56.40	20.30	—	76.70

Type L copper coupling with soft-soldered joints

Description	Craft@Hrs	Unit	Material $	Labor $	Equipment $	Total $
½"	P1@.107	Ea	.38	4.76	—	5.14
¾"	P1@.150	Ea	.90	6.67	—	7.57
1"	P1@.193	Ea	2.41	8.58	—	10.99
1¼"	P1@.236	Ea	4.20	10.50	—	14.70
1½"	P1@.278	Ea	6.07	12.40	—	18.47
2"	P1@.371	Ea	8.93	16.50	—	25.43
2½"	P1@.457	Ea	21.10	20.30	—	41.40
3"	P1@.543	Ea	32.10	24.10	—	56.20
4"	P1@.714	Ea	63.00	31.70	—	94.70

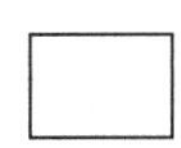

Class 125 bronze body gate valve, soft-soldered joints

Description	Craft@Hrs	Unit	Material $	Labor $	Equipment $	Total $
½"	P1@.200	Ea	22.80	8.89	—	31.69
¾"	P1@.249	Ea	28.50	11.10	—	39.60
1"	P1@.299	Ea	40.10	13.30	—	53.40
1¼"	P1@.398	Ea	50.40	17.70	—	68.10
1½"	P1@.448	Ea	68.00	19.90	—	87.90
2"	P1@.498	Ea	114.00	22.10	—	136.10
2½"	P1@.830	Ea	188.00	36.90	—	224.90
3"	P1@1.24	Ea	270.00	55.10	—	325.10

Class 125 iron body gate valve, flanged ends

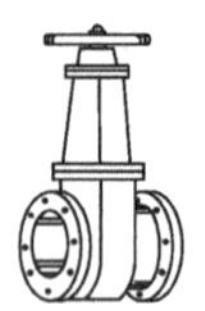

Description	Craft@Hrs	Unit	Material $	Labor $	Equipment $	Total $
2"	P1@.500	Ea	435.00	22.20	—	457.20
2½"	P1@.600	Ea	586.00	26.70	—	612.70
3"	P1@.750	Ea	636.00	33.30	—	669.30
4"	P1@1.35	Ea	934.00	60.00	—	994.00
5"	P1@2.00	Ea	1,780.00	88.90	—	1,868.90
6"	P1@2.50	Ea	1,780.00	111.00	—	1,891.00

Class 125 bronze body globe valve, with soft-soldered joints

Description	Craft@Hrs	Unit	Material $	Labor $	Equipment $	Total $
½"	P1@.200	Ea	42.20	8.89	—	51.09
¾"	P1@.249	Ea	56.20	11.10	—	67.30
1"	P1@.299	Ea	96.90	13.30	—	110.20
1¼"	P1@.398	Ea	136.00	17.70	—	153.70
1½"	P1@.448	Ea	185.00	19.90	—	204.90
2"	P1@.498	Ea	299.00	22.10	—	321.10

Class 125 iron body globe valve, flanged ends

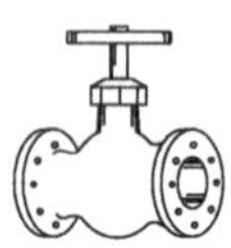

Description	Craft@Hrs	Unit	Material $	Labor $	Equipment $	Total $
2½"	P1@.600	Ea	493.00	26.70	—	519.70
3"	P1@.750	Ea	591.00	33.30	—	624.30
4"	P1@1.35	Ea	798.00	60.00	—	858.00

200 PSIG iron body butterfly valve, lug-type, lever operated

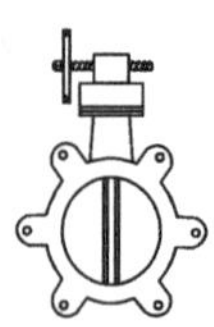

Description	Craft@Hrs	Unit	Material $	Labor $	Equipment $	Total $
2"	P1@.450	Ea	179.00	20.00	—	199.00
2½"	P1@.450	Ea	186.00	20.00	—	206.00
3"	P1@.550	Ea	196.00	24.40	—	220.40
4"	P1@.550	Ea	244.00	24.40	—	268.40

200 PSIG iron body butterfly valve, wafer-type, lever operated

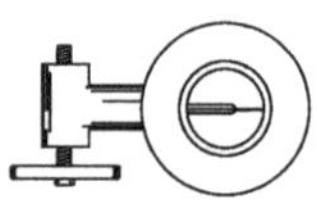

Description	Craft@Hrs	Unit	Material $	Labor $	Equipment $	Total $
2"	P1@.450	Ea	164.00	20.00	—	184.00
2½"	P1@.450	Ea	167.00	20.00	—	187.00
3"	P1@.550	Ea	179.00	24.40	—	203.40
4"	P1@.550	Ea	217.00	24.40	—	241.40

Description	Craft@Hrs	Unit	Material $	Labor $	Equipment $	Total $

Class 125 bronze body 2-piece ball valve, soft-soldered joints

Description	Craft@Hrs	Unit	Material $	Labor $	Equipment $	Total $
½"	P1@.200	Ea	11.60	8.89	—	20.49
¾"	P1@.249	Ea	15.40	11.10	—	26.50
1"	P1@.299	Ea	26.70	13.30	—	40.00
1¼"	P1@.398	Ea	44.60	17.70	—	62.30
1½"	P1@.448	Ea	60.30	19.90	—	80.20
2"	P1@.498	Ea	76.40	22.10	—	98.50
3"	P1@1.24	Ea	524.00	55.10	—	579.10
4"	P1@1.45	Ea	684.00	64.50	—	748.50

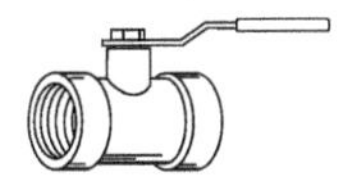

Class 125 bronze body swing check valve, soft-soldered joints

Description	Craft@Hrs	Unit	Material $	Labor $	Equipment $	Total $
½"	P1@.200	Ea	23.90	8.89	—	32.79
¾"	P1@.249	Ea	34.10	11.10	—	45.20
1"	P1@.299	Ea	44.50	13.30	—	57.80
1¼"	P1@.398	Ea	63.60	17.70	—	81.30
1½"	P1@.448	Ea	89.60	19.90	—	109.50
2"	P1@.498	Ea	148.00	22.10	—	170.10
2½"	P1@.830	Ea	299.00	36.90	—	335.90
3"	P1@1.24	Ea	430.00	55.10	—	485.10

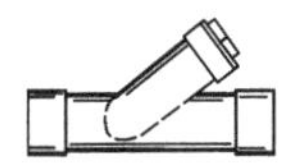

Class 125 iron body swing check valve, flanged ends

Description	Craft@Hrs	Unit	Material $	Labor $	Equipment $	Total $
2"	P1@.500	Ea	209.00	22.20	—	231.20
2½"	P1@.600	Ea	264.00	26.70	—	290.70
3"	P1@.750	Ea	329.00	33.30	—	362.30
4"	P1@1.35	Ea	483.00	60.00	—	543.00

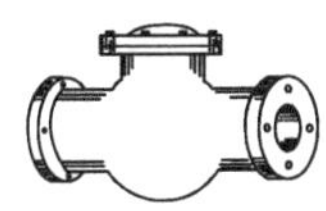

Class 125 iron body silent check valve, wafer-type

Description	Craft@Hrs	Unit	Material $	Labor $	Equipment $	Total $
2"	P1@.500	Ea	153.00	22.20	—	175.20
2½"	P1@.600	Ea	170.00	26.70	—	196.70
3"	P1@.750	Ea	195.00	33.30	—	228.30
4"	P1@1.35	Ea	260.00	60.00	—	320.00

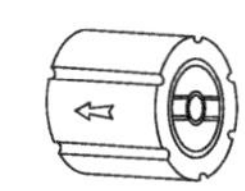

Class 125 bronze body strainer, threaded ends

Description	Craft@Hrs	Unit	Material $	Labor $	Equipment $	Total $
½"	P1@.230	Ea	37.40	10.20	—	47.60
¾"	P1@.260	Ea	49.10	11.60	—	60.70
1"	P1@.330	Ea	60.20	14.70	—	74.90
1¼"	P1@.440	Ea	84.20	19.60	—	103.80
1½"	P1@.495	Ea	126.00	22.00	—	148.00
2"	P1@.550	Ea	220.00	24.40	—	244.40

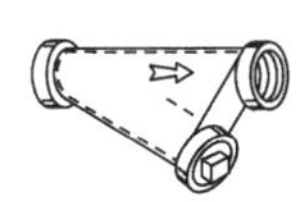

Class 125 iron body strainer, flanged ends

Description	Craft@Hrs	Unit	Material $	Labor $	Equipment $	Total $
2"	P1@.500	Ea	168.00	22.20	—	190.20
2½"	P1@.600	Ea	187.00	26.70	—	213.70
3"	P1@.750	Ea	218.00	33.30	—	251.30
4"	P1@1.35	Ea	367.00	60.00	—	427.00

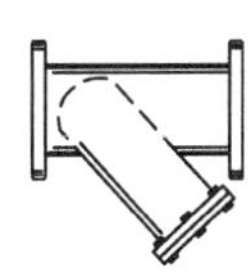

Copper, Type L with Soft-Soldered Joints

Description	Craft@Hrs	Unit	Material $	Labor $	Equipment $	Total $

Installation of 2-way control valve, threaded joints

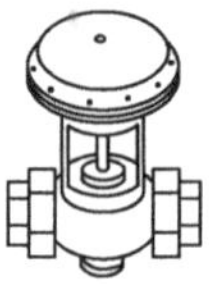

Description	Craft@Hrs	Unit	Material $	Labor $	Equipment $	Total $
½"	P1@.210	Ea	—	9.33	—	9.33
¾"	P1@.275	Ea	—	12.20	—	12.20
1"	P1@.350	Ea	—	15.60	—	15.60
1¼"	P1@.430	Ea	—	19.10	—	19.10
1½"	P1@.505	Ea	—	22.40	—	22.40
2"	P1@.675	Ea	—	30.00	—	30.00
2½"	P1@.830	Ea	—	36.90	—	36.90
3"	P1@.990	Ea	—	44.00	—	44.00

Installation of 3-way control valve, threaded joints

Description	Craft@Hrs	Unit	Material $	Labor $	Equipment $	Total $
½"	P1@.260	Ea	—	11.60	—	11.60
¾"	P1@.365	Ea	—	16.20	—	16.20
1"	P1@.475	Ea	—	21.10	—	21.10
1¼"	P1@.575	Ea	—	25.60	—	25.60
1½"	P1@.680	Ea	—	30.20	—	30.20
2"	P1@.910	Ea	—	40.40	—	40.40
2½"	P1@1.12	Ea	—	49.80	—	49.80
3"	P1@1.33	Ea	—	59.10	—	59.10

Companion flange, 150 pound cast brass

Description	Craft@Hrs	Unit	Material $	Labor $	Equipment $	Total $
2"	P1@.290	Ea	330.00	12.90	—	342.90
2½"	P1@.380	Ea	410.00	16.90	—	426.90
3"	P1@.460	Ea	410.00	20.40	—	430.40
4"	P1@.600	Ea	655.00	26.70	—	681.70

Bolt and gasket sets

Description	Craft@Hrs	Unit	Material $	Labor $	Equipment $	Total $
2"	P1@.500	Ea	4.64	22.20	—	26.84
2½"	P1@.650	Ea	5.42	28.90	—	34.32
3"	P1@.750	Ea	9.19	33.30	—	42.49
4"	P1@1.00	Ea	15.40	44.50	—	59.90

Thermometer with well

Description	Craft@Hrs	Unit	Material $	Labor $	Equipment $	Total $
7"	P1@.250	Ea	186.00	11.10	—	197.10
9"	P1@.250	Ea	192.00	11.10	—	203.10

Dial-type pressure gauge

Description	Craft@Hrs	Unit	Material $	Labor $	Equipment $	Total $
2½"	P1@.200	Ea	38.50	8.89	—	47.39
3½"	P1@.200	Ea	51.20	8.89	—	60.09

Pressure/temperature tap

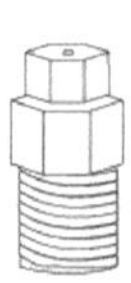

Description	Craft@Hrs	Unit	Material $	Labor $	Equipment $	Total $
Tap	P1@.150	Ea	17.70	6.67	—	24.37

Description	Craft@Hrs	Unit	Material $	Labor $	Equipment $	Total $

Hanger with swivel assembly

½"	P1@.250	Ea	4.50	11.10	—	15.60
¾"	P1@.250	Ea	4.98	11.10	—	16.08
1"	P1@.250	Ea	5.20	11.10	—	16.30
1¼"	P1@.300	Ea	5.34	13.30	—	18.64
1½"	P1@.300	Ea	5.74	13.30	—	19.04
2"	P1@.300	Ea	5.99	13.30	—	19.29
2½"	P1@.350	Ea	8.15	15.60	—	23.75
3"	P1@.350	Ea	10.10	15.60	—	25.70
4"	P1@.350	Ea	11.10	15.60	—	26.70

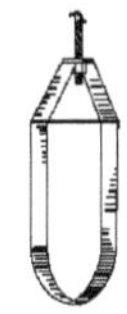

Riser clamp

½"	P1@.100	Ea	3.30	4.45	—	7.75
¾"	P1@.100	Ea	4.87	4.45	—	9.32
1"	P1@.100	Ea	4.91	4.45	—	9.36
1¼"	P1@.105	Ea	5.90	4.67	—	10.57
1½"	P1@.110	Ea	6.22	4.89	—	11.11
2"	P1@.115	Ea	6.62	5.11	—	11.73
2½"	P1@.120	Ea	6.93	5.33	—	12.26
3"	P1@.120	Ea	7.53	5.33	—	12.86
4"	P1@.125	Ea	9.61	5.56	—	15.17

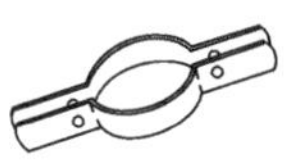

Copper, Type M with Brazed Joints

Type M hard-drawn copper pipe with wrought copper fittings and brazed joints is used in a wide variety of plumbing and HVAC systems such as potable water, heating hot water and chilled water.

Brazed joints are those made with silver or other alloy filler metals having melting points at, or above, 1,000 degrees F. Maximum working pressure/temperature relationships for brazed joints are approximately as follows:

Maximum Working Pressure (PSIG)*			
Water Temperature (degrees F.)	Nominal Pipe Size (inches)		
	Up to 1	1¼ to 2	2½ to 4
Up to 350	270	190	150

**For copper pipe and solder-type fittings using brazing alloys melting at, or above, 1,000 degrees F.*

This section has been arranged to save the estimator's time by including all normally-used system components such as pipe, fittings, valves, hanger assemblies, riser clamps and miscellaneous items under one heading. Additional items can be found under "Plumbing and Piping Specialties." The cost estimates in this section are based on the conditions, limitations and wage rates described in the section "How to Use This Book" beginning on page 5.

Type M copper pipe with brazed joints installed horizontally. Complete installation including six to twelve wrought copper tees and six to twelve wrought copper elbows every 100', and hangers spaced to meet plumbing code *(a tee & elbow every 8.3' for ½" dia., a tee and elbow every 16.6' for 4" dia.).*

Description	Craft@Hrs	Unit	Material $	Labor $	Equipment $	Total $
½"	P1@.113	LF	2.51	5.02	—	7.53
¾"	P1@.123	LF	3.87	5.47	—	9.34
1"	P1@.143	LF	8.03	6.36	—	14.39
1¼"	P1@.163	LF	13.60	7.25	—	20.85
1½"	P1@.183	LF	15.90	8.13	—	24.03
2"	P1@.213	LF	33.60	9.47	—	43.07
2½"	P1@.243	LF	49.30	10.80	—	60.10
3"	P1@.273	LF	61.70	12.10	—	73.80
4"	P1@.313	LF	90.90	13.90	—	104.80

Type M copper pipe with brazed joints, installed risers. Complete installation including a wrought copper reducing tee every floor and a riser clamp every other floor.

Description	Craft@Hrs	Unit	Material $	Labor $	Equipment $	Total $
½"	P1@.062	LF	2.13	2.76	—	4.89
¾"	P1@.072	LF	3.45	3.20	—	6.65
1"	P1@.077	LF	6.75	3.42	—	10.17
1¼"	P1@.082	LF	9.95	3.64	—	13.59
1½"	P1@.102	LF	13.20	4.53	—	17.73
2"	P1@.132	LF	17.40	5.87	—	23.27
2½"	P1@.152	LF	37.90	6.76	—	44.66
3"	P1@.172	LF	46.40	7.65	—	54.05
4"	P1@.192	LF	68.80	8.53	—	77.33

Description	Craft@Hrs	Unit	Material $	Labor $	Equipment $	Total $

Type M copper pipe with brazed joints, pipe only

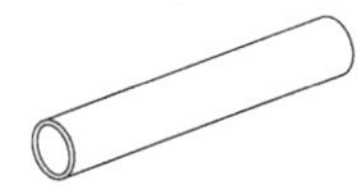

½"	P1@.032	LF	1.67	1.42	—	3.09
¾"	P1@.035	LF	2.72	1.56	—	4.28
1"	P1@.038	LF	4.25	1.69	—	5.94
1¼"	P1@.042	LF	6.40	1.87	—	8.27
1½"	P1@.046	LF	8.95	2.04	—	10.99
2"	P1@.053	LF	16.60	2.36	—	18.96
2½"	P1@.060	LF	22.80	2.67	—	25.47
3"	P1@.066	LF	28.50	2.93	—	31.43
4"	P1@.080	LF	45.90	3.56	—	49.46

Type M copper 45-degree ell C x C with brazed joints

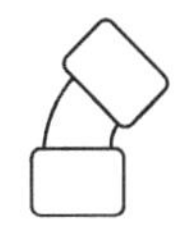

½"	P1@.129	Ea	.97	5.73	—	6.70
¾"	P1@.181	Ea	1.86	8.05	—	9.91
1"	P1@.232	Ea	5.49	10.30	—	15.79
1¼"	P1@.284	Ea	7.43	12.60	—	20.03
1½"	P1@.335	Ea	11.10	14.90	—	26.00
2"	P1@.447	Ea	17.50	19.90	—	37.40
2½"	P1@.550	Ea	44.80	24.40	—	69.20
3"	P1@.654	Ea	59.20	29.10	—	88.30
4"	P1@.860	Ea	114.00	38.20	—	152.20

Type M copper 90-degree ell C x C with brazed joints

½"	P1@.129	Ea	.56	5.73	—	6.29
¾"	P1@.181	Ea	1.17	8.05	—	9.22
1"	P1@.232	Ea	4.19	10.30	—	14.49
1¼"	P1@.284	Ea	6.76	12.60	—	19.36
1½"	P1@.335	Ea	10.20	14.90	—	25.10
2"	P1@.447	Ea	17.70	19.90	—	37.60
2½"	P1@.550	Ea	37.00	24.40	—	61.40
3"	P1@.654	Ea	46.90	29.10	—	76.00
4"	P1@.860	Ea	99.20	38.20	—	137.40

Type M copper 90-degree ell ftg. x C with brazed joints

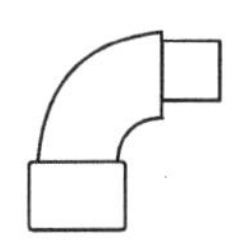

½"	P1@.129	Ea	.92	5.73	—	6.65
¾"	P1@.181	Ea	1.98	8.05	—	10.03
1"	P1@.232	Ea	5.93	10.30	—	16.23
1¼"	P1@.284	Ea	8.26	12.60	—	20.86
1½"	P1@.335	Ea	12.30	14.90	—	27.20
2"	P1@.447	Ea	24.20	19.90	—	44.10
2½"	P1@.550	Ea	50.40	24.40	—	74.80
3"	P1@.654	Ea	56.70	29.10	—	85.80
4"	P1@.860	Ea	114.00	38.20	—	152.20

Description	Craft@Hrs	Unit	Material $	Labor $	Equipment $	Total $

Type M copper tee C x C x C with brazed joints

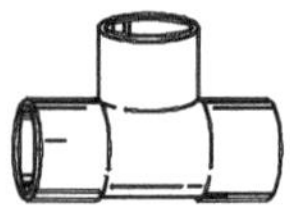

Description	Craft@Hrs	Unit	Material $	Labor $	Equipment $	Total $
½"	P1@.156	Ea	.92	6.93	—	7.85
¾"	P1@.204	Ea	2.48	9.07	—	11.55
1"	P1@.263	Ea	9.34	11.70	—	21.04
1¼"	P1@.322	Ea	14.90	14.30	—	29.20
1½"	P1@.380	Ea	21.80	16.90	—	38.70
2"	P1@.507	Ea	33.60	22.50	—	56.10
2½"	P1@.624	Ea	68.10	27.70	—	95.80
3"	P1@.741	Ea	105.00	32.90	—	137.90
4"	P1@.975	Ea	176.00	43.30	—	219.30

Type M copper reducing branch tee C x C x C with brazed joints

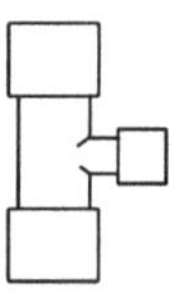

Description	Craft@Hrs	Unit	Material $	Labor $	Equipment $	Total $
½" x 3/8"	P1@.129	Ea	1.87	5.73	—	7.60
¾" x ½"	P1@.182	Ea	3.97	8.09	—	12.06
1" x ½"	P1@.209	Ea	7.78	9.29	—	17.07
1" x ¾"	P1@.234	Ea	7.78	10.40	—	18.18
1¼" x ½"	P1@.241	Ea	11.70	10.70	—	22.40
1¼" x ¾"	P1@.264	Ea	11.70	11.70	—	23.40
1¼" x 1"	P1@.287	Ea	11.70	12.80	—	24.50
1½" x ½"	P1@.291	Ea	14.40	12.90	—	27.30
1½" x ¾"	P1@.307	Ea	14.40	13.60	—	28.00
1½" x 1"	P1@.323	Ea	14.40	14.40	—	28.80
1½" x 1¼"	P1@.339	Ea	14.40	15.10	—	29.50
2" x ½"	P1@.348	Ea	18.80	15.50	—	34.30
2" x ¾"	P1@.373	Ea	18.80	16.60	—	35.40
2" x 1"	P1@.399	Ea	18.80	17.70	—	36.50
2" x 1¼"	P1@.426	Ea	18.80	18.90	—	37.70
2" x 1½"	P1@.452	Ea	18.80	20.10	—	38.90
2½" x ½"	P1@.468	Ea	56.80	20.80	—	77.60
2½" x ¾"	P1@.485	Ea	56.80	21.60	—	78.40
2½" x 1"	P1@.502	Ea	56.80	22.30	—	79.10
2½" x 1½"	P1@.537	Ea	56.80	23.90	—	80.70
2½" x 2"	P1@.555	Ea	56.80	24.70	—	81.50
3" x 1¼"	P1@.562	Ea	64.20	25.00	—	89.20
3" x 2"	P1@.626	Ea	66.40	27.80	—	94.20
3" x 2½"	P1@.660	Ea	66.40	29.30	—	95.70
4" x 1¼"	P1@.669	Ea	116.00	29.70	—	145.70
4" x 1½"	P1@.718	Ea	116.00	31.90	—	147.90
4" x 2"	P1@.768	Ea	116.00	34.10	—	150.10
4" x 2½"	P1@.819	Ea	116.00	36.40	—	152.40
4" x 3"	P1@.868	Ea	116.00	38.60	—	154.60

Description	Craft@Hrs	Unit	Material $	Labor $	Equipment $	Total $

Type M copper reducer with brazed joints

Description	Craft@Hrs	Unit	Material $	Labor $	Equipment $	Total $
½" x 3/8"	P1@.148	Ea	.67	6.58	—	7.25
¾" x ½"	P1@.155	Ea	1.26	6.89	—	8.15
1" x ¾"	P1@.206	Ea	2.28	9.16	—	11.44
1¼" x ½"	P1@.206	Ea	3.82	9.16	—	12.98
1¼" x ¾"	P1@.232	Ea	3.82	10.30	—	14.12
1¼" x 1"	P1@.258	Ea	3.82	11.50	—	15.32
1½" x ½"	P1@.232	Ea	5.83	10.30	—	16.13
1½" x ¾"	P1@.257	Ea	5.83	11.40	—	17.23
1½" x 1"	P1@.283	Ea	5.83	12.60	—	18.43
1½" x 1¼"	P1@.308	Ea	5.83	13.70	—	19.53
2" x ½"	P1@.287	Ea	8.79	12.80	—	21.59
2" x ¾"	P1@.277	Ea	8.79	12.30	—	21.09
2" x 1"	P1@.338	Ea	8.79	15.00	—	23.79
2" x 1¼"	P1@.365	Ea	8.79	16.20	—	24.99
2" x 1½"	P1@.390	Ea	8.79	17.30	—	26.09
2½" x 1"	P1@.390	Ea	23.90	17.30	—	41.20
2½" x 1¼"	P1@.416	Ea	23.90	18.50	—	42.40
2½" x 1½"	P1@.442	Ea	23.90	19.60	—	43.50
2½" x 2"	P1@.497	Ea	23.90	22.10	—	46.00
3" x 1¼"	P1@.468	Ea	28.00	20.80	—	48.80
3" x 1½"	P1@.493	Ea	28.00	21.90	—	49.90
3" x 2"	P1@.548	Ea	28.00	24.40	—	52.40
3" x 2½"	P1@.600	Ea	28.00	26.70	—	54.70
4" x 2"	P1@.652	Ea	53.90	29.00	—	82.90
4" x 2½"	P1@.704	Ea	53.90	31.30	—	85.20
4" x 3"	P1@.755	Ea	53.90	33.60	—	87.50

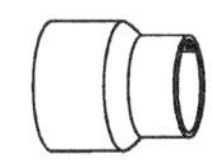

Type M copper adapter C x MPT with brazed joint

Description	Craft@Hrs	Unit	Material $	Labor $	Equipment $	Total $
½"	P1@.090	Ea	1.65	4.00	—	5.65
¾"	P1@.126	Ea	2.59	5.60	—	8.19
1"	P1@.162	Ea	6.44	7.20	—	13.64
1¼"	P1@.198	Ea	12.00	8.80	—	20.80
1½"	P1@.234	Ea	13.60	10.40	—	24.00
2"	P1@.312	Ea	16.90	13.90	—	30.80
2½"	P1@.384	Ea	54.90	17.10	—	72.00
3"	P1@.456	Ea	91.00	20.30	—	111.30

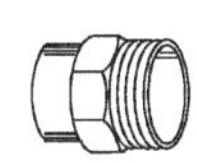

Type M copper adapter C x FPT with brazed joint

Description	Craft@Hrs	Unit	Material $	Labor $	Equipment $	Total $
½"	P1@.090	Ea	2.76	4.00	—	6.76
¾"	P1@.126	Ea	5.01	5.60	—	10.61
1"	P1@.162	Ea	7.56	7.20	—	14.76
1¼"	P1@.198	Ea	16.00	8.80	—	24.80
1½"	P1@.234	Ea	19.20	10.40	—	29.60
2"	P1@.312	Ea	28.60	13.90	—	42.50
2½"	P1@.384	Ea	72.00	17.10	—	89.10
3"	P1@.456	Ea	99.00	20.30	—	119.30

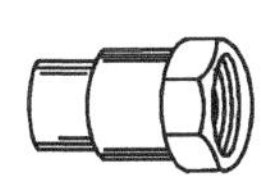

Description	Craft@Hrs	Unit	Material $	Labor $	Equipment $	Total $

Type M copper flush bushing with brazed joints

½" x 3/8"	P1@.148	Ea	1.67	6.58	—	8.25
¾" x ½"	P1@.155	Ea	2.87	6.89	—	9.76
1" x ¾"	P1@.206	Ea	7.43	9.16	—	16.59
1" x ½"	P1@.206	Ea	8.48	9.16	—	17.64
1¼" x 1"	P1@.258	Ea	8.81	11.50	—	20.31
1½" x 1¼"	P1@.308	Ea	11.20	13.70	—	24.90
2" x 1½"	P1@.390	Ea	20.50	17.30	—	37.80

Type M copper union with brazed joints

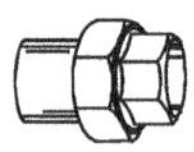

½"	P1@.146	Ea	6.75	6.49	—	13.24
¾"	P1@.205	Ea	8.05	9.11	—	17.16
1"	P1@.263	Ea	18.10	11.70	—	29.80
1¼"	P1@.322	Ea	32.90	14.30	—	47.20
1½"	P1@.380	Ea	40.40	16.90	—	57.30
2"	P1@.507	Ea	58.20	22.50	—	80.70

Type M copper dielectric union with brazed joints

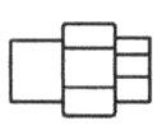

½"	P1@.146	Ea	6.32	6.49	—	12.81
¾"	P1@.205	Ea	6.32	9.11	—	15.43
1"	P1@.263	Ea	13.10	11.70	—	24.80
1¼"	P1@.322	Ea	22.30	14.30	—	36.60
1½"	P1@.380	Ea	30.70	16.90	—	47.60
2"	P1@.507	Ea	41.00	22.50	—	63.50

Type M copper cap with brazed joints

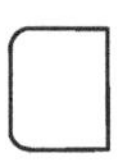

½"	P1@.083	Ea	.56	3.69	—	4.25
¾"	P1@.116	Ea	1.12	5.16	—	6.28
1"	P1@.149	Ea	2.54	6.62	—	9.16
1¼"	P1@.182	Ea	3.64	8.09	—	11.73
1½"	P1@.215	Ea	5.31	9.56	—	14.87
2"	P1@.286	Ea	9.24	12.70	—	21.94
2½"	P1@.352	Ea	23.50	15.60	—	39.10
3"	P1@.418	Ea	36.70	18.60	—	55.30
4"	P1@.550	Ea	56.40	24.40	—	80.80

Type M copper coupling with brazed joints

½"	P1@.129	Ea	.36	5.73	—	6.09
¾"	P1@.181	Ea	.90	8.05	—	8.95
1"	P1@.232	Ea	2.41	10.30	—	12.71
1¼"	P1@.284	Ea	4.20	12.60	—	16.80
1½"	P1@.335	Ea	6.07	14.90	—	20.97
2"	P1@.447	Ea	8.93	19.90	—	28.83
2½"	P1@.550	Ea	21.10	24.40	—	45.50
3"	P1@.654	Ea	32.10	29.10	—	61.20
4"	P1@.860	Ea	63.00	38.20	—	101.20

Class 125 bronze body gate valve, brazed joints

Description	Craft@Hrs	Unit	Material $	Labor $	Equipment $	Total $
½"	P1@.240	Ea	22.80	10.70	—	33.50
¾"	P1@.300	Ea	28.50	13.30	—	41.80
1"	P1@.360	Ea	40.10	16.00	—	56.10
1¼"	P1@.480	Ea	50.40	21.30	—	71.70
1½"	P1@.540	Ea	68.00	24.00	—	92.00
2"	P1@.600	Ea	114.00	26.70	—	140.70
2½"	P1@1.00	Ea	188.00	44.50	—	232.50
3"	P1@1.50	Ea	270.00	66.70	—	336.70

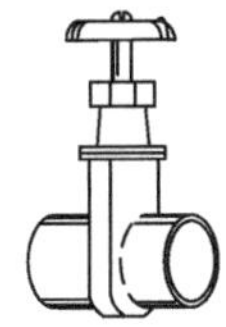

Class 125 iron body gate valve, flanged ends

Description	Craft@Hrs	Unit	Material $	Labor $	Equipment $	Total $
2"	P1@.500	Ea	435.00	22.20	—	457.20
2½"	P1@.600	Ea	586.00	26.70	—	612.70
3"	P1@.750	Ea	636.00	33.30	—	669.30
4"	P1@1.35	Ea	934.00	60.00	—	994.00
5"	P1@2.00	Ea	1,780.00	88.90	—	1,868.90
6"	P1@2.50	Ea	1,780.00	111.00	—	1,891.00

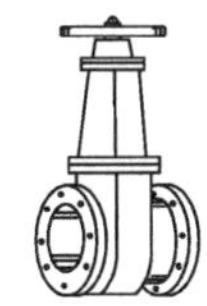

Class 125 bronze body globe valve, brazed joints

Description	Craft@Hrs	Unit	Material $	Labor $	Equipment $	Total $
½"	P1@.240	Ea	42.20	10.70	—	52.90
¾"	P1@.300	Ea	56.20	13.30	—	69.50
1"	P1@.360	Ea	96.90	16.00	—	112.90
1¼"	P1@.480	Ea	136.00	21.30	—	157.30
1½"	P1@.540	Ea	185.00	24.00	—	209.00
2"	P1@.600	Ea	299.00	26.70	—	325.70

Class 125 iron body globe valve, flanged ends

Description	Craft@Hrs	Unit	Material $	Labor $	Equipment $	Total $
2½"	P1@.650	Ea	493.00	28.90	—	521.90
3"	P1@.750	Ea	591.00	33.30	—	624.30
4"	P1@1.35	Ea	798.00	60.00	—	858.00

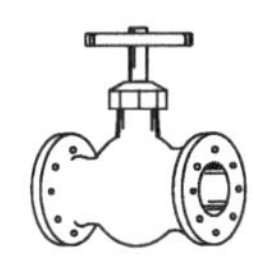

200 PSIG iron body butterfly valve, lug-type, lever operated

Description	Craft@Hrs	Unit	Material $	Labor $	Equipment $	Total $
2"	P1@.450	Ea	179.00	20.00	—	199.00
2½"	P1@.450	Ea	186.00	20.00	—	206.00
3"	P1@.550	Ea	196.00	24.40	—	220.40
4"	P1@.550	Ea	244.00	24.40	—	268.40

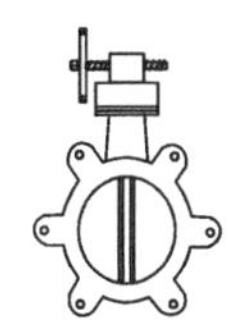

200 PSIG iron body butterfly valve, wafer-type, lever operated

Description	Craft@Hrs	Unit	Material $	Labor $	Equipment $	Total $
2"	P1@.450	Ea	164.00	20.00	—	184.00
2½"	P1@.450	Ea	167.00	20.00	—	187.00
3"	P1@.550	Ea	179.00	24.40	—	203.40
4"	P1@.550	Ea	217.00	24.40	—	241.40

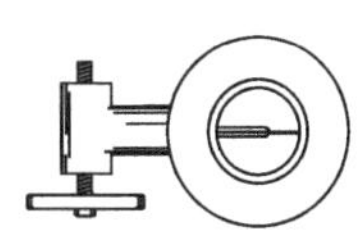

Description	Craft@Hrs	Unit	Material $	Labor $	Equipment $	Total $

Class 125 bronze body 2-piece ball valve, brazed joints

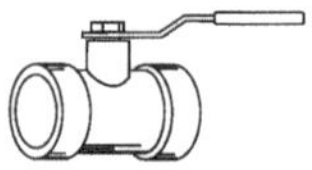

Description	Craft@Hrs	Unit	Material $	Labor $	Equipment $	Total $
½"	P1@.240	Ea	11.60	10.70	—	22.30
¾"	P1@.300	Ea	15.40	13.30	—	28.70
1"	P1@.360	Ea	26.70	16.00	—	42.70
1¼"	P1@.480	Ea	44.60	21.30	—	65.90
1½"	P1@.540	Ea	60.30	24.00	—	84.30
2"	P1@.600	Ea	76.40	26.70	—	103.10
3"	P1@1.50	Ea	524.00	66.70	—	590.70
4"	P1@1.80	Ea	684.00	80.00	—	764.00

Class 125 bronze body swing check valve, brazed joints

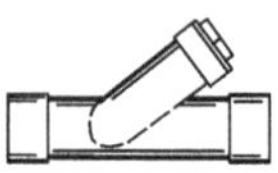

Description	Craft@Hrs	Unit	Material $	Labor $	Equipment $	Total $
½"	P1@.240	Ea	23.90	10.70	—	34.60
¾"	P1@.300	Ea	34.10	13.30	—	47.40
1"	P1@.360	Ea	44.50	16.00	—	60.50
1¼"	P1@.480	Ea	63.60	21.30	—	84.90
1½"	P1@.540	Ea	89.60	24.00	—	113.60
2"	P1@.600	Ea	148.00	26.70	—	174.70
2½"	P1@1.00	Ea	299.00	44.50	—	343.50
3"	P1@1.50	Ea	430.00	66.70	—	496.70

Class 125 iron body swing check valve, flanged ends

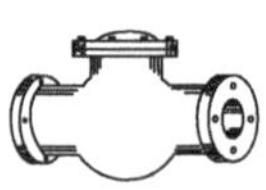

Description	Craft@Hrs	Unit	Material $	Labor $	Equipment $	Total $
2"	P1@.500	Ea	209.00	22.20	—	231.20
2½"	P1@.600	Ea	264.00	26.70	—	290.70
3"	P1@.750	Ea	329.00	33.30	—	362.30
4"	P1@1.35	Ea	483.00	60.00	—	543.00

Class 125 iron body silent check valve, wafer-type

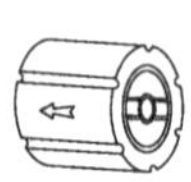

Description	Craft@Hrs	Unit	Material $	Labor $	Equipment $	Total $
2"	P1@.500	Ea	153.00	22.20	—	175.20
2½"	P1@.600	Ea	170.00	26.70	—	196.70
3"	P1@.750	Ea	195.00	33.30	—	228.30
4"	P1@1.35	Ea	260.00	60.00	—	320.00

Class 125 bronze body strainer, threaded ends

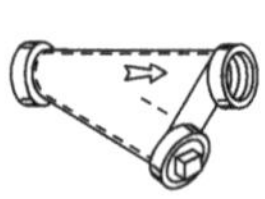

Description	Craft@Hrs	Unit	Material $	Labor $	Equipment $	Total $
½"	P1@.230	Ea	37.40	10.20	—	47.60
¾"	P1@.260	Ea	49.10	11.60	—	60.70
1"	P1@.330	Ea	60.20	14.70	—	74.90
1¼"	P1@.440	Ea	84.20	19.60	—	103.80
1½"	P1@.495	Ea	126.00	22.00	—	148.00
2"	P1@.550	Ea	220.00	24.40	—	244.40

Class 125 iron body strainer, flanged ends

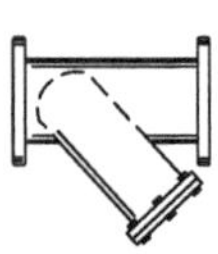

Description	Craft@Hrs	Unit	Material $	Labor $	Equipment $	Total $
2"	P1@.500	Ea	168.00	22.20	—	190.20
2½"	P1@.600	Ea	187.00	26.70	—	213.70
3"	P1@.750	Ea	218.00	33.30	—	251.30
4"	P1@1.35	Ea	367.00	60.00	—	427.00

Installation of 2-way control valve, threaded joints

Description	Craft@Hrs	Unit	Material $	Labor $	Equipment $	Total $
½"	P1@.210	Ea	—	9.33	—	9.33
¾"	P1@.275	Ea	—	12.20	—	12.20
1"	P1@.350	Ea	—	15.60	—	15.60
1¼"	P1@.430	Ea	—	19.10	—	19.10
1½"	P1@.505	Ea	—	22.40	—	22.40
2"	P1@.675	Ea	—	30.00	—	30.00
2½"	P1@.830	Ea	—	36.90	—	36.90
3"	P1@.990	Ea	—	44.00	—	44.00

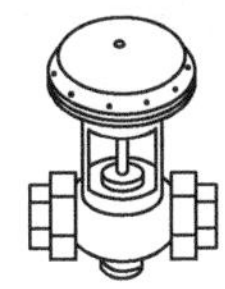

Installation of 3-way control valve, threaded joints

Description	Craft@Hrs	Unit	Material $	Labor $	Equipment $	Total $
½"	P1@.260	Ea	—	11.60	—	11.60
¾"	P1@.365	Ea	—	16.20	—	16.20
1"	P1@.475	Ea	—	21.10	—	21.10
1¼"	P1@.575	Ea	—	25.60	—	25.60
1½"	P1@.680	Ea	—	30.20	—	30.20
2"	P1@.910	Ea	—	40.40	—	40.40
2½"	P1@1.12	Ea	—	49.80	—	49.80
3"	P1@1.33	Ea	—	59.10	—	59.10

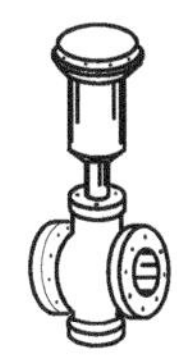

Companion flange, 150 pound cast brass

Description	Craft@Hrs	Unit	Material $	Labor $	Equipment $	Total $
2"	P1@.290	Ea	330.00	12.90	—	342.90
2½"	P1@.380	Ea	410.00	16.90	—	426.90
3"	P1@.460	Ea	410.00	20.40	—	430.40
4"	P1@.600	Ea	655.00	26.70	—	681.70

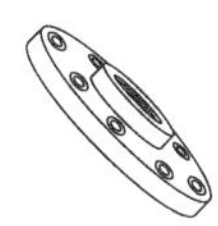

Bolt and gasket sets

Description	Craft@Hrs	Unit	Material $	Labor $	Equipment $	Total $
2"	P1@.500	Ea	4.64	22.20	—	26.84
2½"	P1@.650	Ea	5.42	28.90	—	34.32
3"	P1@.750	Ea	9.19	33.30	—	42.49
4"	P1@1.00	Ea	15.40	44.50	—	59.90

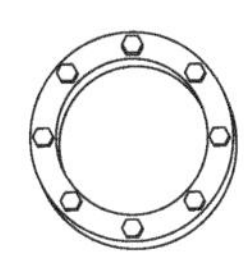

Thermometer with well

Description	Craft@Hrs	Unit	Material $	Labor $	Equipment $	Total $
7"	P1@.250	Ea	186.00	11.10	—	197.10
9"	P1@.250	Ea	192.00	11.10	—	203.10

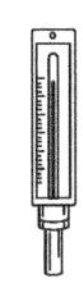

Dial-type pressure gauge

Description	Craft@Hrs	Unit	Material $	Labor $	Equipment $	Total $
2½"	P1@.200	Ea	38.50	8.89	—	47.39
3½"	P1@.200	Ea	51.20	8.89	—	60.09

Pressure/temperature tap

Description	Craft@Hrs	Unit	Material $	Labor $	Equipment $	Total $
Tap	P1@.150	Ea	17.70	6.67	—	24.37

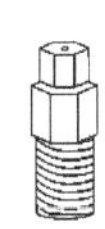

Copper, Type M with Soft-Soldered Joints

Type M hard-drawn copper pipe with wrought copper fittings and soft-soldered joints is used in a wide variety of plumbing and HVAC systems such as potable water, heating hot water, chilled water, and A.C. condensate.

Soft-soldered joints are those made with solders having melting points in the 350 degree F. to 500 degree F. range. Maximum working pressure/temperature relationships for soft-soldered joints are approximately as follows:

Maximum Working Pressures (PSIG)*				
Soft-solder Type	Water Temperature (degrees F.)	Nominal Pipe Size (inches)		
		Up to 1	1¼ to 2	2½ x 4
50-50 tin-lead**	100	200	175	150
	150	150	125	100
	200	100	90	75
	250	85	75	50
95-5 tin-antimony	100	500	400	300
	150	400	350	275
	200	300	250	200
	250	200	175	150

**For copper pipe and solder-type fittings using soft-solders melting at approximately 350 degrees F. to 500 degrees F.*

***The use of any solder containing lead is not allowed in potable water systems.*

This section has been arranged to save the estimator's time by including all normally-used system components such as pipe, fittings, valves, hanger assemblies, riser clamps and miscellaneous items under one heading. Additional items can be found under "Plumbing and Piping Specialties." The cost estimates in this section are based on the conditions, limitations and wage rates described in the section "How to Use This Book" beginning on page 5.

Type M copper pipe with soft-soldered joints installed horizontally. Complete installation including six to twelve wrought copper tees and six to twelve wrought copper elbows every 100', and hangers spaced to meet plumbing code *(a tee & elbow every 8.3' for ½" dia., a tee and elbow every 16.6' for 4" dia.).*

Description	Craft@Hrs	Unit	Material $	Labor $	Equipment $	Total $
½"	P1@.110	LF	2.51	4.89	—	7.40
¾"	P1@.120	LF	3.87	5.33	—	9.20
1"	P1@.140	LF	8.03	6.22	—	14.25
1¼"	P1@.160	LF	13.60	7.11	—	20.71
1½"	P1@.180	LF	15.90	8.00	—	23.90
2"	P1@.210	LF	33.60	9.33	—	42.93
2½"	P1@.240	LF	49.30	10.70	—	60.00
3"	P1@.270	LF	61.70	12.00	—	73.70
4"	P1@.312	LF	90.90	13.90	—	104.80

Description	Craft@Hrs	Unit	Material $	Labor $	Equipment $	Total $

Type M copper pipe with soft-soldered joints, installed risers. Complete installation including a wrought copper reducing tee every floor and a riser clamp every other floor.

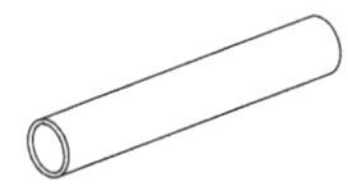

Description	Craft@Hrs	Unit	Material $	Labor $	Equipment $	Total $
½"	P1@.060	LF	2.13	2.67	—	4.80
¾"	P1@.070	LF	3.45	3.11	—	6.56
1"	P1@.075	LF	6.75	3.33	—	10.08
1¼"	P1@.080	LF	9.95	3.56	—	13.51
1½"	P1@.100	LF	13.20	4.45	—	17.65
2"	P1@.130	LF	17.40	5.78	—	23.18
2½"	P1@.150	LF	37.90	6.67	—	44.57
3"	P1@.170	LF	46.40	7.56	—	53.96
4"	P1@.190	LF	68.80	8.45	—	77.25

Type M copper pipe with soft-soldered joints, pipe only

Description	Craft@Hrs	Unit	Material $	Labor $	Equipment $	Total $
½"	P1@.032	LF	1.67	1.42	—	3.09
¾"	P1@.035	LF	2.73	1.56	—	4.29
1"	P1@.038	LF	4.24	1.69	—	5.93
1¼"	P1@.042	LF	6.40	1.87	—	8.27
1½"	P1@.046	LF	8.95	2.04	—	10.99
2"	P1@.053	LF	16.60	2.36	—	18.96
2½"	P1@.060	LF	22.80	2.67	—	25.47
3"	P1@.066	LF	28.50	2.93	—	31.43
4"	P1@.080	LF	45.90	3.56	—	49.46

Type M copper 45-degree ell C x C with soft-soldered joints

Description	Craft@Hrs	Unit	Material $	Labor $	Equipment $	Total $
½"	P1@.107	Ea	.97	4.76	—	5.73
¾"	P1@.150	Ea	1.70	6.67	—	8.37
1"	P1@.193	Ea	5.49	8.58	—	14.07
1¼"	P1@.236	Ea	6.18	10.50	—	16.68
1½"	P1@.278	Ea	11.10	12.40	—	23.50
2"	P1@.371	Ea	17.50	16.50	—	34.00
2½"	P1@.457	Ea	44.80	20.30	—	65.10
3"	P1@.543	Ea	59.20	24.10	—	83.30
4"	P1@.714	Ea	114.00	31.70	—	145.70

Type M copper 90-degree ell C x C with soft-soldered joints

Description	Craft@Hrs	Unit	Material $	Labor $	Equipment $	Total $
½"	P1@.107	Ea	.53	4.76	—	5.29
¾"	P1@.150	Ea	1.09	6.67	—	7.76
1"	P1@.193	Ea	4.19	8.58	—	12.77
1¼"	P1@.236	Ea	6.76	10.50	—	17.26
1½"	P1@.278	Ea	10.20	12.40	—	22.60
2"	P1@.371	Ea	17.70	16.50	—	34.20
2½"	P1@.457	Ea	37.00	20.30	—	57.30
3"	P1@.543	Ea	46.90	24.10	—	71.00
4"	P1@.714	Ea	99.20	31.70	—	130.90

Copper, Type M with Soft-Soldered Joints

Type M copper 90-degree ell ftg. x C with soft-soldered joint

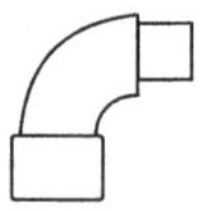

Description	Craft@Hrs	Unit	Material $	Labor $	Equipment $	Total $
½"	P1@.107	Ea	.92	4.76	—	5.68
¾"	P1@.150	Ea	1.98	6.67	—	8.65
1"	P1@.193	Ea	5.93	8.58	—	14.51
1¼"	P1@.236	Ea	8.26	10.50	—	18.76
1½"	P1@.278	Ea	12.30	12.40	—	24.70
2"	P1@.371	Ea	24.20	16.50	—	40.70
2½"	P1@.457	Ea	50.40	20.30	—	70.70
3"	P1@.543	Ea	56.70	24.10	—	80.80
4"	P1@.714	Ea	114.00	31.70	—	145.70

Type M copper tee C x C x C with soft-soldered joints

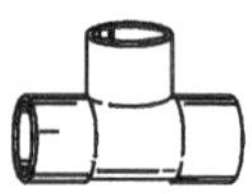

Description	Craft@Hrs	Unit	Material $	Labor $	Equipment $	Total $
½"	P1@.129	Ea	.84	5.73	—	6.57
¾"	P1@.181	Ea	2.26	8.05	—	10.31
1"	P1@.233	Ea	9.34	10.40	—	19.74
1¼"	P1@.285	Ea	14.90	12.70	—	27.60
1½"	P1@.337	Ea	21.80	15.00	—	36.80
2"	P1@.449	Ea	33.60	20.00	—	53.60
2½"	P1@.552	Ea	68.10	24.50	—	92.60
3"	P1@.656	Ea	105.00	29.20	—	134.20
4"	P1@.863	Ea	132.00	38.40	—	170.40

Type M copper branch reducing tee C x C x C with soft-soldered joints

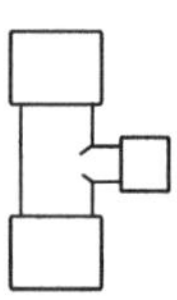

Description	Craft@Hrs	Unit	Material $	Labor $	Equipment $	Total $
½" x 3/8"	P1@.129	Ea	1.87	5.73	—	7.60
¾" x ½"	P1@.182	Ea	3.97	8.09	—	12.06
1" x ½"	P1@.195	Ea	7.78	8.67	—	16.45
1" x ¾"	P1@.219	Ea	7.78	9.73	—	17.51
1¼" x ½"	P1@.225	Ea	11.70	10.00	—	21.70
1¼" x ¾"	P1@.247	Ea	11.70	11.00	—	22.70
1¼" x 1"	P1@.268	Ea	11.70	11.90	—	23.60
1½" x ½"	P1@.272	Ea	14.40	12.10	—	26.50
1½" x ¾"	P1@.287	Ea	14.40	12.80	—	27.20
1½" x 1"	P1@.302	Ea	14.40	13.40	—	27.80
1½" x 1¼"	P1@.317	Ea	14.40	14.10	—	28.50
2" x ½"	P1@.325	Ea	18.80	14.40	—	33.20
2" x ¾"	P1@.348	Ea	18.80	15.50	—	34.30
2" x 1"	P1@.373	Ea	18.80	16.60	—	35.40
2" x 1¼"	P1@.398	Ea	18.80	17.70	—	36.50
2" x 1½"	P1@.422	Ea	18.80	18.80	—	37.60
2½" x ½"	P1@.437	Ea	56.80	19.40	—	76.20
2½" x ¾"	P1@.453	Ea	56.80	20.10	—	76.90
2½" x 1"	P1@.469	Ea	56.80	20.80	—	77.60
2½" x 1½"	P1@.502	Ea	56.80	22.30	—	79.10
2½" x 2"	P1@.519	Ea	56.80	23.10	—	79.90
3" x 1¼"	P1@.525	Ea	64.20	23.30	—	87.50
3" x 2"	P1@.585	Ea	66.40	26.00	—	92.40
3" x 2½"	P1@.617	Ea	66.40	27.40	—	93.80
4" x 1¼"	P1@.625	Ea	116.00	27.80	—	143.80
4" x 1½"	P1@.671	Ea	116.00	29.80	—	145.80
4" x 2"	P1@.718	Ea	116.00	31.90	—	147.90
4" x 2½"	P1@.765	Ea	116.00	34.00	—	150.00
4" x 3"	P1@.811	Ea	116.00	36.00	—	152.00

Type M copper reducer with soft-soldered joints

Description	Craft@Hrs	Unit	Material $	Labor $	Equipment $	Total $
½" x 3/8"	P1@.124	Ea	.67	5.51	—	6.18
¾" x ½"	P1@.129	Ea	1.26	5.73	—	6.99
1" x ¾"	P1@.172	Ea	2.28	7.65	—	9.93
1¼" x ½"	P1@.172	Ea	3.82	7.65	—	11.47
1¼" x ¾"	P1@.193	Ea	3.82	8.58	—	12.40
1¼" x 1"	P1@.215	Ea	3.82	9.56	—	13.38
1½" x ½"	P1@.193	Ea	5.83	8.58	—	14.41
1½" x ¾"	P1@.214	Ea	5.83	9.51	—	15.34
1½" x 1"	P1@.236	Ea	5.83	10.50	—	16.33
1½" x 1¼"	P1@.257	Ea	5.83	11.40	—	17.23
2" x ½"	P1@.239	Ea	8.79	10.60	—	19.39
2" x ¾"	P1@.261	Ea	8.79	11.60	—	20.39
2" x 1"	P1@.282	Ea	8.79	12.50	—	21.29
2" x 1¼"	P1@.304	Ea	8.79	13.50	—	22.29
2" x 1½"	P1@.325	Ea	8.79	14.40	—	23.19
2½" x 1"	P1@.325	Ea	23.90	14.40	—	38.30
2½" x 1¼"	P1@.347	Ea	23.90	15.40	—	39.30
2½" x 1½"	P1@.368	Ea	23.90	16.40	—	40.30
2½" x 2"	P1@.414	Ea	23.90	18.40	—	42.30
3" x 2"	P1@.457	Ea	28.00	20.30	—	48.30
3" x 2½"	P1@.500	Ea	28.00	22.20	—	50.20
4" x 2"	P1@.543	Ea	53.90	24.10	—	78.00
4" x 2½"	P1@.586	Ea	53.90	26.00	—	79.90
4" x 3"	P1@.629	Ea	53.90	28.00	—	81.90

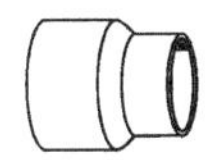

Type M copper adapter C x MPT with soft-soldered joint

Description	Craft@Hrs	Unit	Material $	Labor $	Equipment $	Total $
½"	P1@.075	Ea	1.65	3.33	—	4.98
¾"	P1@.105	Ea	2.59	4.67	—	7.26
1"	P1@.134	Ea	6.44	5.96	—	12.40
1¼"	P1@.164	Ea	12.00	7.29	—	19.29
1½"	P1@.194	Ea	13.60	8.62	—	22.22
2"	P1@.259	Ea	16.90	11.50	—	28.40
2½"	P1@.319	Ea	54.90	14.20	—	69.10
3"	P1@.378	Ea	91.00	16.80	—	107.80

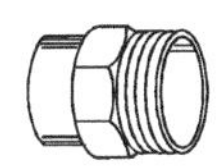

Type M copper adapter C x FPT with soft-soldered joint

Description	Craft@Hrs	Unit	Material $	Labor $	Equipment $	Total $
½"	P1@.075	Ea	2.76	3.33	—	6.09
¾"	P1@.105	Ea	5.01	4.67	—	9.68
1"	P1@.134	Ea	7.56	5.96	—	13.52
1¼"	P1@.164	Ea	16.00	7.29	—	23.29
1½"	P1@.194	Ea	19.20	8.62	—	27.82
2"	P1@.259	Ea	28.60	11.50	—	40.10
2½"	P1@.319	Ea	72.00	14.20	—	86.20
3"	P1@.378	Ea	99.00	16.80	—	115.80

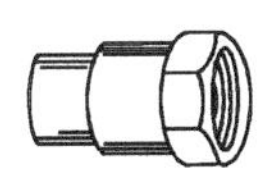

Type M copper flush bushing with soft-soldered joints

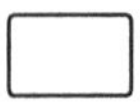

Description	Craft@Hrs	Unit	Material $	Labor $	Equipment $	Total $
½" x 3/8"	P1@.124	Ea	1.67	5.51	—	7.18
¾" x ½"	P1@.129	Ea	2.87	5.73	—	8.60
1" x ¾"	P1@.172	Ea	7.43	7.65	—	15.08
1" x ½"	P1@.172	Ea	8.48	7.65	—	16.13
1¼" x 1"	P1@.215	Ea	8.81	9.56	—	18.37
1½" x 1¼"	P1@.257	Ea	11.20	11.40	—	22.60
2" x 1½"	P1@.325	Ea	20.50	14.40	—	34.90

Type M copper union with soft-soldered joint

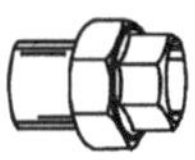

Description	Craft@Hrs	Unit	Material $	Labor $	Equipment $	Total $
½"	P1@.121	Ea	6.75	5.38	—	12.13
¾"	P1@.170	Ea	8.05	7.56	—	15.61
1"	P1@.218	Ea	18.10	9.69	—	27.79
1¼"	P1@.267	Ea	32.90	11.90	—	44.80
1½"	P1@.315	Ea	40.40	14.00	—	54.40
2"	P1@.421	Ea	58.20	18.70	—	76.90

Type M copper dielectric union with soft-soldered joint

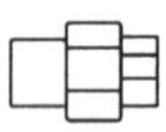

Description	Craft@Hrs	Unit	Material $	Labor $	Equipment $	Total $
½"	P1@.121	Ea	6.32	5.38	—	11.70
¾"	P1@.170	Ea	6.32	7.56	—	13.88
1"	P1@.218	Ea	13.10	9.69	—	22.79
1¼"	P1@.267	Ea	22.30	11.90	—	34.20
1½"	P1@.315	Ea	30.70	14.00	—	44.70
2"	P1@.421	Ea	41.00	18.70	—	59.70

Type M copper cap with soft-soldered joint

Description	Craft@Hrs	Unit	Material $	Labor $	Equipment $	Total $
½"	P1@.069	Ea	.56	3.07	—	3.63
¾"	P1@.096	Ea	1.12	4.27	—	5.39
1"	P1@.124	Ea	2.54	5.51	—	8.05
1¼"	P1@.151	Ea	3.64	6.71	—	10.35
1½"	P1@.178	Ea	5.31	7.91	—	13.22
2"	P1@.237	Ea	9.24	10.50	—	19.74
2½"	P1@.292	Ea	23.50	13.00	—	36.50
3"	P1@.347	Ea	36.70	15.40	—	52.10
4"	P1@.457	Ea	56.40	20.30	—	76.70

Type M copper coupling with soft-soldered joints

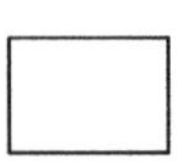

Description	Craft@Hrs	Unit	Material $	Labor $	Equipment $	Total $
½"	P1@.107	Ea	.36	4.76	—	5.12
¾"	P1@.150	Ea	.90	6.67	—	7.57
1"	P1@.193	Ea	2.41	8.58	—	10.99
1¼"	P1@.236	Ea	4.20	10.50	—	14.70
1½"	P1@.278	Ea	6.07	12.40	—	18.47
2"	P1@.371	Ea	8.93	16.50	—	25.43
2½"	P1@.457	Ea	21.10	20.30	—	41.40
3"	P1@.543	Ea	32.10	24.10	—	56.20
4"	P1@.714	Ea	63.00	31.70	—	94.70

Description	Craft@Hrs	Unit	Material $	Labor $	Equipment $	Total $

Class 125 bronze body gate valve with soft-soldered joints

Description	Craft@Hrs	Unit	Material $	Labor $	Equipment $	Total $
½"	P1@.200	Ea	22.80	8.89	—	31.69
¾"	P1@.249	Ea	28.50	11.10	—	39.60
1"	P1@.299	Ea	40.10	13.30	—	53.40
1¼"	P1@.398	Ea	50.40	17.70	—	68.10
1½"	P1@.448	Ea	68.00	19.90	—	87.90
2"	P1@.498	Ea	114.00	22.10	—	136.10
2½"	P1@.830	Ea	188.00	36.90	—	224.90
3"	P1@1.24	Ea	270.00	55.10	—	325.10

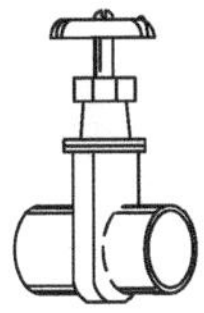

Class 125 iron body gate valve with flanged ends

Description	Craft@Hrs	Unit	Material $	Labor $	Equipment $	Total $
2"	P1@.500	Ea	435.00	22.20	—	457.20
2½"	P1@.600	Ea	586.00	26.70	—	612.70
3"	P1@.750	Ea	636.00	33.30	—	669.30
4"	P1@1.35	Ea	934.00	60.00	—	994.00
5"	P1@2.00	Ea	1,780.00	88.90	—	1,868.90
6"	P1@2.50	Ea	1,780.00	111.00	—	1,891.00

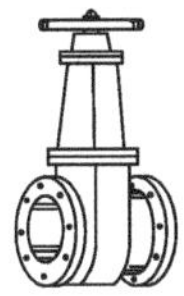

Class 125 bronze body globe valve with soft-soldered joints

Description	Craft@Hrs	Unit	Material $	Labor $	Equipment $	Total $
½"	P1@.200	Ea	42.20	8.89	—	51.09
¾"	P1@.249	Ea	56.20	11.10	—	67.30
1"	P1@.299	Ea	96.90	13.30	—	110.20
1¼"	P1@.398	Ea	136.00	17.70	—	153.70
1½"	P1@.448	Ea	185.00	19.90	—	204.90
2"	P1@.498	Ea	299.00	22.10	—	321.10

Class 125 iron body globe valve with flanged ends

Description	Craft@Hrs	Unit	Material $	Labor $	Equipment $	Total $
2½"	P1@.600	Ea	493.00	26.70	—	519.70
3"	P1@.750	Ea	591.00	33.30	—	624.30
4"	P1@1.35	Ea	798.00	60.00	—	858.00

200 PSIG iron body butterfly valve, lug-type, lever operated

Description	Craft@Hrs	Unit	Material $	Labor $	Equipment $	Total $
2"	P1@.450	Ea	179.00	20.00	—	199.00
2½"	P1@.450	Ea	186.00	20.00	—	206.00
3"	P1@.550	Ea	196.00	24.40	—	220.40
4"	P1@.550	Ea	244.00	24.40	—	268.40

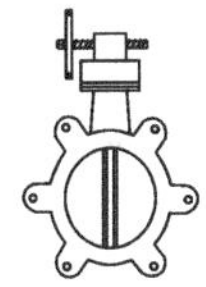

200 PSIG iron body butterfly valve, wafer-type, lever operated

Description	Craft@Hrs	Unit	Material $	Labor $	Equipment $	Total $
2"	P1@.450	Ea	164.00	20.00	—	184.00
2½"	P1@.450	Ea	167.00	20.00	—	187.00
3"	P1@.550	Ea	179.00	24.40	—	203.40
4"	P1@.550	Ea	217.00	24.40	—	241.40

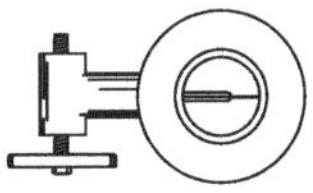

Description	Craft@Hrs	Unit	Material $	Labor $	Equipment $	Total $

Class 125 bronze body 2-piece ball valve with soft-soldered joints

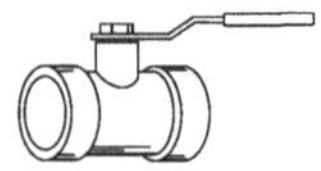

½"	P1@.200	Ea	11.60	8.89	—	20.49
¾"	P1@.249	Ea	15.40	11.10	—	26.50
1"	P1@.299	Ea	26.70	13.30	—	40.00
1¼"	P1@.398	Ea	44.60	17.70	—	62.30
1½"	P1@.448	Ea	60.30	19.90	—	80.20
2"	P1@.498	Ea	76.40	22.10	—	98.50
3"	P1@1.24	Ea	524.00	55.10	—	579.10
4"	P1@1.45	Ea	684.00	64.50	—	748.50

Class 125 bronze body swing check valve with soft-soldered joints

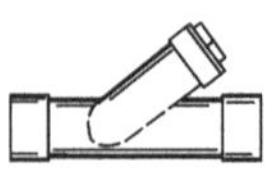

½"	P1@.200	Ea	23.90	8.89	—	32.79
¾"	P1@.249	Ea	34.10	11.10	—	45.20
1"	P1@.299	Ea	44.50	13.30	—	57.80
1¼"	P1@.398	Ea	63.60	17.70	—	81.30
1½"	P1@.448	Ea	89.60	19.90	—	109.50
2"	P1@.498	Ea	148.00	22.10	—	170.10
2½"	P1@.830	Ea	299.00	36.90	—	335.90
3"	P1@1.24	Ea	430.00	55.10	—	485.10

Class 125 iron body swing check valve, flanged ends

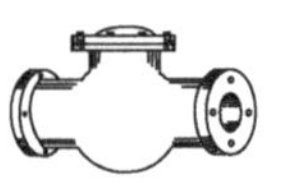

2"	P1@.500	Ea	209.00	22.20	—	231.20
2½"	P1@.600	Ea	264.00	26.70	—	290.70
3"	P1@.750	Ea	329.00	33.30	—	362.30
4"	P1@1.35	Ea	483.00	60.00	—	543.00

Class 125 iron body silent check valve, wafer-type

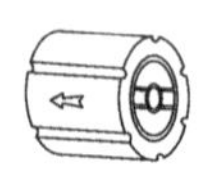

2"	P1@.500	Ea	153.00	22.20	—	175.20
2½"	P1@.600	Ea	170.00	26.70	—	196.70
3"	P1@.750	Ea	195.00	33.30	—	228.30
4"	P1@1.35	Ea	260.00	60.00	—	320.00

Class 125 bronze body strainer, threaded ends

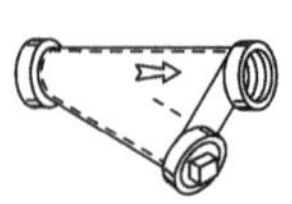

½"	P1@.230	Ea	37.40	10.20	—	47.60
¾"	P1@.260	Ea	49.10	11.60	—	60.70
1"	P1@.330	Ea	60.20	14.70	—	74.90
1¼"	P1@.440	Ea	84.20	19.60	—	103.80
1½"	P1@.495	Ea	126.00	22.00	—	148.00
2"	P1@.550	Ea	220.00	24.40	—	244.40

Class 125 iron body strainer, flanged ends

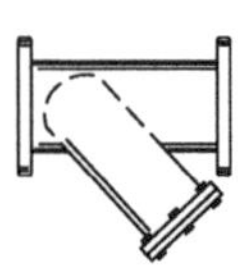

2"	P1@.500	Ea	168.00	22.20	—	190.20
2½"	P1@.600	Ea	187.00	26.70	—	213.70
3"	P1@.750	Ea	218.00	33.30	—	251.30
4"	P1@1.35	Ea	367.00	60.00	—	427.00

Description	Craft@Hrs	Unit	Material $	Labor $	Equipment $	Total $

Installation of 2-way control valve, threaded joints

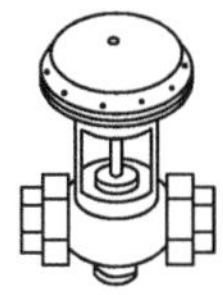

Description	Craft@Hrs	Unit	Material $	Labor $	Equipment $	Total $
½"	P1@.210	Ea	—	9.33	—	9.33
¾"	P1@.275	Ea	—	12.20	—	12.20
1"	P1@.350	Ea	—	15.60	—	15.60
1¼"	P1@.430	Ea	—	19.10	—	19.10
1½"	P1@.505	Ea	—	22.40	—	22.40
2"	P1@.675	Ea	—	30.00	—	30.00
2½"	P1@.830	Ea	—	36.90	—	36.90
3"	P1@.990	Ea	—	44.00	—	44.00

Installation of 3-way control valve, threaded joints

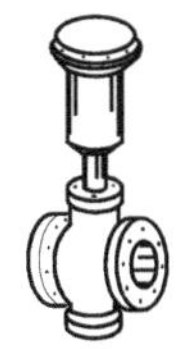

Description	Craft@Hrs	Unit	Material $	Labor $	Equipment $	Total $
½"	P1@.260	Ea	—	11.60	—	11.60
¾"	P1@.365	Ea	—	16.20	—	16.20
1"	P1@.475	Ea	—	21.10	—	21.10
1¼"	P1@.575	Ea	—	25.60	—	25.60
1½"	P1@.680	Ea	—	30.20	—	30.20
2"	P1@.910	Ea	—	40.40	—	40.40
2½"	P1@1.12	Ea	—	49.80	—	49.80
3"	P1@1.33	Ea	—	59.10	—	59.10

Companion flange, 150 pound cast brass

Description	Craft@Hrs	Unit	Material $	Labor $	Equipment $	Total $
2"	P1@.290	Ea	330.00	12.90	—	342.90
2½"	P1@.380	Ea	410.00	16.90	—	426.90
3"	P1@.460	Ea	410.00	20.40	—	430.40
4"	P1@.600	Ea	655.00	26.70	—	681.70

Bolt and gasket sets

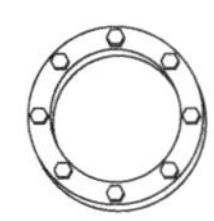

Description	Craft@Hrs	Unit	Material $	Labor $	Equipment $	Total $
2"	P1@.500	Ea	4.68	22.20	—	26.88
2½"	P1@.650	Ea	5.18	28.90	—	34.08
3"	P1@.750	Ea	5.41	33.30	—	38.71
4"	P1@1.00	Ea	5.55	44.50	—	50.05

Thermometer with well

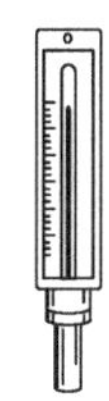

Description	Craft@Hrs	Unit	Material $	Labor $	Equipment $	Total $
7"	P1@.250	Ea	186.00	11.10	—	197.10
9"	P1@.250	Ea	192.00	11.10	—	203.10

Dial-type pressure gauge

Description	Craft@Hrs	Unit	Material $	Labor $	Equipment $	Total $
2½"	P1@.200	Ea	38.50	8.89	—	47.39
3½"	P1@.200	Ea	51.20	8.89	—	60.09

Pressure/temperature tap

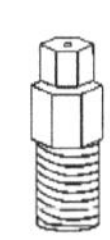

Description	Craft@Hrs	Unit	Material $	Labor $	Equipment $	Total $
Tap	P1@.150	Ea	17.70	6.67	—	24.37

Copper, Press Fit Fittings

Press Fit (Pressfast) Copper Pipe Fittings. For joining ½" to 4" type K, L or M hard drawn or annealed copper pipe and tube (ASTM 888). No brazing required. Includes a square cut and de-burr of the pipe ends and joining with a hand-held electric press fit tool. Equipment cost, when shown, is $1.50 (to 1-1/2" dia.) to $3.53 (2" and greater dia.) per hour for pressfit tool. Fittings include an O-ring for each joint.

Description	Craft@Hrs	Unit	Material $	Labor $	Equipment $	Total $

45-degree Copper press fit ell (P x P)

Description	Craft@Hrs	Unit	Material $	Labor $	Equipment $	Total $
½"	P1@.170	Ea	3.38	7.56	1.02	11.96
¾"	P1@.170	Ea	4.39	7.56	1.02	12.97
1"	P1@.180	Ea	14.00	8.00	1.08	23.08
1¼"	P1@.190	Ea	20.10	8.45	1.14	29.69
1½"	P1@.200	Ea	32.50	8.89	1.20	42.59
2"	P1@.200	Ea	45.50	8.89	2.82	57.21
2½"	P1@.225	Ea	100.00	10.00	3.17	113.17
3"	P1@.240	Ea	140.00	10.70	3.39	154.09
4"	P1@.260	Ea	200.00	11.60	3.67	215.27

90-degree Copper press fit ell (P x P)

Description	Craft@Hrs	Unit	Material $	Labor $	Equipment $	Total $
½"	P1@.170	Ea	2.79	7.56	1.02	11.37
¾"	P1@.170	Ea	4.69	7.56	1.02	13.27
1"	P1@.180	Ea	10.40	8.00	1.08	19.48
1¼"	P1@.190	Ea	20.90	8.45	1.14	30.49
1½"	P1@.200	Ea	39.40	8.89	1.20	49.49
2"	P1@.200	Ea	55.00	8.89	2.82	66.71
2½"	P1@.225	Ea	116.00	10.00	3.17	129.17
3"	P1@.240	Ea	217.00	10.70	3.39	231.09
4"	P1@.260	Ea	265.00	11.60	3.67	280.27

Coupling, Copper, press fit (P x P)

Description	Craft@Hrs	Unit	Material $	Labor $	Equipment $	Total $
½"	P1@.170	Ea	2.51	7.56	1.02	11.09
¾"	P1@.170	Ea	3.81	7.56	1.02	12.39
1"	P1@.180	Ea	7.65	8.00	1.08	16.73
1¼"	P1@.190	Ea	9.49	8.45	1.14	19.08
1½"	P1@.200	Ea	17.70	8.89	1.20	27.79
2"	P1@.200	Ea	22.50	8.89	2.82	34.21
2½"	P1@.225	Ea	74.90	10.00	3.17	88.07
3"	P1@.240	Ea	96.60	10.70	3.39	110.69
4"	P1@.260	Ea	133.00	11.60	3.67	148.27

Description	Craft@Hrs	Unit	Material $	Labor $	Equipment $	Total $

Tee, Copper, press fit (P x P)

Description	Craft@Hrs	Unit	Material $	Labor $	Equipment $	Total $
½"	P1@.170	Ea	3.38	7.56	1.02	11.96
¾"	P1@.170	Ea	4.39	7.56	1.02	12.97
1"	P1@.180	Ea	14.00	8.00	1.08	23.08
1¼"	P1@.190	Ea	20.10	8.45	1.14	29.69
1½"	P1@.200	Ea	32.50	8.89	1.20	42.59
2"	P1@.200	Ea	45.50	8.89	2.82	57.21
2½"	P1@.225	Ea	100.00	10.00	3.17	113.17
3"	P1@.240	Ea	140.00	10.70	3.39	154.09
4"	P1@.260	Ea	200.00	11.60	3.67	215.27

Reducing tee, Copper press fit (P x P x P)

Description	Craft@Hrs	Unit	Material $	Labor $	Equipment $	Total $
¾ x ½"	P1@.200	Ea	10.20	8.89	1.20	20.29
1 x ¾"	P1@.200	Ea	13.00	8.89	1.20	23.09
1¼" x ¾"	P1@.240	Ea	26.40	10.70	1.44	38.54
1¼" x 1"	P1@.240	Ea	18.80	10.70	1.44	30.94
1½" x ¾"	P1@.255	Ea	40.20	11.30	1.53	53.03
1½ x 1"	P1@.255	Ea	41.80	11.30	1.53	54.63
2" x ¾"	P1@.270	Ea	65.10	12.00	3.81	80.91
2" x 1"	P1@.270	Ea	67.40	12.00	3.81	83.21
2" x 1½"	P1@.280	Ea	51.00	12.40	3.95	67.35
3" x 1½"	P1@.380	Ea	254.00	16.90	5.36	276.26
3" x 2"	P1@.400	Ea	225.00	17.80	5.64	248.44
4" x 1½"	P1@.400	Ea	303.00	17.80	5.64	326.44
4" x 2"	P1@.420	Ea	286.00	18.70	5.93	310.63
4" x 3"	P1@.450	Ea	265.00	20.00	6.35	291.35

Union, Copper press fit (P x M)

Description	Craft@Hrs	Unit	Material $	Labor $	Equipment $	Total $
½"	P1@.220	Ea	22.80	9.78	1.32	33.90
¾"	P1@.220	Ea	29.90	9.78	1.32	41.00
1"	P1@.260	Ea	47.90	11.60	1.56	61.06
1¼"	P1@.280	Ea	68.30	12.40	1.68	82.38
1½"	P1@.280	Ea	93.20	12.40	1.68	107.28
2"	P1@.300	Ea	152.00	13.30	4.23	169.53
2½"	P1@.320	Ea	232.00	14.20	4.52	250.72
3"	P1@.380	Ea	251.00	16.90	5.36	273.26
4"	P1@.420	Ea	289.00	18.70	5.93	313.63

Description	Craft@Hrs	Unit	Material $	Labor $	Equipment $	Total $

Male threaded Copper press fit adapter (P x M)

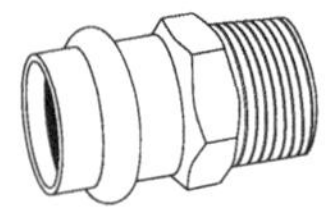

Description	Craft@Hrs	Unit	Material $	Labor $	Equipment $	Total $
½"	P1@.220	Ea	3.60	9.78	1.32	14.70
¾" x ½"	P1@.220	Ea	7.35	9.78	1.32	18.45
¾"	P1@.220	Ea	6.50	9.78	1.32	17.60
¾" x 1"	P1@.240	Ea	35.50	10.70	1.44	47.64
1 x ¾"	P1@.250	Ea	18.80	11.10	1.50	31.40
1"	P1@.260	Ea	11.90	11.60	1.56	25.06
1¼"	P1@.280	Ea	25.80	12.40	1.68	39.88
1½"	P1@.280	Ea	37.40	12.40	1.68	51.48
2"	P1@.300	Ea	70.10	13.30	4.23	87.63
2½"	P1@.320	Ea	155.00	14.20	4.52	173.72
3"	P1@.380	Ea	194.00	16.90	5.36	216.26
4"	P1@.420	Ea	235.00	18.70	5.93	259.63

Female threaded Copper press fit adapter (P x F)

Description	Craft@Hrs	Unit	Material $	Labor $	Equipment $	Total $
½"	P1@.200	Ea	4.22	8.89	1.20	14.31
¾"	P1@.220	Ea	6.89	9.78	1.32	17.99
¾" x ½"	P1@.220	Ea	7.68	9.78	1.32	18.78
¾"	P1@.220	Ea	6.50	9.78	1.32	17.60
1" x ½"	P1@.240	Ea	22.60	10.70	1.44	34.74
1 x ¾"	P1@.240	Ea	21.40	10.70	1.44	33.54
1"	P1@.260	Ea	12.50	11.60	1.56	25.66
1¼"	P1@.280	Ea	28.60	12.40	1.68	42.68
1½"	P1@.280	Ea	41.70	12.40	3.95	58.05
2"	P1@.300	Ea	72.70	13.30	4.23	90.23
3"	P1@.380	Ea	194.00	16.90	5.36	216.26
4"	P1@.420	Ea	236.00	18.70	5.93	260.63

Ball Valve, Brass press fit (P x P)

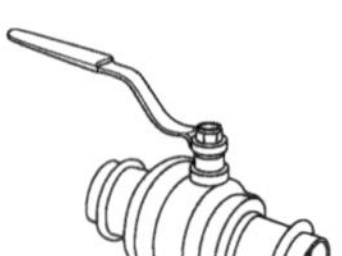

Description	Craft@Hrs	Unit	Material $	Labor $	Equipment $	Total $
½"	P1@.220	Ea	33.30	9.78	1.32	44.40
¾"	P1@.225	Ea	44.20	10.00	1.35	55.55
1"	P1@.275	Ea	57.80	12.20	1.65	71.65
1¼"	P1@.295	Ea	113.00	13.10	1.76	127.86
1½"	P1@.295	Ea	141.00	13.10	1.76	155.86
2"	P1@.330	Ea	231.00	14.70	4.66	250.36
3"	P1@.395	Ea	493.00	17.60	5.57	516.17
4"	P1@.435	Ea	625.00	19.30	6.14	650.44

Hard drawn rigid copper roll-grooved pipe with factory grooved copper fittings is commonly used for domestic or potable water distribution systems.

Maximum operating pressure for fittings is 300 psi (2065kPa). Operating temperature for fittings ranges from −30 F. to 250 F. (−34 C. – 121 C.).

Consult manufacturer for pipe pressure and temperature ratings and for the various combinations of pipe, fittings and system applications.

The cost estimates in this section are based on the conditions, limitations and wage rates described in the section "How to Use This Book" beginning on page 5.

Description	Craft@Hrs	Unit	Material $	Labor $	Equipment $	Total $
Type K Copper pipe only, roll-grooved, no hangers or fittings						
2"	P1@.080	LF	21.70	3.56	—	25.26
2½"	P1@.090	LF	31.60	4.00	—	35.60
3"	P1@.105	LF	42.90	4.67	—	47.57
4"	P1@.145	LF	67.00	6.45	—	73.45
6"	P1@.195	LF	102.00	8.67	—	110.67
Type L Copper pipe only, roll-grooved, no hangers or fittings						
2"	P1@.070	LF	19.70	3.11	—	22.81
2½"	P1@.070	LF	28.40	3.11	—	31.51
3"	P1@.090	LF	38.50	4.00	—	42.50
4"	P1@.130	LF	60.20	5.78	—	65.98
6"	P1@.180	LF	92.10	8.00	—	100.10
90-degree copper elbow, roll-grooved, style #610, (Victaulic)						
2"	P1@.410	Ea	50.60	18.20	—	68.80
2½"	P1@.430	Ea	56.10	19.10	—	75.20
3"	P1@.500	Ea	78.60	22.20	—	100.80
4"	P1@.690	Ea	168.00	30.70	—	198.70
6"	P1@1.19	Ea	881.00	52.90	—	933.90
45-degree copper elbow, roll-grooved, style #611, (Victaulic)						
2"	P1@.410	Ea	44.70	18.20	—	62.90
2½"	P1@.430	Ea	48.70	19.10	—	67.80
3"	P1@.500	Ea	67.10	22.20	—	89.30
4"	P1@.690	Ea	154.00	30.70	—	184.70
6"	P1@1.19	Ea	752.00	52.90	—	804.90
Tee, copper, roll-grooved, style #620, (Victaulic)						
2"	P1@.510	Ea	86.60	22.70	—	109.30
2½"	P1@.540	Ea	91.60	24.00	—	115.60
3"	P1@.630	Ea	138.00	28.00	—	166.00
4"	P1@.860	Ea	305.00	38.20	—	343.20
6"	P1@1.49	Ea	1,010.00	66.20	—	1,076.20

Description	Craft@Hrs	Unit	Material $	Labor $	Equipment $	Total $

Reducing tee, copper, roll-grooved, style #625, (Victaulic)

2½" x 2"	P1@.540	Ea	141.00	24.00	—	165.00
3" x 2½"	P1@.630	Ea	151.00	28.00	—	179.00
4" x 3"	P1@.860	Ea	355.00	38.20	—	393.20

Reducing tee, copper, roll-grooved, style #626, (Victaulic)

2½" x 1"	P1@.500	Ea	179.00	22.20	—	201.20
2½" x 1½"	P1@.520	Ea	179.00	23.10	—	202.10
3" x 1½"	P1@.600	Ea	189.00	26.70	—	215.70

Reducer, copper, roll-grooved, style #650, (Victaulic)

2½" x 2"	P1@.430	Ea	75.60	19.10	—	94.70
3" x 2"	P1@.500	Ea	75.60	22.20	—	97.80
3" x 2½"	P1@.520	Ea	75.60	23.10	—	98.70
4" x 2"	P1@.690	Ea	149.00	30.70	—	179.70
4" x 3"	P1@.720	Ea	149.00	32.00	—	181.00
6" x 3"	P1@1.20	Ea	458.00	53.30	—	511.30

Roll-grooved copper coupling with gasket, style #650, (Victaulic)

2"	P1@.300	Ea	30.50	13.30	—	43.80
2½"	P1@.350	Ea	34.50	15.60	—	50.10
3"	P1@.400	Ea	37.30	17.80	—	55.10
4"	P1@.500	Ea	47.90	22.20	—	70.10
5"	P1@.600	Ea	96.70	26.70	—	123.40
6"	P1@.700	Ea	111.00	31.10	—	142.10
8"	P1@.900	Ea	171.00	40.00	—	211.00
10"	P1@1.10	Ea	253.00	48.90	—	301.90
12"	P1@1.35	Ea	379.00	60.00	—	439.00

Flange adapter, copper, roll-grooved, style #641, (Victaulic)

2½"	P1@.300	Ea	150.00	13.30	—	163.30
3"	P1@.400	Ea	160.00	17.80	—	177.80
4"	P1@.550	Ea	220.00	24.40	—	244.40

Butterfly valve, brass, roll-grooved, lever handle, style #608, (Victaulic)

2½"	P1@.450	Ea	637.00	20.00	—	657.00
3"	P1@.450	Ea	820.00	20.00	—	840.00
4"	P1@.550	Ea	1,100.00	24.40	—	1,124.40

Type L soft-drawn (annealed) is used for potable water, service and distribution, fire protection, solar, fuel and oil, compressed air LPG gas etc.

Coiled Pipe

Tubing: ASTM B88, Seamless copper water and gas tube, NSF/ANSI 61 Annex G, Safe drinking water (3rd Party cert.)

Allow additional costs for permits, inspections and application.

Follow all Local, Regional and National codes/Utility guide lines were applicable, when installing gas/propane and potable water services.

The cost estimates in this section are based on the conditions, limitations and wage rates described in the section "How to Use This Book" beginning on page 5.

Description	Craft@Hrs	Unit	Material $	Labor $	Equipment $	Total $
Type L Soft Copper Pipe (coil). Above grade application. Pipe only. Make additional allowances per intended application.						
¼"	P1@.035	LF	1.18	1.56	—	2.74
3/8"	P1@.035	LF	1.66	1.56	—	3.22
½"	P1@.038	LF	5.20	1.69	—	6.89
¾"	P1@.044	LF	8.42	1.96	—	10.38
1"	P1@.056	LF	10.70	2.49	—	13.19
Type L Soft Copper Pipe (coil). Buried application. Make additional allowances for trenching/backfill, Permits and Code requirements.						
¼"	P1@.032	LF	1.18	1.42	.43	3.03
3/8"	P1@.032	LF	1.66	1.42	.53	3.61
½"	P1@.036	LF	5.20	1.60	.67	7.47
¾"	P1@.042	LF	8.42	1.87	.76	11.05
1"	P1@.054	LF	10.70	2.40	.84	13.94
Brass Corporation Coupling – Copper Pipe.						
¼"	P1@.100	LF	13.50	4.45	.29	18.24
3/8"	P1@.100	LF	21.70	4.45	.29	26.44
½"	P1@.125	LF	31.70	5.56	.29	37.55
¾"	P1@.140	LF	37.50	6.22	.33	44.05
1"	P1@.165	LF	42.30	7.33	.41	50.04
Brass Corporation Adapter (MIP/FIP) – Copper.						
¼"	P1@.115	LF	16.00	5.11	.29	21.40
3/8"	P1@.115	LF	24.70	5.11	.29	30.10
½"	P1@.135	LF	35.60	6.00	.29	41.89
¾"	P1@.145	LF	40.60	6.45	.33	47.38
1"	P1@.175	LF	45.20	7.78	.41	53.39

Corrugated Stainless Steel Tubing (CSST) CSST supplies natural gas or liquefied petroleum gas (LP gas) to appliances. Can be installed in all modes of construction; residential, commercial and industrial; new construction retrofits. The reusable brass components provide a reliable metal-to-metal gas tight seal. Tubing: ASTM A240 Type 304 Stainless Steel (annealed). Jacket: UV resistant. Polyethylene complying with requirement of ASTM E84 25/50. Allow additional costs for permits, inspections and applications. Follow all Local, Regional and National codes/Utility guidelines where applicable, when installing gas/propane services.

The cost estimates in this section are based on the conditions, limitations and wage rates described in the section "How to Use This Book" beginning on page 5.

Description	Craft@Hrs	Unit	Material $	Labor $	Equipment $	Total $
CSST with Arc/UV resistant Jacket – Flex Gas Pipe. Above grade application. Pipe only. Make additional allowances per intended application.						
½"	P1@.035	LF	3.15	1.56	—	4.71
¾"	P1@.040	LF	4.34	1.78	—	6.12
1"	P1@.050	LF	6.30	2.22	—	8.52
1¼"	P1@.065	LF	7.35	2.89	—	10.24
1½"	P1@.080	LF	13.10	3.56	—	16.66
2"	P1@.090	LF	18.40	4.00	—	22.40
CSST with Arc/UV resistant Jacket – Flex Gas Pipe. Buried application. Make additional allowances for trenching/backfill, Permits and Code requirements.						
½"	P1@.030	LF	3.15	1.33	.43	4.91
¾"	P1@.035	LF	4.34	1.56	.53	6.43
1"	P1@.045	LF	6.30	2.00	.67	8.97
1¼"	P1@.060	LF	7.35	2.67	.76	10.78
1½"	P1@.075	LF	13.10	3.33	.84	17.27
2"	P1@.085	LF	18.40	3.78	1.02	23.20
CSST Brass Coupling – Gas Pipe.						
½"	P1@.100	LF	18.90	4.45	.29	23.64
¾"	P1@.100	LF	25.20	4.45	.33	29.98
1"	P1@.160	LF	38.30	7.11	.41	45.82
1¼"	P1@.220	LF	84.00	9.78	.45	94.23
1½"	P1@.280	LF	173.00	12.40	.49	185.89
2"	P1@.330	LF	315.00	14.70	.58	330.28
CSST Brass Adapter (MIP/FIP) – Gas Pipe.						
½"	P1@.115	LF	9.45	5.11	.29	14.85
¾"	P1@.115	LF	12.60	5.11	.33	18.04
1"	P1@.175	LF	19.30	7.78	.41	27.49
1¼"	P1@.235	LF	43.40	10.40	.45	54.25
1½"	P1@.305	LF	91.70	13.60	.49	105.79
2"	P1@.355	LF	161.00	15.80	.58	177.38
CSST Brass Reducing Tee – Gas Pipe.						
½"	P1@.128	LF	26.60	5.69	.29	32.58
¾"	P1@.128	LF	40.60	5.69	.33	46.62
1"	P1@.188	LF	61.60	8.36	.41	70.37
1¼"	P1@.248	LF	116.00	11.00	.45	127.45
1½"	P1@.328	LF	203.00	14.60	.49	218.09
2"	P1@.368	LF	354.00	16.40	.58	370.98

PVC (Polyvinyl Chloride) Schedule 40 pipe with Schedule 40 socket-type fittings and solvent-welded joints is widely used in process, chemical, A.C. condensate, potable water and irrigation piping systems. PVC is available in Type I (normal impact) and Type II (high impact). Type I has a maximum temperature rating of 150 degrees F., and Type II is rated at 140 degrees F.

Consult the manufacturers for maximum pressure ratings and recommended joint solvents for specific applications.

Because of the current unreliability of many plastic valves, most engineers specify standard bronze-body and iron-body valves for use in PVC water piping systems. Plastic valves, however, have to be used in systems conveying liquids that may be injurious to metallic valves.

This section has been arranged to save the estimator's time by including all normally-used system components such as pipe, fittings, valves, hanger assemblies, riser clamps and miscellaneous items under one heading. Additional items can be found under "Plumbing and Piping Specialties." The cost estimates in this section are based on the conditions, limitations and wage rates described in the section "How to Use This Book" beginning on page 5.

Equipment cost, where shown, is $4.57 per hour for a 2-ton chain hoist.

Schedule 40 PVC pipe assembly with solvent-weld joints installed horizontally. Complete installation including three to eight tees and three to eight elbows every 100' and hangers spaced to meet plumbing code *(a tee and elbow every 12' for 1/2" dia., a tee and elbow every 40' for 8" dia.).*

Description	Craft@Hrs	Unit	Material $	Labor $	Equipment $	Total $
½"	P1@.084	LF	2.24	3.73	—	5.97
¾"	P1@.091	LF	2.67	4.04	—	6.71
1"	P1@.099	LF	2.80	4.40	—	7.20
1¼"	P1@.114	LF	2.93	5.07	—	8.00
1½"	P1@.121	LF	3.32	5.38	—	8.70
2"	P1@.130	LF	3.42	5.78	—	9.20
2½"	P1@.134	LF	5.53	5.96	—	11.49
3"	P1@.146	LF	8.66	6.49	—	15.15
4"	P1@.177	LF	11.00	7.87	—	18.87
6"	P1@.231	LF	20.20	10.30	—	30.50
8"	P1@.236	LF	35.90	10.50	—	46.40

Schedule 40 PVC pipe assembly with solvent-weld joints, installed risers. Complete installation including a reducing tee every floor and a riser clamp every other.

Description	Craft@Hrs	Unit	Material $	Labor $	Equipment $	Total $
½"	P1@.040	LF	1.10	1.78	—	2.88
¾"	P1@.047	LF	1.45	2.09	—	3.54
1"	P1@.052	LF	1.85	2.31	—	4.16
1¼"	P1@.061	LF	2.32	2.71	—	5.03
1½"	P1@.069	LF	2.85	3.07	—	5.92
2"	P1@.077	LF	2.99	3.42	—	6.41
2½"	P1@.088	LF	4.75	3.91	—	8.66
3"	P1@.099	LF	7.22	4.40	—	11.62
4"	P1@.125	LF	9.29	5.56	—	14.85
6"	P1@.202	LF	21.50	8.98	—	30.48
8"	P1@.266	LF	39.60	11.80	—	51.40

Schedule 40 PVC pipe with solvent-weld joints

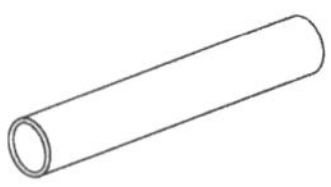

Description	Craft@Hrs	Unit	Material $	Labor $	Equipment $	Total $
½"	P1@.020	LF	.83	.89	—	1.72
¾"	P1@.025	LF	1.23	1.11	—	2.34
1"	P1@.030	LF	1.67	1.33	—	3.00
1¼"	P1@.035	LF	2.03	1.56	—	3.59
1½"	P1@.040	LF	2.40	1.78	—	4.18
2"	P1@.045	LF	2.49	2.00	—	4.49
2½"	P1@.050	LF	3.92	2.22	—	6.14
3"	P1@.055	LF	6.29	2.44	—	8.73
4"	P1@.070	LF	7.33	3.11	—	10.44
5"	P1@.095	LF	13.80	4.22	—	18.02
6"	P1@.120	LF	13.80	5.33	—	19.13
8"	P1@.160	LF	19.50	7.11	—	26.61

Schedule 40 PVC 45-degree ell S x S with solvent-weld joints

Description	Craft@Hrs	Unit	Material $	Labor $	Equipment $	Total $
½"	P1@.100	Ea	.41	4.45	—	4.86
¾"	P1@.115	Ea	.48	5.11	—	5.59
1"	P1@.120	Ea	.76	5.33	—	6.09
1¼"	P1@.160	Ea	1.42	7.11	—	8.53
1½"	P1@.180	Ea	1.51	8.00	—	9.51
2"	P1@.200	Ea	2.38	8.89	—	11.27
2½"	P1@.250	Ea	7.11	11.10	—	18.21
3"	P1@.300	Ea	8.57	13.30	—	21.87
4"	P1@.500	Ea	15.30	22.20	—	37.50
5"	P1@.550	Ea	39.50	24.40	—	63.90
6"	P1@.600	Ea	49.40	26.70	—	76.10
8"	P1@.800	Ea	125.00	35.60	—	160.60

Schedule 40 PVC 90-degree ell S x S with solvent-weld joints

Description	Craft@Hrs	Unit	Material $	Labor $	Equipment $	Total $
½"	P1@.100	Ea	.40	4.45	—	4.85
¾"	P1@.115	Ea	.45	5.11	—	5.56
1"	P1@.120	Ea	.76	5.33	—	6.09
1¼"	P1@.160	Ea	1.40	7.11	—	8.51
1½"	P1@.180	Ea	1.47	8.00	—	9.47
2"	P1@.200	Ea	2.32	8.89	—	11.21
2½"	P1@.250	Ea	7.09	11.10	—	18.19
3"	P1@.300	Ea	8.47	13.30	—	21.77
4"	P1@.500	Ea	15.30	22.20	—	37.50
5"	P1@.550	Ea	39.20	24.40	—	63.60
6"	P1@.600	Ea	48.40	26.70	—	75.10
8"	P1@.800	Ea	124.00	35.60	—	159.60

Schedule 40 PVC tee S x S x S with solvent-weld joints

Description	Craft@Hrs	Unit	Material $	Labor $	Equipment $	Total $
½"	P1@.130	Ea	.54	5.78	—	6.32
¾"	P1@.140	Ea	.58	6.22	—	6.80
1"	P1@.170	Ea	1.07	7.56	—	8.63
1¼"	P1@.220	Ea	1.69	9.78	—	11.47
1½"	P1@.250	Ea	1.87	11.10	—	12.97
2"	P1@.280	Ea	2.41	12.40	—	14.81
2½"	P1@.350	Ea	9.94	15.60	—	25.54
3"	P1@.420	Ea	12.70	18.70	—	31.40
4"	P1@.560	Ea	23.60	24.90	—	48.50
5"	P1@.710	Ea	56.80	31.60	—	88.40
6"	P1@.840	Ea	82.30	37.30	—	119.60
8"	P1@1.12	Ea	182.00	49.80	—	231.80

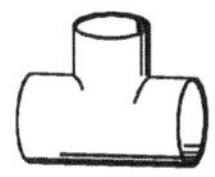

Schedule 40 PVC reducing tee S x S x S with solvent-weld joints

Description	Craft@Hrs	Unit	Material $	Labor $	Equipment $	Total $
¾" x ½"	P1@.140	Ea	.65	6.22	—	6.87
1" x ½"	P1@.170	Ea	1.19	7.56	—	8.75
1" x ¾"	P1@.170	Ea	1.78	7.56	—	9.34
1¼" x ½"	P1@.220	Ea	1.78	9.78	—	11.56
1¼" x ¾"	P1@.220	Ea	1.78	9.78	—	11.56
1¼" x 1"	P1@.220	Ea	1.78	9.78	—	11.56
1½" x ½"	P1@.250	Ea	3.08	11.10	—	14.18
1½" x ¾"	P1@.250	Ea	3.08	11.10	—	14.18
1½" x 1"	P1@.250	Ea	3.08	11.10	—	14.18
1½" x 1¼"	P1@.250	Ea	3.08	11.10	—	14.18
2" x ½"	P1@.280	Ea	4.38	12.40	—	16.78
2" x ¾"	P1@.280	Ea	4.38	12.40	—	16.78
2" x 1"	P1@.280	Ea	4.38	12.40	—	16.78
2" x 1¼"	P1@.280	Ea	4.38	12.40	—	16.78
2" x 1½"	P1@.280	Ea	4.38	12.40	—	16.78
2½" x 1"	P1@.350	Ea	9.11	15.60	—	24.71
2½" x 1¼"	P1@.350	Ea	9.11	15.60	—	24.71
2½" x 1½"	P1@.350	Ea	9.11	15.60	—	24.71
2½" x 2"	P1@.350	Ea	9.11	15.60	—	24.71
3" x 2"	P1@.420	Ea	13.50	18.70	—	32.20
3" x 2½"	P1@.420	Ea	13.50	18.70	—	32.20
4" x 2"	P1@.560	Ea	22.70	24.90	—	47.60
4" x 2½"	P1@.560	Ea	22.70	24.90	—	47.60
4" x 3"	P1@.560	Ea	22.70	24.90	—	47.60
5" x 3"	P1@.710	Ea	40.30	31.60	—	71.90
5" x 4"	P1@.710	Ea	40.30	31.60	—	71.90
6" x 3"	P1@.840	Ea	66.10	37.30	—	103.40
6" x 4"	P1@.840	Ea	66.10	37.30	—	103.40
6" x 5"	P1@.840	Ea	66.10	37.30	—	103.40
8" x 4"	P1@1.12	Ea	146.00	49.80	—	195.80
8" x 5"	P1@1.12	Ea	146.00	49.80	—	195.80
8" x 6"	P1@1.12	Ea	146.00	49.80	—	195.80

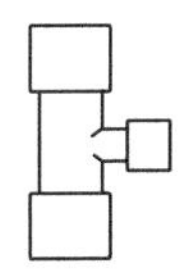

Schedule 40 PVC reducing bushing SPIG x S with solvent-weld joints

Description	Craft@Hrs	Unit	Material $	Labor $	Equipment $	Total $
¾" x ½"	P1@.050	Ea	.42	2.22	—	2.64
1" x ½"	P1@.060	Ea	.74	2.67	—	3.41
1" x ¾"	P1@.060	Ea	.74	2.67	—	3.41
1¼" x ½"	P1@.080	Ea	.99	3.56	—	4.55
1¼" x ¾"	P1@.080	Ea	.99	3.56	—	4.55
1¼" x 1"	P1@.080	Ea	.99	3.56	—	4.55
1½" x ½"	P1@.100	Ea	1.03	4.45	—	5.48
1½" x ¾"	P1@.100	Ea	1.03	4.45	—	5.48
1½" x 1"	P1@.100	Ea	1.03	4.45	—	5.48
1½" x 1¼"	P1@.100	Ea	1.03	4.45	—	5.48
2" x 1"	P1@.110	Ea	1.73	4.89	—	6.62
2" x 1¼"	P1@.110	Ea	1.73	4.89	—	6.62
2" x 1½"	P1@.110	Ea	1.73	4.89	—	6.62
2½" x 1¼"	P1@.130	Ea	2.76	5.78	—	8.54
2½" x 1½"	P1@.130	Ea	2.76	5.78	—	8.54
2½" x 2"	P1@.130	Ea	2.76	5.78	—	8.54
3" x 2"	P1@.150	Ea	4.12	6.67	—	10.79
3" x 2½"	P1@.150	Ea	4.12	6.67	—	10.79
4" x 3"	P1@.200	Ea	9.20	8.89	—	18.09
5" x 3"	P1@.250	Ea	18.10	11.10	—	29.20
5" x 4"	P1@.250	Ea	18.10	11.10	—	29.20
6" x 4"	P1@.300	Ea	26.50	13.30	—	39.80
6" x 5"	P1@.300	Ea	26.50	13.30	—	39.80
8" x 4"	P1@.400	Ea	79.60	17.80	—	97.40
8" x 5"	P1@.400	Ea	79.60	17.80	—	97.40

Schedule 40 PVC adapter MPT x S with solvent-weld joints

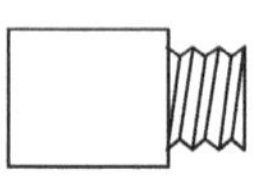

Description	Craft@Hrs	Unit	Material $	Labor $	Equipment $	Total $
½"	P1@.060	Ea	.36	2.67	—	3.03
¾"	P1@.065	Ea	.40	2.89	—	3.29
1"	P1@.070	Ea	.70	3.11	—	3.81
1¼"	P1@.100	Ea	.86	4.45	—	5.31
1½"	P1@.110	Ea	1.16	4.89	—	6.05
2"	P1@.120	Ea	3.79	5.33	—	9.12
2½"	P1@.150	Ea	4.44	6.67	—	11.11
3"	P1@.180	Ea	6.45	8.00	—	14.45
4"	P1@.240	Ea	8.22	10.70	—	18.92
5"	P1@.370	Ea	14.00	16.40	—	30.40
6"	P1@.420	Ea	21.40	18.70	—	40.10

Schedule 40 PVC adapter FPT x S with solvent-weld joints

Description	Craft@Hrs	Unit	Material $	Labor $	Equipment $	Total $
½"	P1@.110	Ea	.45	4.89	—	5.34
¾"	P1@.120	Ea	.56	5.33	—	5.89
1"	P1@.130	Ea	.65	5.78	—	6.43
1¼"	P1@.170	Ea	1.00	7.56	—	8.56
1½"	P1@.210	Ea	1.16	9.33	—	10.49
2"	P1@.290	Ea	1.56	12.90	—	14.46
2½"	P1@.330	Ea	3.89	14.70	—	18.59
3"	P1@.470	Ea	7.76	20.90	—	28.66
4"	P1@.630	Ea	14.30	28.00	—	42.30
5"	P1@.890	Ea	22.70	39.60	—	62.30
6"	P1@.970	Ea	31.90	43.10	—	75.00

Description	Craft@Hrs	Unit	Material $	Labor $	Equipment $	Total $

Schedule 40 PVC cap S with solvent-weld joints

Description	Craft@Hrs	Unit	Material $	Labor $	Equipment $	Total $
½"	P1@.060	Ea	.36	2.67	—	3.03
¾"	P1@.065	Ea	.42	2.89	—	3.31
1"	P1@.070	Ea	.65	3.11	—	3.76
1¼"	P1@.100	Ea	.99	4.45	—	5.44
1½"	P1@.110	Ea	1.00	4.89	—	5.89
2"	P1@.120	Ea	1.20	5.33	—	6.53
2½"	P1@.150	Ea	3.79	6.67	—	10.46
3"	P1@.180	Ea	4.18	8.00	—	12.18
4"	P1@.240	Ea	9.57	10.70	—	20.27
5"	P1@.370	Ea	15.90	16.40	—	32.30
6"	P1@.420	Ea	22.80	18.70	—	41.50

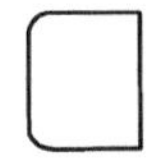

Schedule 40 PVC plug MPT with solvent-weld joints

Description	Craft@Hrs	Unit	Material $	Labor $	Equipment $	Total $
½"	P1@.150	Ea	1.05	6.67	—	7.72
¾"	P1@.160	Ea	1.15	7.11	—	8.26
1"	P1@.180	Ea	1.87	8.00	—	9.87
1¼"	P1@.210	Ea	1.93	9.33	—	11.26
1½"	P1@.230	Ea	2.08	10.20	—	12.28
2"	P1@.240	Ea	2.72	10.70	—	13.42
2½"	P1@.260	Ea	3.74	11.60	—	15.34
3"	P1@.300	Ea	5.66	13.30	—	18.96
4"	P1@.400	Ea	13.00	17.80	—	30.80

Schedule 40 PVC coupling S x S with solvent-weld joints

Description	Craft@Hrs	Unit	Material $	Labor $	Equipment $	Total $
½"	P1@.100	Ea	.24	4.45	—	4.69
¾"	P1@.115	Ea	.36	5.11	—	5.47
1"	P1@.120	Ea	.63	5.33	—	5.96
1¼"	P1@.160	Ea	.86	7.11	—	7.97
1½"	P1@.180	Ea	.90	8.00	—	8.90
2"	P1@.200	Ea	1.41	8.89	—	10.30
2½"	P1@.250	Ea	3.08	11.10	—	14.18
3"	P1@.300	Ea	4.85	13.30	—	18.15
4"	P1@.500	Ea	6.96	22.20	—	29.16
5"	P1@.550	Ea	13.00	24.40	—	37.40
6"	P1@.600	Ea	22.00	26.70	—	48.70
8"	P1@.800	Ea	41.20	35.60	—	76.80

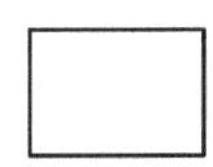

Schedule 40 PVC union S x S with solvent-weld joints

Description	Craft@Hrs	Unit	Material $	Labor $	Equipment $	Total $
1/2"	P1@.125	Ea	3.24	5.56	—	8.80
3/4"	P1@.140	Ea	4.02	6.22	—	10.24
1"	P1@.160	Ea	4.18	7.11	—	11.29
1¼"	P1@.200	Ea	8.62	8.89	—	17.51
1½"	P1@.225	Ea	9.82	10.00	—	19.82
2"	P1@.250	Ea	15.30	11.10	—	26.40
2½"	P1@.275	Ea	22.70	12.20	—	34.90
3"	P1@.300	Ea	27.70	13.30	—	41.00

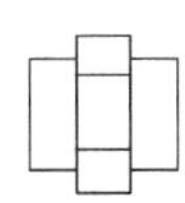

Description	Craft@Hrs	Unit	Material $	Labor $	Equipment $	Total $
Class 125 bronze body gate valve, threaded ends						
½"	P1@.210	Ea	22.80	9.33	—	32.13
¾"	P1@.250	Ea	28.50	11.10	—	39.60
1"	P1@.300	Ea	40.10	13.30	—	53.40
1¼"	P1@.400	Ea	50.40	17.80	—	68.20
1½"	P1@.450	Ea	68.00	20.00	—	88.00
2"	P1@.500	Ea	114.00	22.20	—	136.20
2½"	P1@.750	Ea	188.00	33.30	—	221.30
3"	P1@.950	Ea	270.00	42.20	—	312.20
Class 125 iron body gate valve, flanged ends						
2"	P1@.500	Ea	435.00	22.20	—	457.20
2½"	P1@.600	Ea	586.00	26.70	—	612.70
3"	P1@.750	Ea	636.00	33.30	—	669.30
4"	P1@1.35	Ea	934.00	60.00	—	994.00
5"	ER@2.00	Ea	1,780.00	94.80	1.34	1,876.14
6"	ER@2.50	Ea	1,780.00	119.00	1.67	1,900.67
8"	ER@3.00	Ea	2,940.00	142.00	2.01	3,084.01
Class 125 bronze body globe valve, threaded ends						
½"	P1@.210	Ea	42.20	9.33	—	51.53
¾"	P1@.250	Ea	56.20	11.10	—	67.30
1"	P1@.300	Ea	96.90	13.30	—	110.20
1¼"	P1@.400	Ea	136.00	17.80	—	153.80
1½"	P1@.450	Ea	185.00	20.00	—	205.00
2"	P1@.500	Ea	299.00	22.20	—	321.20
200 PSIG iron body butterfly valve, lug-type, lever operated						
2"	P1@.450	Ea	179.00	20.00	—	199.00
2½"	P1@.450	Ea	186.00	20.00	—	206.00
3"	P1@.550	Ea	196.00	24.40	—	220.40
4"	P1@.550	Ea	244.00	24.40	—	268.40
5"	ER@.800	Ea	321.00	37.90	.54	359.44
6"	ER@.800	Ea	392.00	37.90	.54	430.44
8"	ER@.900	Ea	536.00	42.70	.60	579.30
200 PSIG iron body butterfly valve, wafer-type, lever operated						
2"	P1@.450	Ea	164.00	20.00	—	184.00
2½"	P1@.450	Ea	167.00	20.00	—	187.00
3"	P1@.550	Ea	179.00	24.40	—	203.40
4"	P1@.550	Ea	217.00	24.40	—	241.40
5"	ER@.800	Ea	290.00	37.90	.54	328.44
6"	ER@.800	Ea	363.00	37.90	.54	401.44
8"	ER@.900	Ea	503.00	42.70	.60	546.30

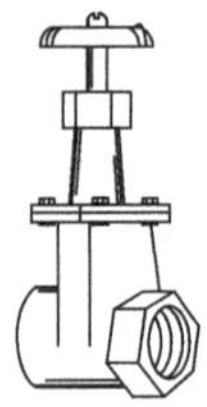
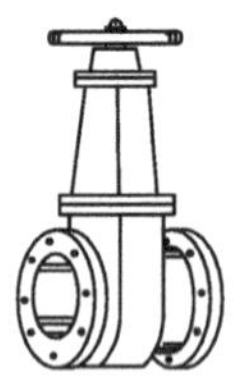
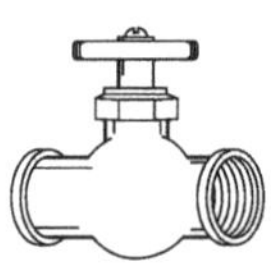

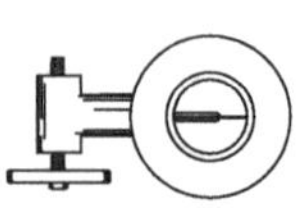

Description	Craft@Hrs	Unit	Material $	Labor $	Equipment $	Total $
Class 125 bronze body 2-piece ball valve, threaded ends						
½"	P1@.210	Ea	11.60	9.33	—	20.93
¾"	P1@.250	Ea	15.40	11.10	—	26.50
1"	P1@.300	Ea	26.70	13.30	—	40.00
1¼"	P1@.400	Ea	44.60	17.80	—	62.40
1½"	P1@.450	Ea	60.30	20.00	—	80.30
2"	P1@.500	Ea	76.40	22.20	—	98.60
3"	P1@.625	Ea	524.00	27.80	—	551.80
4"	P1@.690	Ea	684.00	30.70	—	714.70
PVC ball valve, solid body, solvent weld joints, EDPM (female socket)						
½"	P1@.170	Ea	7.46	7.56	—	15.02
¾"	P1@.200	Ea	8.44	8.89	—	17.33
1"	P1@.220	Ea	12.30	9.78	—	22.08
1¼"	P1@.280	Ea	14.10	12.40	—	26.50
1½"	P1@.315	Ea	23.80	14.00	—	37.80
2"	P1@.385	Ea	29.70	17.10	—	46.80
PVC ball valve, solid body, threaded joints, EDPM (female pipe thread)						
½"	P1@.200	Ea	7.46	8.89	—	16.35
¾"	P1@.240	Ea	8.44	10.70	—	19.14
1"	P1@.290	Ea	12.30	12.90	—	25.20
1¼"	P1@.390	Ea	14.10	17.30	—	31.40
1½"	P1@.420	Ea	23.80	18.70	—	42.50
2"	P1@.490	Ea	29.70	21.80	—	51.50
PVC ball valve, tru-union, threaded joints, EDPM (female pipe thread)						
2"	P1@.490	Ea	79.30	21.80	—	101.10
2½"	P1@.520	Ea	91.70	23.10	—	114.80
3"	P1@.590	Ea	141.00	26.20	—	167.20
4"	P1@.630	Ea	217.00	28.00	—	245.00

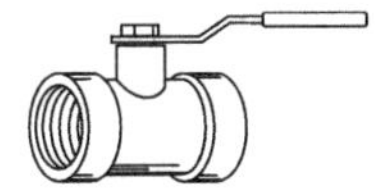

Description	Craft@Hrs	Unit	Material $	Labor $	Equipment $	Total $
PVC ball valve, union type body, solvent-weld joints (female socket)						
½"	P1@.185	Ea	16.00	8.22	—	24.22
¾"	P1@.225	Ea	20.00	10.00	—	30.00
1"	P1@.270	Ea	30.40	12.00	—	42.40
1¼"	P1@.360	Ea	39.20	16.00	—	55.20
1½"	P1@.400	Ea	52.00	17.80	—	69.80
2"	P1@.465	Ea	79.30	20.70	—	100.00
2½"	P1@.490	Ea	91.70	21.80	—	113.50
3"	P1@.540	Ea	141.00	24.00	—	165.00
4"	P1@.600	Ea	217.00	26.70	—	243.70
Class 125 bronze body swing check valve, threaded						
½"	P1@.210	Ea	23.90	9.33	—	33.23
¾"	P1@.250	Ea	34.10	11.10	—	45.20
1"	P1@.300	Ea	44.50	13.30	—	57.80
1¼"	P1@.400	Ea	63.60	17.80	—	81.40
1½"	P1@.450	Ea	89.60	20.00	—	109.60
2"	P1@.500	Ea	148.00	22.20	—	170.20
Class 125 iron body swing check valve, flanged ends						
2"	P1@.500	Ea	209.00	22.20	—	231.20
2½"	P1@.600	Ea	264.00	26.70	—	290.70
3"	P1@.750	Ea	329.00	33.30	—	362.30
4"	P1@1.35	Ea	483.00	60.00	—	543.00
5"	ER@2.00	Ea	930.00	94.80	1.34	1,026.14
6"	ER@2.50	Ea	930.00	119.00	1.67	1,050.67
8"	ER@3.00	Ea	1,690.00	142.00	2.01	1,834.01
Class 125 iron body silent check valve, wafer-type						
2"	P1@.500	Ea	153.00	22.20	—	175.20
2½"	P1@.600	Ea	170.00	26.70	—	196.70
3"	P1@.750	Ea	195.00	33.30	—	228.30
4"	P1@1.35	Ea	260.00	60.00	—	320.00
5"	ER@2.00	Ea	399.00	94.80	1.34	495.14
6"	ER@2.50	Ea	544.00	119.00	1.67	664.67
8"	ER@3.00	Ea	930.00	142.00	2.01	1,074.01
Class 125 bronze body strainer, threaded ends						
½"	P1@.210	Ea	37.40	9.33	—	46.73
¾"	P1@.250	Ea	49.10	11.10	—	60.20
1"	P1@.300	Ea	60.20	13.30	—	73.50
1¼"	P1@.400	Ea	84.20	17.80	—	102.00
1½"	P1@.450	Ea	126.00	20.00	—	146.00
2"	P1@.500	Ea	220.00	22.20	—	242.20

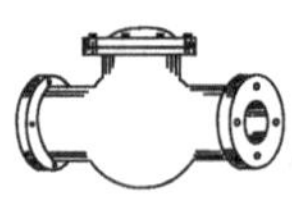
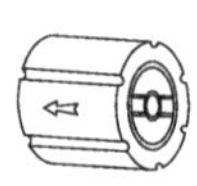
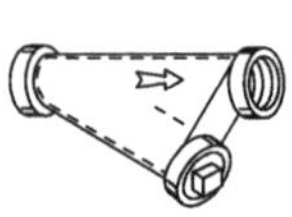

Description	Craft@Hrs	Unit	Material $	Labor $	Equipment $	Total $

Class 125 iron body strainer, flanged ends

Description	Craft@Hrs	Unit	Material $	Labor $	Equipment $	Total $
2"	P1@.500	Ea	168.00	22.20	—	190.20
2½"	P1@.600	Ea	187.00	26.70	—	213.70
3"	P1@.750	Ea	218.00	33.30	—	251.30
4"	P1@1.35	Ea	367.00	60.00	—	427.00
5"	ER@2.00	Ea	749.00	94.80	1.34	845.14
6"	ER@2.50	Ea	749.00	119.00	1.67	869.67
8"	ER@3.00	Ea	1,260.00	142.00	2.01	1,404.01

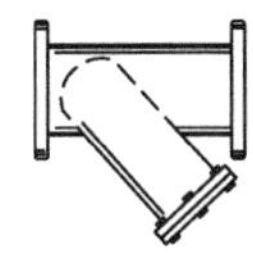

Installation of 2-way control valve, threaded joints

Description	Craft@Hrs	Unit	Material $	Labor $	Equipment $	Total $
½"	P1@.210	Ea	—	9.33	—	9.33
¾"	P1@.275	Ea	—	12.20	—	12.20
1"	P1@.350	Ea	—	15.60	—	15.60
1¼"	P1@.430	Ea	—	19.10	—	19.10
1½"	P1@.505	Ea	—	22.40	—	22.40
2"	P1@.675	Ea	—	30.00	—	30.00
2½"	P1@.830	Ea	—	36.90	—	36.90
3"	P1@.990	Ea	—	44.00	—	44.00

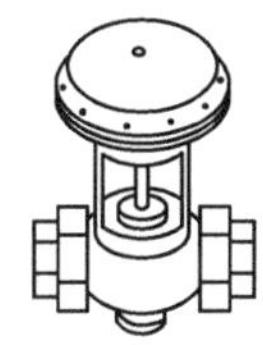

Installation of 3-way control valve, threaded joints

Description	Craft@Hrs	Unit	Material $	Labor $	Equipment $	Total $
½"	P1@.260	Ea	—	11.60	—	11.60
¾"	P1@.365	Ea	—	16.20	—	16.20
1"	P1@.475	Ea	—	21.10	—	21.10
1¼"	P1@.575	Ea	—	25.60	—	25.60
1½"	P1@.680	Ea	—	30.20	—	30.20
2"	P1@.910	Ea	—	40.40	—	40.40
2½"	P1@1.12	Ea	—	49.80	—	49.80
3"	P1@1.33	Ea	—	59.10	—	59.10

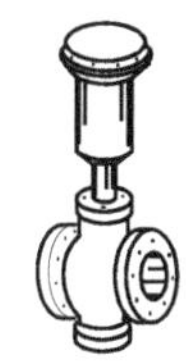

PVC companion flange

Description	Craft@Hrs	Unit	Material $	Labor $	Equipment $	Total $
2"	P1@.140	Ea	14.60	6.22	—	20.82
2½"	P1@.180	Ea	44.10	8.00	—	52.10
3"	P1@.210	Ea	45.50	9.33	—	54.83
4"	P1@.280	Ea	46.70	12.40	—	59.10
5"	P1@.310	Ea	105.00	13.80	—	118.80
6"	P1@.420	Ea	105.00	18.70	—	123.70
8"	P1@.560	Ea	157.00	24.90	—	181.90

Bolt and gasket sets

Description	Craft@Hrs	Unit	Material $	Labor $	Equipment $	Total $
2"	P1@.500	Ea	4.64	22.20	—	26.84
2½"	P1@.650	Ea	5.42	28.90	—	34.32
3"	P1@.750	Ea	9.19	33.30	—	42.49
4"	P1@1.00	Ea	15.40	44.50	—	59.90
5"	P1@1.10	Ea	29.60	48.90	—	78.50
6"	P1@1.20	Ea	29.60	53.30	—	82.90
8"	P1@1.25	Ea	33.00	55.60	—	88.60

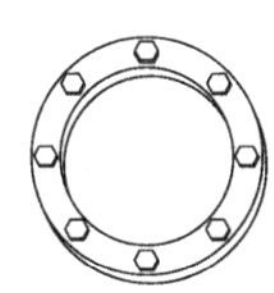

Description	Craft@Hrs	Unit	Material $	Labor $	Equipment $	Total $

Thermometer with well

Description	Craft@Hrs	Unit	Material $	Labor $	Equipment $	Total $
7"	P1@.250	Ea	186.00	11.10	—	197.10
9"	P1@.250	Ea	192.00	11.10	—	203.10

Dial-type pressure gauge

Description	Craft@Hrs	Unit	Material $	Labor $	Equipment $	Total $
2½"	P1@.200	Ea	38.50	8.89	—	47.39
3½"	P1@.200	Ea	51.20	8.89	—	60.09

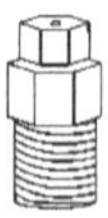

Pressure/temperature tap

Description	Craft@Hrs	Unit	Material $	Labor $	Equipment $	Total $
Tap	P1@.150	Ea	17.60	6.67	—	24.27

Hanger with swivel assembly

Description	Craft@Hrs	Unit	Material $	Labor $	Equipment $	Total $
½"	P1@.250	Ea	4.50	11.10	—	15.60
¾"	P1@.250	Ea	4.98	11.10	—	16.08
1"	P1@.250	Ea	5.20	11.10	—	16.30
1¼"	P1@.300	Ea	5.34	13.30	—	18.64
1½"	P1@.300	Ea	5.74	13.30	—	19.04
2"	P1@.300	Ea	5.99	13.30	—	19.29
2½"	P1@.350	Ea	8.15	15.60	—	23.75
3"	P1@.350	Ea	10.10	15.60	—	25.70
4"	P1@.350	Ea	11.10	15.60	—	26.70
5"	P1@.450	Ea	15.80	20.00	—	35.80
6"	P1@.450	Ea	17.90	20.00	—	37.90
8"	P1@.450	Ea	22.40	20.00	—	42.40

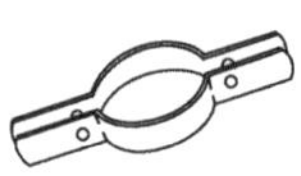

Riser clamp

Description	Craft@Hrs	Unit	Material $	Labor $	Equipment $	Total $
½"	P1@.100	Ea	3.33	4.45	—	7.78
¾"	P1@.100	Ea	4.92	4.45	—	9.37
1"	P1@.100	Ea	4.96	4.45	—	9.41
1¼"	P1@.105	Ea	5.96	4.67	—	10.63
1½"	P1@.110	Ea	6.28	4.89	—	11.17
2"	P1@.115	Ea	6.69	5.11	—	11.80
2½"	P1@.120	Ea	7.00	5.33	—	12.33
3"	P1@.120	Ea	7.60	5.33	—	12.93
4"	P1@.125	Ea	9.70	5.56	—	15.26
5"	P1@.180	Ea	13.70	8.00	—	21.70
6"	P1@.200	Ea	16.50	8.89	—	25.39
8"	P1@.200	Ea	26.90	8.89	—	35.79

PVC (Polyvinyl Chloride) Schedule 80 pipe with Schedule 80 socket-type fittings and solvent-welded joints is widely used in process, chemical, A.C. condensate, potable water and irrigation piping systems. PVC is available in Type I (normal impact) and Type II (high impact). Type I has a maximum temperature rating of 150 degrees F., and Type II is rated at 140 degrees F.

Schedule 80 PVC can be threaded, but it is more economical to use solvent-welded joints.

Manufacturers should be consulted for maximum pressure ratings and recommended joints solvents for specific applications.

Because of the current unreliability of many plastic valves, most engineers specify standard bronze-body and iron-body valves for use in PVC water piping systems. Plastic valves, however, have to be used in systems conveying liquids that may be injurious to metallic valves.

This section has been arranged to save the estimator's time by including all normally-used system components such as pipe, fittings, valves, hanger assemblies, riser clamps and miscellaneous items under one heading. Additional items can be found under "Plumbing and Piping Specialties." The cost estimates in this section are based on the conditions, limitations and wage rates described in the section "How to Use This Book" beginning on page 5.

Equipment cost, where shown, is $4.57 per hour for a 2-ton chain hoist.

Schedule 80 PVC pipe assembly with solvent-weld joints installed horizontally. Complete installation including three to eight tees and three to eight elbows every 100' and hangers spaced to meet plumbing code *(a tee and elbow every 12' for 1/2" dia., a tee and elbow every 40' for 8"dia.).*

Description	Craft@Hrs	Unit	Material $	Labor $	Equipment $	Total $
½"	P1@.085	LF	3.03	3.78	—	6.81
¾"	P1@.093	LF	3.79	4.13	—	7.92
1"	P1@.099	LF	4.17	4.40	—	8.57
1¼"	P1@.115	LF	4.88	5.11	—	9.99
1½"	P1@.117	LF	5.23	5.20	—	10.43
2"	P1@.120	LF	5.51	5.33	—	10.84
2½"	P1@.125	LF	9.09	5.56	—	14.65
3"	P1@.129	LF	12.30	5.73	—	18.03
4"	P1@.161	LF	17.70	7.16	—	24.86
6"	P1@.253	LF	29.40	11.20	—	40.60
8"	P1@.299	LF	49.40	13.30	—	62.70

Description	Craft@Hrs	Unit	Material $	Labor $	Equipment $	Total $

Schedule 80 PVC pipe assembly with solvent-weld joints, installed risers. Complete installation including a reducing tee every floor and a riser clamp every other.

½"	P1@.041	LF	1.50	1.82	—	3.32
¾"	P1@.048	LF	2.16	2.13	—	4.29
1"	P1@.054	LF	2.62	2.40	—	5.02
1¼"	P1@.063	LF	3.07	2.80	—	5.87
1½"	P1@.071	LF	4.02	3.16	—	7.18
2"	P1@.079	LF	4.30	3.51	—	7.81
2½"	P1@.090	LF	7.03	4.00	—	11.03
3"	P1@.101	LF	9.11	4.49	—	13.60
4"	P1@.129	LF	13.00	5.73	—	18.73
6"	P1@.242	LF	25.60	10.80	—	36.40
8"	P1@.296	LF	43.20	13.20	—	56.40

Schedule 80 PVC pipe with solvent-weld joints

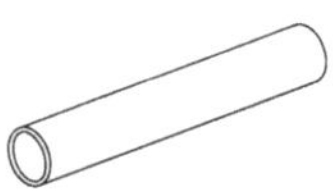

½"	P1@.021	LF	1.24	.93	—	2.17
¾"	P1@.026	LF	1.66	1.16	—	2.82
1"	P1@.032	LF	2.18	1.42	—	3.60
1¼"	P1@.037	LF	2.58	1.64	—	4.22
1½"	P1@.042	LF	3.46	1.87	—	5.33
2"	P1@.047	LF	3.58	2.09	—	5.67
2½"	P1@.052	LF	5.71	2.31	—	8.02
3"	P1@.057	LF	7.63	2.53	—	10.16
4"	P1@.074	LF	11.30	3.29	—	14.59
6"	P1@.160	LF	23.60	7.11	—	30.71
8"	P1@.190	LF	35.80	8.45	—	44.25

Schedule 80 PVC 45-degree ell S x S with solvent-weld joints

½"	P1@.105	Ea	3.22	4.67	—	7.89
¾"	P1@.120	Ea	4.86	5.33	—	10.19
1"	P1@.125	Ea	7.33	5.56	—	12.89
1¼"	P1@.170	Ea	9.38	7.56	—	16.94
1½"	P1@.190	Ea	11.00	8.45	—	19.45
2"	P1@.210	Ea	14.30	9.33	—	23.63
2½"	P1@.262	Ea	30.00	11.60	—	41.60
3"	P1@.315	Ea	36.50	14.00	—	50.50
4"	P1@.520	Ea	65.70	23.10	—	88.80
6"	P1@.630	Ea	82.70	28.00	—	110.70
8"	P1@.840	Ea	179.00	37.30	—	216.30

Description	Craft@Hrs	Unit	Material $	Labor $	Equipment $	Total $

Schedule 80 PVC 90-degree ell S x S with solvent-weld joints

Description	Craft@Hrs	Unit	Material $	Labor $	Equipment $	Total $
½"	P1@.105	Ea	1.75	4.67	—	6.42
¾"	P1@.120	Ea	2.21	5.33	—	7.54
1"	P1@.125	Ea	3.47	5.56	—	9.03
1¼"	P1@.170	Ea	4.74	7.56	—	12.30
1½"	P1@.190	Ea	5.05	8.45	—	13.50
2"	P1@.210	Ea	6.05	9.33	—	15.38
2½"	P1@.262	Ea	14.10	11.60	—	25.70
3"	P1@.315	Ea	15.90	14.00	—	29.90
4"	P1@.520	Ea	24.20	23.10	—	47.30
6"	P1@.630	Ea	69.10	28.00	—	97.10
8"	P1@.840	Ea	189.00	37.30	—	226.30

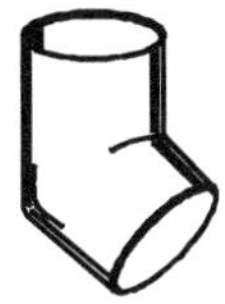

Schedule 80 PVC tee S x S x S with solvent-weld joints

Description	Craft@Hrs	Unit	Material $	Labor $	Equipment $	Total $
½"	P1@.135	Ea	4.82	6.00	—	10.82
¾"	P1@.145	Ea	5.05	6.45	—	11.50
1"	P1@.180	Ea	6.25	8.00	—	14.25
1¼"	P1@.230	Ea	17.30	10.20	—	27.50
1½"	P1@.265	Ea	17.30	11.80	—	29.10
2"	P1@.295	Ea	21.50	13.10	—	34.60
2½"	P1@.368	Ea	23.50	16.40	—	39.90
3"	P1@.440	Ea	29.50	19.60	—	49.10
4"	P1@.585	Ea	34.10	26.00	—	60.10
6"	P1@.880	Ea	115.00	39.10	—	154.10
8"	P1@1.18	Ea	269.00	52.50	—	321.50

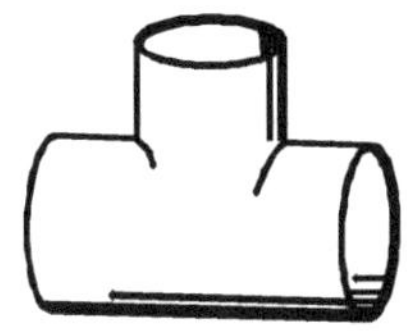

Schedule 80 PVC reducing tee S x S x S with solvent-weld joints

Description	Craft@Hrs	Unit	Material $	Labor $	Equipment $	Total $
¾" x ½"	P1@.145	Ea	5.45	6.45	—	11.90
1" x ½"	P1@.175	Ea	5.88	7.78	—	13.66
1" x ¾"	P1@.175	Ea	5.88	7.78	—	13.66
1¼" x ½"	P1@.230	Ea	19.30	10.20	—	29.50
1¼" x ¾"	P1@.230	Ea	19.30	10.20	—	29.50
1¼" x 1"	P1@.230	Ea	19.30	10.20	—	29.50
1½" x ¾"	P1@.260	Ea	20.60	11.60	—	32.20
1½" x 1"	P1@.260	Ea	20.60	11.60	—	32.20
1½" x 1¼"	P1@.260	Ea	20.60	11.60	—	32.20
2" x ½"	P1@.290	Ea	25.30	12.90	—	38.20
2" x ¾"	P1@.290	Ea	25.30	12.90	—	38.20
2" x 1"	P1@.290	Ea	25.30	12.90	—	38.20
2" x 1½"	P1@.290	Ea	25.30	12.90	—	38.20
3" x 2"	P1@.430	Ea	35.90	19.10	—	55.00
4" x 2"	P1@.570	Ea	40.80	25.30	—	66.10
4" x 3"	P1@.570	Ea	40.80	25.30	—	66.10
6" x 4"	P1@.850	Ea	138.00	37.80	—	175.80
8" x 4"	P1@1.16	Ea	309.00	51.60	—	360.60
8" x 5"	P1@1.16	Ea	309.00	51.60	—	360.60
8" x 6"	P1@1.16	Ea	309.00	51.60	—	360.60

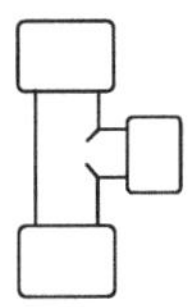

Schedule 80 PVC reducing bushing SPIG x S with solvent-weld joints

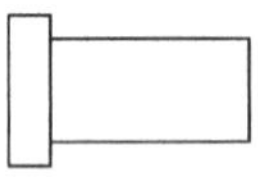

Description	Craft@Hrs	Unit	Material $	Labor $	Equipment $	Total $
¾" x ½"	P1@.053	Ea	.99	2.36	—	3.35
1" x ½"	P1@.063	Ea	2.84	2.80	—	5.64
1" x ¾"	P1@.063	Ea	2.84	2.80	—	5.64
1¼" x ½"	P1@.084	Ea	4.55	3.73	—	8.28
1¼" x ¾"	P1@.084	Ea	4.55	3.73	—	8.28
1¼" x 1"	P1@.084	Ea	4.55	3.73	—	8.28
1½" x ½"	P1@.105	Ea	6.05	4.67	—	10.72
1½" x ¾"	P1@.105	Ea	6.05	4.67	—	10.72
1½" x 1"	P1@.105	Ea	6.05	4.67	—	10.72
1½" x 1¼"	P1@.105	Ea	6.05	4.67	—	10.72
2" x 1"	P1@.116	Ea	8.66	5.16	—	13.82
2" x 1¼"	P1@.116	Ea	8.66	5.16	—	13.82
2" x 1½"	P1@.116	Ea	8.66	5.16	—	13.82
3" x 2"	P1@.158	Ea	23.90	7.02	—	30.92
4" x 2½"	P1@.210	Ea	33.00	9.33	—	42.33
4" x 3"	P1@.210	Ea	33.00	9.33	—	42.33
6" x 3"	P1@.315	Ea	45.90	14.00	—	59.90
6" x 4"	P1@.315	Ea	45.90	14.00	—	59.90
8" x 4"	P1@.420	Ea	110.00	18.70	—	128.70
8" x 6"	P1@.420	Ea	110.00	18.70	—	128.70

Schedule 80 PVC adapter MPT x S with solvent-weld joints

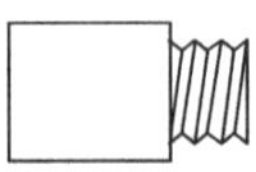

Description	Craft@Hrs	Unit	Material $	Labor $	Equipment $	Total $
½"	P1@.063	Ea	3.61	2.80	—	6.41
¾"	P1@.068	Ea	3.95	3.02	—	6.97
1"	P1@.074	Ea	6.79	3.29	—	10.08
1¼"	P1@.105	Ea	8.04	4.67	—	12.71
1½"	P1@.116	Ea	11.50	5.16	—	16.66
2"	P1@.126	Ea	16.60	5.60	—	22.20
2½"	P1@.158	Ea	19.20	7.02	—	26.22
3"	P1@.189	Ea	21.00	8.40	—	29.40
4"	P1@.252	Ea	37.10	11.20	—	48.30
5"	P1@.385	Ea	51.90	17.10	—	69.00
6"	P1@.435	Ea	119.00	19.30	—	138.30

Schedule 80 PVC adapter FPT x S with solvent-weld joints

Description	Craft@Hrs	Unit	Material $	Labor $	Equipment $	Total $
½"	P1@.116	Ea	2.90	5.16	—	8.06
¾"	P1@.126	Ea	4.30	5.60	—	9.90
1"	P1@.137	Ea	6.28	6.09	—	12.37
1¼"	P1@.179	Ea	10.20	7.96	—	18.16
1½"	P1@.221	Ea	12.30	9.82	—	22.12
2"	P1@.305	Ea	21.70	13.60	—	35.30
2½"	P1@.345	Ea	34.30	15.30	—	49.60
3"	P1@.490	Ea	38.60	21.80	—	60.40
4"	P1@.655	Ea	66.40	29.10	—	95.50
5"	P1@.920	Ea	108.00	40.90	—	148.90
6"	P1@1.10	Ea	132.00	48.90	—	180.90

Description	Craft@Hrs	Unit	Material $	Labor $	Equipment $	Total $

Schedule 80 PVC cap S with solvent-weld joints

Description	Craft@Hrs	Unit	Material $	Labor $	Equipment $	Total $
½"	P1@.063	Ea	3.04	2.80	—	5.84
¾"	P1@.070	Ea	3.14	3.11	—	6.25
1"	P1@.075	Ea	5.60	3.33	—	8.93
1¼"	P1@.105	Ea	6.79	4.67	—	11.46
1½"	P1@.115	Ea	6.79	5.11	—	11.90
2"	P1@.125	Ea	13.30	5.56	—	18.86
2½"	P1@.158	Ea	27.10	7.02	—	34.12
3"	P1@.189	Ea	32.20	8.40	—	40.60
4"	P1@.250	Ea	54.30	11.10	—	65.40
5"	P1@.390	Ea	111.00	17.30	—	128.30
6"	P1@.450	Ea	134.00	20.00	—	154.00

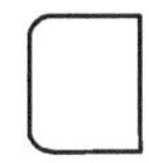

Schedule 80 PVC plug MPT with solvent-weld joints

Description	Craft@Hrs	Unit	Material $	Labor $	Equipment $	Total $
½"	P1@.155	Ea	2.94	6.89	—	9.83
¾"	P1@.165	Ea	3.04	7.33	—	10.37
1"	P1@.190	Ea	3.70	8.45	—	12.15
1¼"	P1@.220	Ea	5.45	9.78	—	15.23
1½"	P1@.240	Ea	6.58	10.70	—	17.28
2"	P1@.255	Ea	6.78	11.30	—	18.08
2½"	P1@.275	Ea	17.00	12.20	—	29.20
3"	P1@.320	Ea	22.10	14.20	—	36.30
4"	P1@.425	Ea	43.00	18.90	—	61.90

Schedule 80 PVC coupling S x S with solvent-weld joints

Description	Craft@Hrs	Unit	Material $	Labor $	Equipment $	Total $
½"	P1@.105	Ea	3.09	4.67	—	7.76
¾"	P1@.120	Ea	4.14	5.33	—	9.47
1"	P1@.125	Ea	4.29	5.56	—	9.85
1¼"	P1@.170	Ea	6.55	7.56	—	14.11
1½"	P1@.190	Ea	7.02	8.45	—	15.47
2"	P1@.210	Ea	7.51	9.33	—	16.84
2½"	P1@.262	Ea	18.50	11.60	—	30.10
3"	P1@.315	Ea	21.30	14.00	—	35.30
4"	P1@.520	Ea	26.90	23.10	—	50.00
6"	P1@.630	Ea	57.30	28.00	—	85.30
8"	P1@.840	Ea	78.80	37.30	—	116.10

Schedule 80 PVC union S x S with solvent-weld joints

Description	Craft@Hrs	Unit	Material $	Labor $	Equipment $	Total $
½"	P1@.125	Ea	6.19	5.56	—	11.75
¾"	P1@.140	Ea	8.53	6.22	—	14.75
1"	P1@.160	Ea	8.99	7.11	—	16.10
1¼"	P1@.200	Ea	17.90	8.89	—	26.79
1½"	P1@.225	Ea	20.30	10.00	—	30.30
2"	P1@.250	Ea	27.40	11.10	—	38.50
2½"	P1@.275	Ea	43.40	12.20	—	55.60
3"	P1@.300	Ea	51.10	13.30	—	64.40

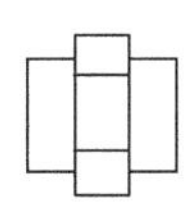

Description	Craft@Hrs	Unit	Material $	Labor $	Equipment $	Total $

Class 125 bronze body gate valve, threaded ends

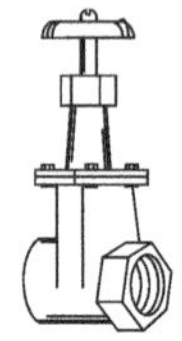

½"	P1@.210	Ea	22.80	9.33	—	32.13
¾"	P1@.250	Ea	28.50	11.10	—	39.60
1"	P1@.300	Ea	40.10	13.30	—	53.40
1¼"	P1@.400	Ea	50.40	17.80	—	68.20
1½"	P1@.450	Ea	68.00	20.00	—	88.00
2"	P1@.500	Ea	114.00	22.20	—	136.20
2½"	P1@.750	Ea	188.00	33.30	—	221.30
3"	P1@.950	Ea	270.00	42.20	—	312.20

Class 125 iron body gate valve, flanged ends

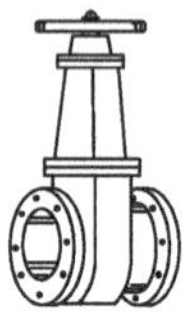

2"	P1@.500	Ea	435.00	22.20	—	457.20
2½"	P1@.600	Ea	586.00	26.70	—	612.70
3"	P1@.750	Ea	636.00	33.30	—	669.30
4"	P1@1.35	Ea	934.00	60.00	—	994.00
5"	ER@2.00	Ea	1,780.00	94.80	1.34	1,876.14
6"	ER@2.50	Ea	1,780.00	119.00	1.67	1,900.67
8"	ER@3.00	Ea	2,940.00	142.00	2.01	3,084.01

Class 125 bronze body globe valve, threaded ends

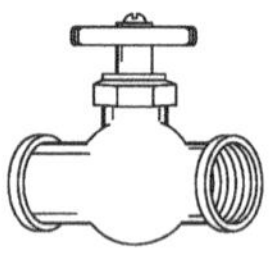

½"	P1@.210	Ea	42.20	9.33	—	51.53
¾"	P1@.250	Ea	56.20	11.10	—	67.30
1"	P1@.300	Ea	96.90	13.30	—	110.20
1¼"	P1@.400	Ea	136.00	17.80	—	153.80
1½"	P1@.450	Ea	185.00	20.00	—	205.00
2"	P1@.500	Ea	299.00	22.20	—	321.20

200 PSIG iron body butterfly valve, lug-type, lever operated

2"	P1@.450	Ea	179.00	20.00	—	199.00
2½"	P1@.450	Ea	186.00	20.00	—	206.00
3"	P1@.550	Ea	196.00	24.40	—	220.40
4"	P1@.550	Ea	244.00	24.40	—	268.40
5"	ER@.800	Ea	321.00	37.90	.54	359.44
6"	ER@.800	Ea	392.00	37.90	.54	430.44
8"	ER@.900	Ea	536.00	42.70	.60	579.30

200 PSIG iron body butterfly valve, wafer-type, lever operated

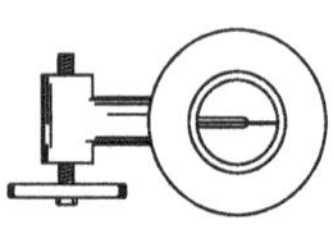

2"	P1@.450	Ea	164.00	20.00	—	184.00
2½"	P1@.450	Ea	167.00	20.00	—	187.00
3"	P1@.550	Ea	179.00	24.40	—	203.40
4"	P1@.550	Ea	217.00	24.40	—	241.40
5"	ER@.800	Ea	290.00	37.90	.54	328.44
6"	ER@.800	Ea	363.00	37.90	.54	401.44
8"	ER@.900	Ea	503.00	42.70	.60	546.30

Description	Craft@Hrs	Unit	Material $	Labor $	Equipment $	Total $
Class 125 bronze body 2-piece ball valve, threaded ends						
½"	P1@.210	Ea	11.60	9.33	—	20.93
¾"	P1@.250	Ea	15.40	11.10	—	26.50
1"	P1@.300	Ea	26.70	13.30	—	40.00
1¼"	P1@.400	Ea	44.60	17.80	—	62.40
1½"	P1@.450	Ea	60.30	20.00	—	80.30
2"	P1@.500	Ea	76.40	22.20	—	98.60
3"	P1@.625	Ea	524.00	27.80	—	551.80
4"	P1@.690	Ea	684.00	30.70	—	714.70
PVC ball valve, solid body, solvent-weld joints, EDPM (female socket)						
½"	P1@.170	Ea	7.46	7.56	—	15.02
¾"	P1@.200	Ea	8.44	8.89	—	17.33
1"	P1@.220	Ea	12.30	9.78	—	22.08
1¼"	P1@.280	Ea	14.10	12.40	—	26.50
1½"	P1@.315	Ea	23.80	14.00	—	37.80
2"	P1@.385	Ea	29.70	17.10	—	46.80
PVC ball valve, solid body, threaded joints, EDPM (female pipe thread)						
½"	P1@.200	Ea	7.46	8.89	—	16.35
¾"	P1@.240	Ea	8.44	10.70	—	19.14
1"	P1@.290	Ea	12.30	12.90	—	25.20
1¼"	P1@.390	Ea	14.10	17.30	—	31.40
1½"	P1@.420	Ea	23.80	18.70	—	42.50
2"	P1@.490	Ea	29.70	21.80	—	51.50
PVC ball valve, tru-union, threaded joints, EDPM (female pipe thread)						
2"	P1@.490	Ea	79.30	21.80	—	101.10
2½"	P1@.520	Ea	91.70	23.10	—	114.80
3"	P1@.590	Ea	141.00	26.20	—	167.20
4"	P1@.630	Ea	217.00	28.00	—	245.00
PVC ball valve, union-type body, solvent-weld joints (female socket)						
½"	P1@.185	Ea	16.00	8.22	—	24.22
¾"	P1@.225	Ea	20.00	10.00	—	30.00
1"	P1@.270	Ea	30.40	12.00	—	42.40
1¼"	P1@.360	Ea	39.20	16.00	—	55.20
1½"	P1@.400	Ea	52.00	17.80	—	69.80
2"	P1@.465	Ea	79.30	20.70	—	100.00
2½"	P1@.490	Ea	91.70	21.80	—	113.50
3"	P1@.540	Ea	141.00	24.00	—	165.00
4"	P1@.600	Ea	217.00	26.70	—	243.70

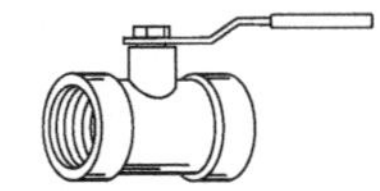

Description	Craft@Hrs	Unit	Material $	Labor $	Equipment $	Total $

Class 125 bronze body swing check valve, threaded ends

Description	Craft@Hrs	Unit	Material $	Labor $	Equipment $	Total $
½"	P1@.210	Ea	23.90	9.33	—	33.23
¾"	P1@.250	Ea	34.10	11.10	—	45.20
1"	P1@.300	Ea	44.50	13.30	—	57.80
1¼"	P1@.400	Ea	63.60	17.80	—	81.40
1½"	P1@.450	Ea	89.60	20.00	—	109.60
2"	P1@.500	Ea	148.00	22.20	—	170.20

Class 125 iron body swing check valve, flanged ends

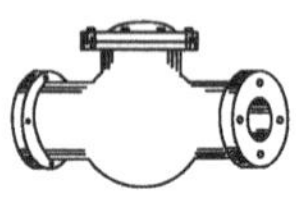

Description	Craft@Hrs	Unit	Material $	Labor $	Equipment $	Total $
2"	P1@.500	Ea	209.00	22.20	—	231.20
2½"	P1@.600	Ea	264.00	26.70	—	290.70
3"	P1@.750	Ea	329.00	33.30	—	362.30
4"	P1@1.35	Ea	483.00	60.00	—	543.00
5"	ER@2.00	Ea	930.00	94.80	1.34	1,026.14
6"	ER@2.50	Ea	930.00	119.00	1.67	1,050.67
8"	ER@3.00	Ea	1,690.00	142.00	2.01	1,834.01

Class 125 iron body silent check valve, wafer-type

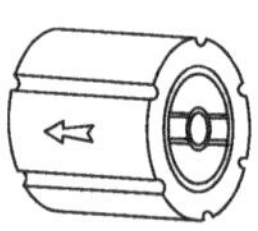

Description	Craft@Hrs	Unit	Material $	Labor $	Equipment $	Total $
2"	P1@.500	Ea	153.00	22.20	—	175.20
2½"	P1@.600	Ea	170.00	26.70	—	196.70
3"	P1@.750	Ea	195.00	33.30	—	228.30
4"	P1@1.35	Ea	260.00	60.00	—	320.00
5"	ER@2.00	Ea	399.00	94.80	1.34	495.14
6"	ER@2.50	Ea	544.00	119.00	1.67	664.67
8"	ER@3.00	Ea	930.00	142.00	2.01	1,074.01

Class 125 bronze body strainer, threaded ends

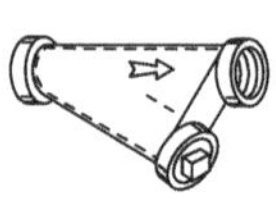

Description	Craft@Hrs	Unit	Material $	Labor $	Equipment $	Total $
½"	P1@.210	Ea	37.40	9.33	—	46.73
¾"	P1@.250	Ea	49.10	11.10	—	60.20
1"	P1@.300	Ea	60.20	13.30	—	73.50
1¼"	P1@.400	Ea	84.20	17.80	—	102.00
1½"	P1@.450	Ea	126.00	20.00	—	146.00
2"	P1@.500	Ea	220.00	22.20	—	242.20

Class 125 iron body strainer, flanged ends

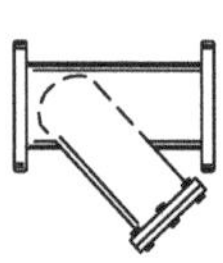

Description	Craft@Hrs	Unit	Material $	Labor $	Equipment $	Total $
2"	P1@.500	Ea	168.00	22.20	—	190.20
2½"	P1@.600	Ea	187.00	26.70	—	213.70
3"	P1@.750	Ea	218.00	33.30	—	251.30
4"	P1@1.35	Ea	367.00	60.00	—	427.00
5"	ER@2.00	Ea	998.00	94.80	1.34	1,094.14
6"	ER@2.50	Ea	998.00	119.00	1.67	1,118.67
8"	ER@3.00	Ea	1,700.00	142.00	2.01	1,844.01

Installation of 2-way control valve, threaded joints

Description	Craft@Hrs	Unit	Material $	Labor $	Equipment $	Total $
½"	P1@.210	Ea	—	9.33	—	9.33
¾"	P1@.275	Ea	—	12.20	—	12.20
1"	P1@.350	Ea	—	15.60	—	15.60
1¼"	P1@.430	Ea	—	19.10	—	19.10
1½"	P1@.505	Ea	—	22.40	—	22.40
2"	P1@.675	Ea	—	30.00	—	30.00
2½"	P1@.830	Ea	—	36.90	—	36.90
3"	P1@.990	Ea	—	44.00	—	44.00

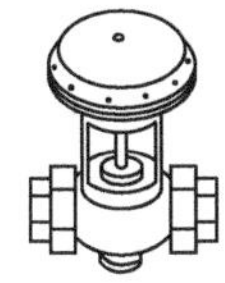

Installation of 3-way control valve, threaded joints

Description	Craft@Hrs	Unit	Material $	Labor $	Equipment $	Total $
½"	P1@.260	Ea	—	11.60	—	11.60
¾"	P1@.365	Ea	—	16.20	—	16.20
1"	P1@.475	Ea	—	21.10	—	21.10
1¼"	P1@.575	Ea	—	25.60	—	25.60
1½"	P1@.680	Ea	—	30.20	—	30.20
2"	P1@.910	Ea	—	40.40	—	40.40
2½"	P1@1.12	Ea	—	49.80	—	49.80
3"	P1@1.33	Ea	—	59.10	—	59.10

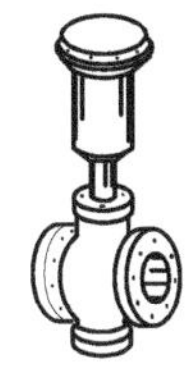

PVC companion flange

Description	Craft@Hrs	Unit	Material $	Labor $	Equipment $	Total $
2"	P1@.140	Ea	15.40	6.22	—	21.62
2½"	P1@.180	Ea	45.80	8.00	—	53.80
3"	P1@.210	Ea	47.80	9.33	—	57.13
4"	P1@.280	Ea	49.50	12.40	—	61.90
5"	P1@.310	Ea	110.00	13.80	—	123.80
6"	P1@.420	Ea	110.00	18.70	—	128.70
8"	P1@.560	Ea	164.00	24.90	—	188.90

Bolt and gasket set

Description	Craft@Hrs	Unit	Material $	Labor $	Equipment $	Total $
2"	P1@.500	Ea	4.64	22.20	—	26.84
2½"	P1@.650	Ea	5.42	28.90	—	34.32
3"	P1@.750	Ea	9.19	33.30	—	42.49
4"	P1@1.00	Ea	15.40	44.50	—	59.90
5"	P1@1.10	Ea	29.10	48.90	—	78.00
6"	P1@1.20	Ea	29.10	53.30	—	82.40
8"	P1@1.25	Ea	32.60	55.60	—	88.20

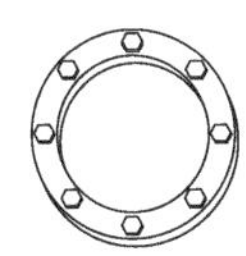

Thermometer with well

Description	Craft@Hrs	Unit	Material $	Labor $	Equipment $	Total $
7"	P1@.250	Ea	186.00	11.10	—	197.10
9"	P1@.250	Ea	192.00	11.10	—	203.10

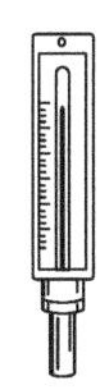

Dial-type pressure gauge

Description	Craft@Hrs	Unit	Material $	Labor $	Equipment $	Total $
2½"	P1@.200	Ea	38.50	8.89	—	47.39
3½"	P1@.200	Ea	51.20	8.89	—	60.09

Description	Craft@Hrs	Unit	Material $	Labor $	Equipment $	Total $

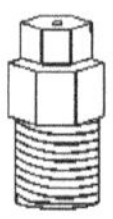

Pressure/temperature tap

Description	Craft@Hrs	Unit	Material $	Labor $	Equipment $	Total $
Tap	P1@.150	Ea	17.60	6.67	—	24.27

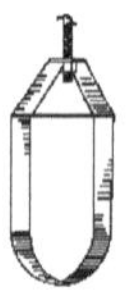

Hanger with swivel assembly

Description	Craft@Hrs	Unit	Material $	Labor $	Equipment $	Total $
½"	P1@.250	Ea	4.50	11.10	—	15.60
¾"	P1@.250	Ea	4.98	11.10	—	16.08
1"	P1@.250	Ea	5.20	11.10	—	16.30
1¼"	P1@.300	Ea	5.34	13.30	—	18.64
1½"	P1@.300	Ea	5.74	13.30	—	19.04
2"	P1@.300	Ea	5.99	13.30	—	19.29
2½"	P1@.350	Ea	8.15	15.60	—	23.75
3"	P1@.350	Ea	10.10	15.60	—	25.70
4"	P1@.350	Ea	11.10	15.60	—	26.70
5"	P1@.450	Ea	15.80	20.00	—	35.80
6"	P1@.450	Ea	17.90	20.00	—	37.90
8"	P1@.450	Ea	22.40	20.00	—	42.40

Riser clamp

Description	Craft@Hrs	Unit	Material $	Labor $	Equipment $	Total $
½"	P1@.100	Ea	3.33	4.45	—	7.78
¾"	P1@.100	Ea	4.92	4.45	—	9.37
1"	P1@.100	Ea	4.96	4.45	—	9.41
1¼"	P1@.105	Ea	5.96	4.67	—	10.63
1½"	P1@.110	Ea	6.28	4.89	—	11.17
2"	P1@.115	Ea	6.69	5.11	—	11.80
2½"	P1@.120	Ea	7.00	5.33	—	12.33
3"	P1@.120	Ea	7.60	5.33	—	12.93
4"	P1@.125	Ea	9.70	5.56	—	15.26
5"	P1@.180	Ea	13.70	8.00	—	21.70
6"	P1@.200	Ea	16.50	8.89	—	25.39
8"	P1@.200	Ea	26.90	8.89	—	35.79

HDPE – High Density Polyethylene pipe, or PEX pipe, is cross-linked at the molecular level. An additional external layer of an EVOH polymer creates an oxygen barrier and restricts oxygen permeability. It is manufactured and sold in various coiled lengths (200' to 1,000' rolls). The jointing method can be either crimped or compression connections. It will not rust or corrode and, because of its flexibility, installations require approximately 40% less fittings. Expansion rates are similar to copper, but its thermal coefficient (heat loss) is over 800 times less than copper. It is also chemically resistant to most acids, salt solutions, alkalis, fats and oils.

Common applications include ice or snow melting systems, radiant floor heating systems, water service tubing, hot & cold domestic water service, chilled water systems, compressed air systems, solar and process piping applications. Ice or snow melting systems require an oxygen barrier type PEX pipe selection.

Maximum temperature	Maximum pressure	Pipe selection
73 degrees F	160 psi	PEX
140 degrees F	120 psi	PEX
180 degrees F	100 psi	PEX

PEX is the designation for HDPE pipe

Description	Craft@Hrs	Unit	Material $	Labor $	Equipment $	Total $

Cross linked HD Polyethylene-PEX pipe with crimped joints (PEX), pipe only

Description	Craft@Hrs	Unit	Material $	Labor $	Equipment $	Total $
3/8"	P1@.028	LF	.40	1.24	—	1.64
1/2"	P1@.028	LF	.59	1.24	—	1.83
5/8"	P1@.030	LF	.82	1.33	—	2.15
3/4"	P1@.030	LF	1.16	1.33	—	2.49
1"	P1@.034	LF	1.86	1.51	—	3.37
Add for oxygen barrier type		%	25.0	—	—	—

90-degree brass ell with crimped joints (PEX)

Description	Craft@Hrs	Unit	Material $	Labor $	Equipment $	Total $
1/2"	P1@.080	Ea	1.50	3.56	—	5.06
5/8"	P1@.085	Ea	1.80	3.78	—	5.58
3/4"	P1@.090	Ea	3.48	4.00	—	7.48
1"	P1@.105	Ea	3.78	4.67	—	8.45

90-degree brass ell with crimped joints x male pipe thread (PEX x MPT)

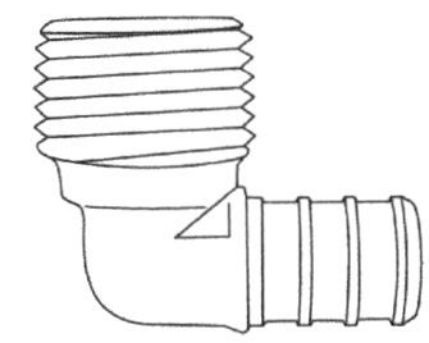

Description	Craft@Hrs	Unit	Material $	Labor $	Equipment $	Total $
1/2" x 1/2"	P1@.085	Ea	1.64	3.78	—	5.42
1/2" x 3/8"	P1@.095	Ea	2.07	4.22	—	6.29

Description	Craft@Hrs	Unit	Material $	Labor $	Equipment $	Total $

90-degree brass ell with crimped joints x copper socket (PEX x C)

1/2"	P1@.085	Ea	1.64	3.78	—	5.42
1/2" x 3/4"	P1@.090	Ea	2.07	4.00	—	6.07
5/8" x 3/4"	P1@.095	Ea	2.22	4.22	—	6.44
3/4"	P1@.095	Ea	1.82	4.22	—	6.04

90-degree brass wingback ell with crimped joints x female pipe thread (PEX x FPT)

1/2"	P1@.085	Ea	3.19	3.78	—	6.97
3/4"	P1@.095	Ea	3.78	4.22	—	8.00

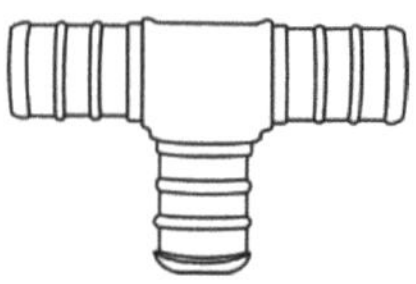

Tee (brass) with crimped joints (PEX)

1/2"	P1@.095	Ea	1.61	4.22	—	5.83
5/8"	P1@.105	Ea	1.87	4.67	—	6.54
3/4"	P1@.110	Ea	3.78	4.89	—	8.67
1"	P1@.120	Ea	4.64	5.33	—	9.97

Reducing tee (brass) with crimped joints (PEX)

1/2" x 1/2" x 5/8"	P1@.110	Ea	1.87	4.89	—	6.76
1/2" x 1/2" x 3/4"	P1@.115	Ea	2.18	5.11	—	7.29
5/8" x 1/2" x 1/2"	P1@.115	Ea	1.87	5.11	—	6.98
5/8" x 1/2" x 5/8"	P1@.115	Ea	1.87	5.11	—	6.98
5/8" x 5/8" x 1/2"	P1@.115	Ea	1.87	5.11	—	6.98
3/4" x 1/2" x 1/2"	P1@.115	Ea	2.18	5.11	—	7.29
3/4" x 1/2" x 3/4"	P1@.115	Ea	2.33	5.11	—	7.44
3/4" x 3/4" x 1/2"	P1@.115	Ea	2.33	5.11	—	7.44
3/4" x 3/4" x 5/8"	P1@.115	Ea	2.33	5.11	—	7.44
3/4" x 3/4" x 1"	P1@.120	Ea	4.49	5.33	—	9.82
1" x 1/2" x 1/2"	P1@.120	Ea	4.37	5.33	—	9.70
1" x 1/2" x 3/4"	P1@.120	Ea	4.07	5.33	—	9.40
1" x 1/2" x 1"	P1@.120	Ea	4.49	5.33	—	9.82
1" x 3/4" x 1/2"	P1@.120	Ea	4.49	5.33	—	9.82
1" x 3/4" x 3/4"	P1@.120	Ea	4.49	5.33	—	9.82
1" x 3/4" x 1"	P1@.120	Ea	4.49	5.33	—	9.82
1" x 1" x 1/2"	P1@.120	Ea	4.05	5.33	—	9.38
1" x 1" x 3/4"	P1@.120	Ea	4.64	5.33	—	9.97

Coupling (brass) with crimped joints (PEX)

1/2"	P1@.080	Ea	.73	3.56	—	4.29
5/8"	P1@.085	Ea	.92	3.78	—	4.70
3/4"	P1@.090	Ea	2.05	4.00	—	6.05
1"	P1@.105	Ea	2.09	4.67	—	6.76

Description	Craft@Hrs	Unit	Material $	Labor $	Equipment $	Total $

Reducing coupling (brass) with crimped joints (PEX)

Description	Craft@Hrs	Unit	Material $	Labor $	Equipment $	Total $
1/2"	P1@.080	Ea	.92	3.56	—	4.48
5/8"	P1@.085	Ea	1.31	3.78	—	5.09
3/4"	P1@.090	Ea	1.40	4.00	—	5.40
1"	P1@.105	Ea	1.70	4.67	—	6.37

Adapter (brass) with crimped joints x male pipe thread (PEX x MPT)

Description	Craft@Hrs	Unit	Material $	Labor $	Equipment $	Total $
1/2" x 1/2"	P1@.080	Ea	1.09	3.56	—	4.65
1/2" x 3/4"	P1@.085	Ea	1.40	3.78	—	5.18
5/8" x 3/4"	P1@.085	Ea	1.39	3.78	—	5.17
3/4" x 1/2"	P1@.090	Ea	1.87	4.00	—	5.87
3/4" x 3/4"	P1@.090	Ea	1.99	4.00	—	5.99
3/4" x 1"	P1@.105	Ea	3.19	4.67	—	7.86
1" x 1"	P1@.105	Ea	3.48	4.67	—	8.15

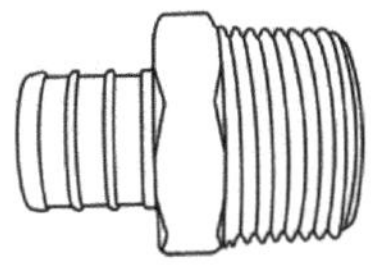

Adapter (brass) with crimped joints x female pipe thread (PEX x FPT)

Description	Craft@Hrs	Unit	Material $	Labor $	Equipment $	Total $
1/2" x 1/2"	P1@.080	Ea	1.13	3.56	—	4.69
1/2" x 3/4"	P1@.085	Ea	1.61	3.78	—	5.39
5/8" x 3/4"	P1@.085	Ea	1.61	3.78	—	5.39
3/4" x 1/2"	P1@.090	Ea	1.87	4.00	—	5.87
3/4" x 3/4"	P1@.090	Ea	2.07	4.00	—	6.07
3/4" x 1"	P1@.105	Ea	3.45	4.67	—	8.12
1" x 1"	P1@.105	Ea	3.78	4.67	—	8.45

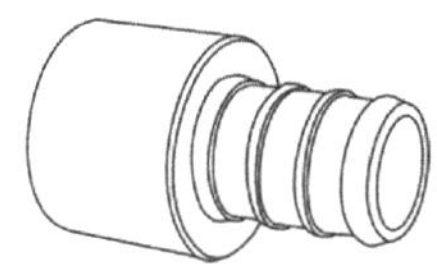

Adapter (brass) with crimped joints x copper fitting (spigot) (PEX x C ftg)

Description	Craft@Hrs	Unit	Material $	Labor $	Equipment $	Total $
1/2" x 1/2"	P1@.080	Ea	1.31	3.56	—	4.87
1/2" x 3/4"	P1@.085	Ea	1.62	3.78	—	5.40
5/8" x 3/4"	P1@.085	Ea	1.61	3.78	—	5.39
3/4" x 3/4"	P1@.090	Ea	3.19	4.00	—	7.19
1" x 1"	P1@.105	Ea	3.78	4.67	—	8.45

Adapter (brass) with crimped joints x copper socket (PEX x C)

Description	Craft@Hrs	Unit	Material $	Labor $	Equipment $	Total $
1/2" x 1/2"	P1@.080	Ea	1.09	3.56	—	4.65
1/2" x 3/4"	P1@.085	Ea	1.63	3.78	—	5.41
5/8" x 3/4"	P1@.085	Ea	1.69	3.78	—	5.47
3/4" x 1/2"	P1@.090	Ea	1.80	4.00	—	5.80
3/4" x 3/4"	P1@.090	Ea	2.62	4.00	—	6.62
3/4" x 1"	P1@.105	Ea	3.78	4.67	—	8.45
1" x 1"	P1@.105	Ea	4.37	4.67	—	9.04

Description	Craft@Hrs	Unit	Material $	Labor $	Equipment $	Total $

Cap (brass) with crimped joints (PEX)

1/2"	P1@.060	Ea	.50	2.67	—	3.17
5/8"	P1@.065	Ea	.70	2.89	—	3.59
3/4"	P1@.070	Ea	.92	3.11	—	4.03
1"	P1@.085	Ea	1.13	3.78	—	4.91

Mini ball valve (brass) with crimped joints (PEX)

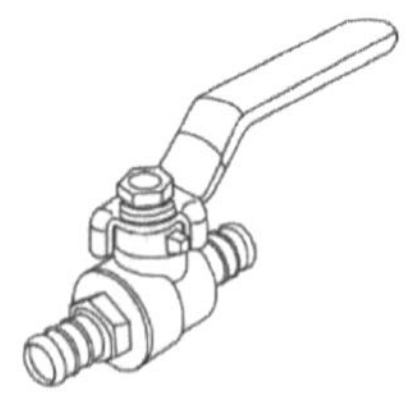

1/2" x 3/8" straight	P1@.150	Ea	8.54	6.67	—	15.21
1/2" x 3/8" angle	P1@.150	Ea	8.54	6.67	—	15.21
1/2" x copper	P1@.150	Ea	9.48	6.67	—	16.15
1/2" x MPT	P1@.150	Ea	9.48	6.67	—	16.15
1/2"	P1@.150	Ea	9.48	6.67	—	16.15
1/2" x comp	P1@.150	Ea	9.48	6.67	—	16.15
5/8"	P1@.165	Ea	14.30	7.33	—	21.63
5/8" x 3/4" copper	P1@.170	Ea	14.30	7.56	—	21.86
5/8" x 3/4" MPT	P1@.170	Ea	14.30	7.56	—	21.86
3/4"	P1@.170	Ea	15.30	7.56	—	22.86
3/4" x copper	P1@.170	Ea	15.30	7.56	—	22.86

Balancing valve (brass) with crimped joints (PEX)

1/2" x copper	P1@.150	Ea	51.90	6.67	—	58.57
1/2" x MPT	P1@.150	Ea	60.00	6.67	—	66.67

Manifolds (headers) copper with crimped joints (PEX)

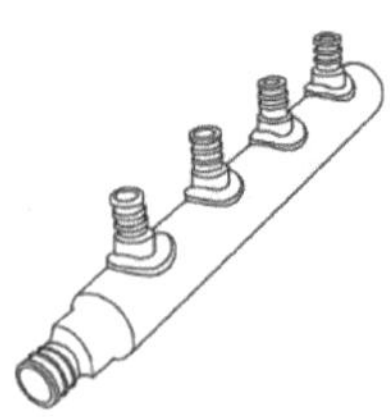

3 - 1/2" outlets	P1@.750	Ea	41.80	33.30	—	75.10
3 - 5/8" outlets	P1@.750	Ea	47.30	33.30	—	80.60
4 - 1/2" outlets	P1@.850	Ea	53.70	37.80	—	91.50
4 - 5/8" outlets	P1@.850	Ea	62.10	37.80	—	99.90
5 - 1/2" outlets	P1@1.00	Ea	68.70	44.50	—	113.20
6 - 1/2" outlets	P1@1.20	Ea	78.10	53.30	—	131.40
8 - 1/2" outlets	P1@1.45	Ea	112.00	64.50	—	176.50
10 - 1/2" outlets	P1@1.65	Ea	144.00	73.30	—	217.30
12 - 1/2" outlets	P1@1.85	Ea	174.00	82.20	—	256.20

Manifolds (headers) copper with mini ball valves and crimped joints (PEX)

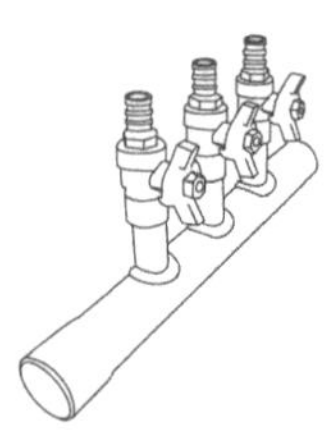

3 - 1/2" outlets	P1@.750	Ea	156.00	33.30	—	189.30
3 - 5/8" outlets	P1@.750	Ea	189.00	33.30	—	222.30
4 - 1/2" outlets	P1@.850	Ea	204.00	37.80	—	241.80
4 - 5/8" outlets	P1@.850	Ea	248.00	37.80	—	285.80
5 - 1/2" outlets	P1@1.00	Ea	259.00	44.50	—	303.50
6 - 1/2" outlets	P1@1.20	Ea	297.00	53.30	—	350.30

Description	Craft@Hrs	Unit	Material $	Labor $	Equipment $	Total $
Crimp rings, nickel-plated soft copper						
1/2"	—	Ea	.21	—	—	.21
5/8"	—	Ea	.22	—	—	.22
3/4"	—	Ea	.25	—	—	.25
1"	—	Ea	.37	—	—	.37
Replacement O-rings						
1/2"	—	Ea	.12	—	—	.12
5/8"	—	Ea	.12	—	—	.12
3/4"	—	Ea	.19	—	—	.19
1"	—	Ea	.19	—	—	.19
Pipe hangers (polypropylene)						
1/2"	P1@.050	Ea	.06	2.22	—	2.28
5/8"	P1@.050	Ea	.06	2.22	—	2.28
3/4"	P1@.050	Ea	.06	2.22	—	2.28
1"	P1@.050	Ea	.08	2.22	—	2.30
Pipe clips (nail clips)						
1/2"	P1@.030	Ea	.13	1.33	—	1.46
5/8"	P1@.030	Ea	.23	1.33	—	1.56
1"	P1@.030	Ea	.52	1.33	—	1.85
Miscellaneous tools						
Crimp tool	—	Ea	299.00	—	—	299.00
Beveling tool	—	Ea	6.05	—	—	6.05
Reaming tool	—	Ea	36.20	—	—	36.20
Pipe bender kit	—	Ea	534.00	—	—	534.00
Bending spring	—	Ea	44.80	—	—	44.80
Pipe cutter	—	Ea	21.70	—	—	21.70

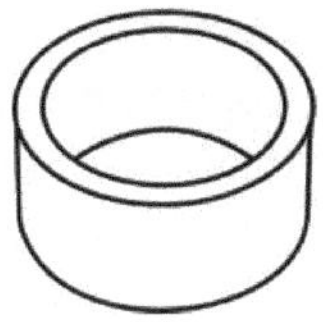
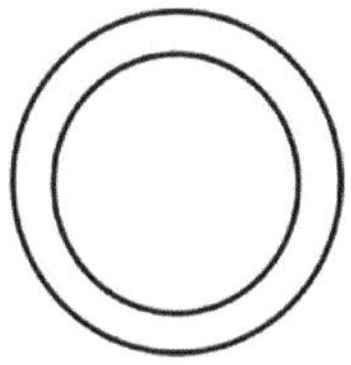

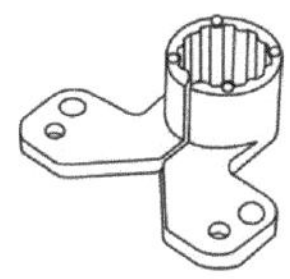
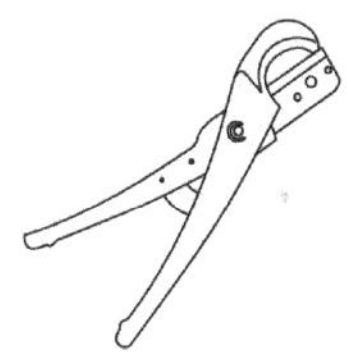

Polyethylene Pipe with Compression Joints

HDPE – High Density Polyethylene pipe, is cross-linked at the molecular level. An additional external layer of an EVOH polymer creates an oxygen barrier and restricts oxygen permeability. It is manufactured and sold in various coiled lengths (200' to 1,000' rolls). The jointing method can be either crimped or compression connections. It will not rust or corrode and, because of its flexibility, installations require approximately 40% less fittings. Expansion rates are similar to copper, but its thermal coefficient (heat loss) is over 800 times less than copper. It is also chemically resistant to most acids, salt solutions, alkalis, fats and oils.

Common applications include ice or snow melting systems, radiant floor heating systems, water service tubing, hot & cold domestic water service, chilled water systems, compressed air systems, solar and process piping applications. Ice or snow melting systems require an oxygen barrier type selection.

Maximum temperature	Maximum pressure	Pipe selection
73 degrees F	160 psi	PEX
140 degrees F	120 psi	PEX
180 degrees F	100 psi	PEX

PEX is the designation for HDPE pipe

Description	Craft@Hrs	Unit	Material $	Labor $	Equipment $	Total $

Cross linked HD Polyurethane pipe with compression joints, pipe only

3/8"	P1@.028	LF	.36	1.24	—	1.60
1/2"	P1@.028	LF	.55	1.24	—	1.79
5/8"	P1@.030	LF	.74	1.33	—	2.07
3/4"	P1@.030	LF	1.03	1.33	—	2.36
1"	P1@.034	LF	1.66	1.51	—	3.17

90-degree brass ell with compression joints

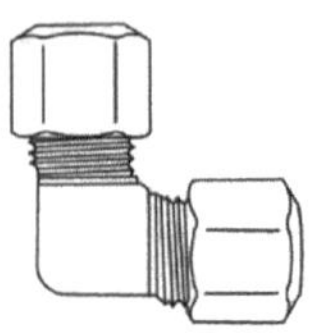

1/2"	P1@.092	Ea	3.41	4.09	—	7.50
5/8"	P1@.098	Ea	4.73	4.36	—	9.09
3/4"	P1@.104	Ea	6.90	4.62	—	11.52
1"	P1@.121	Ea	12.00	5.38	—	17.38

90-degree brass ell with compression joints x male pipe thread (MPT)

½" x ½"	P1@.098	Ea	5.31	4.36	—	9.67
½" x ¾"	P1@.105	Ea	6.44	4.67	—	11.11
¾" x ¾"	P1@.109	Ea	5.94	4.85	—	10.79

Description	Craft@Hrs	Unit	Material $	Labor $	Equipment $	Total $
90-degree brass ell with compression joints x copper socket (C)						
1/2"	P1@.098	Ea	2.90	4.36	—	7.26
1/2" x 3/4"	P1@.109	Ea	3.09	4.85	—	7.94
5/8" x 3/4"	P1@.112	Ea	3.68	4.98	—	8.66
3/4"	P1@.115	Ea	5.58	5.11	—	10.69
90-degree brass wingback ell with compression joints x female pipe thread (FPT)						
1/2"	P1@.098	Ea	2.52	4.36	—	6.88
5/8"	P1@.105	Ea	4.33	4.67	—	9.00
3/4"	P1@.115	Ea	6.63	5.11	—	11.74
Tee (brass) with compression joints						
1/2"	P1@.109	Ea	4.16	4.85	—	9.01
5/8"	P1@.121	Ea	6.68	5.38	—	12.06
3/4"	P1@.127	Ea	9.88	5.65	—	15.53
1"	P1@.141	Ea	16.40	6.27	—	22.67
Reducing tee (brass) with compression joints						
5/8" x 5/8" x 1/2"	P1@.138	Ea	6.13	6.13	—	12.26
3/4" x 1/2" x 3/4"	P1@.138	Ea	8.69	6.13	—	14.82
3/4" x 3/4" x 1/2"	P1@.138	Ea	8.89	6.13	—	15.02
3/4" x 1/2" x 3/4"	P1@.138	Ea	7.86	6.13	—	13.99
3/4" x 3/4" x 5/8"	P1@.138	Ea	9.00	6.13	—	15.13
1" x 1/2" x 1/2"	P1@.142	Ea	13.40	6.31	—	19.71
1" x 1" x 1/2"	P1@.142	Ea	17.20	6.31	—	23.51
1" x 3/4" x 3/4"	P1@.142	Ea	17.20	6.31	—	23.51
1" x 1" x 3/4"	P1@.142	Ea	17.20	6.31	—	23.51
Coupling (brass) with compression joints						
3/8"	P1@.090	Ea	2.87	4.00	—	6.87
1/2"	P1@.092	Ea	3.13	4.09	—	7.22
5/8"	P1@.098	Ea	4.27	4.36	—	8.63
3/4"	P1@.104	Ea	6.32	4.62	—	10.94
1"	P1@.121	Ea	11.30	5.38	—	16.68
Reducing coupling (brass) with compression joints						
5/8" x 1/2"	P1@.092	Ea	3.75	4.09	—	7.84
3/4" x 1/2"	P1@.098	Ea	5.67	4.36	—	10.03
3/4" x 5/8"	P1@.104	Ea	5.49	4.62	—	10.11
1" x 3/4"	P1@.121	Ea	7.63	5.38	—	13.01
Cap (brass) with compression joints						
1/2"	P1@.069	Ea	2.11	3.07	—	5.18
5/8"	P1@.075	Ea	2.90	3.33	—	6.23
3/4"	P1@.081	Ea	3.81	3.60	—	7.41
1"	P1@.098	Ea	5.67	4.36	—	10.03

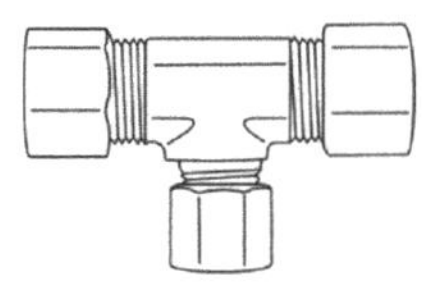

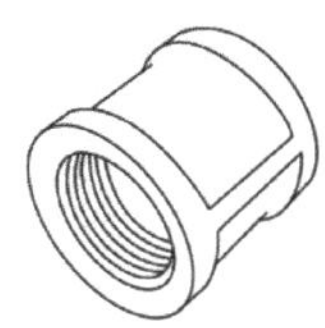

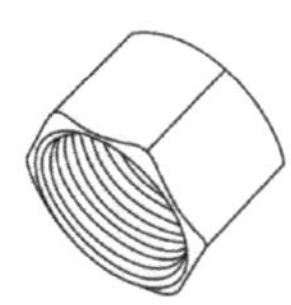

Polyethylene Pipe with Compression Joints

Mini ball valve (brass) with compression joints

Description	Craft@Hrs	Unit	Material $	Labor $	Equipment $	Total $
1/2" x 3/8" straight	P1@.173	Ea	8.54	7.69	—	16.23
1/2" x 3/8" angle	P1@.173	Ea	8.54	7.69	—	16.23
1/2" x copper	P1@.173	Ea	9.48	7.69	—	17.17
1/2" x MPT	P1@.173	Ea	9.48	7.69	—	17.17
1/2"	P1@.173	Ea	9.48	7.69	—	17.17
5/8"	P1@.173	Ea	9.48	7.69	—	17.17
5/8" x 3/4" copper	P1@.196	Ea	14.30	8.71	—	23.01
5/8" x 3/4" MPT	P1@.196	Ea	14.30	8.71	—	23.01
3/4"	P1@.196	Ea	15.30	8.71	—	24.01
3/4" x copper	P1@.196	Ea	15.30	8.71	—	24.01

Balancing valve (brass) with compression joints

1/2" x copper	P1@.173	Ea	57.40	7.69	—	65.09
1/2" x MPT	P1@.173	Ea	63.90	7.69	—	71.59

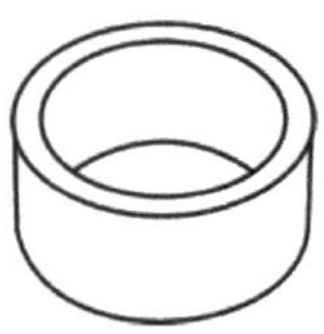

Crimp rings, nickel-plated soft copper

1/2"	—	Ea	.21	—	—	.21
5/8"	—	Ea	.22	—	—	.22
3/4"	—	Ea	.25	—	—	.25
1"	—	Ea	.37	—	—	.37

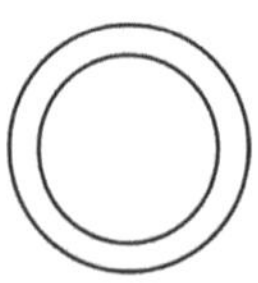

Replacement O-rings

1/2"	—	Ea	.13	—	—	.13
5/8"	—	Ea	.13	—	—	.13
3/4"	—	Ea	.20	—	—	.20
1"	—	Ea	.20	—	—	.20

Pipe hangers (polypropylene)

1/2"	P1@.050	Ea	.06	2.22	—	2.28
5/8"	P1@.050	Ea	.06	2.22	—	2.28
3/4"	P1@.050	Ea	.06	2.22	—	2.28
1"	P1@.050	Ea	.08	2.22	—	2.30

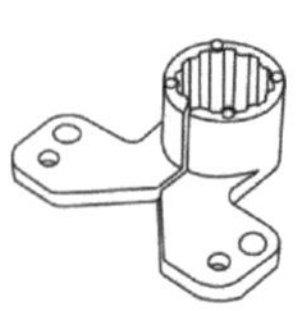

Pipe clips (nail clips)

1/2"	P1@.030	Ea	.13	1.33	—	1.46
5/8"	P1@.030	Ea	.23	1.33	—	1.56
1"	P1@.030	Ea	.52	1.33	—	1.85

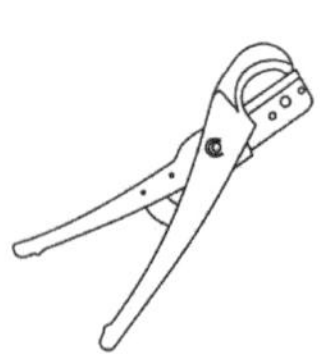

Miscellaneous tools

Crimp tool	—	Ea	299.00	—	—	299.00
Beveling tool	—	Ea	6.05	—	—	6.05
Reaming tool	—	Ea	36.20	—	—	36.20
Pipe bender kit	—	Ea	534.00	—	—	534.00
Bending spring	—	Ea	44.80	—	—	44.80
Pipe cutter	—	Ea	21.70	—	—	21.70

Description	Craft@Hrs	Unit	Material $	Labor $	Equipment $	Total $
Water meters, turbine-type. Including brass connection unions						
½"	P1@.500	Ea	201.00	22.20	—	223.20
¾"	P1@.500	Ea	326.00	22.20	—	348.20
1"	P1@.650	Ea	436.00	28.90	—	464.90
1½"	P1@.700	Ea	1,170.00	31.10	—	1,201.10
2"	P1@.800	Ea	1,750.00	35.60	—	1,785.60
3"	P1@.900	Ea	2,690.00	40.00	—	2,730.00
4"	P1@.975	Ea	5,010.00	43.30	—	5,053.30
Water meters, compound-type. Including brass connection unions						
2"	P1@.800	Ea	4,090.00	35.60	—	4,125.60
3"	P1@.900	Ea	5,330.00	40.00	—	5,370.00
4"	P1@.975	Ea	8,400.00	43.30	—	8,443.30
Water meter by-pass and connection assembly. Includes three isolation ball valves, two tees, two 90-degree elbows, 10' of Type L copper pipe and standard support devices. Make additional allowances for the cost of the water meter.						
½"	P1@1.00	Ea	98.40	44.50	—	142.90
¾"	P1@1.50	Ea	149.00	66.70	—	215.70
1"	P1@1.95	Ea	257.00	86.70	—	343.70
1½"	P1@2.45	Ea	473.00	109.00	—	582.00
2"	P1@3.25	Ea	697.00	144.00	—	841.00
3"	P1@5.50	Ea	2,080.00	244.00	—	2,324.00
4"	P1@6.75	Ea	3,530.00	300.00	—	3,830.00
Backflow preventers – reduced pressure. Including valves and test ports						
¾"	P1@1.00	Ea	390.00	44.50	—	434.50
1"	P1@1.25	Ea	483.00	55.60	—	538.60
1½"	P1@2.00	Ea	776.00	88.90	—	864.90
2"	P1@2.15	Ea	964.00	95.60	—	1,059.60
2½"	P1@4.00	Ea	3,500.00	178.00	—	3,678.00
3"	P1@4.75	Ea	4,610.00	211.00	—	4,821.00
4"	P1@6.00	Ea	5,250.00	267.00	—	5,517.00
6"	P1@7.95	Ea	5,740.00	353.00	—	6,093.00
Backflow preventers – reduced pressure. Including integral ball valves and test ports						
½"	P1@.900	Ea	237.00	40.00	—	277.00
¾"	P1@1.00	Ea	280.00	44.50	—	324.50
1"	P1@1.25	Ea	351.00	55.60	—	406.60
1¼"	P1@1.50	Ea	515.00	66.70	—	581.70
1½"	P1@2.00	Ea	560.00	88.90	—	648.90
2"	P1@2.15	Ea	630.00	95.60	—	725.60

Description	Craft@Hrs	Unit	Material $	Labor $	Equipment $	Total $

Backflow preventers, double-check valve assembly. Including integral ball valves and test ports

½"	P1@1.00	Ea	178.00	44.50	—	222.50
¾"	P1@1.00	Ea	197.00	44.50	—	241.50
1"	P1@1.25	Ea	237.00	55.60	—	292.60
1¼"	P1@1.95	Ea	398.00	86.70	—	484.70
1½"	P1@2.00	Ea	413.00	88.90	—	501.90
2"	P1@2.15	Ea	501.00	95.60	—	596.60
2½"	P1@2.95	Ea	2,060.00	131.00	—	2,191.00
3"	P1@3.45	Ea	2,630.00	153.00	—	2,783.00
4"	P1@3.95	Ea	3,570.00	176.00	—	3,746.00
6"	P1@5.75	Ea	6,020.00	256.00	—	6,276.00

Vacuum breakers, atmospheric. Female pipe thread

¼"	P1@.350	Ea	58.40	15.60	—	74.00
½"	P1@.350	Ea	81.50	15.60	—	97.10
¾"	P1@.350	Ea	99.60	15.60	—	115.20

Vacuum breakers, hose connection. Male/female hose thread, polished brass

½"	P1@.150	Ea	24.10	6.67	—	30.77
¾"	P1@.150	Ea	29.50	6.67	—	36.17

Suction diffusers

2"	P1@1.25	Ea	402.00	55.60	—	457.60
3"	P1@2.50	Ea	674.00	111.00	—	785.00
4"	P1@3.25	Ea	858.00	144.00	—	1,002.00
6"	P1@4.80	Ea	1,230.00	213.00	11.20	1,454.20
8"	P1@5.95	Ea	2,280.00	264.00	13.90	2,557.90
10"	P1@6.75	Ea	3,090.00	300.00	15.80	3,405.80

Triple-duty valves

2"	P1@1.25	Ea	521.00	55.60	—	576.60
3"	P1@2.50	Ea	707.00	111.00	—	818.00
4"	P1@3.25	Ea	1,280.00	144.00	—	1,424.00
6"	P1@4.80	Ea	2,110.00	213.00	11.20	2,334.20
8"	P1@5.95	Ea	3,080.00	264.00	13.90	3,357.90
10"	P1@6.75	Ea	4,530.00	300.00	15.80	4,845.80

In-line circulating pump, all bronze. 115 volt, including flange kit

1/25 HP	P1@1.50	Ea	618.00	66.70	—	684.70
1/16 HP	P1@1.55	Ea	779.00	68.90	—	847.90
1/12 HP	P1@1.60	Ea	1,070.00	71.10	—	1,141.10
1/6 HP	P1@1.75	Ea	1,650.00	77.80	—	1,727.80
1/4 HP	P1@1.95	Ea	2,730.00	86.70	—	2,816.70
1 HP	P1@2.25	Ea	3,900.00	100.00	—	4,000.00
1½ HP	P1@2.85	Ea	4,710.00	127.00	—	4,837.00

Description	Craft@Hrs	Unit	Material $	Labor $	Equipment $	Total $

In-line circulating pump, iron body. 115 volt, including flange kit

1/25 HP	P1@1.50	Ea	293.00	66.70	—	359.70
1/16 HP	P1@1.55	Ea	352.00	68.90	—	420.90
1/12 HP	P1@1.60	Ea	662.00	71.10	—	733.10
1/6 HP	P1@1.75	Ea	1,120.00	77.80	—	1,197.80
1/4 HP	P1@1.95	Ea	1,210.00	86.70	—	1,296.70
1 HP	P1@2.25	Ea	2,070.00	100.00	—	2,170.00
1½ HP	P1@2.85	Ea	2,700.00	127.00	—	2,827.00

Primed steel cam-lock non-fire-rated access doors

8" x 8"	P1@.400	Ea	69.70	17.80	—	87.50
12" x 12"	P1@.500	Ea	80.30	22.20	—	102.50
16" x 16"	P1@.600	Ea	98.70	26.70	—	125.40
18" x 18"	P1@.800	Ea	113.00	35.60	—	148.60
24" x 24"	P1@1.25	Ea	168.00	55.60	—	223.60
36" x 36"	P1@1.60	Ea	310.00	71.10	—	381.10
Add for Allen key lock		Ea	4.97	—	—	4.97
Add for cylinder lock		Ea	12.10	—	—	12.10

Primed steel cam-lock 1½ hour fire-rated access doors

8" x 8"	P1@.400	Ea	257.00	17.80	—	274.80
12" x 12"	P1@.500	Ea	270.00	22.20	—	292.20
16" x 16"	P1@.600	Ea	319.00	26.70	—	345.70
18" x 18"	P1@.800	Ea	324.00	35.60	—	359.60
24" x 24"	P1@1.25	Ea	455.00	55.60	—	510.60
24" x 36"	P1@1.50	Ea	587.00	66.70	—	653.70
48" x 48"	P1@2.00	Ea	908.00	88.90	—	996.90
Add for cylinder lock		Ea	12.20	—	—	12.20

Threaded automatic air vent

Float type	P1@.150	Ea	13.90	6.67	—	20.57
35 PSIG	P1@.150	Ea	88.10	6.67	—	94.77
150 PSIG	P1@.150	Ea	143.00	6.67	—	149.67

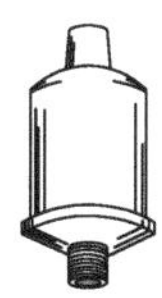

Threaded manual air vent

175 PSIG	P1@.150	Ea	8.51	6.67	—	15.18

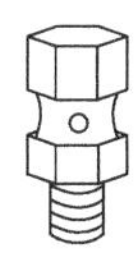

150 pound forged steel slip-on companion flange, welding-type

2½"	P1@.610	Ea	42.60	27.10	—	69.70
3"	P1@.730	Ea	42.60	32.40	—	75.00
4"	P1@.980	Ea	58.60	43.60	—	102.20
6"	ER@1.47	Ea	77.40	69.70	.98	148.08
8"	ER@1.77	Ea	143.00	83.90	1.18	228.08
10"	ER@2.20	Ea	240.00	104.00	1.47	345.47
12"	ER@2.64	Ea	358.00	125.00	1.77	484.77

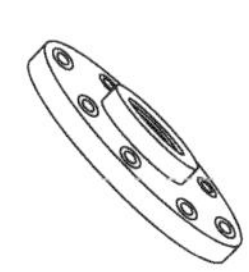

Description	Craft@Hrs	Unit	Material $	Labor $	Equipment $	Total $

150 pound forged steel companion flange, threaded

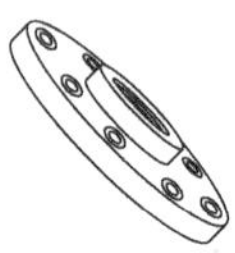

2"	P1@.290	Ea	45.60	12.90	—	58.50
2½"	P1@.380	Ea	68.70	16.90	—	85.60
3"	P1@.460	Ea	76.30	20.40	—	96.70
4"	P1@.600	Ea	102.00	26.70	—	128.70
6"	ER@.680	Ea	151.00	32.20	.46	183.66
8"	ER@.760	Ea	207.00	36.00	.51	243.51

300 pound forged steel slip-on companion flange, welding-type

2½"	P1@.810	Ea	48.90	36.00	—	84.90
3"	P1@.980	Ea	48.90	43.60	—	92.50
4"	P1@1.30	Ea	71.40	57.80	—	129.20
6"	ER@1.95	Ea	131.00	92.40	1.31	224.71
8"	ER@2.35	Ea	185.00	111.00	1.57	297.57
10"	ER@2.90	Ea	304.00	137.00	1.94	442.94
12"	ER@3.50	Ea	484.00	166.00	2.34	652.34

PVC companion flange

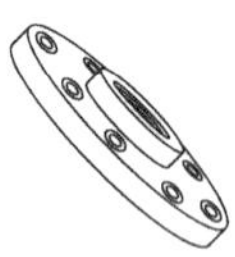

2"	P1@.140	Ea	14.50	6.22	—	20.72
2½"	P1@.180	Ea	43.70	8.00	—	51.70
3"	P1@.210	Ea	45.10	9.33	—	54.43
4"	P1@.280	Ea	46.20	12.40	—	58.60
5"	P1@.310	Ea	104.00	13.80	—	117.80
6"	P1@.420	Ea	104.00	18.70	—	122.70
8"	P1@.560	Ea	155.00	24.90	—	179.90

Bolt and gasket set

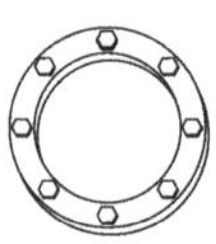

2"	P1@.500	Ea	4.64	22.20	—	26.84
2½"	P1@.650	Ea	5.42	28.90	—	34.32
3"	P1@.750	Ea	9.19	33.30	—	42.49
4"	P1@1.00	Ea	15.40	44.50	—	59.90
5"	P1@1.10	Ea	21.30	48.90	—	70.20
6"	P1@1.20	Ea	21.30	53.30	—	74.60
8"	P1@1.25	Ea	27.60	55.60	—	83.20
10"	P1@1.70	Ea	33.10	75.60	—	108.70
12"	P1@2.20	Ea	44.50	97.80	—	142.30

Dielectric union, brazed joints

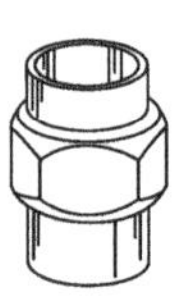

1/2"	P1@.146	Ea	5.81	6.49	—	12.30
3/4"	P1@.205	Ea	5.81	9.11	—	14.92
1"	P1@.263	Ea	12.00	11.70	—	23.70
1¼"	P1@.322	Ea	20.40	14.30	—	34.70
1½"	P1@.380	Ea	28.20	16.90	—	45.10
2"	P1@.507	Ea	37.60	22.50	—	60.10

Description	Craft@Hrs	Unit	Material $	Labor $	Equipment $	Total $

125/150# galvanized ASME expansion tank

15 gallon	P1@1.75	Ea	357.00	77.80	—	434.80
30 gallon	P1@2.00	Ea	643.00	88.90	—	731.90
40 gallon	P1@2.35	Ea	712.00	104.00	—	816.00
60 gallon	P1@2.65	Ea	967.00	118.00	—	1,085.00
80 gallon	P1@2.95	Ea	1,780.00	131.00	—	1,911.00
100 gallon	P1@3.25	Ea	2,140.00	144.00	—	2,284.00
120 gallon	P1@3.50	Ea	2,300.00	156.00	—	2,456.00

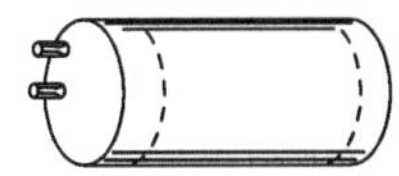

Expansion tank fitting

¾"	P1@.300	Ea	80.70	13.30	—	94.00

Lead roof pipe flashing

1½"	P1@.250	Ea	36.60	11.10	—	47.70
2"	P1@.250	Ea	48.20	11.10	—	59.30
3"	P1@.250	Ea	55.80	11.10	—	66.90
4"	P1@.250	Ea	61.50	11.10	—	72.60
6"	P1@.350	Ea	78.80	15.60	—	94.40
8"	P1@.400	Ea	99.60	17.80	—	117.40

Roof flashing, 20" x 20"

Pitched roof, neoprene						
all pipe	P1@.300	Ea	22.40	13.30	—	35.70
Pitched roof, aluminum/neoprene						
all pipe	P1@.300	Ea	23.70	13.30	—	37.00
Flat roof, aluminum/neoprene						
all pipe	P1@.300	Ea	45.10	13.30	—	58.40

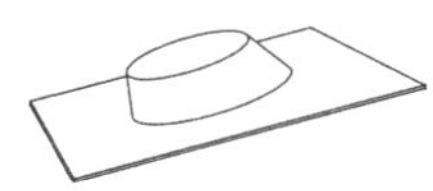

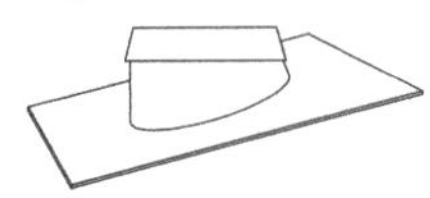

Braided stainless steel pipe connector, solder ends, brazed joints

½"	P1@.240	Ea	47.30	10.70	—	58.00
¾"	P1@.300	Ea	62.70	13.30	—	76.00
1"	P1@.360	Ea	93.90	16.00	—	109.90
1¼"	P1@.480	Ea	109.00	21.30	—	130.30
1½"	P1@.540	Ea	125.00	24.00	—	149.00
2"	P1@.600	Ea	182.00	26.70	—	208.70

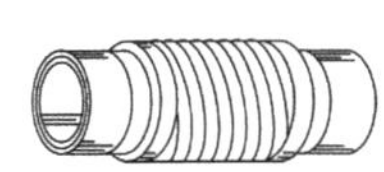

Braided stainless steel pipe connector, threaded joints

½"	P1@.210	Ea	53.40	9.33	—	62.73
¾"	P1@.250	Ea	57.70	11.10	—	68.80
1"	P1@.300	Ea	71.90	13.30	—	85.20
1¼"	P1@.400	Ea	73.30	17.80	—	91.10
1½"	P1@.450	Ea	99.50	20.00	—	119.50
2"	P1@.500	Ea	142.00	22.20	—	164.20

Braided stainless steel pipe connector, flanged joints

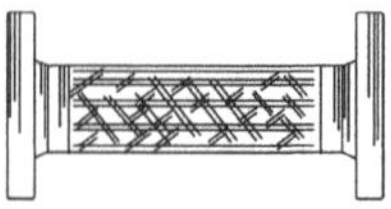

Description	Craft@Hrs	Unit	Material $	Labor $	Equipment $	Total $
2"	P1@.500	Ea	126.00	22.20	—	148.20
2½"	P1@.630	Ea	181.00	28.00	—	209.00
3"	P1@.750	Ea	235.00	33.30	—	268.30
4"	P1@1.50	Ea	307.00	66.70	—	373.70
6"	P1@2.50	Ea	610.00	111.00	—	721.00
8"	P1@3.00	Ea	1,370.00	133.00	—	1,503.00

Galvanized steel band pipe hanger

Description	Craft@Hrs	Unit	Material $	Labor $	Equipment $	Total $
½"	P1@.250	Ea	2.14	11.10	—	13.24
¾"	P1@.250	Ea	2.14	11.10	—	13.24
1"	P1@.250	Ea	2.14	11.10	—	13.24
1¼"	P1@.300	Ea	2.27	13.30	—	15.57
1½"	P1@.300	Ea	2.41	13.30	—	15.71
2"	P1@.300	Ea	2.53	13.30	—	15.83
2½"	P1@.350	Ea	3.10	15.60	—	18.70
3"	P1@.350	Ea	3.56	15.60	—	19.16
4"	P1@.350	Ea	4.52	15.60	—	20.12
6"	P1@.450	Ea	7.11	20.00	—	27.11
8"	P1@.450	Ea	10.60	20.00	—	30.60

Trapeze pipe hanger

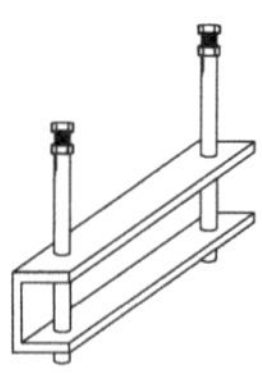

Description	Craft@Hrs	Unit	Material $	Labor $	Equipment $	Total $
24"	P1@1.00	Ea	48.00	44.50	—	92.50
36"	P1@1.25	Ea	61.10	55.60	—	116.70
48"	P1@1.50	Ea	93.00	66.70	—	159.70

Wall bracket support with anchors and bolts

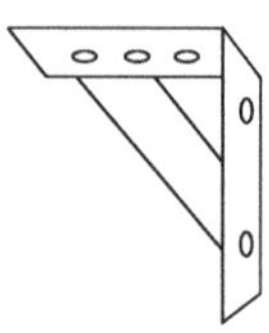

Description	Craft@Hrs	Unit	Material $	Labor $	Equipment $	Total $
6" x 6"	P1@.600	Ea	22.30	26.70	—	49.00
8" x 8"	P1@.650	Ea	25.80	28.90	—	54.70
10" x 10"	P1@.700	Ea	29.10	31.10	—	60.20
12" x 12"	P1@.800	Ea	35.10	35.60	—	70.70
15" x 15"	P1@.950	Ea	40.40	42.20	—	82.60
18" x 18"	P1@1.10	Ea	57.10	48.90	—	106.00
24" x 24"	P1@1.60	Ea	82.60	71.10	—	153.70

Galvanized U-bolts with nuts

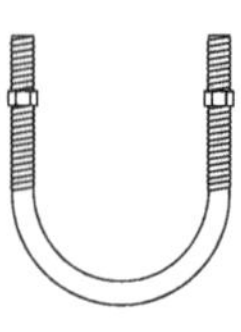

Description	Craft@Hrs	Unit	Material $	Labor $	Equipment $	Total $
½"	P1@.160	Ea	1.94	7.11	—	9.05
¾"	P1@.165	Ea	1.94	7.33	—	9.27
1"	P1@.170	Ea	2.11	7.56	—	9.67
1¼"	P1@.175	Ea	2.41	7.78	—	10.19
1½"	P1@.180	Ea	2.92	8.00	—	10.92
2"	P1@.190	Ea	3.12	8.45	—	11.57
2½"	P1@.210	Ea	4.82	9.33	—	14.15
3"	P1@.250	Ea	5.57	11.10	—	16.67
4"	P1@.300	Ea	6.30	13.30	—	19.60
6"	P1@.400	Ea	9.86	17.80	—	27.66

Description	Craft@Hrs	Unit	Material $	Labor $	Equipment $	Total $
Nail-on wire pipe hooks						
½"	P1@.100	Ea	.09	4.45	—	4.54
¾"	P1@.100	Ea	.11	4.45	—	4.56
1"	P1@.100	Ea	.12	4.45	—	4.57
Galvanized steel pipe sleeves						
1"	P1@.120	Ea	4.52	5.33	—	9.85
1¼"	P1@.125	Ea	5.00	5.56	—	10.56
1½"	P1@.125	Ea	5.66	5.56	—	11.22
2"	P1@.130	Ea	6.11	5.78	—	11.89
2½"	P1@.150	Ea	6.20	6.67	—	12.87
3"	P1@.180	Ea	6.34	8.00	—	14.34
4"	P1@.220	Ea	7.21	9.78	—	16.99
5"	P1@.250	Ea	8.99	11.10	—	20.09
6"	P1@.270	Ea	9.72	12.00	—	21.72
8"	P1@.270	Ea	11.20	12.00	—	23.20
10"	P1@.290	Ea	13.00	12.90	—	25.90
12"	P1@.310	Ea	17.70	13.80	—	31.50
14"	P1@.330	Ea	32.90	14.70	—	47.60

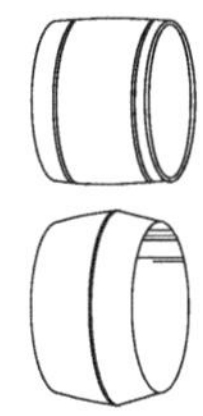

Description	Craft@Hrs	Unit	Material $	Labor $	Equipment $	Total $
Dial-type pressure gauge						
2½"	P1@.200	Ea	38.50	8.89	—	47.39
3½"	P1@.200	Ea	51.20	8.89	—	60.09

Description	Craft@Hrs	Unit	Material $	Labor $	Equipment $	Total $
Riser clamp						
½"	P1@.100	Ea	3.30	4.45	—	7.75
¾"	P1@.100	Ea	4.87	4.45	—	9.32
1"	P1@.100	Ea	4.91	4.45	—	9.36
1¼"	P1@.105	Ea	5.90	4.67	—	10.57
1½"	P1@.110	Ea	6.22	4.89	—	11.11
2"	P1@.115	Ea	6.62	5.11	—	11.73
2½"	P1@.120	Ea	6.93	5.33	—	12.26
3"	P1@.120	Ea	7.53	5.33	—	12.86
4"	P1@.125	Ea	9.61	5.56	—	15.17
5"	P1@.180	Ea	13.30	8.00	—	21.30
6"	P1@.200	Ea	15.90	8.89	—	24.79
8"	P1@.200	Ea	25.90	8.89	—	34.79
10"	P1@.250	Ea	38.40	11.10	—	49.50
12"	P1@.250	Ea	45.50	11.10	—	56.60

Description	Craft@Hrs	Unit	Material $	Labor $	Equipment $	Total $
15 PSIG float and thermostatic steam trap, threaded						
¾"	P1@.250	Ea	362.00	11.10	—	373.10
1"	P1@.300	Ea	362.00	13.30	—	375.30
1¼"	P1@.400	Ea	437.00	17.80	—	454.80
1½"	P1@.450	Ea	820.00	20.00	—	840.00
2"	P1@.500	Ea	1,050.00	22.20	—	1,072.20

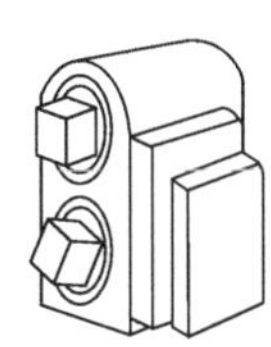

Description	Craft@Hrs	Unit	Material $	Labor $	Equipment $	Total $

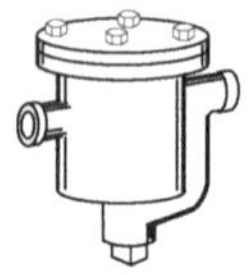

15 PSIG inverted bucket steam trap, threaded joints

Description	Craft@Hrs	Unit	Material $	Labor $	Equipment $	Total $
½"	P1@.210	Ea	210.00	9.33	—	219.33
¾"	P1@.250	Ea	516.00	11.10	—	527.10
1"	P1@.300	Ea	739.00	13.30	—	752.30
1¼"	P1@.400	Ea	1,230.00	17.80	—	1,247.80
1½"	P1@.450	Ea	1,230.00	20.00	—	1,250.00
2"	P1@.500	Ea	1,850.00	22.20	—	1,872.20

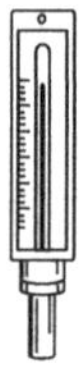

Thermometer with well

Description	Craft@Hrs	Unit	Material $	Labor $	Equipment $	Total $
7"	P1@.250	Ea	186.00	11.10	—	197.10
9"	P1@.250	Ea	192.00	11.10	—	203.10

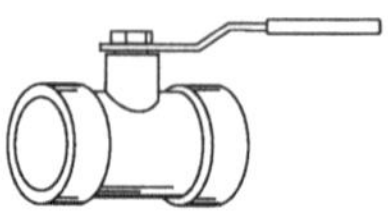

Class 125 bronze body 2-piece ball valve, brazed joints

Description	Craft@Hrs	Unit	Material $	Labor $	Equipment $	Total $
½"	P1@.240	Ea	11.60	10.70	—	22.30
¾"	P1@.300	Ea	15.40	13.30	—	28.70
1"	P1@.360	Ea	26.70	16.00	—	42.70
1¼"	P1@.480	Ea	44.60	21.30	—	65.90
1½"	P1@.540	Ea	60.30	24.00	—	84.30
2"	P1@.600	Ea	76.40	26.70	—	103.10
3"	P1@1.50	Ea	524.00	66.70	—	590.70
4"	P1@1.80	Ea	684.00	80.00	—	764.00

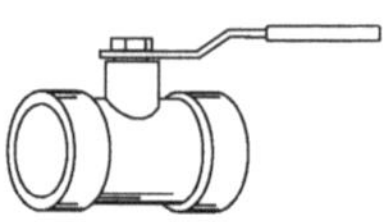

Class 125 bronze body 2-piece ball valve, soft-soldered joints

Description	Craft@Hrs	Unit	Material $	Labor $	Equipment $	Total $
½"	P1@.200	Ea	11.60	8.89	—	20.49
¾"	P1@.249	Ea	15.40	11.10	—	26.50
1"	P1@.299	Ea	26.70	13.30	—	40.00
1¼"	P1@.308	Ea	44.60	13.70	—	58.30
1½"	P1@.498	Ea	60.30	22.10	—	82.40
2"	P1@.498	Ea	76.40	22.10	—	98.50
3"	P1@1.24	Ea	524.00	55.10	—	579.10
4"	P1@1.45	Ea	684.00	64.50	—	748.50

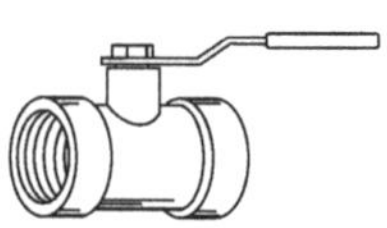

Class 125 bronze body 2-piece ball valve, threaded joints

Description	Craft@Hrs	Unit	Material $	Labor $	Equipment $	Total $
½"	P1@.210	Ea	11.60	9.33	—	20.93
¾"	P1@.250	Ea	15.40	11.10	—	26.50
1"	P1@.300	Ea	26.70	13.30	—	40.00
1¼"	P1@.400	Ea	44.60	17.80	—	62.40
1½"	P1@.450	Ea	60.30	20.00	—	80.30
2"	P1@.500	Ea	76.40	22.20	—	98.60
3"	P1@.625	Ea	524.00	27.80	—	551.80
4"	P1@.690	Ea	684.00	30.70	—	714.70

Description	Craft@Hrs	Unit	Material $	Labor $	Equipment $	Total $

Class 150, 600 pound W.O.G. bronze body 2-piece ball valve, threaded joints

½"	P1@.210	Ea	11.60	9.33	—	20.93
¾"	P1@.250	Ea	15.40	11.10	—	26.50
1"	P1@.300	Ea	26.70	13.30	—	40.00
1¼"	P1@.400	Ea	44.60	17.80	—	62.40
1½"	P1@.450	Ea	60.30	20.00	—	80.30
2"	P1@.500	Ea	76.40	22.20	—	98.60

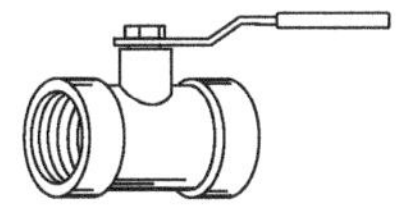

Bronze body circuit balance valve

½"	P1@.210	Ea	83.60	9.33	—	92.93
¾"	P1@.250	Ea	88.90	11.10	—	100.00
1"	P1@.300	Ea	103.00	13.30	—	116.30
1¼"	P1@.400	Ea	128.00	17.80	—	145.80
1½"	P1@.400	Ea	163.00	17.80	—	180.80
2"	P1@.500	Ea	227.00	22.20	—	249.20
2½"	P1@1.75	Ea	862.00	77.80	—	939.80
3"	P1@2.50	Ea	1,330.00	111.00	—	1,441.00
4"	P1@3.15	Ea	2,260.00	140.00	—	2,400.00

200 PSIG iron body butterfly valve, lug-type, lever operated

2"	P1@.450	Ea	179.00	20.00	—	199.00
2½"	P1@.450	Ea	186.00	20.00	—	206.00
3"	P1@.550	Ea	196.00	24.40	—	220.40
4"	P1@.505	Ea	244.00	22.40	—	266.40
6"	ER@.800	Ea	392.00	37.90	.54	430.44
8"	ER@.800	Ea	536.00	37.90	.54	574.44
10"	ER@.900	Ea	745.00	42.70	.60	788.30
12"	ER@1.00	Ea	975.00	47.40	.67	1,023.07

200 PSIG iron body butterfly valve, wafer-type, lever operated

2"	P1@.450	Ea	164.00	20.00	—	184.00
2½"	P1@.450	Ea	167.00	20.00	—	187.00
3"	P1@.550	Ea	179.00	24.40	—	203.40
4"	P1@.550	Ea	217.00	24.40	—	241.40
6"	ER@.800	Ea	363.00	37.90	.54	401.44
8"	ER@.800	Ea	503.00	37.90	.54	541.44
10"	ER@.900	Ea	704.00	42.70	.60	747.30
12"	ER@1.00	Ea	926.00	47.40	.67	974.07

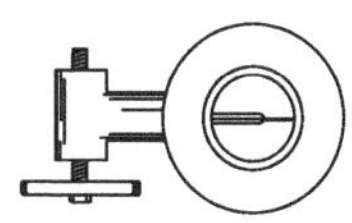

Class 125 bronze body swing check valve, with brazed joints

½"	P1@.240	Ea	23.90	10.70	—	34.60
¾"	P1@.300	Ea	34.10	13.30	—	47.40
1"	P1@.360	Ea	44.50	16.00	—	60.50
1¼"	P1@.480	Ea	63.60	21.30	—	84.90
1½"	P1@.540	Ea	89.60	24.00	—	113.60
2"	P1@.600	Ea	148.00	26.70	—	174.70
2½"	P1@1.00	Ea	299.00	44.50	—	343.50
3"	P1@1.50	Ea	430.00	66.70	—	496.70

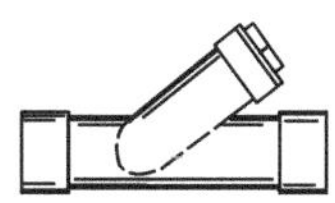

Description	Craft@Hrs	Unit	Material $	Labor $	Equipment $	Total $
Class 125 bronze body swing check valve with soft-soldered joints						
½"	P1@.200	Ea	23.90	8.89	—	32.79
¾"	P1@.249	Ea	34.10	11.10	—	45.20
1"	P1@.299	Ea	44.50	13.30	—	57.80
1¼"	P1@.398	Ea	63.60	17.70	—	81.30
1½"	P1@.448	Ea	89.60	19.90	—	109.50
2"	P1@.498	Ea	148.00	22.10	—	170.10
2½"	P1@.830	Ea	299.00	36.90	—	335.90
3"	P1@1.24	Ea	430.00	55.10	—	485.10

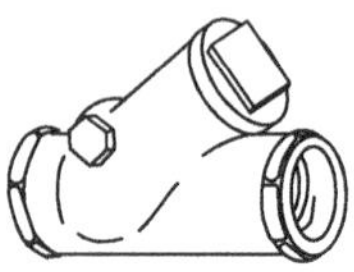

Description	Craft@Hrs	Unit	Material $	Labor $	Equipment $	Total $
Class 125 bronze body swing check valve, threaded joints						
½"	P1@.210	Ea	23.90	9.33	—	33.23
¾"	P1@.250	Ea	34.10	11.10	—	45.20
1"	P1@.300	Ea	44.50	13.30	—	57.80
1¼"	P1@.400	Ea	63.60	17.80	—	81.40
1½"	P1@.450	Ea	89.60	20.00	—	109.60
2"	P1@.500	Ea	148.00	22.20	—	170.20

Description	Craft@Hrs	Unit	Material $	Labor $	Equipment $	Total $
Class 300 bronze body swing check valve, threaded joints						
½"	P1@.210	Ea	80.40	9.33	—	89.73
¾"	P1@.250	Ea	95.10	11.10	—	106.20
1"	P1@.300	Ea	116.00	13.30	—	129.30
1¼"	P1@.400	Ea	169.00	17.80	—	186.80
1½"	P1@.450	Ea	242.00	20.00	—	262.00
2"	P1@.500	Ea	336.00	22.20	—	358.20

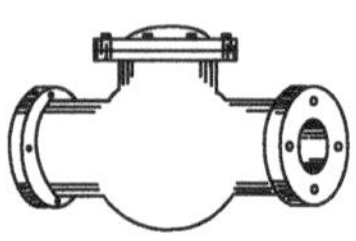

Description	Craft@Hrs	Unit	Material $	Labor $	Equipment $	Total $
Class 125 iron body swing check valve, flanged joints						
2"	P1@.500	Ea	209.00	22.20	—	231.20
2½"	P1@.600	Ea	264.00	26.70	—	290.70
3"	P1@.750	Ea	329.00	33.30	—	362.30
4"	P1@1.35	Ea	483.00	60.00	—	543.00
6"	ER@2.50	Ea	937.00	119.00	1.67	1,057.67
8"	ER@3.00	Ea	1,690.00	142.00	2.01	1,834.01
10"	ER@4.00	Ea	2,740.00	190.00	2.68	2,932.68
12"	ER@4.50	Ea	3,760.00	213.00	3.01	3,976.01

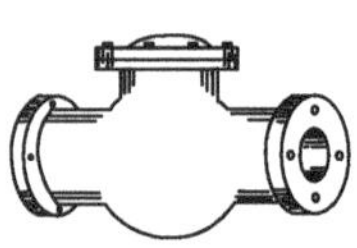

Description	Craft@Hrs	Unit	Material $	Labor $	Equipment $	Total $
Class 250 iron body swing check valve, flanged joints						
2"	P1@.500	Ea	538.00	22.20	—	560.20
2½"	P1@.600	Ea	628.00	26.70	—	654.70
3"	P1@.750	Ea	779.00	33.30	—	812.30
4"	P1@1.35	Ea	956.00	60.00	—	1,016.00
6"	ER@2.50	Ea	1,840.00	119.00	1.67	1,960.67
8"	ER@3.00	Ea	3,410.00	142.00	2.01	3,554.01

Description	Craft@Hrs	Unit	Material $	Labor $	Equipment $	Total $
Class 125 iron body silent check valve, wafer-type						
2"	P1@.500	Ea	153.00	22.20	—	175.20
2½"	P1@.600	Ea	170.00	26.70	—	196.70
3"	P1@.750	Ea	195.00	33.30	—	228.30
4"	P1@1.35	Ea	260.00	60.00	—	320.00
5"	P1@2.00	Ea	399.00	88.90	—	487.90
6"	P1@2.50	Ea	544.00	111.00	—	655.00
8"	P1@3.00	Ea	930.00	133.00	—	1,063.00
Class 125 iron body silent check valve, flanged joints						
2"	P1@.500	Ea	153.00	22.20	—	175.20
2½"	P1@.600	Ea	176.00	26.70	—	202.70
3"	P1@.750	Ea	198.00	33.30	—	231.30
4"	P1@1.35	Ea	259.00	60.00	—	319.00
6"	ER@2.50	Ea	466.00	119.00	1.67	586.67
8"	ER@3.00	Ea	840.00	142.00	2.01	984.01
10"	ER@4.00	Ea	1,300.00	190.00	2.68	1,492.68
Class 250 iron body silent check valve, flanged joints						
2"	P1@.500	Ea	225.00	22.20	—	247.20
2½"	P1@.600	Ea	259.00	26.70	—	285.70
3"	P1@.750	Ea	288.00	33.30	—	321.30
4"	P1@1.35	Ea	391.00	60.00	—	451.00
6"	ER@2.50	Ea	687.00	119.00	1.67	807.67
8"	ER@3.00	Ea	1,250.00	142.00	2.01	1,394.01
10"	ER@4.00	Ea	1,980.00	190.00	2.68	2,172.68
Class 125 bronze body gate valve, solder ends						
½"	P1@.240	Ea	22.80	10.70	—	33.50
¾"	P1@.300	Ea	28.50	13.30	—	41.80
1"	P1@.360	Ea	40.10	16.00	—	56.10
1¼"	P1@.480	Ea	50.40	21.30	—	71.70
1½"	P1@.540	Ea	68.00	24.00	—	92.00
2"	P1@.600	Ea	114.00	26.70	—	140.70
2½"	P1@1.00	Ea	188.00	44.50	—	232.50
3"	P1@1.50	Ea	270.00	66.70	—	336.70
Class 125 bronze body gate valve, threaded joints						
½"	P1@.210	Ea	22.80	9.33	—	32.13
¾"	P1@.250	Ea	28.50	11.10	—	39.60
1"	P1@.300	Ea	40.10	13.30	—	53.40
1¼"	P1@.400	Ea	50.40	17.80	—	68.20
1½"	P1@.450	Ea	68.00	20.00	—	88.00
2"	P1@.500	Ea	114.00	22.20	—	136.20
2½"	P1@.750	Ea	188.00	33.30	—	221.30
3"	P1@.950	Ea	270.00	42.20	—	312.20

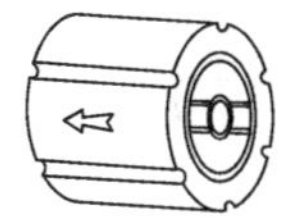
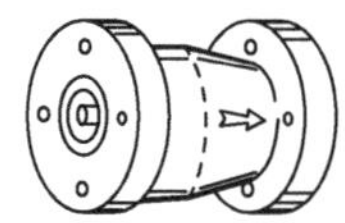
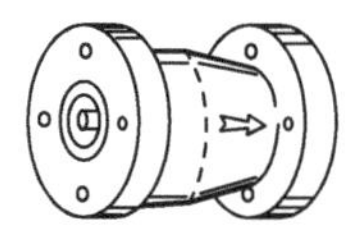
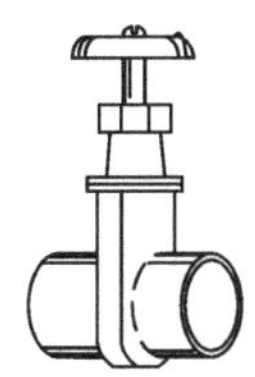
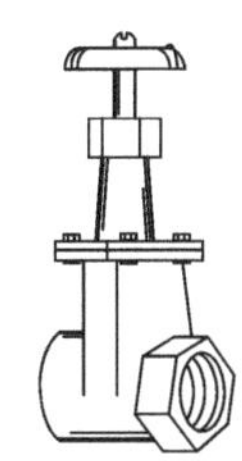

Class 300 bronze body gate valve, threaded joints

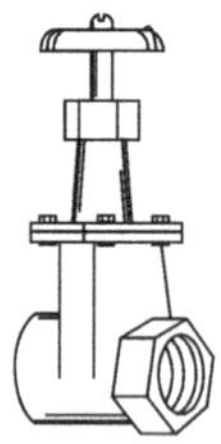

Description	Craft@Hrs	Unit	Material $	Labor $	Equipment $	Total $
½"	P1@.210	Ea	47.60	9.33	—	56.93
¾"	P1@.250	Ea	62.20	11.10	—	73.30
1"	P1@.300	Ea	75.30	13.30	—	88.60
1¼"	P1@.400	Ea	103.00	17.80	—	120.80
1½"	P1@.450	Ea	130.00	20.00	—	150.00
2"	P1@.500	Ea	191.00	22.20	—	213.20
2½"	P1@.750	Ea	389.00	33.30	—	422.30
3"	P1@.950	Ea	552.00	42.20	—	594.20

Class 125 iron body gate valve, flanged joints

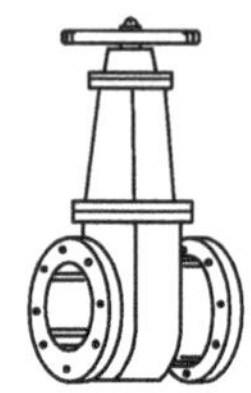

Description	Craft@Hrs	Unit	Material $	Labor $	Equipment $	Total $
2"	P1@.500	Ea	435.00	22.20	—	457.20
2½"	P1@.600	Ea	586.00	26.70	—	612.70
3"	P1@.750	Ea	636.00	33.30	—	669.30
4"	P1@1.35	Ea	934.00	60.00	—	994.00
6"	ER@2.50	Ea	1,780.00	119.00	1.67	1,900.67
8"	ER@3.00	Ea	1,780.00	142.00	2.01	1,924.01
10"	ER@4.00	Ea	4,120.00	190.00	2.68	4,312.68
12"	ER@4.50	Ea	5,680.00	213.00	3.01	5,896.01

Class 250 iron body gate valve, flanged joints

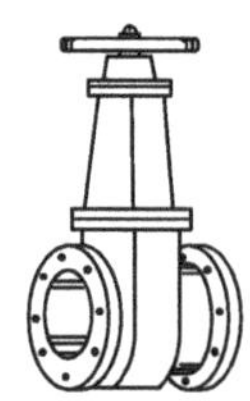

Description	Craft@Hrs	Unit	Material $	Labor $	Equipment $	Total $
2"	P1@.500	Ea	651.00	22.20	—	673.20
2½"	P1@.600	Ea	470.00	26.70	—	496.70
3"	P1@.750	Ea	594.00	33.30	—	627.30
4"	P1@1.35	Ea	1,010.00	60.00	—	1,070.00
6"	ER@2.50	Ea	1,880.00	119.00	1.67	2,000.67
8"	ER@3.00	Ea	2,900.00	142.00	2.01	3,044.01
10"	ER@4.00	Ea	5,160.00	190.00	2.68	5,352.68
12"	ER@4.50	Ea	8,410.00	213.00	3.01	8,626.01

Class 125 bronze body globe valve, brazed joints

Description	Craft@Hrs	Unit	Material $	Labor $	Equipment $	Total $
½"	P1@.240	Ea	42.20	10.70	—	52.90
¾"	P1@.300	Ea	56.20	13.30	—	69.50
1"	P1@.360	Ea	96.90	16.00	—	112.90
1¼"	P1@.480	Ea	136.00	21.30	—	157.30
1½"	P1@.540	Ea	185.00	24.00	—	209.00
2"	P1@.600	Ea	299.00	26.70	—	325.70

Class 125 bronze body globe valve, soft-soldered joints

Description	Craft@Hrs	Unit	Material $	Labor $	Equipment $	Total $
½"	P1@.200	Ea	42.20	8.89	—	51.09
¾"	P1@.249	Ea	56.20	11.10	—	67.30
1"	P1@.299	Ea	96.90	13.30	—	110.20
1¼"	P1@.398	Ea	136.00	17.70	—	153.70
1½"	P1@.448	Ea	185.00	19.90	—	204.90
2"	P1@.498	Ea	299.00	22.10	—	321.10

Description	Craft@Hrs	Unit	Material $	Labor $	Equipment $	Total $
Class 125 bronze body globe valve, threaded ends						
½"	P1@.210	Ea	42.20	9.33	—	51.53
¾"	P1@.250	Ea	56.20	11.10	—	67.30
1"	P1@.300	Ea	96.90	13.30	—	110.20
1¼"	P1@.400	Ea	136.00	17.80	—	153.80
1½"	P1@.450	Ea	185.00	20.00	—	205.00
2"	P1@.500	Ea	299.00	22.20	—	321.20
Class 300 bronze body globe valve, threaded joints						
½"	P1@.210	Ea	81.30	9.33	—	90.63
¾"	P1@.250	Ea	103.00	11.10	—	114.10
1"	P1@.300	Ea	146.00	13.30	—	159.30
1¼"	P1@.400	Ea	212.00	17.80	—	229.80
1½"	P1@.450	Ea	285.00	20.00	—	305.00
2"	P1@.500	Ea	392.00	22.20	—	414.20
Class 125 iron body globe valve, flanged joints						
2"	P1@.500	Ea	299.00	22.20	—	321.20
2½"	P1@.600	Ea	493.00	26.70	—	519.70
3"	P1@.750	Ea	591.00	33.30	—	624.30
4"	P1@1.35	Ea	798.00	60.00	—	858.00
6"	ER@2.50	Ea	1,310.00	119.00	1.67	1,430.67
8"	ER@3.00	Ea	2,000.00	142.00	2.01	2,144.01
10"	ER@4.00	Ea	3,060.00	190.00	2.68	3,252.68
Class 250 iron body globe valve, flanged joints						
2"	P1@.500	Ea	654.00	22.20	—	676.20
2½"	P1@.600	Ea	815.00	26.70	—	841.70
3"	P1@.750	Ea	969.00	33.30	—	1,002.30
4"	P1@1.35	Ea	1,340.00	60.00	—	1,400.00
6"	ER@2.50	Ea	2,630.00	119.00	1.67	2,750.67
8"	ER@3.00	Ea	5,850.00	142.00	2.01	5,994.01
Class 125 bronze body strainer, threaded joints						
½"	P1@.230	Ea	37.40	10.20	—	47.60
¾"	P1@.260	Ea	49.10	11.60	—	60.70
1"	P1@.330	Ea	60.20	14.70	—	74.90
1¼"	P1@.440	Ea	84.20	19.60	—	103.80
1½"	P1@.495	Ea	126.00	22.00	—	148.00
2"	P1@.550	Ea	220.00	24.40	—	244.40
Class 125 iron body strainer, flanged joints						
2"	P1@.500	Ea	168.00	22.20	—	190.20
2½"	P1@.600	Ea	187.00	26.70	—	213.70
3"	P1@.750	Ea	218.00	33.30	—	251.30
4"	P1@1.35	Ea	367.00	60.00	—	427.00
6"	ER@2.50	Ea	687.00	119.00	1.67	807.67
8"	ER@3.00	Ea	1,160.00	142.00	2.01	1,304.01

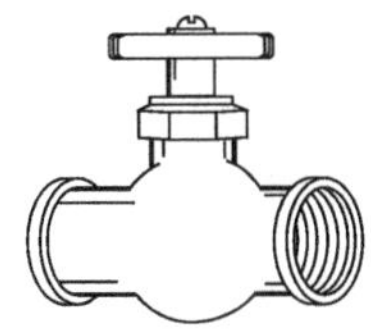
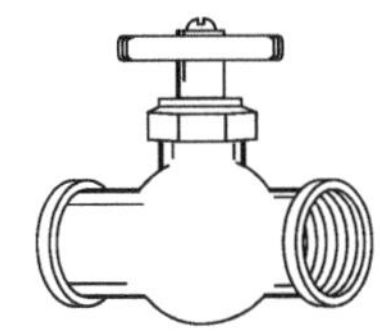
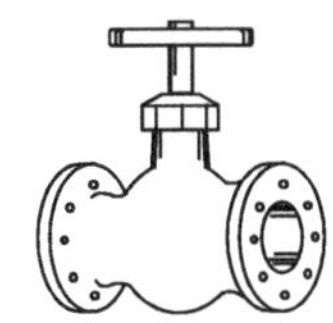
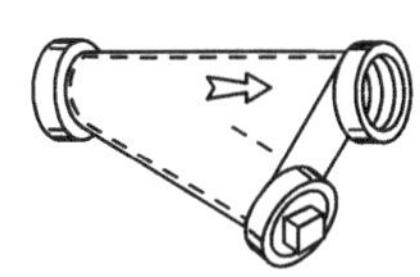
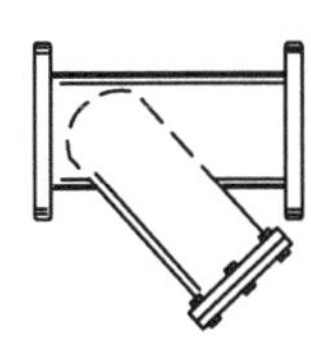

Description	Craft@Hrs	Unit	Material $	Labor $	Equipment $	Total $

Class 250 iron body strainer, flanged joints

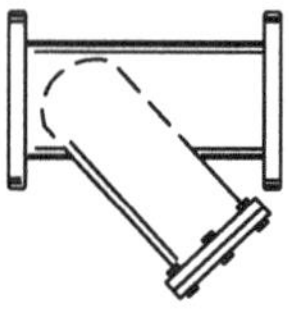

2"	P1@.500	Ea	262.00	22.20	—	284.20
2½"	P1@.600	Ea	287.00	26.70	—	313.70
3"	P1@.750	Ea	337.00	33.30	—	370.30
4"	P1@1.35	Ea	595.00	60.00	—	655.00
6"	ER@2.50	Ea	953.00	119.00	1.67	1,073.67
8"	ER@3.00	Ea	1,740.00	142.00	2.01	1,884.01

Installation of 2-way control valve, threaded joints

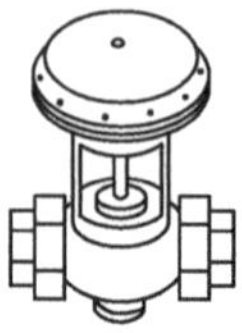

½"	P1@.210	Ea	—	9.33	—	9.33
¾"	P1@.250	Ea	—	11.10	—	11.10
1"	P1@.300	Ea	—	13.30	—	13.30
1¼"	P1@.400	Ea	—	17.80	—	17.80
1½"	P1@.450	Ea	—	20.00	—	20.00
2"	P1@.500	Ea	—	22.20	—	22.20
2½"	P1@.830	Ea	—	36.90	—	36.90
3"	P1@.990	Ea	—	44.00	—	44.00

Installation of 2-way control valve, flanged joints

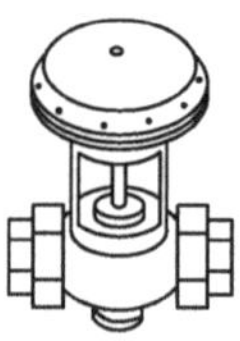

2"	P1@.500	Ea	—	22.20	—	22.20
2½"	P1@.600	Ea	—	26.70	—	26.70
3"	P1@.750	Ea	—	33.30	—	33.30
4"	P1@1.35	Ea	—	60.00	—	60.00
6"	P1@2.50	Ea	—	111.00	5.86	116.86
8"	P1@3.00	Ea	—	133.00	7.03	140.03
10"	P1@4.00	Ea	—	178.00	9.37	187.37
12"	P1@4.50	Ea	—	200.00	10.50	210.50

Installation of 3-way control valve, threaded joints

½"	P1@.260	Ea	—	11.60	—	11.60
¾"	P1@.365	Ea	—	16.20	—	16.20
1"	P1@.475	Ea	—	21.10	—	21.10
1¼"	P1@.575	Ea	—	25.60	—	25.60
1½"	P1@.680	Ea	—	30.20	—	30.20
2"	P1@.910	Ea	—	40.40	—	40.40
2½"	P1@1.12	Ea	—	49.80	—	49.80
3"	P1@1.33	Ea	—	59.10	—	59.10

Description	Craft@Hrs	Unit	Material $	Labor $	Equipment $	Total $

Installation of 3-way control valve, flanged joints

2"	P1@.910	Ea	—	40.40	—	40.40
2½"	P1@1.12	Ea	—	49.80	—	49.80
3"	P1@1.33	Ea	—	59.10	—	59.10
4"	P1@2.00	Ea	—	88.90	—	88.90
6"	P1@3.70	Ea	—	164.00	8.67	172.67
8"	P1@4.40	Ea	—	196.00	10.30	206.30
10"	P1@5.90	Ea	—	262.00	13.80	275.80
12"	P1@6.50	Ea	—	289.00	15.20	304.20

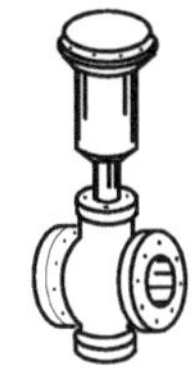

Lever-handled gas valve, threaded joints

½"	P1@.210	Ea	35.20	9.33	—	44.53
¾"	P1@.250	Ea	39.00	11.10	—	50.10
1"	P1@.300	Ea	48.30	13.30	—	61.60
1¼"	P1@.400	Ea	64.80	17.80	—	82.60
1½"	P1@.450	Ea	87.30	20.00	—	107.30
2"	P1@.500	Ea	127.00	22.20	—	149.20

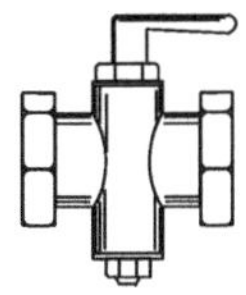

Plug-type gas valve, flanged joints, lubricated

2"	P1@.500	Ea	264.00	22.20	—	286.20
2½"	P1@.600	Ea	428.00	26.70	—	454.70
3"	P1@.750	Ea	601.00	33.30	—	634.30
4"	P1@1.35	Ea	823.00	60.00	—	883.00
6"	P1@2.50	Ea	1,470.00	111.00	—	1,581.00

¾" hose bibb

Low quality	P1@.350	Ea	12.70	15.60	—	28.30
Med. quality	P1@.350	Ea	18.30	15.60	—	33.90
High quality	P1@.350	Ea	24.10	15.60	—	39.70

Water pressure reducing valve, threaded joints, 25-75 PSI

½"	P1@.210	Ea	124.00	9.33	—	133.33
¾"	P1@.250	Ea	160.00	11.10	—	171.10
1"	P1@.300	Ea	260.00	13.30	—	273.30
1¼"	P1@.400	Ea	317.00	17.80	—	334.80
1½"	P1@.450	Ea	452.00	20.00	—	472.00
2"	P1@.500	Ea	715.00	22.20	—	737.20

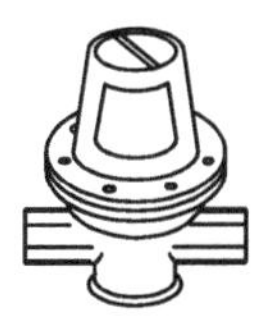

Water hammer arrester, threaded joints

½"	P1@.160	Ea	55.10	7.11	—	62.21
¾"	P1@.190	Ea	64.00	8.45	—	72.45
1"	P1@.230	Ea	162.00	10.20	—	172.20
1¼"	P1@.300	Ea	189.00	13.30	—	202.30
1½"	P1@.340	Ea	271.00	15.10	—	286.10
2"	P1@.380	Ea	413.00	16.90	—	429.90

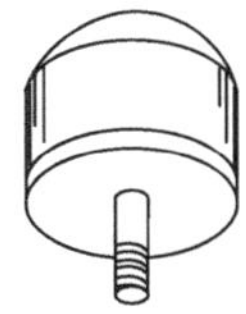

Description	Craft@Hrs	Unit	Material $	Labor $	Equipment $	Total $

Carbon steel weldolet, Schedule 40

Description	Craft@Hrs	Unit	Material $	Labor $	Equipment $	Total $
½"	P1@.330	Ea	17.00	14.70	1.07	32.77
¾"	P1@.500	Ea	17.60	22.20	1.62	41.42
1"	P1@.670	Ea	18.50	29.80	2.17	50.47
1¼"	P1@.830	Ea	22.20	36.90	2.69	61.79
1½"	P1@1.00	Ea	22.20	44.50	3.24	69.94
2"	P1@1.33	Ea	24.60	59.10	4.31	88.01
2½"	P1@1.67	Ea	54.00	74.20	5.42	133.62
3"	P1@2.00	Ea	61.40	88.90	6.49	156.79
4"	P1@2.67	Ea	79.20	119.00	8.66	206.86
6"	P1@4.00	Ea	327.00	178.00	13.00	518.00
8"	P1@4.80	Ea	480.00	213.00	15.60	708.60
10"	P1@6.00	Ea	665.00	267.00	19.50	951.50
12"	P1@7.20	Ea	817.00	320.00	23.30	1,160.30

Carbon steel weldolet, Schedule 80

Description	Craft@Hrs	Unit	Material $	Labor $	Equipment $	Total $
½"	P1@.440	Ea	18.90	19.60	1.43	39.93
¾"	P1@.670	Ea	19.30	29.80	2.17	51.27
1"	P1@.890	Ea	21.10	39.60	2.89	63.59
1¼"	P1@1.11	Ea	26.60	49.30	3.60	79.50
1½"	P1@1.33	Ea	26.60	59.10	4.31	90.01
2"	P1@1.77	Ea	28.70	78.70	5.74	113.14
2½"	P1@2.22	Ea	64.10	98.70	7.20	170.00
3"	P1@2.66	Ea	65.90	118.00	8.63	192.53
4"	P1@3.55	Ea	146.00	158.00	11.50	315.50
6"	P1@5.32	Ea	459.00	236.00	17.30	712.30
8"	P1@6.38	Ea	614.00	284.00	20.70	918.70

Carbon steel threadolet, Schedule 40

Description	Craft@Hrs	Unit	Material $	Labor $	Equipment $	Total $
¾"	P1@.330	Ea	10.60	14.70	—	25.30
1"	P1@.440	Ea	11.70	19.60	—	31.30
1¼"	P1@.560	Ea	16.40	24.90	—	41.30
1½"	P1@.670	Ea	18.50	29.80	—	48.30
2"	P1@.890	Ea	21.50	39.60	—	61.10
2½"	P1@1.11	Ea	66.20	49.30	—	115.50

Carbon steel threadolet, Schedule 80

Description	Craft@Hrs	Unit	Material $	Labor $	Equipment $	Total $
½"	P1@.300	Ea	15.90	13.30	—	29.20
¾"	P1@.440	Ea	17.50	19.60	—	37.10
1"	P1@.590	Ea	21.60	26.20	—	47.80
1¼"	P1@.740	Ea	97.00	32.90	—	129.90
1½"	P1@.890	Ea	97.00	39.60	—	136.60
2"	P1@1.18	Ea	126.00	52.50	—	178.50

Service weight cast iron soil pipe with hub-less coupled joints is widely used for non-pressurized underground and in-building drain, waste and vent (DWV) systems.

This section has been arranged to save the estimator's time by including all normally-used system components such as pipe, fittings, hanger assemblies and riser clamps under one heading. Additional items can be found under "Plumbing and Piping Specialties." The cost estimates in this section are based on the conditions, limitations and wage rates described in the section "How to Use This Book" beginning on page 5.

Equipment cost, where shown, is $4.57 per hour for a 2-ton chain hoist.

Description	Craft@Hrs	Unit	Material $	Labor $	Equipment $	Total $

DWV cast iron no-hub (MJ) horizontal pipe assembly. Horizontally hung in a building. Assembly includes fittings, couplings, hangers and rod. Based on a reducing wye and a 45-degree elbow every 15 feet and a hanger every 6 feet for 2" pipe. A reducing wye and a 45-degree elbow every 40 feet and a hanger every 8 feet for 10" pipe. *(Distance between fittings in the pipe assembly increases as pipe diameter increases.)* Use these figures for preliminary estimates.

2"	P1@.130	LF	11.80	5.78	—	17.58
3"	P1@.160	LF	13.00	7.11	—	20.11
4"	P1@.190	LF	15.20	8.45	—	23.65
6"	ER@.220	LF	29.60	10.40	.15	40.15
8"	ER@.260	LF	56.80	12.30	.17	69.27
10"	ER@.310	LF	87.90	14.70	.21	102.81

DWV cast iron no-hub (MJ), pipe only

2"	P1@.050	LF	6.90	2.22	—	9.12
3"	P1@.060	LF	8.22	2.67	—	10.89
4"	P1@.080	LF	10.70	3.56	—	14.26
6"	ER@.110	LF	20.10	5.22	.15	25.47
8"	ER@.130	LF	37.10	6.16	.17	43.43
10"	ER@.140	LF	64.60	6.64	.21	71.45

DWV cast iron no-hub 1/8 bend

2"	P1@.250	Ea	6.70	11.10	—	17.80
3"	P1@.350	Ea	8.72	15.60	—	24.32
4"	P1@.500	Ea	11.00	22.20	—	33.20
6"	ER@.650	Ea	35.40	30.80	.44	66.64
8"	ER@.710	Ea	94.90	33.70	.48	129.08
10"	ER@.800	Ea	141.00	37.90	.54	179.44

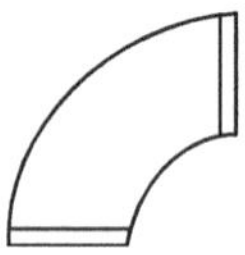

DWV cast iron no-hub ¼ bend

Description	Craft@Hrs	Unit	Material $	Labor $	Equipment $	Total $
2"	P1@.250	Ea	8.17	11.10	—	19.27
3"	P1@.350	Ea	10.10	15.60	—	25.70
4"	P1@.500	Ea	15.40	22.20	—	37.60
6"	ER@.650	Ea	37.50	30.80	.44	68.74
8"	ER@.710	Ea	145.00	33.70	.48	179.18
10"	ER@.710	Ea	159.00	33.70	.48	193.18

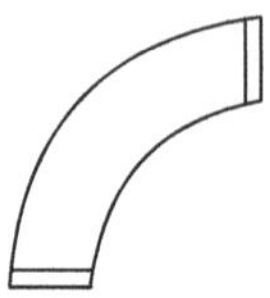

DWV cast iron no-hub long sweep ¼ bend

Description	Craft@Hrs	Unit	Material $	Labor $	Equipment $	Total $
2"	P1@.250	Ea	14.20	11.10	—	25.30
3"	P1@.350	Ea	17.20	15.60	—	32.80
4"	P1@.500	Ea	27.20	22.20	—	49.40
5"	ER@.580	Ea	49.90	27.50	.39	77.79
6"	ER@.650	Ea	60.70	30.80	.44	91.94

DWV cast iron no-hub low-heel outlet ¼ bend

Description	Craft@Hrs	Unit	Material $	Labor $	Equipment $	Total $
3"	P1@.470	Ea	14.40	20.90	—	35.30
4"	P1@.650	Ea	17.70	28.90	—	46.60

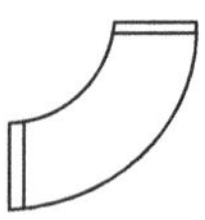

DWV cast iron no-hub closet bend

Description	Craft@Hrs	Unit	Material $	Labor $	Equipment $	Total $
3"	P1@.350	Ea	33.30	15.60	—	48.90
4"	P1@.500	Ea	34.20	22.20	—	56.40

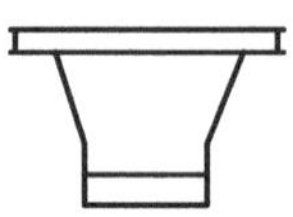

DWV cast iron no-hub closet flange

Description	Craft@Hrs	Unit	Material $	Labor $	Equipment $	Total $
4" x 3"	P1@.500	Ea	48.70	22.20	—	70.90

DWV cast iron no-hub P-trap

Description	Craft@Hrs	Unit	Material $	Labor $	Equipment $	Total $
2"	P1@.250	Ea	13.30	11.10	—	24.40
3"	P1@.350	Ea	41.10	15.60	—	56.70
4"	P1@.500	Ea	50.80	22.20	—	73.00
6"	ER@.650	Ea	181.00	30.80	.44	212.24

DWV cast iron no-hub wye

Description	Craft@Hrs	Unit	Material $	Labor $	Equipment $	Total $
2"	P1@.380	Ea	10.40	16.90	—	27.30
3"	P1@.520	Ea	14.50	23.10	—	37.60
4"	P1@.700	Ea	22.90	31.10	—	54.00
6"	ER@.900	Ea	61.60	42.70	.60	104.90
8"	ER@1.15	Ea	156.00	54.50	.77	211.27
10"	ER@1.50	Ea	375.00	71.10	1.00	447.10

DWV cast iron no-hub reducing wye

Description	Craft@Hrs	Unit	Material $	Labor $	Equipment $	Total $
3" x 2"	P1@.470	Ea	13.90	20.90	—	34.80
4" x 2"	P1@.600	Ea	18.70	26.70	—	45.40
4" x 3"	P1@.650	Ea	19.60	28.90	—	48.50
6" x 2"	ER@.770	Ea	50.00	36.50	.52	87.02
6" x 3"	ER@.770	Ea	54.30	36.50	.52	91.32
6" x 4"	ER@.800	Ea	53.30	37.90	.54	91.74
8" x 2"	ER@1.00	Ea	128.00	47.40	.67	176.07
8" x 3"	ER@1.00	Ea	130.00	47.40	.67	178.07
8" x 4"	ER@1.00	Ea	132.00	47.40	.67	180.07
8" x 6"	ER@1.10	Ea	136.00	52.20	.74	188.94

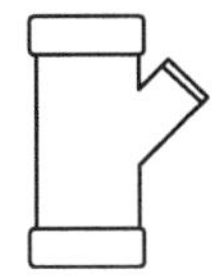

DWV cast iron no-hub combination upright wye and 1/8 bend

Description	Craft@Hrs	Unit	Material $	Labor $	Equipment $	Total $
2"	P1@.380	Ea	25.80	16.90	—	42.70
3"	P1@.520	Ea	41.70	23.10	—	64.80
4"	P1@.700	Ea	55.00	31.10	—	86.10

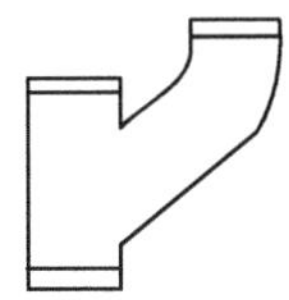

DWV cast iron no-hub combination upright reducing wye and 1/8 bend

Description	Craft@Hrs	Unit	Material $	Labor $	Equipment $	Total $
3" x 2"	P1@.470	Ea	45.20	20.90	—	66.10
4" x 2"	P1@.600	Ea	53.80	26.70	—	80.50
4" x 3"	P1@.650	Ea	56.30	28.90	—	85.20

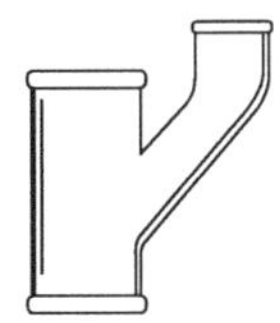

DWV cast iron no-hub double wye

Description	Craft@Hrs	Unit	Material $	Labor $	Equipment $	Total $
2"	P1@.500	Ea	32.10	22.20	—	54.30
3"	P1@.700	Ea	32.50	31.10	—	63.60
4"	P1@.950	Ea	50.50	42.20	—	92.70
6"	ER@1.10	Ea	176.00	52.20	.74	228.94
8"	ER@1.45	Ea	320.00	68.70	.97	389.67

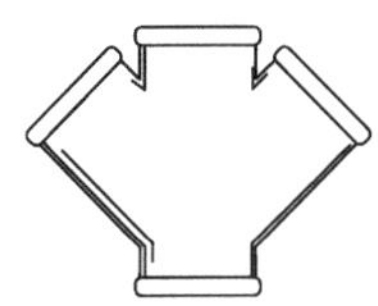

DWV cast iron no-hub double reducing wye

Description	Craft@Hrs	Unit	Material $	Labor $	Equipment $	Total $
3" x 2"	P1@.520	Ea	32.10	23.10	—	55.20
4" x 2"	P1@.740	Ea	44.70	32.90	—	77.60
4" x 3"	P1@1.00	Ea	46.00	44.50	—	90.50
6" x 4"	ER@1.75	Ea	173.00	83.00	1.17	257.17

DWV cast iron no-hub combination wye and 1/8 bend

Description	Craft@Hrs	Unit	Material $	Labor $	Equipment $	Total $
2"	P1@.380	Ea	20.00	16.90	—	36.90
3"	P1@.520	Ea	28.60	23.10	—	51.70
4"	P1@.700	Ea	45.60	31.10	—	76.70
6"	ER@.900	Ea	155.00	42.70	.60	198.30
8"	ER@1.15	Ea	312.00	54.50	.77	367.27

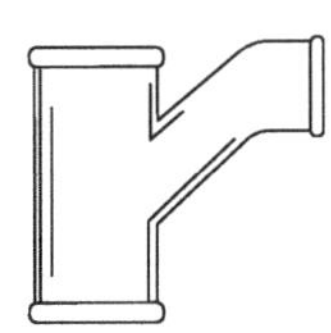

DWV cast iron no-hub combination reducing wye and 1/8 bend

Description	Craft@Hrs	Unit	Material $	Labor $	Equipment $	Total $
3" x 2"	P1@.470	Ea	24.30	20.90	—	45.20
4" x 2"	P1@.600	Ea	36.90	26.70	—	63.60
4" x 3"	P1@.650	Ea	39.60	28.90	—	68.50
6" x 2"	ER@.770	Ea	177.00	36.50	.52	214.02
6" x 3"	ER@.770	Ea	186.00	36.50	.52	223.02
6" x 4"	ER@.800	Ea	191.00	37.90	.54	229.44
8" x 4"	ER@1.20	Ea	347.00	56.90	.80	404.70

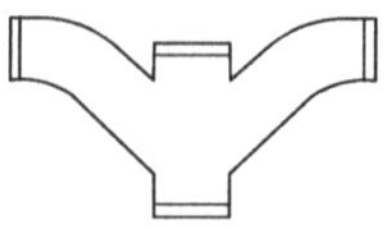

DWV cast iron no-hub double combination wye and 1/8 bend

2"	P1@.500	Ea	34.80	22.20	—	57.00
3"	P1@.700	Ea	55.80	31.10	—	86.90
4"	P1@.950	Ea	68.70	42.20	—	110.90

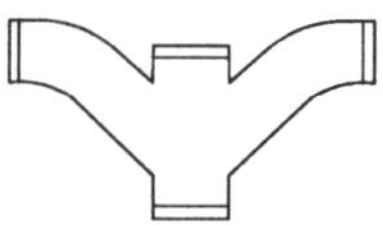

DWV cast iron no-hub double combination reducing wye and 1/8 bend

3" x 2"	P1@.600	Ea	39.40	26.70	—	66.10
4" x 2"	P1@1.25	Ea	62.10	55.60	—	117.70
4" x 3"	P1@1.35	Ea	64.90	60.00	—	124.90

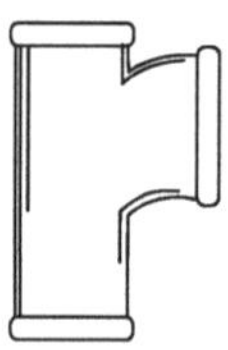

DWV cast iron no-hub sanitary tee (TY)

2"	P1@.380	Ea	12.20	16.90	—	29.10
3"	P1@.520	Ea	14.50	23.10	—	37.60
4"	P1@.700	Ea	24.10	31.10	—	55.20
6"	ER@.900	Ea	73.00	42.70	.60	116.30
8"	ER@1.15	Ea	193.00	54.50	.77	248.27

DWV cast iron no-hub reducing sanitary tee (TY)

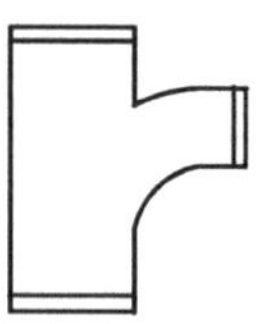

3" x 2"	P1@.470	Ea	13.80	20.90	—	34.70
4" x 2"	P1@.600	Ea	18.50	26.70	—	45.20
4" x 3"	P1@.650	Ea	21.20	28.90	—	50.10
6" x 2"	ER@.770	Ea	57.90	36.50	.52	94.92
6" x 3"	ER@.770	Ea	54.70	36.50	.52	91.72
6" x 4"	ER@.800	Ea	59.30	37.90	.54	97.74
8" x 3"	ER@1.00	Ea	152.00	47.40	.67	200.07
8" x 4"	ER@1.00	Ea	148.00	47.40	.67	196.07
8" x 6"	ER@1.10	Ea	156.00	52.20	.74	208.94

DWV cast iron no-hub tapped sanitary tee (tapped TY)

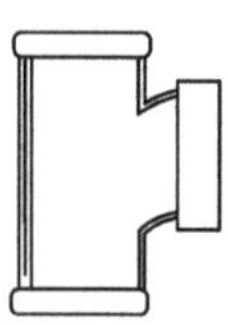

2" x 1½"	P1@.200	Ea	34.00	8.89	—	42.89
2" x 2"	P1@.250	Ea	34.00	11.10	—	45.10
3" x 2"	P1@.350	Ea	43.30	15.60	—	58.90
4" x 2"	P1@.425	Ea	51.30	18.90	—	70.20
6" x 2"	ER@.700	Ea	133.00	33.20	.47	166.67

Cast Iron, DWV, Service Weight, No-Hub with Coupled Joints

DWV cast iron no-hub test tee

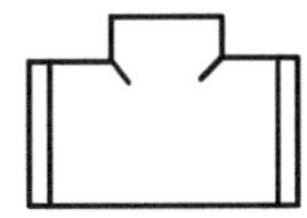

Description	Craft@Hrs	Unit	Material $	Labor $	Equipment $	Total $
2"	P1@.370	Ea	43.00	16.40	—	59.40
3"	P1@.520	Ea	60.70	23.10	—	83.80
4"	P1@.750	Ea	77.40	33.30	—	110.70
6"	ER@1.12	Ea	277.00	53.10	.75	330.85
8"	ER@1.25	Ea	503.00	59.30	.84	563.14

DWV cast iron no-hub cross (double TY)

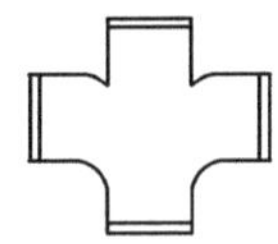

Description	Craft@Hrs	Unit	Material $	Labor $	Equipment $	Total $
2"	P1@.500	Ea	34.20	22.20	—	56.40
3"	P1@.700	Ea	36.30	31.10	—	67.40
4"	P1@.950	Ea	52.50	42.20	—	94.70
6"	ER@1.30	Ea	194.00	61.60	.87	256.47

DWV cast iron no-hub reducing cross (double TY)

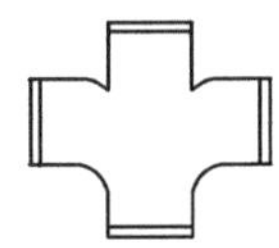

Description	Craft@Hrs	Unit	Material $	Labor $	Equipment $	Total $
3" x 2"	P1@.600	Ea	32.20	26.70	—	58.90
4" x 2"	P1@.900	Ea	44.60	40.00	—	84.60
4" x 3"	P1@.900	Ea	45.60	40.00	—	85.60
6" x 4"	ER@1.20	Ea	179.00	56.90	.80	236.70
8" x 4"	ER@1.50	Ea	449.00	71.10	1.00	521.10

DWV cast iron no-hub reducer

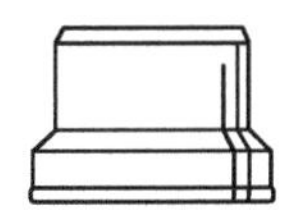

Description	Craft@Hrs	Unit	Material $	Labor $	Equipment $	Total $
2"	P1@.120	Ea	6.72	5.33	—	12.05
3"	P1@.130	Ea	7.24	5.78	—	13.02
4"	P1@.150	Ea	10.70	6.67	—	17.37
6"	ER@.200	Ea	28.70	9.48	.13	38.31
8"	ER@.400	Ea	73.70	19.00	.27	92.97
10"	ER@.650	Ea	114.00	30.80	.44	145.24

DWV cast iron no-hub cap

Description	Craft@Hrs	Unit	Material $	Labor $	Equipment $	Total $
2"	P1@.125	Ea	4.10	5.56	—	9.66
3"	P1@.175	Ea	7.03	7.78	—	14.81
4"	P1@.250	Ea	9.16	11.10	—	20.26
6"	ER@.375	Ea	17.40	17.80	.25	35.45
8"	ER@.500	Ea	32.10	23.70	.33	56.13
10"	ER@.750	Ea	49.60	35.60	.50	85.70

MJ (mechanical joint) coupling with gasket (labor included with fittings)

Description	Craft@Hrs	Unit	Material $	Labor $	Equipment $	Total $
2"	—	Ea	5.77	—	—	5.77
3"	—	Ea	6.31	—	—	6.31
4"	—	Ea	6.75	—	—	6.75
6"	—	Ea	18.40	—	—	18.40
8"	—	Ea	34.30	—	—	34.30
10"	—	Ea	47.60	—	—	47.60

Description	Craft@Hrs	Unit	Material $	Labor $	Equipment $	Total $

Hanger with swivel assembly

Description	Craft@Hrs	Unit	Material $	Labor $	Equipment $	Total $
2"	P1@.400	Ea	5.99	17.80	—	23.79
3"	P1@.450	Ea	8.15	20.00	—	28.15
4"	P1@.450	Ea	10.10	20.00	—	30.10
5"	P1@.500	Ea	14.70	22.20	—	36.90
6"	P1@.500	Ea	16.60	22.20	—	38.80
8"	P1@.550	Ea	21.70	24.40	—	46.10
10"	P1@.650	Ea	28.60	28.90	—	57.50

Riser clamp

Description	Craft@Hrs	Unit	Material $	Labor $	Equipment $	Total $
2"	P1@.115	Ea	6.62	5.11	—	11.73
3"	P1@.120	Ea	6.93	5.33	—	12.26
4"	P1@.125	Ea	7.53	5.56	—	13.09
5"	P1@.180	Ea	11.40	8.00	—	19.40
6"	P1@.200	Ea	13.80	8.89	—	22.69
8"	P1@.200	Ea	22.50	8.89	—	31.39
10"	P1@.250	Ea	33.30	11.10	—	44.40

Cast Iron, DWV, Service Weight, Hub & Spigot with Gasketed Joints

Service weight hub and spigot cast iron soil pipe with gasketed joints is widely used for non-pressurized underground and in-building drain, waste and vent (DWV) systems.

This section has been arranged to save the estimator's time by including all normally-used system components such as pipe, fittings, valves, hanger assemblies and riser clamps under one heading. Additional items can be found under "Plumbing and Piping Specialties." The cost estimates in this section are based on the conditions, limitations and wage rates described in the section "How to Use This Book" beginning on page 5.

Equipment cost, where shown, is $4.57 per hour for a 2-ton chain hoist.

DWV cast iron pipe, hub and spigot, gasketed joints

Description	Craft@Hrs	Unit	Material $	Labor $	Equipment $	Total $
2"	P1@.060	LF	11.80	2.67	—	14.47
3"	P1@.070	LF	13.00	3.11	—	16.11
4"	P1@.090	LF	17.10	4.00	—	21.10
6"	ER@.110	LF	32.50	5.22	.07	37.79
8"	ER@.140	LF	54.60	6.64	.09	61.33
10"	ER@.180	LF	87.20	8.53	.12	95.85
12"	ER@.230	LF	113.00	10.90	.15	124.05
15"	ER@.280	LF	170.00	13.30	.19	183.49

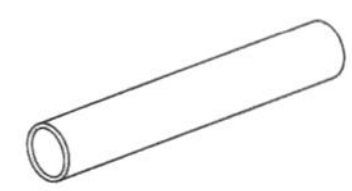

DWV cast iron 1/8 bend, hub and spigot, gasketed joints

Description	Craft@Hrs	Unit	Material $	Labor $	Equipment $	Total $
2"	P1@.300	Ea	13.30	13.30	—	26.60
3"	P1@.400	Ea	18.00	17.80	—	35.80
4"	P1@.550	Ea	25.30	24.40	—	49.70
6"	ER@.700	Ea	63.70	33.20	.47	97.37
8"	ER@1.15	Ea	151.00	54.50	.77	206.27
10"	ER@1.25	Ea	204.00	59.30	.84	264.14
12"	ER@2.25	Ea	471.00	107.00	1.51	579.51
15"	ER@2.50	Ea	789.00	119.00	1.67	909.67

DWV cast iron ¼ bend, hub and spigot, gasketed joints

Description	Craft@Hrs	Unit	Material $	Labor $	Equipment $	Total $
2"	P1@.300	Ea	14.70	13.30	—	28.00
3"	P1@.400	Ea	22.60	17.80	—	40.40
4"	P1@.550	Ea	39.90	24.40	—	64.30
6"	ER@.900	Ea	75.60	42.70	.60	118.90
8"	ER@1.15	Ea	203.00	54.50	.77	258.27
10"	ER@1.25	Ea	280.00	59.30	.84	340.14
12"	ER@2.25	Ea	587.00	107.00	1.51	695.51
15"	ER@2.50	Ea	1,180.00	119.00	1.67	1,300.67

DWV cast iron heel inlet ¼ bend, hub and spigot, gasketed joints

Description	Craft@Hrs	Unit	Material $	Labor $	Equipment $	Total $
3" x 2"	P1@.550	Ea	27.70	24.40	—	52.10
4" x 2"	P1@.700	Ea	33.70	31.10	—	64.80

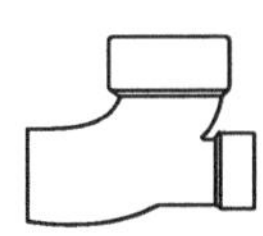

Cast Iron, DWV, Service Weight, Hub & Spigot with Gasketed Joints

DWV cast iron 1/16 bend, hub and spigot, gasketed joints

Description	Craft@Hrs	Unit	Material $	Labor $	Equipment $	Total $
2"	P1@.300	Ea	8.33	13.30	—	21.63
3"	P1@.400	Ea	14.10	17.80	—	31.90
4"	P1@.550	Ea	19.50	24.40	—	43.90
6"	ER@.700	Ea	32.40	33.20	.47	66.07
8"	ER@1.15	Ea	114.00	54.50	.77	169.27

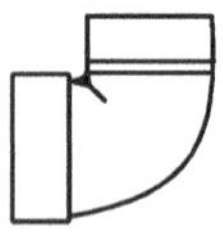

DWV cast iron closet bend, hub and spigot, gasketed joints

Description	Craft@Hrs	Unit	Material $	Labor $	Equipment $	Total $
4" x 4"	P1@.400	Ea	48.40	17.80	—	66.20

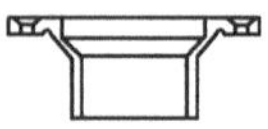

DWV cast iron closet flange, hub and spigot, gasketed joints

Description	Craft@Hrs	Unit	Material $	Labor $	Equipment $	Total $
4" x 2"	P1@.075	Ea	15.70	3.33	—	19.03
4" x 3"	P1@.100	Ea	21.80	4.45	—	26.25
4" x 4"	P1@.125	Ea	28.00	5.56	—	33.56

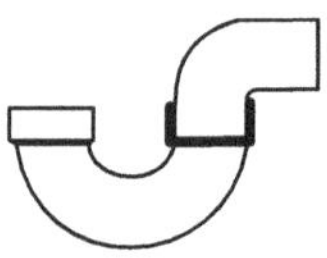

DWV cast iron regular P-trap, hub and spigot, gasketed joints

Description	Craft@Hrs	Unit	Material $	Labor $	Equipment $	Total $
2"	P1@.300	Ea	56.10	13.30	—	69.40
3"	P1@.400	Ea	77.80	17.80	—	95.60
4"	P1@.550	Ea	120.00	24.40	—	144.40
6"	ER@.700	Ea	264.00	33.20	.47	297.67
8"	ER@1.15	Ea	778.00	54.50	.77	833.27
10"	ER@2.00	Ea	1,070.00	94.80	1.34	1,166.14

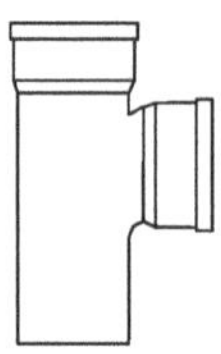

DWV cast iron sanitary tee (TY), hub and spigot, gasketed joints

Description	Craft@Hrs	Unit	Material $	Labor $	Equipment $	Total $
2"	P1@.400	Ea	26.90	17.80	—	44.70
3"	P1@.600	Ea	47.10	26.70	—	73.80
4"	P1@.750	Ea	94.70	33.30	—	128.00
6"	ER@.900	Ea	218.00	42.70	.60	261.30
8"	ER@1.50	Ea	314.00	71.10	1.00	386.10

DWV cast iron sanitary reducing tee (TY), hub and spigot, gasketed joints

Description	Craft@Hrs	Unit	Material $	Labor $	Equipment $	Total $
3" x 2"	P1@.400	Ea	38.20	17.80	—	56.00
4" x 2"	P1@.650	Ea	99.40	28.90	—	128.30
4" x 3"	P1@.700	Ea	99.40	31.10	—	130.50
6" x 2"	ER@.750	Ea	135.00	35.60	.50	171.10
6" x 3"	ER@.800	Ea	135.00	37.90	.54	173.44
6" x 4"	ER@.850	Ea	135.00	40.30	.57	175.87

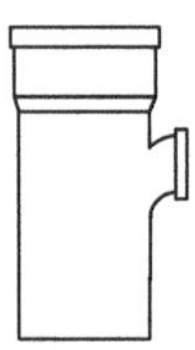

DWV cast iron sanitary tapped tee (tapped TY), hub and spigot, gasketed joints

Description	Craft@Hrs	Unit	Material $	Labor $	Equipment $	Total $
2"	P1@.300	Ea	23.10	13.30	—	36.40
3"	P1@.400	Ea	33.90	17.80	—	51.70
4"	P1@.550	Ea	34.60	24.40	—	59.00

Description	Craft@Hrs	Unit	Material $	Labor $	Equipment $	Total $
DWV cast iron combination wye and 1/8 bend, hub and spigot, gasketed joints						
2"	P1@.400	Ea	23.80	17.80	—	41.60
3"	P1@.600	Ea	36.00	26.70	—	62.70
4"	P1@.750	Ea	50.00	33.30	—	83.30
5"	ER@.850	Ea	93.80	40.30	.57	134.67
6"	ER@.900	Ea	119.00	42.70	.60	162.30

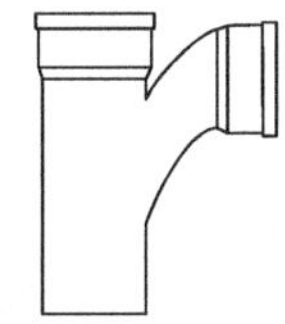

DWV cast iron combination reducing wye and 1/8 bend, hub and spigot, gasketed joints

Description	Craft@Hrs	Unit	Material $	Labor $	Equipment $	Total $
3" x 2"	P1@.550	Ea	26.90	24.40	—	51.30
4" x 2"	P1@.650	Ea	36.90	28.90	—	65.80
4" x 3"	P1@.700	Ea	42.80	31.10	—	73.90
5" x 2"	ER@.700	Ea	76.00	33.20	.47	109.67
5" x 4"	ER@.800	Ea	89.90	37.90	.54	128.34
6" x 2"	ER@.750	Ea	77.00	35.60	.50	113.10
6" x 3"	ER@.750	Ea	81.80	35.60	.50	117.90
6" x 4"	ER@.800	Ea	85.90	37.90	.54	124.34

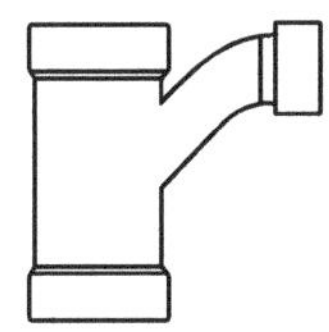

DWV cast iron combination double wye and 1/8 bend, hub and spigot, gasketed joints

Description	Craft@Hrs	Unit	Material $	Labor $	Equipment $	Total $
2"	P1@.500	Ea	45.70	22.20	—	67.90
3"	P1@.750	Ea	62.40	33.30	—	95.70
4"	P1@.950	Ea	88.80	42.20	—	131.00
6"	ER@1.10	Ea	305.00	52.20	.74	357.94

DWV cast iron combination double reducing wye and 1/8 bend, hub and spigot, gasketed joints

Description	Craft@Hrs	Unit	Material $	Labor $	Equipment $	Total $
3" x 2"	P1@.550	Ea	50.20	24.40	—	74.60
4" x 2"	P1@.650	Ea	63.10	28.90	—	92.00
4" x 3"	P1@.700	Ea	71.50	31.10	—	102.60
6" x 4"	ER@.800	Ea	202.00	37.90	.54	240.44

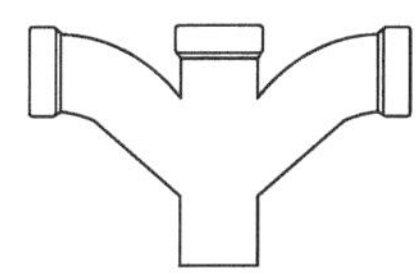

DWV cast iron wye, hub and spigot, gasketed joints

Description	Craft@Hrs	Unit	Material $	Labor $	Equipment $	Total $
2"	P1@.400	Ea	22.00	17.80	—	39.80
3"	P1@.600	Ea	37.90	26.70	—	64.60
4"	P1@.750	Ea	50.70	33.30	—	84.00
6"	ER@.900	Ea	122.00	42.70	.60	165.30
8"	ER@1.50	Ea	272.00	71.10	1.00	344.10
10"	ER@2.00	Ea	457.00	94.80	1.34	553.14
12"	ER@2.50	Ea	1,030.00	119.00	1.67	1,150.67
15"	ER@3.00	Ea	2,040.00	142.00	2.01	2,184.01

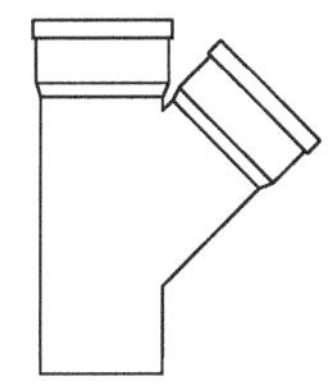

DWV cast iron reducing wye, hub and spigot, gasketed joints

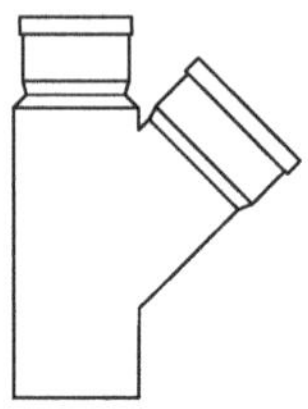

Description	Craft@Hrs	Unit	Material $	Labor $	Equipment $	Total $
4" x 2"	P1@.650	Ea	38.20	28.90	—	67.10
4" x 3"	P1@.700	Ea	46.80	31.10	—	77.90
6" x 2"	ER@.750	Ea	97.50	35.60	.50	133.60
6" x 3"	ER@.750	Ea	97.50	35.60	.50	133.60
6" x 4"	ER@.800	Ea	98.50	37.90	.54	136.94
8" x 4"	ER@1.35	Ea	226.00	64.00	.90	290.90
8" x 6"	ER@1.45	Ea	241.00	68.70	.97	310.67
10" x 4"	ER@1.85	Ea	390.00	87.70	1.24	478.94
10" x 6"	ER@1.90	Ea	401.00	90.10	1.27	492.37
10" x 8"	ER@1.95	Ea	429.00	92.40	1.31	522.71
12" x 6"	ER@2.35	Ea	726.00	111.00	1.57	838.57
12" x 8"	ER@2.40	Ea	797.00	114.00	1.61	912.61
12" x 10"	ER@2.45	Ea	859.00	116.00	1.64	976.64
15" x 6"	ER@2.85	Ea	1,170.00	135.00	1.91	1,306.91
15" x 8"	ER@2.85	Ea	1,180.00	135.00	1.91	1,316.91
15" x 10"	ER@2.90	Ea	1,340.00	137.00	1.94	1,478.94
15" x 12"	ER@2.95	Ea	1,350.00	140.00	1.97	1,491.97

DWV cast iron double wye, hub and spigot, gasketed joints

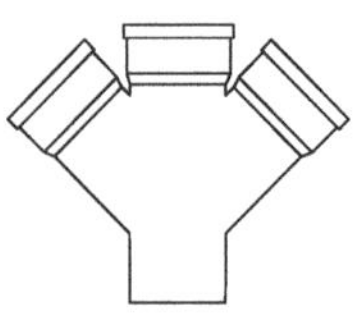

Description	Craft@Hrs	Unit	Material $	Labor $	Equipment $	Total $
2"	P1@.500	Ea	28.60	22.20	—	50.80
3"	P1@.750	Ea	49.60	33.30	—	82.90
4"	P1@.950	Ea	68.80	42.20	—	111.00
6"	ER@1.10	Ea	159.00	52.20	.74	211.94
8"	ER@1.85	Ea	395.00	87.70	1.24	483.94
10"	ER@2.50	Ea	551.00	119.00	1.67	671.67
12"	ER@2.75	Ea	798.00	130.00	1.84	929.84
15"	ER@3.50	Ea	2,410.00	166.00	2.34	2,578.34

DWV cast iron reducing double wye, hub and spigot, gasketed joints

Description	Craft@Hrs	Unit	Material $	Labor $	Equipment $	Total $
3" x 2"	P1@.550	Ea	41.80	24.40	—	66.20
4" x 2"	P1@.650	Ea	51.20	28.90	—	80.10
4" x 3"	P1@.700	Ea	56.80	31.10	—	87.90
6" x 4"	ER@.800	Ea	120.00	37.90	.54	158.44

DWV cast iron reducer, hub and spigot, gasketed joints

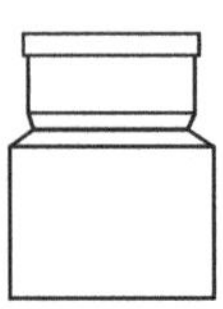

Description	Craft@Hrs	Unit	Material $	Labor $	Equipment $	Total $
3"	P1@.150	Ea	14.70	6.67	—	21.37
4"	P1@.200	Ea	38.20	8.89	—	47.09
6"	ER@.320	Ea	59.60	15.20	.21	75.01
8"	ER@.350	Ea	126.00	16.60	.23	142.83
10"	ER@.570	Ea	266.00	27.00	.38	293.38
12"	ER@.620	Ea	353.00	29.40	.42	382.82
15"	ER@1.12	Ea	654.00	53.10	.75	707.85

Description	Craft@Hrs	Unit	Material $	Labor $	Equipment $	Total $

Gaskets for hub and spigot joints

Description	Craft@Hrs	Unit	Material $	Labor $	Equipment $	Total $
2"	—	Ea	3.73	—	—	3.73
3"	—	Ea	4.93	—	—	4.93
4"	—	Ea	6.28	—	—	6.28
5"	—	Ea	9.63	—	—	9.63
6"	—	Ea	10.10	—	—	10.10
8"	—	Ea	21.90	—	—	21.90
10"	—	Ea	34.20	—	—	34.20
12"	—	Ea	43.50	—	—	43.50
15"	—	Ea	52.00	—	—	52.00

Hanger with swivel assembly

Description	Craft@Hrs	Unit	Material $	Labor $	Equipment $	Total $
2"	P1@.400	Ea	5.99	17.80	—	23.79
3"	P1@.450	Ea	8.15	20.00	—	28.15
4"	P1@.450	Ea	10.10	20.00	—	30.10
5"	P1@.500	Ea	14.70	22.20	—	36.90
6"	P1@.500	Ea	16.60	22.20	—	38.80
8"	P1@.550	Ea	21.70	24.40	—	46.10
10"	P1@.650	Ea	28.60	28.90	—	57.50
12"	P1@.750	Ea	39.40	33.30	—	72.70
15"	P1@.850	Ea	52.30	37.80	—	90.10

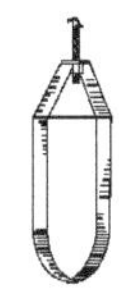

Riser clamp

Description	Craft@Hrs	Unit	Material $	Labor $	Equipment $	Total $
2"	P1@.115	Ea	3.30	5.11	—	8.41
3"	P1@.120	Ea	4.87	5.33	—	10.20
4"	P1@.125	Ea	4.91	5.56	—	10.47
5"	P1@.180	Ea	11.40	8.00	—	19.40
6"	P1@.200	Ea	13.80	8.89	—	22.69
8"	P1@.200	Ea	22.50	8.89	—	31.39
10"	P1@.250	Ea	33.30	11.10	—	44.40
12"	P1@.250	Ea	41.40	11.10	—	52.50

Copper, DWV with Soft-Soldered Joints

Copper drainage tubing is used for non-pressurized above-grade, interior building drain, waste and vent (DWV) systems.

Because of its thin walls, it must be handled carefully and protected from damage during building construction.

This section has been arranged to save the estimator's time by including all normally-used system components such as pipe, fittings, hanger assemblies and riser clamps under one heading. Additional items can be found under "Plumbing and Piping Specialties." The cost estimates in this section are based on the conditions, limitations and wage rates described in the section "How to Use This Book" beginning on page 5.

DWV copper pipe assembly, installed horizontally. Assembly includes fittings, couplings, hangers and rod. Based on a reducing wye and a 45-degree elbow every 12 feet for 1¼" pipe, a reducing wye and a 45-degree elbow every 25 feet for 4" pipe, and hangers spaced to meet plumbing code.

Description	Craft@Hrs	Unit	Material $	Labor $	Equipment $	Total $
1¼"	P1@.130	LF	13.50	5.78	—	19.28
1½"	P1@.132	LF	13.70	5.87	—	19.57
2"	P1@.137	LF	17.70	6.09	—	23.79
3"	P1@.179	LF	31.60	7.96	—	39.56
4"	P1@.202	LF	59.20	8.98	—	68.18

DWV copper pipe assembly, installed risers. Complete installation, including a reducing tee every floor and a riser clamp every other.

Description	Craft@Hrs	Unit	Material $	Labor $	Equipment $	Total $
1¼"	P1@.069	LF	10.60	3.07	—	13.67
1½"	P1@.074	LF	11.90	3.29	—	15.19
2"	P1@.086	LF	16.30	3.82	—	20.12
3"	P1@.110	LF	31.70	4.89	—	36.59
4"	P1@.134	LF	52.90	5.96	—	58.86

DWV copper pipe, soft-soldered joints

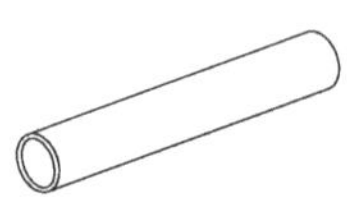

Description	Craft@Hrs	Unit	Material $	Labor $	Equipment $	Total $
1¼"	P1@.040	LF	5.82	1.78	—	7.60
1½"	P1@.045	LF	7.39	2.00	—	9.39
2"	P1@.050	LF	9.72	2.22	—	11.94
3"	P1@.060	LF	15.30	2.67	—	17.97
4"	P1@.070	LF	19.60	3.11	—	22.71

DWV copper 1/8 bend, soft-soldered joints

Description	Craft@Hrs	Unit	Material $	Labor $	Equipment $	Total $
1¼"	P1@.170	Ea	6.60	7.56	—	14.16
1½"	P1@.200	Ea	6.06	8.89	—	14.95
2"	P1@.270	Ea	11.00	12.00	—	23.00
3"	P1@.400	Ea	27.10	17.80	—	44.90
4"	P1@.540	Ea	111.00	24.00	—	135.00

DWV copper ¼ bend, soft-soldered joints

Description	Craft@Hrs	Unit	Material $	Labor $	Equipment $	Total $
1¼"	P1@.170	Ea	7.41	7.56	—	14.97
1½"	P1@.200	Ea	8.46	8.89	—	17.35
2"	P1@.270	Ea	13.60	12.00	—	25.60
3"	P1@.400	Ea	35.80	17.80	—	53.60
4"	P1@.540	Ea	173.00	24.00	—	197.00

DWV copper closet flange, soft-soldered joints

Description	Craft@Hrs	Unit	Material $	Labor $	Equipment $	Total $
4" x 3"	P1@.680	Ea	35.30	30.20	—	65.50
4" x 4"	P1@.980	Ea	43.40	43.60	—	87.00

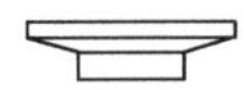

DWV copper P-trap, soft-soldered joints

Description	Craft@Hrs	Unit	Material $	Labor $	Equipment $	Total $
1¼"	P1@.240	Ea	50.00	10.70	—	60.70
1½"	P1@.240	Ea	43.60	10.70	—	54.30
2"	P1@.240	Ea	66.30	10.70	—	77.00
3"	P1@.320	Ea	191.00	14.20	—	205.20

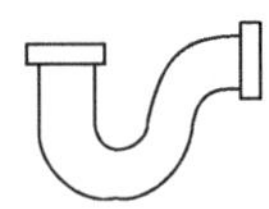

DWV copper sanitary tee (TY), soft-soldered joints

Description	Craft@Hrs	Unit	Material $	Labor $	Equipment $	Total $
1½"	P1@.240	Ea	15.40	10.70	—	26.10
2"	P1@.320	Ea	21.10	14.20	—	35.30
3"	P1@.490	Ea	72.50	21.80	—	94.30
4"	P1@.650	Ea	183.00	28.90	—	211.90

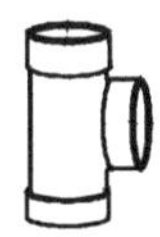

DWV copper sanitary cross, soft-soldered joints

Description	Craft@Hrs	Unit	Material $	Labor $	Equipment $	Total $
1½"	P1@.640	Ea	102.00	28.40	—	130.40
2"	P1@.900	Ea	147.00	40.00	—	187.00
3"	P1@1.36	Ea	416.00	60.50	—	476.50

DWV copper combination wye and 1/8 bend, soft-soldered joints

Description	Craft@Hrs	Unit	Material $	Labor $	Equipment $	Total $
1½"	P1@.240	Ea	32.40	10.70	—	43.10
2"	P1@.320	Ea	55.30	14.20	—	69.50
3"	P1@.490	Ea	183.00	21.80	—	204.80
4"	P1@.650	Ea	412.00	28.90	—	440.90

DWV copper wye, soft-soldered joints

Description	Craft@Hrs	Unit	Material $	Labor $	Equipment $	Total $
1½"	P1@.240	Ea	25.30	10.70	—	36.00
2"	P1@.320	Ea	36.10	14.20	—	50.30
3"	P1@.490	Ea	90.50	21.80	—	112.30
4"	P1@.650	Ea	183.00	28.90	—	211.90

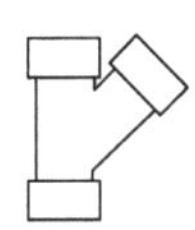

DWV copper cleanout with threaded plug

Description	Craft@Hrs	Unit	Material $	Labor $	Equipment $	Total $
1¼"	P1@.100	Ea	12.80	4.45	—	17.25
1½"	P1@.160	Ea	19.10	7.11	—	26.21
2"	P1@.230	Ea	24.90	10.20	—	35.10

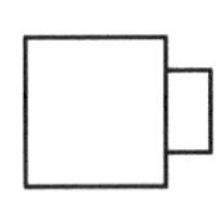

Copper, DWV with Soft-Soldered Joints

DWV copper female adapter

Description	Craft@Hrs	Unit	Material $	Labor $	Equipment $	Total $
1½"	P1@.130	Ea	9.85	5.78	—	15.63
2"	P1@.170	Ea	18.80	7.56	—	26.36
3"	P1@.230	Ea	74.80	10.20	—	85.00
4"	P1@.300	Ea	128.00	13.30	—	141.30

DWV copper male adapter

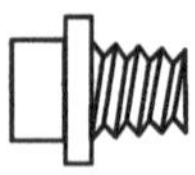

Description	Craft@Hrs	Unit	Material $	Labor $	Equipment $	Total $
1½"	P1@.130	Ea	11.10	5.78	—	16.88
2"	P1@.170	Ea	13.30	7.56	—	20.86
3"	P1@.230	Ea	65.70	10.20	—	75.90
4"	P1@.300	Ea	110.00	13.30	—	123.30

DWV copper soil adapter (copper x MJ)

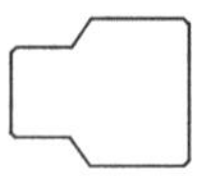

Description	Craft@Hrs	Unit	Material $	Labor $	Equipment $	Total $
2"	P1@.170	Ea	16.00	7.56	—	23.56
3"	P1@.250	Ea	32.80	11.10	—	43.90
4"	P1@.320	Ea	61.20	14.20	—	75.40

DWV copper test cap, soft-soldered joint

Description	Craft@Hrs	Unit	Material $	Labor $	Equipment $	Total $
1½"	P1@.100	Ea	.38	4.45	—	4.83
2"	P1@.130	Ea	.54	5.78	—	6.32
3"	P1@.200	Ea	.69	8.89	—	9.58
4"	P1@.250	Ea	1.43	11.10	—	12.53

DWV copper test tee, soft-soldered joints

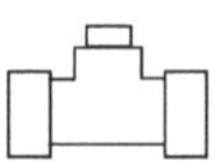

Description	Craft@Hrs	Unit	Material $	Labor $	Equipment $	Total $
1½"	P1@.200	Ea	36.80	8.89	—	45.69
2"	P1@.270	Ea	47.90	12.00	—	59.90
3"	P1@.400	Ea	129.00	17.80	—	146.80

DWV copper fitting reducer, soft-soldered joints

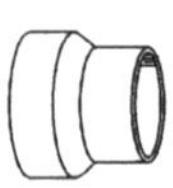

Description	Craft@Hrs	Unit	Material $	Labor $	Equipment $	Total $
1½"	P1@.100	Ea	6.06	4.45	—	10.51
2"	P1@.140	Ea	7.78	6.22	—	14.00
3"	P1@.200	Ea	27.10	8.89	—	35.99
4"	P1@.270	Ea	43.20	12.00	—	55.20

DWV copper coupling, soft-soldered joints

Description	Craft@Hrs	Unit	Material $	Labor $	Equipment $	Total $
1¼"	P1@.170	Ea	3.24	7.56	—	10.80
1½"	P1@.200	Ea	3.73	8.89	—	12.62
2"	P1@.270	Ea	5.31	12.00	—	17.31
3"	P1@.400	Ea	9.85	17.80	—	27.65
4"	P1@.540	Ea	38.70	24.00	—	62.70

Description	Craft@Hrs	Unit	Material $	Labor $	Equipment $	Total $

Hanger with swivel assembly

Description	Craft@Hrs	Unit	Material $	Labor $	Equipment $	Total $
1¼"	P1@.300	Ea	7.71	13.30	—	21.01
1½"	P1@.300	Ea	8.29	13.30	—	21.59
2"	P1@.300	Ea	8.65	13.30	—	21.95
3"	P1@.350	Ea	11.80	15.60	—	27.40
4"	P1@.350	Ea	14.60	15.60	—	30.20

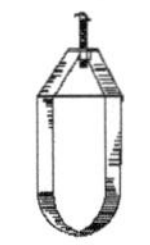

Riser clamp

Description	Craft@Hrs	Unit	Material $	Labor $	Equipment $	Total $
1¼"	P1@.105	Ea	8.53	4.67	—	13.20
1½"	P1@.110	Ea	8.98	4.89	—	13.87
2"	P1@.115	Ea	9.57	5.11	—	14.68
3"	P1@.120	Ea	10.00	5.33	—	15.33
4"	P1@.125	Ea	10.90	5.56	—	16.46

ABS, DWV with Solvent-Weld Joints

ABS (Acrylonitrile Butadiene Styrene) ASTM D 2661-73 is made from various grades of rubber-based resins and is primarily used for non-pressurized drain, waste and vent (DWV) systems.

ABS can also be used for low-temperature (100 degree F. maximum) water distribution systems, but consult the manufacturers for the proper ASTM type, maximum pressure limitations and recommended joint solvent.

The current Uniform Plumbing Code (UPC) states in Sections 401 and 503 that "ABS piping systems shall be limited to those structures where combustible construction is allowed." This does not preclude its use for underground DWV systems.

Not every jurisdiction allows the use of Cell core pipe. Cell core, or Foam core pipe is regulated by ASTM-628. Solid core pipe is regulated by ASTM-6221.

This section has been arranged to save the estimator's time by including all normally-used system components such as pipe, fittings, hanger assemblies and riser clamps under one heading. The cost estimates in this section are based on the conditions, limitations and wage rates described in the section "How to Use This Book" beginning on page 5.

ABS DWV pipe with solvent-weld joints – Solid Core

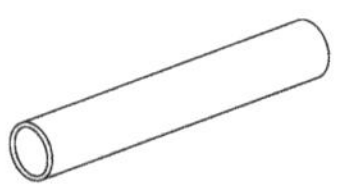

Description	Craft@Hrs	Unit	Material $	Labor $	Equipment $	Total $
1½"	P1@.020	LF	1.33	.89	—	2.22
2"	P1@.025	LF	2.09	1.11	—	3.20
3"	P1@.030	LF	3.33	1.33	—	4.66
4"	P1@.040	LF	5.42	1.78	—	7.20
6"	P1@.055	LF	10.00	2.44	—	12.44

ABS DWV pipe with solvent-weld joints – Cell Core

Description	Craft@Hrs	Unit	Material $	Labor $	Equipment $	Total $
1½"	P1@.020	LF	1.03	.89	—	1.92
2"	P1@.025	LF	1.61	1.11	—	2.72
3"	P1@.030	LF	2.56	1.33	—	3.89
4"	P1@.040	LF	4.17	1.78	—	5.95
6"	P1@.055	LF	7.74	2.44	—	10.18

ABS DWV 1/8 bend with solvent-weld joints

Description	Craft@Hrs	Unit	Material $	Labor $	Equipment $	Total $
1½"	P1@.120	Ea	.92	5.33	—	6.25
2"	P1@.125	Ea	1.64	5.56	—	7.20
3"	P1@.175	Ea	3.97	7.78	—	11.75
4"	P1@.250	Ea	7.50	11.10	—	18.60
6"	P1@.325	Ea	59.50	14.40	—	73.90

ABS DWV 1/8 street bend with solvent-weld joints

Description	Craft@Hrs	Unit	Material $	Labor $	Equipment $	Total $
1½"	P1@.060	Ea	1.00	2.67	—	3.67
2"	P1@.065	Ea	1.92	2.89	—	4.81
3"	P1@.088	Ea	4.21	3.91	—	8.12
4"	P1@.125	Ea	11.60	5.56	—	17.16
6"	P1@.163	Ea	58.30	7.25	—	65.55

ABS DWV ¼ bend with solvent-weld joints

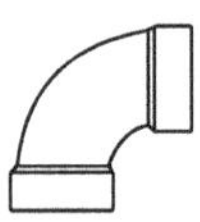

Description	Craft@Hrs	Unit	Material $	Labor $	Equipment $	Total $
1½"	P1@.120	Ea	1.10	5.33	—	6.43
2"	P1@.125	Ea	1.89	5.56	—	7.45
3"	P1@.175	Ea	4.75	7.78	—	12.53
4"	P1@.250	Ea	10.80	11.10	—	21.90
6"	P1@.325	Ea	61.20	14.40	—	75.60

Description	Craft@Hrs	Unit	Material $	Labor $	Equipment $	Total $

ABS DWV closet bend with solvent-weld joints

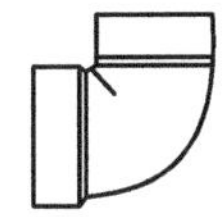

3" x 4"	P1@.250	Ea	16.10	11.10	—	27.20

ABS DWV closet flange with solvent-weld joints

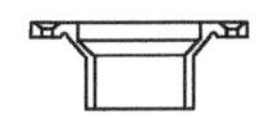

4"	P1@.125	Ea	7.25	5.56	—	12.81

ABS DWV P-trap with solvent-weld joints

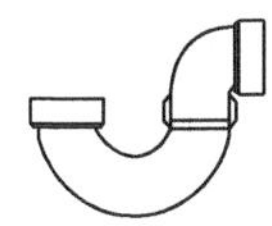

1½"	P1@.150	Ea	3.34	6.67	—	10.01
2"	P1@.190	Ea	8.23	8.45	—	16.68
3"	P1@.260	Ea	19.60	11.60	—	31.20
4"	P1@.350	Ea	37.90	15.60	—	53.50

ABS DWV sanitary tee (TY) with solvent-weld joints

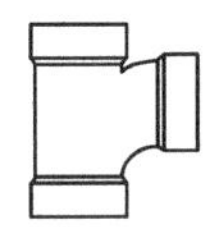

1½"	P1@.150	Ea	1.61	6.67	—	8.28
2"	P1@.190	Ea	3.70	8.45	—	12.15
3"	P1@.260	Ea	7.11	11.60	—	18.71
4"	P1@.350	Ea	15.00	15.60	—	30.60
6"	P1@.450	Ea	68.60	20.00	—	88.60

ABS DWV reducing sanitary tee (TY) with solvent-weld joints

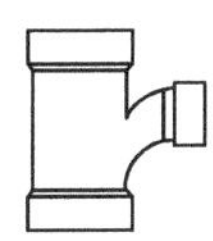

2" x 1½"	P1@.160	Ea	2.68	7.11	—	9.79
3" x 1½"	P1@.210	Ea	4.56	9.33	—	13.89
3" x 2"	P1@.235	Ea	5.09	10.40	—	15.49
4" x 2"	P1@.300	Ea	14.70	13.30	—	28.00
4" x 3"	P1@.325	Ea	17.50	14.40	—	31.90

ABS DWV double sanitary tee (TY) with solvent-weld joints

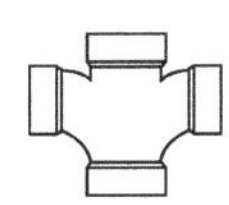

1½"	P1@.210	Ea	4.89	9.33	—	14.22
2"	P1@.250	Ea	7.37	11.10	—	18.47
3"	P1@.350	Ea	11.60	15.60	—	27.20
4"	P1@.475	Ea	45.90	21.10	—	67.00

ABS DWV combination wye and 1/8 bend with solvent-weld joints

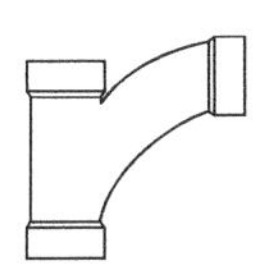

1½"	P1@.150	Ea	9.66	6.67	—	16.33
2"	P1@.190	Ea	12.60	8.45	—	21.05
3"	P1@.260	Ea	16.80	11.60	—	28.40
4"	P1@.350	Ea	29.50	15.60	—	45.10

ABS DWV combination reducing wye and 1/8 bend with solvent-weld joints

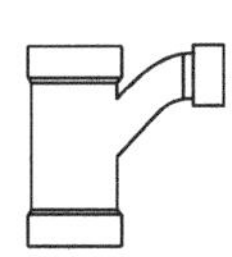

2" x 1½"	P1@.160	Ea	14.00	7.11	—	21.11
3" x 1½"	P1@.210	Ea	19.00	9.33	—	28.33
3" x 2"	P1@.235	Ea	19.00	10.40	—	29.40
4" x 2"	P1@.300	Ea	26.90	13.30	—	40.20
4" x 3"	P1@.325	Ea	31.80	14.40	—	46.20

ABS DWV bushing with solvent-weld joints

Description	Craft@Hrs	Unit	Material $	Labor $	Equipment $	Total $
2" x 1½"	P1@.062	Ea	1.21	2.76	—	3.97
3" x 1½"	P1@.088	Ea	3.69	3.91	—	7.60
3" x 2"	P1@.088	Ea	3.69	3.91	—	7.60
4" x 2"	P1@.125	Ea	6.00	5.56	—	11.56
4" x 3"	P1@.125	Ea	6.00	5.56	—	11.56
6" x 4"	P1@.163	Ea	29.80	7.25	—	37.05

ABS DWV FIP adapter

Description	Craft@Hrs	Unit	Material $	Labor $	Equipment $	Total $
1½"	P1@.080	Ea	1.26	3.56	—	4.82
2"	P1@.085	Ea	2.33	3.78	—	6.11
3"	P1@.155	Ea	5.50	6.89	—	12.39
4"	P1@.220	Ea	10.50	9.78	—	20.28
6"	P1@.285	Ea	40.30	12.70	—	53.00

ABS DWV test cap

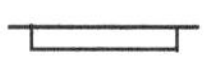

Description	Craft@Hrs	Unit	Material $	Labor $	Equipment $	Total $
1½"	P1@.030	Ea	.63	1.33	—	1.96
2"	P1@.040	Ea	1.02	1.78	—	2.80
3"	P1@.050	Ea	1.62	2.22	—	3.84
4"	P1@.090	Ea	2.53	4.00	—	6.53

ABS DWV cleanout plug

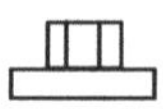

Description	Craft@Hrs	Unit	Material $	Labor $	Equipment $	Total $
1½"	P1@.090	Ea	2.94	4.00	—	6.94
2"	P1@.120	Ea	4.31	5.33	—	9.64
3"	P1@.240	Ea	7.67	10.70	—	18.37
4"	P1@.320	Ea	16.50	14.20	—	30.70
6"	P1@.450	Ea	53.00	20.00	—	73.00

ABS DWV cleanout tee with solvent-weld joints

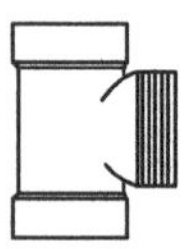

Description	Craft@Hrs	Unit	Material $	Labor $	Equipment $	Total $
1½"	P1@.120	Ea	4.56	5.33	—	9.89
2"	P1@.125	Ea	7.37	5.56	—	12.93
3"	P1@.175	Ea	12.20	7.78	—	19.98
4"	P1@.250	Ea	24.90	11.10	—	36.00

ABS DWV wye with solvent-weld joints

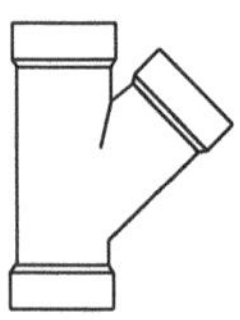

Description	Craft@Hrs	Unit	Material $	Labor $	Equipment $	Total $
1½"	P1@.150	Ea	2.13	6.67	—	8.80
2"	P1@.190	Ea	3.97	8.45	—	12.42
3"	P1@.260	Ea	7.37	11.60	—	18.97
4"	P1@.350	Ea	17.00	15.60	—	32.60
6"	P1@.450	Ea	71.70	20.00	—	91.70

ABS DWV reducing wye with solvent-weld joints

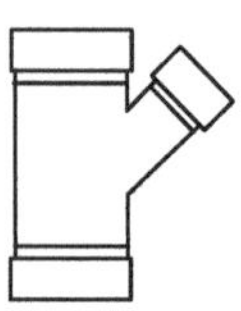

Description	Craft@Hrs	Unit	Material $	Labor $	Equipment $	Total $
2" x 1½"	P1@.160	Ea	3.70	7.11	—	10.81
3" x 1½"	P1@.210	Ea	5.15	9.33	—	14.48
3" x 2"	P1@.235	Ea	5.75	10.40	—	16.15
4" x 2"	P1@.300	Ea	12.30	13.30	—	25.60
4" x 3"	P1@.325	Ea	15.50	14.40	—	29.90
6" x 4"	P1@.400	Ea	59.50	17.80	—	77.30

Description	Craft@Hrs	Unit	Material $	Labor $	Equipment $	Total $

ABS DWV reducer with solvent-weld joints

2" x 1½"	P1@.125	Ea	.84	5.56	—	6.40
3" x 1½"	P1@.150	Ea	3.69	6.67	—	10.36
3" x 2"	P1@.150	Ea	3.69	6.67	—	10.36
4" x 2"	P1@.185	Ea	6.00	8.22	—	14.22
4" x 3"	P1@.210	Ea	5.92	9.33	—	15.25

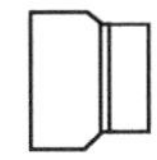

ABS DWV coupling with solvent-weld joints

1½"	P1@.120	Ea	.70	5.33	—	6.03
2"	P1@.125	Ea	1.04	5.56	—	6.60
3"	P1@.175	Ea	2.43	7.78	—	10.21
4"	P1@.250	Ea	4.42	11.10	—	15.52
6"	P1@.325	Ea	26.10	14.40	—	40.50

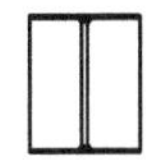

Hanger with swivel assembly

1½"	P1@.300	Ea	5.74	13.30	—	19.04
2"	P1@.300	Ea	5.99	13.30	—	19.29
3"	P1@.350	Ea	8.15	15.60	—	23.75
4"	P1@.350	Ea	10.10	15.60	—	25.70
6"	P1@.450	Ea	11.10	20.00	—	31.10

Riser clamp

1½"	P1@.110	Ea	6.22	4.89	—	11.11
2"	P1@.115	Ea	6.62	5.11	—	11.73
3"	P1@.120	Ea	6.93	5.33	—	12.26
4"	P1@.125	Ea	7.53	5.56	—	13.09
6"	P1@.130	Ea	9.61	5.78	—	15.39

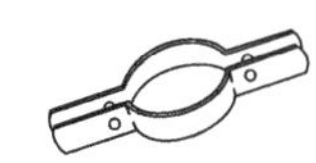

Galvanized steel pipe sleeves

2"	P1@.130	Ea	7.11	5.78	—	12.89
2½"	P1@.150	Ea	7.25	6.67	—	13.92
3"	P1@.180	Ea	7.40	8.00	—	15.40
4"	P1@.220	Ea	8.43	9.78	—	18.21
5"	P1@.250	Ea	10.50	11.10	—	21.60
6"	P1@.270	Ea	11.30	12.00	—	23.30

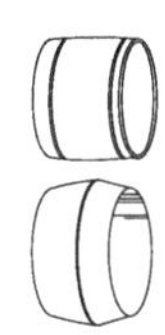

Lead roof pipe flashing

1½"	P1@.250	Ea	35.00	11.10	—	46.10
2"	P1@.250	Ea	46.10	11.10	—	57.20
3"	P1@.250	Ea	53.40	11.10	—	64.50
4"	P1@.250	Ea	58.60	11.10	—	69.70
6"	P1@.350	Ea	75.00	15.60	—	90.60

PVC, DWV with Solvent-Weld Joints

PVC (Polyvinyl Chloride) meets ASTM – D3034, D3212, F1336, F679 & CSA 182.2 (CSA – Canadian Standards Association – similar to UL – Underwriters Laboratories). The applications for PVC DWV have increased in number due to the unique flame-spread characteristics of their compounds, as well as the development of new fire stop devices for use with combustible piping systems.

PVC will not rust, pit or degrade when exposed to moisture and is extremely resistant to a broad range of corrosive agents.

PVC DWV is suitable for sewer, vent and storm water drainage applications in both below- and above-grade applications.

When penetrating a vertical or horizontal fire separation (fire-rated wall or floor) with combustible piping, certified fire stop devices must be used in compliance with local and national codes.

PVC DWV is not permitted in a ceiling space used as a return air plenum and is not permitted in a vertical shaft. PVC DWV has a flame-spread rating of 15.

This section has been arranged to save the estimator's time by including all normally-used system components such as pipe, fittings, hanger assemblies and riser clamps under one heading. The cost estimates in this section are based on the conditions, limitations and wage rates described in the section "How to Use This Book" beginning on page 5.

Description	Craft@Hrs	Unit	Material $	Labor $	Equipment $	Total $
PVC DWV pipe with solvent-weld joints						
1½"	P1@.021	LF	1.15	.93	—	2.08
2"	P1@.026	LF	1.84	1.16	—	3.00
3"	P1@.032	LF	3.91	1.42	—	5.33
4"	P1@.042	LF	4.82	1.87	—	6.69
6"	P1@.058	LF	8.66	2.58	—	11.24
PVC DWV 90-degree elbow with solvent-weld joints						
1½"	P1@.126	Ea	1.76	5.60	—	7.36
2"	P1@.131	Ea	2.65	5.82	—	8.47
3"	P1@.184	Ea	6.95	8.18	—	15.13
4"	P1@.263	Ea	13.70	11.70	—	25.40
6"	P1@.341	Ea	57.70	15.20	—	72.90
PVC DWV 90-degree street elbow with solvent-weld joints						
1½"	P1@.063	Ea	1.85	2.80	—	4.65
2"	P1@.068	Ea	3.15	3.02	—	6.17
3"	P1@.092	Ea	9.22	4.09	—	13.31
4"	P1@.131	Ea	22.30	5.82	—	28.12
6"	P1@.171	Ea	81.70	7.60	—	89.30
PVC DWV 45-degree elbow with solvent-weld joints						
1½"	P1@.126	Ea	1.48	5.60	—	7.08
2"	P1@.131	Ea	1.62	5.82	—	7.44
3"	P1@.184	Ea	4.07	8.18	—	12.25
4"	P1@.263	Ea	10.30	11.70	—	22.00
6"	P1@.341	Ea	62.50	15.20	—	77.70

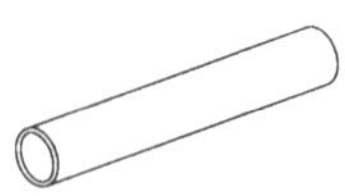

PVC DWV wye with solvent-weld joints

Description	Craft@Hrs	Unit	Material $	Labor $	Equipment $	Total $
1½"	P1@.158	Ea	2.73	7.02	—	9.75
2"	P1@.200	Ea	5.59	8.89	—	14.48
3"	P1@.273	Ea	9.21	12.10	—	21.31
4"	P1@.368	Ea	15.10	16.40	—	31.50
6"	P1@.473	Ea	59.20	21.00	—	80.20

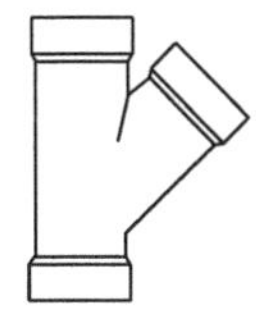

PVC DWV reducing wye with solvent-weld joints

Description	Craft@Hrs	Unit	Material $	Labor $	Equipment $	Total $
2" x 1½"	P1@.168	Ea	4.84	7.47	—	12.31
3" x 1½"	P1@.221	Ea	5.56	9.82	—	15.38
3" x 2"	P1@.247	Ea	7.97	11.00	—	18.97
4" x 2"	P1@.315	Ea	15.50	14.00	—	29.50
4" x 3"	P1@.341	Ea	18.30	15.20	—	33.50
6" x 3"	P1@.473	Ea	67.50	21.00	—	88.50
6" x 4"	P1@.473	Ea	71.80	21.00	—	92.80

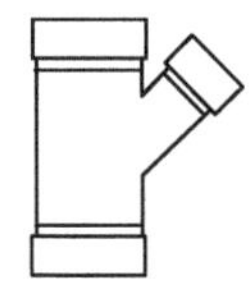

PVC DWV sanitary tee (TY) with solvent-weld joints

Description	Craft@Hrs	Unit	Material $	Labor $	Equipment $	Total $
1½"	P1@.158	Ea	1.94	7.02	—	8.96
2"	P1@.200	Ea	3.31	8.89	—	12.20
3"	P1@.273	Ea	7.10	12.10	—	19.20
4"	P1@.368	Ea	13.60	16.40	—	30.00
6"	P1@.473	Ea	74.40	21.00	—	95.40

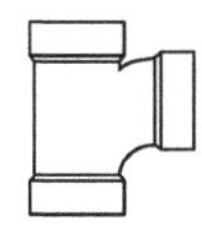

PVC DWV reducing sanitary tee (TY) with solvent-weld joints

Description	Craft@Hrs	Unit	Material $	Labor $	Equipment $	Total $
2" x 1½"	P1@.168	Ea	2.57	7.47	—	10.04
3" x 1½"	P1@.221	Ea	4.43	9.82	—	14.25
3" x 2"	P1@.247	Ea	5.94	11.00	—	16.94
4" x 2"	P1@.315	Ea	13.00	14.00	—	27.00
4" x 3"	P1@.341	Ea	15.00	15.20	—	30.20
6" x 3"	P1@.473	Ea	57.00	21.00	—	78.00
6" x 4"	P1@.473	Ea	57.00	21.00	—	78.00

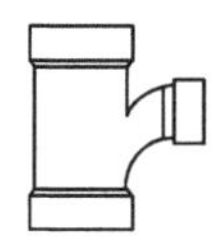

PVC DWV double wye with solvent-weld joints

Description	Craft@Hrs	Unit	Material $	Labor $	Equipment $	Total $
2" x 1½"	P1@.263	Ea	7.27	11.70	—	18.97
3" x 2"	P1@.368	Ea	14.80	16.40	—	31.20
4" x 3"	P1@.473	Ea	48.20	21.00	—	69.20
6" x 4"	P1@.578	Ea	58.20	25.70	—	83.90

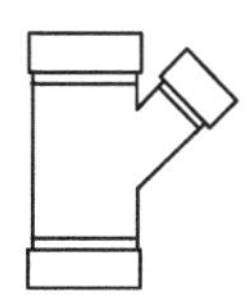

PVC DWV closet flange with solvent-weld joints, adjustable with test plate

Description	Craft@Hrs	Unit	Material $	Labor $	Equipment $	Total $
4" x 3"	P1@.131	Ea	12.50	5.82	—	18.32

PVC DWV closet flange with solvent-weld joints, non-adjustable with test plate

Description	Craft@Hrs	Unit	Material $	Labor $	Equipment $	Total $
4" x 3"	P1@.131	Ea	9.54	5.82	—	15.36

Description	Craft@Hrs	Unit	Material $	Labor $	Equipment $	Total $

PVC DWV closet flange with solvent-weld joints, non-adjustable without test plate

4" x 3"	P1@.131	Ea	8.65	5.82	—	14.47

PVC DWV closet flange with solvent-weld joints, adjustable, offset without test plate

4" x 3"	P1@.170	Ea	18.60	7.56	—	26.16

PVC DWV P-trap with solvent-weld joints

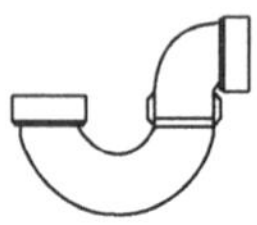

1½"	P1@.160	Ea	4.43	7.11	—	11.54
2"	P1@.200	Ea	9.92	8.89	—	18.81
3"	P1@.285	Ea	23.20	12.70	—	35.90
4"	P1@.370	Ea	45.70	16.40	—	62.10
6"	P1@.580	Ea	176.00	25.80	—	201.80

PVC DWV bushing with solvent-weld joints

2" x 1½"	P1@.062	Ea	.83	2.76	—	3.59
3" x 2"	P1@.088	Ea	2.46	3.91	—	6.37
4" x 3"	P1@.125	Ea	5.15	5.56	—	10.71
6" x 4"	P1@.163	Ea	26.60	7.25	—	33.85

PVC DWV FIP adapter

1½"	P1@.080	Ea	1.21	3.56	—	4.77
2"	P1@.085	Ea	1.98	3.78	—	5.76
3"	P1@.155	Ea	4.78	6.89	—	11.67
4"	P1@.220	Ea	9.35	9.78	—	19.13
6"	P1@.285	Ea	42.00	12.70	—	54.70

PVC DWV MIP adapter

1½"	P1@.080	Ea	1.35	3.56	—	4.91
2"	P1@.085	Ea	2.05	3.78	—	5.83
3"	P1@.155	Ea	4.26	6.89	—	11.15
4"	P1@.220	Ea	11.50	9.78	—	21.28
6"	P1@.285	Ea	20.90	12.70	—	33.60

PVC DWV coupling with solvent-weld joints

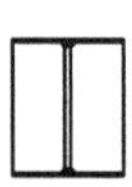

1½"	P1@.120	Ea	.65	5.33	—	5.98
2"	P1@.125	Ea	.93	5.56	—	6.49
3"	P1@.175	Ea	2.18	7.78	—	9.96
4"	P1@.250	Ea	4.57	11.10	—	15.67
6"	P1@.325	Ea	22.30	14.40	—	36.70

Description	Craft@Hrs	Unit	Material $	Labor $	Equipment $	Total $

PVC DWV reducer with solvent-weld joints

2" x 1½"	P1@.125	Ea	1.38	5.56	—	6.94
3" x 2"	P1@.150	Ea	4.16	6.67	—	10.83
4" x 3"	P1@.185	Ea	5.43	8.22	—	13.65
6" x 4"	P1@.210	Ea	55.00	9.33	—	64.33

PVC DWV in-line cleanout with solvent-weld joints

1½"	P1@.120	Ea	5.64	5.33	—	10.97
2"	P1@.125	Ea	8.86	5.56	—	14.42
3"	P1@.175	Ea	13.00	7.78	—	20.78
4"	P1@.250	Ea	27.10	11.10	—	38.20
6" x 4"	P1@.250	Ea	92.10	11.10	—	103.20
6"	P1@.250	Ea	92.10	11.10	—	103.20

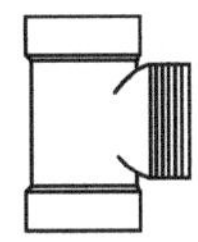

PVC DWV end cleanout

1½"	P1@.090	Ea	2.48	4.00	—	6.48
2"	P1@.120	Ea	3.70	5.33	—	9.03
3"	P1@.240	Ea	7.51	10.70	—	18.21
4"	P1@.320	Ea	13.90	14.20	—	28.10
6"	P1@.450	Ea	48.80	20.00	—	68.80

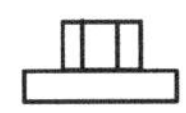

PVC DWV fitting end cleanout

1½"	P1@.090	Ea	2.35	4.00	—	6.35
2"	P1@.120	Ea	4.29	5.33	—	9.62
3"	P1@.240	Ea	5.00	10.70	—	15.70
4"	P1@.320	Ea	10.70	14.20	—	24.90
6"	P1@.450	Ea	55.60	20.00	—	75.60

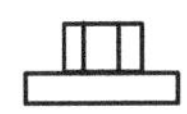

Hanger with swivel assembly

1½"	P1@.300	Ea	5.74	13.30	—	19.04
2"	P1@.300	Ea	5.99	13.30	—	19.29
3"	P1@.350	Ea	8.15	15.60	—	23.75
4"	P1@.350	Ea	10.10	15.60	—	25.70
6"	P1@.450	Ea	11.10	20.00	—	31.10

Riser clamp

1½"	P1@.110	Ea	6.22	4.89	—	11.11
2"	P1@.115	Ea	6.62	5.11	—	11.73
3"	P1@.120	Ea	6.93	5.33	—	12.26
4"	P1@.125	Ea	7.53	5.56	—	13.09
6"	P1@.130	Ea	9.61	5.78	—	15.39

Description	Craft@Hrs	Unit	Material $	Labor $	Equipment $	Total $

Galvanized steel pipe sleeves

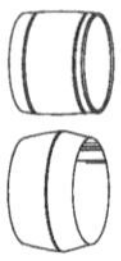

Description	Craft@Hrs	Unit	Material $	Labor $	Equipment $	Total $
2"	P1@.130	Ea	7.11	5.78	—	12.89
2½"	P1@.150	Ea	7.25	6.67	—	13.92
3"	P1@.180	Ea	7.40	8.00	—	15.40
4"	P1@.220	Ea	8.43	9.78	—	18.21
5"	P1@.250	Ea	10.50	11.10	—	21.60
6"	P1@.270	Ea	11.30	12.00	—	23.30

Lead roof pipe flashing

Description	Craft@Hrs	Unit	Material $	Labor $	Equipment $	Total $
1½"	P1@.400	Ea	35.90	17.80	—	53.70
2"	P1@.450	Ea	47.40	20.00	—	67.40
3"	P1@.500	Ea	54.70	22.20	—	76.90
4"	P1@.550	Ea	60.50	24.40	—	84.90

PVC (Polyvinyl Chloride) is polymerized vinyl chloride produced from acetylene and anhydrous hydrochloric acid.

PVC sewer pipe (ASTM D-3034) 10' lay length, is for underground installation and is used primarily to convey sewage and rain water. It is of the bell & spigot type with joints made water-tight by means of an elastomeric gasket (O-ring) factory-installed in the hub end of the pipe and fittings.

This section includes pipe and all commonly-used fittings. Additional items can be found under "Plumbing and Piping Specialties." The cost estimates in this section are based on the conditions, limitations, and wage rates described in the section "How to Use This Book" beginning on page 5.

PVC SDR35 sewer pipe with bell and spigot gasketed joints

Description	Craft@Hrs	Unit	Material $	Labor $	Equipment $	Total $
4"	P1@.050	LF	4.07	2.22	—	6.29
6"	P1@.070	LF	8.95	3.11	—	12.06
8"	P1@.090	LF	14.10	4.00	—	18.10
10"	P1@.110	LF	21.20	4.89	—	26.09
12"	P1@.140	LF	26.70	6.22	—	32.92
15"	P1@.180	LF	39.30	8.00	—	47.30
18"	P1@.230	LF	61.40	10.20	—	71.60

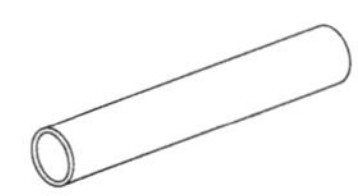

PVC sewer pipe 1/16 bend B x B with bell and spigot gasketed joints

4"	P1@.320	Ea	20.80	14.20	—	35.00
6"	P1@.400	Ea	40.70	17.80	—	58.50
8"	P1@.480	Ea	119.00	21.30	—	140.30
10"	P1@.800	Ea	324.00	35.60	—	359.60
12"	P1@1.06	Ea	427.00	47.10	—	474.10
15"	P1@1.34	Ea	1,120.00	59.60	—	1,179.60
18"	P1@1.60	Ea	1,620.00	71.10	—	1,691.10

PVC sewer pipe 1/16 bend B x S with bell and spigot gasketed joints

4"	P1@.320	Ea	19.60	14.20	—	33.80
6"	P1@.400	Ea	38.70	17.80	—	56.50
8"	P1@.480	Ea	124.00	21.30	—	145.30
10"	P1@.800	Ea	319.00	35.60	—	354.60
12"	P1@1.06	Ea	413.00	47.10	—	460.10
15"	P1@1.34	Ea	900.00	59.60	—	959.60
18"	P1@1.60	Ea	1,380.00	71.10	—	1,451.10

PVC sewer pipe 1/8 bend B x B with bell and spigot gasketed joints

4"	P1@.320	Ea	20.90	14.20	—	35.10
6"	P1@.400	Ea	41.90	17.80	—	59.70
8"	P1@.480	Ea	112.00	21.30	—	133.30
10"	P1@.800	Ea	304.00	35.60	—	339.60
12"	P1@1.06	Ea	448.00	47.10	—	495.10
15"	P1@1.34	Ea	992.00	59.60	—	1,051.60
18"	P1@1.60	Ea	1,580.00	71.10	—	1,651.10

PVC, DWV with Gasketed Bell and Spigot Joints

PVC sewer pipe 1/8 bend B x S with bell and spigot gasketed joints

Description	Craft@Hrs	Unit	Material $	Labor $	Equipment $	Total $
4"	P1@.320	Ea	18.70	14.20	—	32.90
6"	P1@.400	Ea	37.50	17.80	—	55.30
8"	P1@.480	Ea	112.00	21.30	—	133.30
10"	P1@.800	Ea	300.00	35.60	—	335.60
12"	P1@1.06	Ea	434.00	47.10	—	481.10
15"	P1@1.34	Ea	777.00	59.60	—	836.60
18"	P1@1.60	Ea	1,310.00	71.10	—	1,381.10

PVC sewer pipe ¼ bend B x B with bell and spigot gasketed joints

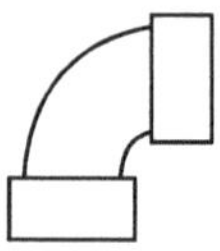

Description	Craft@Hrs	Unit	Material $	Labor $	Equipment $	Total $
4"	P1@.320	Ea	29.90	14.20	—	44.10
6"	P1@.400	Ea	55.40	17.80	—	73.20
8"	P1@.480	Ea	149.00	21.30	—	170.30
10"	P1@.800	Ea	514.00	35.60	—	549.60
12"	P1@1.06	Ea	659.00	47.10	—	706.10
15"	P1@1.34	Ea	1,400.00	59.60	—	1,459.60
18"	P1@1.60	Ea	2,350.00	71.10	—	2,421.10

PVC sewer pipe ¼ bend B x S with bell and spigot gasketed joints

Description	Craft@Hrs	Unit	Material $	Labor $	Equipment $	Total $
4"	P1@.320	Ea	27.30	14.20	—	41.50
6"	P1@.400	Ea	58.20	17.80	—	76.00
8"	P1@.480	Ea	158.00	21.30	—	179.30
10"	P1@.800	Ea	509.00	35.60	—	544.60
12"	P1@1.06	Ea	641.00	47.10	—	688.10
15"	P1@1.34	Ea	1,250.00	59.60	—	1,309.60
18"	P1@1.60	Ea	2,200.00	71.10	—	2,271.10

PVC sewer pipe coupling B x B with bell and spigot gasketed joints

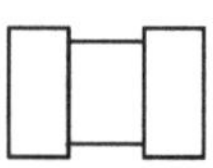

Description	Craft@Hrs	Unit	Material $	Labor $	Equipment $	Total $
4"	P1@.320	Ea	25.60	14.20	—	39.80
6"	P1@.400	Ea	51.40	17.80	—	69.20
8"	P1@.480	Ea	87.50	21.30	—	108.80
10"	P1@.800	Ea	191.00	35.60	—	226.60
12"	P1@1.06	Ea	283.00	47.10	—	330.10
15"	P1@1.34	Ea	582.00	59.60	—	641.60
18"	P1@1.60	Ea	1,130.00	71.10	—	1,201.10

PVC sewer pipe wye B x B x B with bell and spigot gasketed joints

Description	Craft@Hrs	Unit	Material $	Labor $	Equipment $	Total $
4"	P1@.480	Ea	34.80	21.30	—	56.10
6"	P1@.600	Ea	79.70	26.70	—	106.40
8"	P1@.720	Ea	221.00	32.00	—	253.00
10"	P1@1.20	Ea	641.00	53.30	—	694.30
12"	P1@1.59	Ea	909.00	70.70	—	979.70
15"	P1@2.01	Ea	1,580.00	89.30	—	1,669.30
18"	P1@2.40	Ea	2,790.00	107.00	—	2,897.00

Description	Craft@Hrs	Unit	Material $	Labor $	Equipment $	Total $

PVC sewer pipe wye B x S x B with bell and spigot gasketed joints

Description	Craft@Hrs	Unit	Material $	Labor $	Equipment $	Total $
10"	P1@1.20	Ea	633.00	53.30	—	686.30
12"	P1@1.59	Ea	900.00	70.70	—	970.70
15"	P1@2.01	Ea	1,450.00	89.30	—	1,539.30
18"	P1@2.40	Ea	2,560.00	107.00	—	2,667.00

PVC sewer pipe reducing wye B x B x B with bell and spigot gasketed joints

Description	Craft@Hrs	Unit	Material $	Labor $	Equipment $	Total $
6" x 4"	P1@.560	Ea	70.90	24.90	—	95.80
8" x 4"	P1@.640	Ea	106.00	28.40	—	134.40
8" x 6"	P1@.680	Ea	125.00	30.20	—	155.20
10" x 4"	P1@.960	Ea	319.00	42.70	—	361.70
10" x 6"	P1@1.00	Ea	324.00	44.50	—	368.50
10" x 8"	P1@1.04	Ea	514.00	46.20	—	560.20
12" x 4"	P1@1.22	Ea	462.00	54.20	—	516.20
12" x 6"	P1@1.26	Ea	475.00	56.00	—	531.00
12" x 8"	P1@1.30	Ea	709.00	57.80	—	766.80
12" x 10"	P1@1.46	Ea	864.00	64.90	—	928.90
15" x 4"	P1@1.50	Ea	697.00	66.70	—	763.70
15" x 6"	P1@1.54	Ea	824.00	68.50	—	892.50
15" x 8"	P1@1.58	Ea	932.00	70.20	—	1,002.20
15" x 10"	P1@1.74	Ea	1,110.00	77.30	—	1,187.30
15" x 12"	P1@1.87	Ea	1,240.00	83.10	—	1,323.10
18" x 4"	P1@1.76	Ea	1,580.00	78.20	—	1,658.20
18" x 6"	P1@1.80	Ea	1,600.00	80.00	—	1,680.00
18" x 8"	P1@1.84	Ea	1,710.00	81.80	—	1,791.80
18" x 10"	P1@2.00	Ea	1,870.00	88.90	—	1,958.90
18" x 12"	P1@2.13	Ea	2,020.00	94.70	—	2,114.70
18" x 15"	P1@2.27	Ea	2,150.00	101.00	—	2,251.00

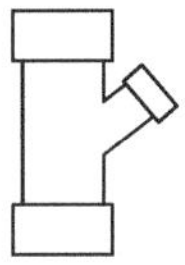

PVC sewer pipe reducing wye B x S x B with bell and spigot gasketed joints

Description	Craft@Hrs	Unit	Material $	Labor $	Equipment $	Total $
10" x 4"	P1@.960	Ea	311.00	42.70	—	353.70
10" x 6"	P1@1.00	Ea	328.00	44.50	—	372.50
10" x 8"	P1@1.04	Ea	388.00	46.20	—	434.20
12" x 4"	P1@1.22	Ea	468.00	54.20	—	522.20
12" x 6"	P1@1.26	Ea	475.00	56.00	—	531.00
12" x 8"	P1@1.30	Ea	548.00	57.80	—	605.80
12" x 10"	P1@1.46	Ea	666.00	64.90	—	730.90
15" x 4"	P1@1.50	Ea	712.00	66.70	—	778.70
15" x 6"	P1@1.54	Ea	805.00	68.50	—	873.50
15" x 8"	P1@1.58	Ea	947.00	70.20	—	1,017.20
15" x 10"	P1@1.74	Ea	1,070.00	77.30	—	1,147.30
15" x 12"	P1@1.87	Ea	1,250.00	83.10	—	1,333.10
18" x 4"	P1@1.76	Ea	1,480.00	78.20	—	1,558.20
18" x 6"	P1@1.80	Ea	1,600.00	80.00	—	1,680.00
18" x 8"	P1@1.84	Ea	1,710.00	81.80	—	1,791.80
18" x 10"	P1@2.00	Ea	1,870.00	88.90	—	1,958.90
18" x 12"	P1@2.13	Ea	2,020.00	94.70	—	2,114.70
18" x 15"	P1@2.27	Ea	2,150.00	101.00	—	2,251.00

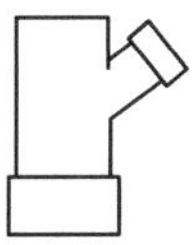

Description	Craft@Hrs	Unit	Material $	Labor $	Equipment $	Total $

PVC sewer pipe tee-wye B x B x B with bell and spigot gasketed joints

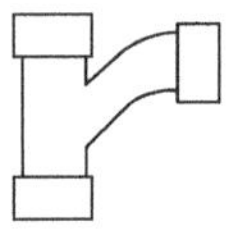

4"	P1@.480	Ea	38.70	21.30	—	60.00
6"	P1@.600	Ea	98.20	26.70	—	124.90
6" x 4"	P1@.560	Ea	83.80	24.90	—	108.70
8" x 4"	P1@.640	Ea	106.00	28.40	—	134.40
8" x 6"	P1@.680	Ea	122.00	30.20	—	152.20

PVC sewer pipe tee B x S x B with bell and spigot gasketed joints

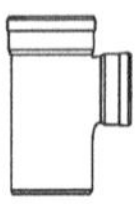

10"	P1@1.20	Ea	535.00	53.30	—	588.30
12"	P1@1.59	Ea	741.00	70.70	—	811.70
15"	P1@2.01	Ea	1,180.00	89.30	—	1,269.30
18"	P1@2.40	Ea	2,220.00	107.00	—	2,327.00

PVC sewer pipe reducing tee B x S x B with bell and spigot gasketed joints

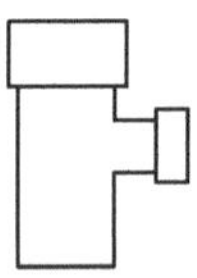

10" x 4"	P1@.960	Ea	339.00	42.70	—	381.70
10" x 6"	P1@1.00	Ea	354.00	44.50	—	398.50
10" x 8"	P1@1.04	Ea	380.00	46.20	—	426.20
12" x 4"	P1@1.22	Ea	416.00	54.20	—	470.20
12" x 6"	P1@1.26	Ea	427.00	56.00	—	483.00
12" x 8"	P1@1.30	Ea	487.00	57.80	—	544.80
12" x 10"	P1@1.46	Ea	617.00	64.90	—	681.90
15" x 4"	P1@1.50	Ea	674.00	66.70	—	740.70
15" x 6"	P1@1.54	Ea	709.00	68.50	—	777.50
15" x 8"	P1@1.58	Ea	726.00	70.20	—	796.20
15" x 10"	P1@1.74	Ea	849.00	77.30	—	926.30
15" x 12"	P1@1.87	Ea	932.00	83.10	—	1,015.10
18" x 4"	P1@1.76	Ea	1,720.00	78.20	—	1,798.20
18" x 6"	P1@1.80	Ea	1,750.00	80.00	—	1,830.00
18" x 8"	P1@1.84	Ea	1,850.00	81.80	—	1,931.80
18" x 10"	P1@2.00	Ea	1,890.00	88.90	—	1,978.90
18" x 12"	P1@2.13	Ea	1,960.00	94.70	—	2,054.70
18" x 15"	P1@2.27	Ea	1,960.00	101.00	—	2,061.00

Description	Craft@Hrs	Unit	Material $	Labor $	Equipment $	Total $

PVC sewer pipe reducer S x B with bell and spigot gasketed joints

Description	Craft@Hrs	Unit	Material $	Labor $	Equipment $	Total $
6" x 4"	P1@.360	Ea	37.50	16.00	—	53.50
8" x 4"	P1@.400	Ea	102.00	17.80	—	119.80
8" x 6"	P1@.440	Ea	114.00	19.60	—	133.60
10" x 4"	P1@.560	Ea	268.00	24.90	—	292.90
10" x 6"	P1@.600	Ea	273.00	26.70	—	299.70
10" x 8"	P1@.640	Ea	330.00	28.40	—	358.40
12" x 4"	P1@.690	Ea	354.00	30.70	—	384.70
12" x 6"	P1@.730	Ea	362.00	32.40	—	394.40
12" x 8"	P1@.770	Ea	416.00	34.20	—	450.20
12" x 10"	P1@.930	Ea	462.00	41.30	—	503.30
15" x 4"	P1@.830	Ea	594.00	36.90	—	630.90
15" x 6"	P1@.870	Ea	617.00	38.70	—	655.70
15" x 8"	P1@.910	Ea	699.00	40.40	—	739.40
15" x 10"	P1@1.07	Ea	699.00	47.60	—	746.60
15" x 12"	P1@1.20	Ea	805.00	53.30	—	858.30
18" x 8"	P1@1.04	Ea	864.00	46.20	—	910.20
18" x 10"	P1@1.20	Ea	875.00	53.30	—	928.30
18" x 12"	P1@1.33	Ea	879.00	59.10	—	938.10
18" x 15"	P1@1.47	Ea	900.00	65.30	—	965.30

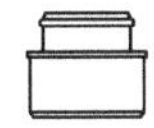

PVC sewer pipe adapter B x S with bell and spigot gasketed joints

Description	Craft@Hrs	Unit	Material $	Labor $	Equipment $	Total $
4"	P1@.320	Ea	22.00	14.20	—	36.20
6"	P1@.400	Ea	40.70	17.80	—	58.50
8"	P1@.480	Ea	104.00	21.30	—	125.30
10"	P1@.800	Ea	128.00	35.60	—	163.60
12"	P1@1.06	Ea	196.00	47.10	—	243.10
15"	P1@1.34	Ea	299.00	59.60	—	358.60
18"	P1@1.60	Ea	653.00	71.10	—	724.10

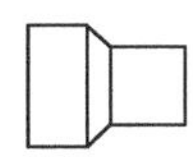

PVC sewer pipe cap with bell and spigot gasketed joints

Description	Craft@Hrs	Unit	Material $	Labor $	Equipment $	Total $
4"	P1@.160	Ea	13.30	7.11	—	20.41
6"	P1@.200	Ea	25.00	8.89	—	33.89
8"	P1@.240	Ea	67.40	10.70	—	78.10
10"	P1@.400	Ea	210.00	17.80	—	227.80
12"	P1@.530	Ea	319.00	23.60	—	342.60
15"	P1@.670	Ea	514.00	29.80	—	543.80
18"	P1@.800	Ea	741.00	35.60	—	776.60

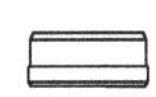

PVC sewer pipe test plug with bell and spigot gasketed joints

Description	Craft@Hrs	Unit	Material $	Labor $	Equipment $	Total $
4"	P1@.160	Ea	10.10	7.11	—	17.21
6"	P1@.200	Ea	15.40	8.89	—	24.29
8"	P1@.240	Ea	55.40	10.70	—	66.10
10"	P1@.400	Ea	184.00	17.80	—	201.80
12"	P1@.530	Ea	220.00	23.60	—	243.60
15"	P1@.670	Ea	434.00	29.80	—	463.80
18"	P1@.800	Ea	601.00	35.60	—	636.60

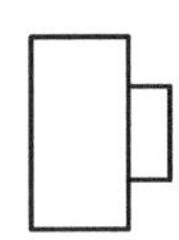

Polypropylene, Schedule 40, with Heat-Fusioned Joints

Polypropylene is used almost exclusively for acid and laboratory drain, waste and vent (DWV) systems.

Joints are made by applying low-voltage current to electrical resistance coils imbedded in propylene collars, which are slipped over the ends of the pipe, then inserted into the hubs of the fittings. Compression clamps are placed around the assemblies and tightened to compress the joints prior to applying electrical current from the power unit to fuse the joint.

This section has been arranged to save the estimator's time by including all normally-used system components such as pipe, fittings, hanger assemblies, and riser clamps under one heading. The cost estimates in this section are based on the conditions, limitations and wage rates described in the section "How to Use This Book" beginning on page 5.

Description	Craft@Hrs	Unit	Material $	Labor $	Equipment $	Total $
Schedule 40 DWV polypropylene pipe with heat-fusioned joints						
1½"	P1@.050	LF	5.73	2.22	—	7.95
2"	P1@.055	LF	7.79	2.44	—	10.23
3"	P1@.060	LF	13.80	2.67	—	16.47
4"	P1@.080	LF	20.00	3.56	—	23.56
6"	P1@.110	LF	35.90	4.89	—	40.79
Schedule 40 DWV polypropylene 1/8 bend with heat-fusioned joints						
1½"	P1@.350	Ea	28.00	15.60	—	43.60
2"	P1@.400	Ea	34.20	17.80	—	52.00
3"	P1@.450	Ea	62.30	20.00	—	82.30
4"	P1@.500	Ea	70.60	22.20	—	92.80
6"	P1@.750	Ea	194.00	33.30	—	227.30
Schedule 40 DWV polypropylene ¼ bend with heat-fusioned joints						
1½"	P1@.350	Ea	28.20	15.60	—	43.80
2"	P1@.400	Ea	34.80	17.80	—	52.60
3"	P1@.450	Ea	59.80	20.00	—	79.80
4"	P1@.500	Ea	94.90	22.20	—	117.10
6"	P1@.750	Ea	234.00	33.30	—	267.30
Schedule 40 DWV polypropylene long sweep ¼ bend with heat-fusioned joints						
1½"	P1@.350	Ea	28.70	15.60	—	44.30
2"	P1@.400	Ea	40.30	17.80	—	58.10
3"	P1@.450	Ea	68.50	20.00	—	88.50
4"	P1@.500	Ea	97.90	22.20	—	120.10
Schedule 40 DWV polypropylene P-trap with heat-fusioned joints						
1½"	P1@.500	Ea	55.40	22.20	—	77.60
2"	P1@.560	Ea	77.00	24.90	—	101.90
3"	P1@.630	Ea	140.00	28.00	—	168.00
4"	P1@.700	Ea	250.00	31.10	—	281.10

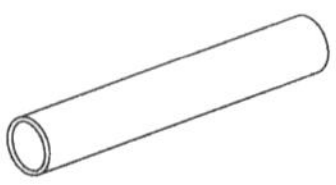

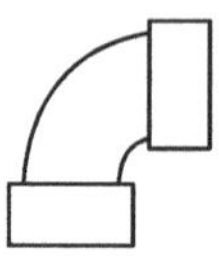
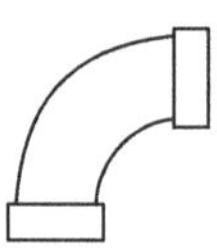
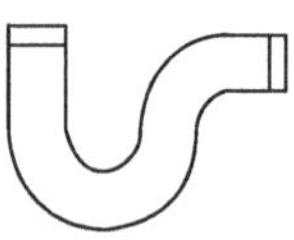

Description	Craft@Hrs	Unit	Material $	Labor $	Equipment $	Total $

Schedule 40 DWV polypropylene sanitary tee (TY) with heat-fusioned joints

1½"	P1@.500	Ea	34.80	22.20	—	57.00
2"	P1@.560	Ea	42.40	24.90	—	67.30
3"	P1@.630	Ea	70.20	28.00	—	98.20
4"	P1@.700	Ea	126.00	31.10	—	157.10

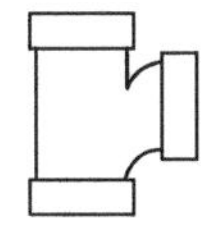

Schedule 40 DWV polypropylene reducing sanitary tee (TY) with heat-fusioned joints

2" x 1½"	P1@.510	Ea	46.40	22.70	—	69.10
3" x 1½"	P1@.530	Ea	80.50	23.60	—	104.10
3" x 2"	P1@.580	Ea	85.90	25.80	—	111.70
4" x 2"	P1@.650	Ea	123.00	28.90	—	151.90
4" x 3"	P1@.700	Ea	129.00	31.10	—	160.10
6" x 4"	P1@.850	Ea	234.00	37.80	—	271.80

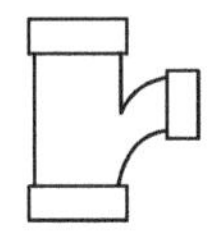

Schedule 40 DWV polypropylene combination wye and 1/8 bend with heat-fusioned joints

1½"	P1@.500	Ea	51.40	22.20	—	73.60
2"	P1@.560	Ea	68.80	24.90	—	93.70
3"	P1@.630	Ea	109.00	28.00	—	137.00
4"	P1@.700	Ea	154.00	31.10	—	185.10

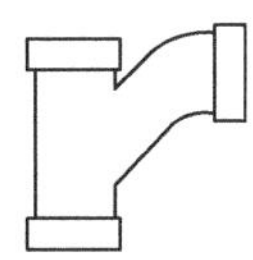

Schedule 40 DWV polypropylene combination reducing wye and 1/8 bend with heat-fusioned joints

2" x 1½"	P1@.500	Ea	61.70	22.20	—	83.90
3" x 1½"	P1@.520	Ea	90.90	23.10	—	114.00
3" x 2"	P1@.570	Ea	98.60	25.30	—	123.90
4" x 2"	P1@.580	Ea	134.00	25.80	—	159.80
4" x 3"	P1@.630	Ea	144.00	28.00	—	172.00

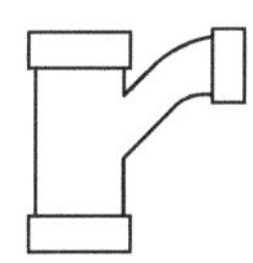

Schedule 40 DWV polypropylene wye with heat-fusioned joints

1½"	P1@.500	Ea	39.50	22.20	—	61.70
2"	P1@.560	Ea	56.10	24.90	—	81.00
3"	P1@.630	Ea	98.60	28.00	—	126.60
4"	P1@.700	Ea	144.00	31.10	—	175.10
6"	P1@1.05	Ea	364.00	46.70	—	410.70

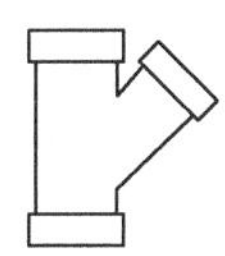

Polypropylene, Schedule 40, with Heat-Fusioned Joints

Schedule 40 DWV polypropylene reducing wye with heat-fusioned joints

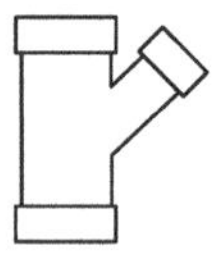

Description	Craft@Hrs	Unit	Material $	Labor $	Equipment $	Total $
2" x 1½"	P1@.510	Ea	54.90	22.70	—	77.60
3" x 2"	P1@.580	Ea	91.50	25.80	—	117.30
4" x 2"	P1@.600	Ea	131.00	26.70	—	157.70
4" x 3"	P1@.650	Ea	135.00	28.90	—	163.90
6" x 2"	P1@.900	Ea	230.00	40.00	—	270.00
6" x 3"	P1@.900	Ea	234.00	40.00	—	274.00
6" x 4"	P1@.950	Ea	237.00	42.20	—	279.20

Schedule 40 DWV polypropylene female adapter with heat-fusioned joints

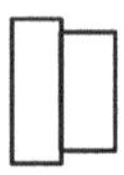

Description	Craft@Hrs	Unit	Material $	Labor $	Equipment $	Total $
1½"	P1@.280	Ea	19.20	12.40	—	31.60
2"	P1@.330	Ea	25.60	14.70	—	40.30
3"	P1@.350	Ea	43.20	15.60	—	58.80
4"	P1@.400	Ea	82.50	17.80	—	100.30

Schedule 40 DWV polypropylene male adapter with heat-fusioned joints

Description	Craft@Hrs	Unit	Material $	Labor $	Equipment $	Total $
1½"	P1@.280	Ea	18.80	12.40	—	31.20
2"	P1@.330	Ea	22.80	14.70	—	37.50
3"	P1@.350	Ea	37.90	15.60	—	53.50
4"	P1@.400	Ea	78.20	17.80	—	96.00

Schedule 40 DWV polypropylene cleanout adapter with heat-fusioned joints

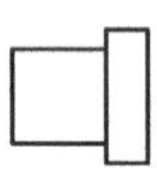

Description	Craft@Hrs	Unit	Material $	Labor $	Equipment $	Total $
1½"	P1@.210	Ea	20.00	9.33	—	29.33
2"	P1@.280	Ea	23.70	12.40	—	36.10
3"	P1@.430	Ea	43.90	19.10	—	63.00
4"	P1@.640	Ea	75.30	28.40	—	103.70
6"	P1@.850	Ea	130.00	37.80	—	167.80

Schedule 40 DWV polypropylene plug with heat-fusioned joints

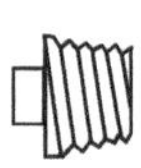

Description	Craft@Hrs	Unit	Material $	Labor $	Equipment $	Total $
1½"	P1@.060	Ea	6.94	2.67	—	9.61
2"	P1@.065	Ea	7.57	2.89	—	10.46
3"	P1@.070	Ea	13.20	3.11	—	16.31
4"	P1@.090	Ea	14.90	4.00	—	18.90

Schedule 40 DWV polypropylene coupling with heat-fusioned joints

Description	Craft@Hrs	Unit	Material $	Labor $	Equipment $	Total $
1½"	P1@.350	Ea	24.50	15.60	—	40.10
2"	P1@.400	Ea	30.10	17.80	—	47.90
3"	P1@.450	Ea	38.50	20.00	—	58.50
4"	P1@.500	Ea	54.90	22.20	—	77.10
6"	P1@.750	Ea	87.60	33.30	—	120.90

Schedule 40 DWV polypropylene reducer with heat-fusioned joints

Description	Craft@Hrs	Unit	Material $	Labor $	Equipment $	Total $
2" x 1½"	P1@.360	Ea	25.60	16.00	—	41.60
3" x 2"	P1@.410	Ea	29.40	18.20	—	47.60
4" x 2"	P1@.430	Ea	65.90	19.10	—	85.00
4" x 3"	P1@.450	Ea	67.70	20.00	—	87.70

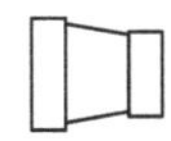

Schedule 40 DWV polypropylene flange with heat-fusioned joints

2"	P1@.350	Ea	84.70	15.60	—	100.30
3"	P1@.380	Ea	175.00	16.90	—	191.90
4"	P1@.430	Ea	228.00	19.10	—	247.10
6"	P1@.630	Ea	300.00	28.00	—	328.00

Schedule 40 DWV polypropylene bolt and gasket set

2"	P1@.500	Ea	8.34	22.20	—	30.54
2½"	P1@.650	Ea	8.83	28.90	—	37.73
3"	P1@.750	Ea	9.28	33.30	—	42.58
4"	P1@1.00	Ea	16.70	44.50	—	61.20

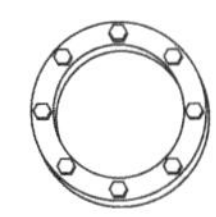

Hanger with swivel assembly

1½"	P1@.300	Ea	5.74	13.30	—	19.04
2"	P1@.300	Ea	5.99	13.30	—	19.29
3"	P1@.350	Ea	8.15	15.60	—	23.75
4"	P1@.350	Ea	10.10	15.60	—	25.70
6"	P1@.450	Ea	11.10	20.00	—	31.10

Riser clamp

1½"	P1@.110	Ea	6.22	4.89	—	11.11
2"	P1@.115	Ea	6.62	5.11	—	11.73
3"	P1@.120	Ea	6.93	5.33	—	12.26
4"	P1@.125	Ea	7.53	5.56	—	13.09
6"	P1@.200	Ea	9.61	8.89	—	18.50

Pipe sleeve

2"	P1@.130	Ea	8.00	5.78	—	13.78
2½"	P1@.150	Ea	8.11	6.67	—	14.78
3"	P1@.180	Ea	8.32	8.00	—	16.32
4"	P1@.220	Ea	9.45	9.78	—	19.23
5"	P1@.250	Ea	11.70	11.10	—	22.80
6"	P1@.270	Ea	12.80	12.00	—	24.80
8"	P1@.270	Ea	14.60	12.00	—	26.60

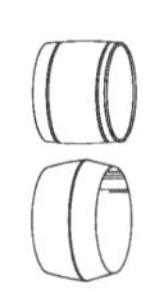

Lead roof pipe flashing

1½"	P1@.250	Ea	36.10	11.10	—	47.20
2"	P1@.250	Ea	47.90	11.10	—	59.00
3"	P1@.250	Ea	55.00	11.10	—	66.10
4"	P1@.250	Ea	60.90	11.10	—	72.00

Floor drain, cast iron, nickel-bronze grate. Add for rough-in (P-trap, fittings, hangers, etc.).

Description	Craft@Hrs	Unit	Material $	Labor $	Equipment $	Total $
3" x 2"	P1@.500	Ea	158.00	22.20	—	180.20
5" x 3"	P1@.500	Ea	195.00	22.20	—	217.20
6" x 4"	P1@.500	Ea	251.00	22.20	—	273.20
8" x 6"	P1@.500	Ea	333.00	22.20	—	355.20

Area drain, cast iron grate. Add for rough-in (P-trap, fittings, hangers, etc.).

Description	Craft@Hrs	Unit	Material $	Labor $	Equipment $	Total $
10" x 3"	P1@.650	Ea	333.00	28.90	—	361.90
12" x 4"	P1@.650	Ea	364.00	28.90	—	392.90
14" x 6"	P1@.700	Ea	392.00	31.10	—	423.10

Roof and overflow drain, cast iron, with plastic dome, flow control weir and deck clamp. Add for rough-in (pipe, fittings, hangers, etc.).

Description	Craft@Hrs	Unit	Material $	Labor $	Equipment $	Total $
8" x 3"	P1@1.50	Ea	449.00	66.70	—	515.70
10" x 4"	P1@1.50	Ea	474.00	66.70	—	540.70
12" x 6"	P1@1.65	Ea	474.00	73.30	—	547.30

Planter drain, cast iron. Add for rough-in (pipe, fittings, hangers, etc.).

Description	Craft@Hrs	Unit	Material $	Labor $	Equipment $	Total $
2"	P1@1.00	Ea	86.10	44.50	—	130.60
3"	P1@1.10	Ea	99.30	48.90	—	148.20
4"	P1@1.15	Ea	120.00	51.10	—	171.10
6"	P1@1.25	Ea	175.00	55.60	—	230.60

Planter drain, plastic. Add for rough-in (pipe, fittings, hangers, etc.).

Description	Craft@Hrs	Unit	Material $	Labor $	Equipment $	Total $
4"	P1@.800	Ea	27.90	35.60	—	63.50
5"	P1@1.00	Ea	29.60	44.50	—	74.10
6"	P1@1.05	Ea	48.40	46.70	—	95.10
8"	P1@1.10	Ea	73.80	48.90	—	122.70

Description	Craft@Hrs	Unit	Material $	Labor $	Equipment $	Total $

Cleanout, in-line, (Barrett) cast iron, (MJ) brass plug. Add for MJ couplings.

2"	P1@.350	Ea	78.10	15.60	—	93.70
3"	P1@.350	Ea	99.90	15.60	—	115.50
4"	P1@.400	Ea	142.00	17.80	—	159.80
6"	P1@.400	Ea	497.00	17.80	—	514.80

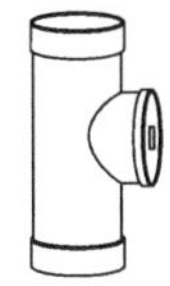

Cleanout, in-line (Barrett), ABS/PVC

2"	P1@.300	Ea	23.10	13.30	—	36.40
3"	P1@.300	Ea	37.20	13.30	—	50.50
4"	P1@.350	Ea	67.30	15.60	—	82.90
6"	P1@.350	Ea	263.00	15.60	—	278.60
8"	P1@.400	Ea	509.00	17.80	—	526.80
10"	P1@.550	Ea	628.00	24.40	—	652.40

Cleanout, end of line (Malcolm), cast iron body, brass top. Add for MJ coupling.

2" raised	P1@.250	Ea	28.10	11.10	—	39.20
2" recessed	P1@.250	Ea	70.10	11.10	—	81.20
3" raised	P1@.250	Ea	32.80	11.10	—	43.90
3" recessed	P1@.300	Ea	103.00	13.30	—	116.30
4" raised	P1@.300	Ea	43.80	13.30	—	57.10
4" recessed	P1@.350	Ea	134.00	15.60	—	149.60
6" raised	P1@.350	Ea	139.00	15.60	—	154.60
6" recessed	P1@.400	Ea	257.00	17.80	—	274.80
8" raised	P1@.450	Ea	298.00	20.00	—	318.00

Cleanout, end-of-line (Malcolm), ABS/PVC

2"	P1@.250	Ea	11.90	11.10	—	23.00
3"	P1@.250	Ea	20.00	11.10	—	31.10
4"	P1@.300	Ea	36.10	13.30	—	49.40
6"	P1@.300	Ea	151.00	13.30	—	164.30
8"	P1@.400	Ea	274.00	17.80	—	291.80
10"	P1@.450	Ea	312.00	20.00	—	332.00
12"	P1@.550	Ea	365.00	24.40	—	389.40

Description	Unit	Material $	Labor $	Equipment $	Total $

Complete wet sprinkler system (square foot costs). System includes subcontractors' overhead and profit. Cost includes design drawings, zone and supervision valves, sprinkler heads, pipe, and connection to water supply service. Make additional allowances if a fire pump is required. Costs are based on ordinary hazard coverage (110 sf per head). Coverage density can vary depending on room sizes and/or hazard rating. Use these costs for preliminary estimates.

Exposed piping to 5,000	SF	—	—	—	5.44
Exposed piping 5,000-15,000	SF	—	—	—	4.58
Exposed piping over 15,000	SF	—	—	—	4.10
Concealed piping to 5,000	SF	—	—	—	5.08
Concealed piping 5,000-15,000	SF	—	—	—	4.38
Concealed piping over 15,000	SF	—	—	—	3.43
Add for dry system	%	—	—	—	20.0

Complete wet sprinkler system (per head costs). Cost includes sprinkler heads, sprinkler mains, branch piping, and supports. Make additional allowances for zone and supervision valves, alarms, connection to water service and fire pump when required. Costs are based on ordinary hazard 155 degree heads. Use these costs for preliminary estimates.

Upright heads to 5,000 SF	Ea	—	—	—	381.00
Upright heads 5,000 to 15,000 SF	Ea	—	—	—	363.00
Upright heads over 15,000 SF	Ea	—	—	—	345.00
Pendent heads to 5,000 SF	Ea	—	—	—	576.00
Pendent heads 5,000 to 15,000 SF	Ea	—	—	—	481.00
Pendent heads over 15,000 SF	Ea	—	—	—	406.00

Description	Craft@Hrs	Unit	Material $	Labor $	Equipment $	Total $

Supervision (zone) valves, flanged. Add for alarm trim if required.

2½" OS&Y gate valve	SL@1.70	Ea	485.00	75.00	—	560.00
3" OS&Y gate vlv	SL@1.80	Ea	514.00	79.40	—	593.40
4" OS&Y gate vlv	SL@2.20	Ea	593.00	97.10	—	690.10
6" OS&Y gate vlv	SL@2.80	Ea	910.00	124.00	—	1,034.00
8" OS&Y gate vlv	SL@3.50	Ea	1,450.00	154.00	—	1,604.00
2½" butterfly valve (lug)	SL@.600	Ea	658.00	26.50	—	684.50
3" butterfly vlv	SL@.700	Ea	665.00	30.90	—	695.90
4" butterfly vlv	SL@.850	Ea	682.00	37.50	—	719.50
6" butterfly vlv	SL@1.15	Ea	944.00	50.70	—	994.70
8" butterfly vlv	SL@1.85	Ea	1,420.00	81.60	—	1,501.60

Description	Craft@Hrs	Unit	Material $	Labor $	Equipment $	Total $
Supervision (zone) valves, grooved. Add for alarm trim if required						
4" butterfly valve	SL@.550	Ea	477.00	24.30	—	501.30
6" butterfly valve	SL@.850	Ea	620.00	37.50	—	657.50
8" butterfly valve	SL@1.25	Ea	939.00	55.20	—	994.20
Alarm valves (flanged or grooved)						
4" alarm valve	SL@3.50	Ea	627.00	154.00	—	781.00
6" alarm valve	SL@4.00	Ea	772.00	176.00	—	948.00
8" alarm valve	SL@4.75	Ea	1,130.00	210.00	—	1,340.00
4" alarm vlv trim	SL@1.25	Ea	451.00	55.20	—	506.20
6" alarm vlv trim	SL@1.25	Ea	451.00	55.20	—	506.20
8" alarm vlv trim	SL@1.25	Ea	451.00	55.20	—	506.20
4" alarm vlv pkg	SL@4.75	Ea	2,110.00	210.00	—	2,320.00
6" alarm vlv pkg	SL@5.50	Ea	2,350.00	243.00	—	2,593.00
8" alarm vlv pkg	SL@6.00	Ea	2,850.00	265.00	—	3,115.00
Dry valves and trim (deluge/pre-action system)						
2" dry valve	SL@2.25	Ea	1,260.00	99.30	—	1,359.30
3" dry valve	SL@2.75	Ea	1,320.00	121.00	—	1,441.00
4" dry valve	SL@3.05	Ea	1,470.00	135.00	—	1,605.00
6" dry valve	SL@3.65	Ea	1,900.00	161.00	—	2,061.00
2" dry valve trim	SL@1.25	Ea	617.00	55.20	—	672.20
3" dry valve trim	SL@1.25	Ea	617.00	55.20	—	672.20
4" dry valve trim	SL@1.25	Ea	617.00	55.20	—	672.20
6" dry valve trim	SL@1.25	Ea	617.00	55.20	—	672.20
Check valves, grooved						
2½"	SL@.450	Ea	236.00	19.90	—	255.90
3"	SL@.500	Ea	260.00	22.10	—	282.10
4"	SL@.550	Ea	280.00	24.30	—	304.30
6"	SL@.850	Ea	459.00	37.50	—	496.50
Check valves, wafer-type, flanged						
2½"	SL@1.65	Ea	309.00	72.80	—	381.80
3"	SL@1.70	Ea	321.00	75.00	—	396.00
4"	SL@2.10	Ea	347.00	92.70	—	439.70
6"	SL@2.60	Ea	567.00	115.00	—	682.00
8"	SL@3.20	Ea	805.00	141.00	—	946.00
Double-check detector valve assembly (flanged)						
2½"	SL@4.00	Ea	3,130.00	176.00	—	3,306.00
3"	SL@4.35	Ea	3,360.00	192.00	—	3,552.00
4"	SL@4.95	Ea	3,650.00	218.00	—	3,868.00
6"	SL@6.25	Ea	5,720.00	276.00	—	5,996.00
8"	SL@6.95	Ea	10,300.00	307.00	—	10,607.00

Description	Craft@Hrs	Unit	Material $	Labor $	Equipment $	Total $
Switches: flow, pressure, monitor, supervisory						
Flow switch	SL@1.00	Ea	205.00	44.10	—	249.10
Pressure switch	SL@1.00	Ea	154.00	44.10	—	198.10
Monitor switch	SL@1.00	Ea	149.00	44.10	—	193.10
Supervisory sw.	SL@1.00	Ea	137.00	44.10	—	181.10
Sprinkler heads, 155 degree						
Pendent, brass	SL@.350	Ea	9.60	15.40	—	25.00
Pendent, chrome	SL@.350	Ea	10.30	15.40	—	25.70
Upright, brass	SL@.350	Ea	9.60	15.40	—	25.00
Upright, chrome	SL@.350	Ea	24.10	15.40	—	39.50
Sidewall, brass	SL@.350	Ea	12.50	15.40	—	27.90
Sidewall, chrome	SL@.350	Ea	13.30	15.40	—	28.70
Sprinkler heads, dry, 165 degree						
Dry pendent	SL@.450	Ea	95.70	19.90	—	115.60
Dry sidewall	SL@.450	Ea	133.00	19.90	—	152.90
Sprinkler heads, 200 degree						
Pendent, brass	SL@.350	Ea	9.60	15.40	—	25.00
Pendent, chrome	SL@.350	Ea	10.30	15.40	—	25.70
Upright, brass	SL@.350	Ea	9.60	15.40	—	25.00
Upright, chrome	SL@.350	Ea	24.10	15.40	—	39.50
Sidewall, brass	SL@.350	Ea	12.50	15.40	—	27.90
Sidewall, chrome	SL@.350	Ea	13.30	15.40	—	28.70
Sprinkler heads, 286 degree						
Pendent, brass	SL@.350	Ea	9.60	15.40	—	25.00
Pendent, chrome	SL@.350	Ea	10.30	15.40	—	25.70
Upright, brass	SL@.350	Ea	9.60	15.40	—	25.00
Upright, chrome	SL@.350	Ea	24.10	15.40	—	39.50
Sidewall, brass	SL@.350	Ea	12.50	15.40	—	27.90
Sidewall, chrome	SL@.350	Ea	13.30	15.40	—	28.70
Sprinkler heads, 360 degree						
Pendent, brass	SL@.350	Ea	6.47	15.40	—	21.87
Upright, brass	SL@.350	Ea	6.47	15.40	—	21.87
Sprinkler heads, 400 degree						
Pendent, brass	SL@.350	Ea	28.50	15.40	—	43.90
Upright, brass	SL@.350	Ea	28.50	15.40	—	43.90

Description	Craft@Hrs	Unit	Material $	Labor $	Equipment $	Total $

Fire hose cabinet, complete assembly. Includes recessed steel cabinet, 2½" and 1½" angle valves, 100' fire hose, hose rack and fog nozzle.

FHC, chrome trim	SL@1.95	Ea	460.00	86.00	—	546.00
FHC, brass trim	SL@1.95	Ea	450.00	86.00	—	536.00
FHC, plastic trim	SL@1.95	Ea	436.00	86.00	—	522.00

Fire hose cabinet components

Recessed cabinet	SL@1.00	Ea	204.00	44.10	—	248.10
Surface cabinet	SL@.850	Ea	236.00	37.50	—	273.50
Cabinet glass	SL@.350	Ea	15.00	15.40	—	30.40
100' fire hose	SL@.250	Ea	191.00	11.00	—	202.00
75' hose rack	SL@.350	Ea	59.00	15.40	—	74.40
100' hose rack	SL@.350	Ea	80.00	15.40	—	95.40
1½" brass fire hose angle valve	SL@.350	Ea	98.70	15.40	—	114.10
1½" chrome fire hose angle valve	SL@.350	Ea	164.00	15.40	—	179.40
2½" brass fire hose angle valve	SL@.450	Ea	190.00	19.90	—	209.90
2½" chrome fire hose angle valve	SL@.450	Ea	74.90	19.90	—	94.80
Brass fog nozzle	SL@.150	Ea	73.70	6.62	—	80.32
Plastic fog nozzle	SL@.150	Ea	19.40	6.62	—	26.02
Chrome fog nozzle	SL@.150	Ea	75.60	6.62	—	82.22

Electric fire pump, skid-mounted with controller and jockey pump

15 KW, 500 gpm	—	Ea	—	—	—	59,400.00
22 KW, 500 gpm	—	Ea	—	—	—	63,600.00
30 KW, 750 gpm	—	Ea	—	—	—	67,300.00
38 KW, 1,000 gpm	—	Ea	—	—	—	74,500.00
56 KW, 1,000 gpm	—	Ea	—	—	—	75,600.00

Diesel fire pump, skid-mounted with controller and jockey pump

250 gpm	—	Ea	—	—	—	88,800.00
500 gpm	—	Ea	—	—	—	92,500.00
750 gpm	—	Ea	—	—	—	94,200.00
1,000 gpm	—	Ea	—	—	—	96,000.00

Description	Craft@Hrs	Unit	Material $	Labor $	Equipment $	Total $

Excess pressure pump

Description	Craft@Hrs	Unit	Material $	Labor $	Equipment $	Total $
1/4 hp	SL@1.35	Ea	670.00	59.60	—	729.60
1/3 hp	SL@1.35	Ea	673.00	59.60	—	732.60
1/2 hp	SL@1.50	Ea	698.00	66.20	—	764.20

Siamese connection, flush, fire dept. connection

Description	Craft@Hrs	Unit	Material $	Labor $	Equipment $	Total $
4" brass	SL@4.00	Ea	888.00	176.00	—	1,064.00
4" chrome	SL@4.00	Ea	957.00	176.00	—	1,133.00
6" brass	SL@4.50	Ea	1,030.00	199.00	—	1,229.00
6" chrome	SL@4.50	Ea	1,050.00	199.00	—	1,249.00

Siamese connection, surface, fire dept. connection

Description	Craft@Hrs	Unit	Material $	Labor $	Equipment $	Total $
4" brass	SL@2.85	Ea	211.00	126.00	—	337.00
6" brass	SL@3.25	Ea	397.00	143.00	—	540.00

Siamese connection, sidewalk, fire dept. connection

Description	Craft@Hrs	Unit	Material $	Labor $	Equipment $	Total $
4" brass	SL@3.25	Ea	802.00	143.00	—	945.00
6" brass	SL@3.75	Ea	1,130.00	165.00	—	1,295.00

Water motor gong

Description	Craft@Hrs	Unit	Material $	Labor $	Equipment $	Total $
Gong	SL@1.50	Ea	376.00	66.20	—	442.20

Multi-purpose fire extinguisher

Description	Craft@Hrs	Unit	Material $	Labor $	Equipment $	Total $
2 lb.	SL@.500	Ea	45.80	22.10	—	67.90
5 lb.	SL@.500	Ea	64.40	22.10	—	86.50
10 lb.	SL@.500	Ea	92.20	22.10	—	114.30
20 lb.	SL@.500	Ea	194.00	22.10	—	216.10

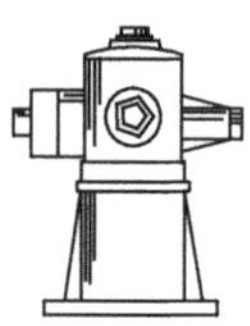

Fire hydrant with spool

Description	Craft@Hrs	Unit	Material $	Labor $	Equipment $	Total $
6"	P1@5.00	Ea	1,730.00	222.00	—	1,952.00

Indicator post

Description	Craft@Hrs	Unit	Material $	Labor $	Equipment $	Total $
6"	P1@6.50	Ea	996.00	289.00	—	1,285.00

Description	Craft@Hrs	Unit	Material $	Labor $	Equipment $	Total $

Black steel pipe (A53), installed horizontal, roll-grooved, Schedule 40. Standard weight, including hangers every 10' and a fitting every 33'. Use these figures for preliminary estimates.

2" (50mm)	SL@.160	LF	8.55	7.06	—	15.61
2½" (65mm)	SL@.180	LF	13.70	7.94	—	21.64
3" (75mm)	SL@.210	LF	17.60	9.27	—	26.87
4" (10cm)	SL@.300	LF	23.70	13.20	—	36.90
6" (15cm)	SK@.450	LF	47.50	21.20	.30	69.00
8" (20cm)	SK@.630	LF	70.50	29.60	.42	100.52
10" (25cm)	SK@.750	LF	137.00	35.30	.50	172.80
12" (30cm)	SK@.800	LF	180.00	37.60	.54	218.14

Black steel pipe (A53), installed riser, roll-grooved, Schedule 40. Standard weight, including a riser clamp every other floor and a tee at every floor. Make additional allowances for sleeving or coring as required. Use these figures for preliminary estimates.

2" (50mm)	SL@.110	LF	8.86	4.85	—	13.71
2½" (65mm)	SL@.120	LF	14.30	5.29	—	19.59
3" (75mm)	SL@.150	LF	17.00	6.62	—	23.62
4" (10cm)	SL@.190	LF	24.10	8.38	—	32.48
6" (15cm)	SK@.300	LF	45.90	14.10	.20	60.20
8" (20cm)	SK@.400	LF	71.40	18.80	.27	90.47
10" (25cm)	SK@.510	LF	111.00	24.00	.34	135.34
12" (30cm)	SK@.560	LF	161.00	26.30	.37	187.67

Black steel pipe (A53), installed horizontal, roll-grooved, thin wall, Schedule 10. Light weight, including hangers every 10' and a fitting every 33'. Use these figures for preliminary estimates.

2" (50mm)	SL@.160	LF	7.11	7.06	—	14.17
2½" (65mm)	SL@.180	LF	10.40	7.94	—	18.34
3" (75mm)	SL@.210	LF	12.10	9.27	—	21.37
4" (10cm)	SL@.300	LF	17.50	13.20	—	30.70
6" (15cm)	SK@.450	LF	33.80	21.20	.30	55.30
8" (20cm)	SK@.630	LF	62.60	29.60	.42	92.62

Black steel pipe (A53), installed riser, roll-grooved, thin wall, Schedule 10. Light weight, including a riser clamp every other floor and a tee at every floor. Make additional allowances for sleeving or coring as required. Use these figures for preliminary estimates.

2" (50mm)	SL@.110	LF	7.98	4.85	—	12.83
2½" (65mm)	SL@.120	LF	11.50	5.29	—	16.79
3" (75mm)	SL@.150	LF	13.50	6.62	—	20.12
4" (10cm)	SL@.190	LF	18.90	8.38	—	27.28
6" (15cm)	SK@.300	LF	36.30	14.10	.20	50.60
8" (20cm)	SK@.400	LF	70.50	18.80	.27	89.57

Description	Craft@Hrs	Unit	Material $	Labor $	Equipment $	Total $

Black steel pipe only (A53), roll-grooved, Schedule 40. Standard weight, no hangers or fittings.

2" (50mm)	SL@.070	LF	3.78	3.09	—	6.87
2½" (65mm)	SL@.070	LF	8.08	3.09	—	11.17
3" (75mm)	SL@.090	LF	9.00	3.97	—	12.97
4" (10cm)	SL@.130	LF	12.60	5.74	—	18.34
6" (15cm)	SK@.160	LF	18.60	7.52	.11	26.23
8" (20cm)	SK@.210	LF	28.20	9.88	.14	38.22
10" (25cm)	SK@.280	LF	40.70	13.20	.19	54.09
12" (30cm)	SK@.340	LF	68.70	16.00	.23	84.93
14" (36cm)	SK@.400	LF	79.30	18.80	.27	98.37
16" (41cm)	SK@.550	LF	92.10	25.90	.37	118.37

Black steel pipe only, (A53) roll-grooved, thin wall, Schedule 10. Light weight, no hangers or fittings.

2" (50mm)	SL@.070	LF	2.83	3.09	—	5.92
2½" (65mm)	SL@.070	LF	5.98	3.09	—	9.07
3" (75mm)	SL@.090	LF	6.63	3.97	—	10.60
4" (10cm)	SL@.130	LF	9.50	5.74	—	15.24
6" (15cm)	SK@.160	LF	13.70	7.52	.11	21.33
8" (20cm)	SK@.210	LF	23.50	9.88	.14	33.52
10" (25cm)	SK@.280	LF	34.20	13.20	.19	47.59
12" (30cm)	SK@.340	LF	57.80	16.00	.23	74.03
14" (36cm)	SK@.400	LF	66.90	18.80	.27	85.97
16" (41cm)	SK@.550	LF	77.80	25.90	.37	104.07

90-degree elbow, roll-grooved, Style #10 (Victaulic)

2"	SL@.410	Ea	42.40	18.10	—	60.50
2½"	SL@.430	Ea	42.40	19.00	—	61.40
3"	SL@.500	Ea	55.80	22.10	—	77.90
4"	SL@.690	Ea	87.30	30.40	—	117.70
6"	SL@1.19	Ea	185.00	52.50	—	237.50
8"	SL@1.69	Ea	312.00	74.60	—	386.60
10"	SL@2.23	Ea	590.00	98.40	—	688.40
12"	SL@2.48	Ea	914.00	109.00	—	1,023.00

45-degree elbow, roll-grooved, Style #11 (Victaulic)

2"	SL@.410	Ea	42.40	18.10	—	60.50
2½"	SL@.430	Ea	42.40	19.00	—	61.40
3"	SL@.500	Ea	55.80	22.10	—	77.90
4"	SL@.690	Ea	87.30	30.40	—	117.70
6"	SL@1.19	Ea	185.00	52.50	—	237.50
8"	SL@1.69	Ea	312.00	74.60	—	386.60
10"	SL@2.23	Ea	590.00	98.40	—	688.40
12"	SL@2.48	Ea	914.00	109.00	—	1,023.00

Description	Craft@Hrs	Unit	Material $	Labor $	Equipment $	Total $
22½-degree elbow, roll-grooved, Style #12 (Victaulic)						
2"	SL@.410	Ea	50.50	18.10	—	68.60
2½"	SL@.430	Ea	50.50	19.00	—	69.50
3"	SL@.500	Ea	52.70	22.10	—	74.80
4"	SL@.690	Ea	99.50	30.40	—	129.90
6"	SL@1.19	Ea	210.00	52.50	—	262.50
8"	SL@1.69	Ea	426.00	74.60	—	500.60
10"	SL@2.23	Ea	593.00	98.40	—	691.40
12"	SL@2.48	Ea	622.00	109.00	—	731.00
Tee, roll-grooved, Style #20 (Victaulic)						
2"	SL@.510	Ea	81.80	22.50	—	104.30
2½"	SL@.540	Ea	81.80	23.80	—	105.60
3"	SL@.630	Ea	98.70	27.80	—	126.50
4"	SL@.860	Ea	137.00	37.90	—	174.90
6"	SL@1.49	Ea	310.00	65.70	—	375.70
8"	SL@2.11	Ea	516.00	93.10	—	609.10
10"	SL@2.79	Ea	897.00	123.00	—	1,020.00
12"	SL@3.10	Ea	1,030.00	137.00	—	1,167.00
Reducing tee, roll-grooved, Style #25 (Victaulic)						
2"	SL@.510	Ea	81.80	22.50	—	104.30
2½"	SL@.540	Ea	107.00	23.80	—	130.80
3"	SL@.630	Ea	97.20	27.80	—	125.00
4"	SL@.860	Ea	164.00	37.90	—	201.90
6"	SL@1.49	Ea	299.00	65.70	—	364.70
8"	SL@2.11	Ea	535.00	93.10	—	628.10
10"	SL@2.79	Ea	696.00	123.00	—	819.00
12"	SL@3.10	Ea	756.00	137.00	—	893.00
Reducer, roll-grooved, Style #50 (Victaulic)						
2"	SL@.410	Ea	27.20	18.10	—	45.30
2½"	SL@.430	Ea	27.20	19.00	—	46.20
3"	SL@.500	Ea	37.80	22.10	—	59.90
4"	SL@.690	Ea	63.40	30.40	—	93.80
6"	SL@1.19	Ea	89.90	52.50	—	142.40
8"	SL@1.69	Ea	155.00	74.60	—	229.60
10"	SL@2.23	Ea	318.00	98.40	—	416.40
Cap, roll-grooved, Style #60 (Victaulic)						
2"	SL@.270	Ea	24.00	11.90	—	35.90
2½"	SL@.280	Ea	24.00	12.40	—	36.40
3"	SL@.330	Ea	30.10	14.60	—	44.70
4"	SL@.450	Ea	34.70	19.90	—	54.60
6"	SL@.770	Ea	66.10	34.00	—	100.10
8"	SL@1.10	Ea	105.00	48.50	—	153.50
10"	SL@1.61	Ea	168.00	71.00	—	239.00

Description	Craft@Hrs	Unit	Material $	Labor $	Equipment $	Total $
Mechanical tee (saddle tee), roll-grooved, Style #920 (Victaulic)						
3"	SL@.750	Ea	82.50	33.10	—	115.60
4"	SL@1.00	Ea	93.70	44.10	—	137.80
6"	SL@1.30	Ea	109.00	57.40	—	166.40
8"	SL@1.50	Ea	169.00	66.20	—	235.20
Coupling, roll-grooved, 800 PSI, Style #77 (Victaulic). Labor included with fittings.						
3"	—	Ea	37.40	—	—	37.40
4"	—	Ea	61.90	—	—	61.90
6"	—	Ea	99.20	—	—	99.20
8"	—	Ea	133.00	—	—	133.00
10"	—	Ea	67.40	—	—	67.40
12"	—	Ea	232.00	—	—	232.00
Coupling, roll-grooved, light weight, Style #75 (Victaulic). Labor included with fittings.						
3"	—	Ea	24.20	—	—	24.20
4"	—	Ea	26.10	—	—	26.10
6"	—	Ea	46.60	—	—	46.60
8"	—	Ea	71.50	—	—	71.50
Coupling, roll-grooved, zeroflex, Style #07 (Victaulic). Labor included with fittings.						
3"	—	Ea	27.90	—	—	27.90
4"	—	Ea	36.60	—	—	36.60
6"	—	Ea	53.90	—	—	53.90
8"	—	Ea	82.00	—	—	82.00
10"	—	Ea	120.00	—	—	120.00
12"	—	Ea	167.00	—	—	167.00
Flange, roll-grooved, Style #741 (Victaulic)						
3"	SL@.500	Ea	105.00	22.10	—	127.10
4"	SL@.690	Ea	143.00	30.40	—	173.40
6"	SL@1.19	Ea	174.00	52.50	—	226.50
8"	SL@1.69	Ea	202.00	74.60	—	276.60
10"	SL@2.23	Ea	301.00	98.40	—	399.40
12"	SL@2.48	Ea	409.00	109.00	—	518.00

Description	Craft@Hrs	Unit	Material $	Labor $	Equipment $	Total $

Schedule 40 carbon steel pipe, threaded and coupled, pipe only. Add for hangers, fittings etc.

1"	P1@.090	LF	2.98	4.00	—	6.98
1¼"	P1@.100	LF	3.90	4.45	—	8.35
1½"	P1@.110	LF	4.63	4.89	—	9.52
2"	P1@.120	LF	6.26	5.33	—	11.59

125# cast iron 90-degree ell, threaded joints

1" x ¾"	P1@.130	Ea	5.28	5.78	—	11.06
1"	P1@.180	Ea	4.90	8.00	—	12.90
1¼"	P1@.240	Ea	6.93	10.70	—	17.63
1½"	P1@.300	Ea	9.66	13.30	—	22.96
2"	P1@.380	Ea	14.70	16.90	—	31.60

125# cast iron 45-degree ell, threaded joints

1"	P1@.180	Ea	6.43	8.00	—	14.43
1¼"	P1@.240	Ea	9.15	10.70	—	19.85
1½"	P1@.300	Ea	11.80	13.30	—	25.10
2"	P1@.380	Ea	15.90	16.90	—	32.80

125# cast iron tee, threaded joints

1"	P1@.230	Ea	6.75	10.20	—	16.95
1¼"	P1@.310	Ea	12.20	13.80	—	26.00
1½"	P1@.390	Ea	15.70	17.30	—	33.00
2"	P1@.490	Ea	21.40	21.80	—	43.20

125# cast iron reducing tee, threaded joints

1" x ¾"	P1@.220	Ea	8.61	9.78	—	18.39
1¼" x ¾"	P1@.290	Ea	11.00	12.90	—	23.90
1¼" x 1"	P1@.290	Ea	12.00	12.90	—	24.90
1½" x ¾"	P1@.370	Ea	15.20	16.40	—	31.60
1½" x 1"	P1@.370	Ea	16.70	16.40	—	33.10
1½" x 1¼"	P1@.370	Ea	18.40	16.40	—	34.80
2" x 1"	P1@.460	Ea	23.80	20.40	—	44.20
2" x 1¼"	P1@.460	Ea	25.60	20.40	—	46.00
2" x 1½"	P1@.460	Ea	27.50	20.40	—	47.90

125# cast iron reducer, threaded joints

1" x ¾"	P1@.200	Ea	6.75	8.89	—	15.64
1¼" x 1"	P1@.210	Ea	8.12	9.33	—	17.45
1½" x 1"	P1@.270	Ea	9.97	12.00	—	21.97
1½" x 1¼"	P1@.270	Ea	10.50	12.00	—	22.50
2" x 1¼"	P1@.340	Ea	15.80	15.10	—	30.90
2" x 1½"	P1@.340	Ea	16.50	15.10	—	31.60

Description	Craft@Hrs	Unit	Material $	Labor $	Equipment $	Total $
125# cast iron cross, threaded joints						
1"	P1@.360	Ea	15.80	16.00	—	31.80
1¼"	P1@.480	Ea	25.90	21.30	—	47.20
1½"	P1@.600	Ea	31.60	26.70	—	58.30
2"	P1@.700	Ea	52.00	31.10	—	83.10
125# cast iron cap, threaded joint						
¾"	P1@.100	Ea	3.32	4.45	—	7.77
1"	P1@.140	Ea	4.02	6.22	—	10.24
1¼"	P1@.180	Ea	5.13	8.00	—	13.13
1½"	P1@.230	Ea	6.50	10.20	—	16.70
2"	P1@.290	Ea	10.20	12.90	—	23.10
125# cast iron plug, threaded joint						
¾"	P1@.100	Ea	2.16	4.45	—	6.61
1"	P1@.140	Ea	2.86	6.22	—	9.08
1¼"	P1@.180	Ea	4.45	8.00	—	12.45
1½"	P1@.230	Ea	5.65	10.20	—	15.85
2"	P1@.290	Ea	7.00	12.90	—	19.90
125# cast iron coupling, threaded joints						
1"	P1@.180	Ea	5.85	8.00	—	13.85
1¼"	P1@.240	Ea	7.21	10.70	—	17.91
1½"	P1@.300	Ea	9.07	13.30	—	22.37
2"	P1@.380	Ea	12.20	16.90	—	29.10

Description	Craft@Hrs	Unit	Material $	Labor $	Equipment $	Total $
Steel pipe nipples — 3" long, standard right-hand thread						
1"	P1@.025	Ea	2.27	1.11	—	3.38
1¼"	P1@.025	Ea	2.91	1.11	—	4.02
1½"	P1@.025	Ea	3.47	1.11	—	4.58
2"	P1@.030	Ea	4.31	1.33	—	5.64
Steel pipe nipples — 4" long, standard right-hand thread						
1"	P1@.025	Ea	2.70	1.11	—	3.81
1¼"	P1@.030	Ea	3.37	1.33	—	4.70
1½"	P1@.030	Ea	4.16	1.33	—	5.49
2"	P1@.030	Ea	5.36	1.33	—	6.69
Steel pipe nipples — 6" long, standard right-hand thread						
1"	P1@.030	Ea	3.47	1.33	—	4.80
1¼"	P1@.030	Ea	4.34	1.33	—	5.67
1½"	P1@.030	Ea	5.28	1.33	—	6.61
2"	P1@.035	Ea	7.10	1.56	—	8.66

Description	Craft@Hrs	Unit	Material $	Labor $	Equipment $	Total $
Chlorinated Polyvinyl Chloride pipe (CPVC), CPVC pipe only, solvent weld. No hangers or fittings.						
¾"	SL@.055	LF	3.89	2.43	—	6.32
1"	SL@.055	LF	5.69	2.43	—	8.12
1¼"	SL@.065	LF	8.90	2.87	—	11.77
1½"	SL@.075	LF	12.50	3.31	—	15.81
2"	SL@.085	LF	18.50	3.75	—	22.25
2½"	SL@.110	LF	28.70	4.85	—	33.55
3"	SL@.125	LF	43.70	5.52	—	49.22
CPVC 90-degree elbow, solvent weld						
¾"	SL@.100	Ea	4.31	4.41	—	8.72
1"	SL@.120	Ea	6.49	5.29	—	11.78
1¼"	SL@.170	Ea	8.30	7.50	—	15.80
1½"	SL@.180	Ea	12.50	7.94	—	20.44
2"	SL@.200	Ea	14.50	8.82	—	23.32
CPVC 45-degree elbow, solvent weld						
¾"	SL@.100	Ea	4.45	4.41	—	8.86
1"	SL@.120	Ea	5.82	5.29	—	11.11
1¼"	SL@.170	Ea	7.66	7.50	—	15.16
1½"	SL@.180	Ea	10.30	7.94	—	18.24
2"	SL@.200	Ea	12.80	8.82	—	21.62
CPVC tee, solvent weld						
¾"	SL@.150	Ea	4.76	6.62	—	11.38
1"	SL@.170	Ea	8.10	7.50	—	15.60
1¼"	SL@.230	Ea	12.70	10.10	—	22.80
1½"	SL@.250	Ea	16.80	11.00	—	27.80
2"	SL@.280	Ea	26.60	12.40	—	39.00
CPVC reducing tee, solvent weld						
1"	SL@.170	Ea	6.97	7.50	—	14.47
1¼"	SL@.230	Ea	12.70	10.10	—	22.80
1½"	SL@.250	Ea	16.90	11.00	—	27.90
2"	SL@.280	Ea	27.10	12.40	—	39.50
3"	SL@.350	Ea	42.60	15.40	—	58.00
CPVC coupling, solvent weld						
¾"	SL@.100	Ea	3.13	4.41	—	7.54
1"	SL@.120	Ea	4.24	5.29	—	9.53
1¼"	SL@.170	Ea	6.70	7.50	—	14.20
1½"	SL@.180	Ea	8.44	7.94	—	16.38
2"	SL@.200	Ea	10.60	8.82	—	19.42
3"	SL@.280	Ea	15.30	12.40	—	27.70

Description	Craft@Hrs	Unit	Material $	Labor $	Equipment $	Total $
CPVC head adapter, solvent weld						
¾" x ½"	SL@.150	Ea	7.09	6.62	—	13.71
1" x ½"	SL@.165	Ea	7.57	7.28	—	14.85
CPVC cap, solvent weld						
¾"	SL@.060	Ea	2.32	2.65	—	4.97
1"	SL@.065	Ea	3.24	2.87	—	6.11
1¼"	SL@.075	Ea	5.57	3.31	—	8.88
1½"	SL@.085	Ea	7.13	3.75	—	10.88
2"	SL@.095	Ea	8.31	4.19	—	12.50
CPVC flange, solvent weld						
1"	SL@.165	Ea	8.31	7.28	—	15.59
1¼"	SL@.175	Ea	10.70	7.72	—	18.42
1½"	SL@.185	Ea	16.70	8.16	—	24.86
2"	SL@.200	Ea	25.10	8.82	—	33.92
3"	SL@.225	Ea	41.80	9.93	—	51.73
CPVC female adapter, solvent weld						
1"	SL@.065	Ea	20.60	2.87	—	23.47
1¼"	SL@.075	Ea	34.20	3.31	—	37.51
1½"	SL@.085	Ea	43.90	3.75	—	47.65
2"	SL@.095	Ea	56.70	4.19	—	60.89

Boilers for industrial, commercial and institutional applications. ASME "H" or "U" stamped, LFUE (Levelized Fuel Utilization Efficiency) certified 90% or better. U.S. Green Building council certified. Cleaver-Brooks or equal. Shipped factory-assembled and skid-mounted. Includes the boiler shell and structure, I-beam skids, combustion chamber, fire tubes, crown sheet, tube sheets (for fire tube), water tubes, mud drum(s), steam drum (for water tube boiler), finned water tubes and check valves (for hot water generators), refractory brick lining, explosion doors, combustion air inlets and stack outlet. Add the cost of the fuel train piping, electrical wiring, refractory, stack, feedwater system, combustion controls, water treatment, expansion tanks (if hot water) or condensate return (if steam) and system start-up. See the sections following packaged boilers. Costs are based on boiler horsepower rating (BHP). Add the cost of commissioning, insurance inspection, permit, permit inspection, hydraulic test, calibration and lockout sealing of emissions monitor, and final run test. Equipment cost assumes a 17-ton hydraulic truck-mounted crane (at $120 per hour). LEED (Leadership in Energy & Environmental Design) certification requires 88% boiler efficiency, recording controls, compliance with SCAQMD 1146.2 emission standards and zone recording thermostats. A LEED-certified boiler with a U.S. Green Building Council rating will add 4% to 15% to the cost of the boiler, as identified in the cost tables that follow.

Hot water or steam cast iron gas boiler, 30 PSIG. Costs shown are based on boiler horsepower rating (BHP). On-site assembly. Includes allowances for associated appurtenances and pipe connections.

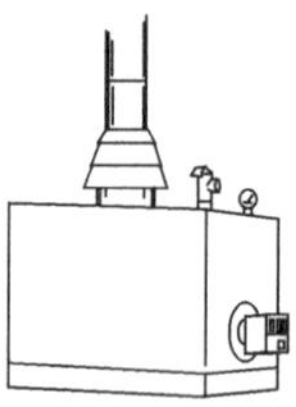

Description	Craft@Hrs	Unit	Material $	Labor $	Equipment $	Total $
4.5 BHP	SN@8.50	Ea	5,550.00	394.00	145.00	6,089.00
5.9 BHP	SN@9.50	Ea	6,920.00	440.00	162.00	7,522.00
9.0 BHP	SN@10.0	Ea	10,300.00	464.00	171.00	10,935.00
15.0 BHP	SN@12.0	Ea	16,100.00	556.00	205.00	16,861.00
20.0 BHP	SN@14.0	Ea	21,300.00	649.00	239.00	22,188.00
30.0 BHP	SN@16.0	Ea	27,400.00	742.00	273.00	28,415.00
45.0 BHP	SN@18.0	Ea	39,800.00	834.00	307.00	40,941.00
60.0 BHP	SN@20.0	Ea	51,700.00	927.00	341.00	52,968.00
Add for LEED-certified boiler	—	%	4.0	—	—	—
Add for LEED inspection	—	Ea	—	—	—	2,060.00
Add for central monitor/ recorder	—	Ea	1,820.00	—	—	1,820.00
Add for recording thermostat	—	Ea	424.00	—	—	424.00

Hot water Scotch marine firetube gas boiler, 30 PSIG. Costs shown are based on boiler horsepower rating (BHP). Set in place only. Make additional allowances for associated appurtenances and pipe connections.

Description	Craft@Hrs	Unit	Material $	Labor $	Equipment $	Total $
30 BHP	SN@10.0	Ea	39,600.00	464.00	171.00	40,235.00
40 BHP	SN@10.0	Ea	41,900.00	464.00	171.00	42,535.00
50 BHP	SN@10.0	Ea	43,700.00	464.00	171.00	44,335.00
60 BHP	SN@12.0	Ea	48,500.00	556.00	205.00	49,261.00
70 BHP	SN@12.0	Ea	52,400.00	556.00	205.00	53,161.00
80 BHP	SN@14.0	Ea	55,800.00	649.00	239.00	56,688.00
100 BHP	SN@14.0	Ea	66,500.00	649.00	239.00	67,388.00
125 BHP	SN@16.0	Ea	71,300.00	742.00	273.00	72,315.00
150 BHP	SN@18.0	Ea	76,600.00	834.00	307.00	77,741.00
200 BHP	SN@18.0	Ea	90,300.00	834.00	307.00	91,441.00
250 BHP	SN@26.0	Ea	99,400.00	1,210.00	444.00	101,054.00
300 BHP	SN@26.0	Ea	107,000.00	1,210.00	444.00	108,654.00
350 BHP	SN@26.0	Ea	122,000.00	1,210.00	444.00	123,654.00
400 BHP	SN@32.0	Ea	125,000.00	1,480.00	546.00	127,026.00
500 BHP	SN@32.0	Ea	148,000.00	1,480.00	546.00	150,026.00
600 BHP	SN@32.0	Ea	167,000.00	1,480.00	546.00	169,026.00
750 BHP	SN@40.0	Ea	174,000.00	1,850.00	683.00	176,533.00
Add for LEED-certified boiler	—	%	4.0	—	—	—
Add for LEED inspection	—	Ea	—	—	—	2,060.00
Add for central monitor/ recorder	—	Ea	4,240.00	—	—	4,240.00
Add for recording thermostat	—	Ea	424.00	—	—	424.00

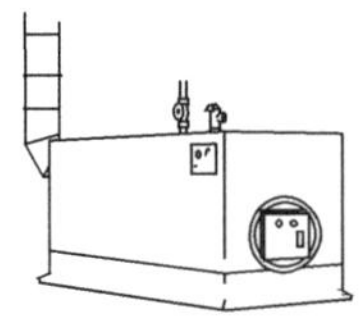

Description	Craft@Hrs	Unit	Material $	Labor $	Equipment $	Total $

Packaged Scotch marine firetube oil or gas fired boiler, 150 PSIG. Costs shown are based on boiler horsepower rating (BHP). Set in place only. Make additional allowances for associated appurtenances and pipe connections.

Description	Craft@Hrs	Unit	Material $	Labor $	Equipment $	Total $
60 BHP	SN@12.0	Ea	55,800.00	556.00	205.00	56,561.00
70 BHP	SN@12.0	Ea	60,000.00	556.00	205.00	60,761.00
80 BHP	SN@14.0	Ea	64,300.00	649.00	239.00	65,188.00
100 BHP	SN@14.0	Ea	77,100.00	649.00	239.00	77,988.00
125 BHP	SN@16.0	Ea	81,700.00	742.00	273.00	82,715.00
150 BHP	SN@18.0	Ea	87,300.00	834.00	307.00	88,441.00
200 BHP	SN@18.0	Ea	116,000.00	834.00	307.00	117,141.00
250 BHP	SN@26.0	Ea	123,000.00	1,210.00	444.00	124,654.00
300 BHP	SN@26.0	Ea	139,000.00	1,210.00	444.00	140,654.00
350 BHP	SN@26.0	Ea	143,000.00	1,210.00	444.00	144,654.00
400 BHP	SN@32.0	Ea	171,000.00	1,480.00	546.00	173,026.00
500 BHP	SN@32.0	Ea	189,000.00	1,480.00	546.00	191,026.00
600 BHP	SN@32.0	Ea	194,000.00	1,480.00	546.00	196,026.00
700 BHP	SN@40.0	Ea	201,000.00	1,850.00	683.00	203,533.00
Add for LEED-certified boiler	—	%	4.0	—	—	—
Add for LEED inspection	—	Ea	—	—	—	2,060.00
Add for central monitor/ recorder	—	Ea	4,240.00	—	—	4,240.00
Add for recording thermostat	—	Ea	424.00	—	—	424.00

Packaged hot water firebox oil or gas fired boiler. Costs shown are based on boiler horsepower rating (BHP). Set in place only. Make additional allowances for associated appurtenances and pipe connections.

Description	Craft@Hrs	Unit	Material $	Labor $	Equipment $	Total $
13 BHP	SN@8.00	Ea	21,000.00	371.00	137.00	21,508.00
16 BHP	SN@8.00	Ea	21,300.00	371.00	137.00	21,808.00
19 BHP	SN@8.00	Ea	22,000.00	371.00	137.00	22,508.00
22 BHP	SN@8.00	Ea	22,600.00	371.00	137.00	23,108.00
28 BHP	SN@8.00	Ea	23,600.00	371.00	137.00	24,108.00
34 BHP	SN@8.00	Ea	27,100.00	371.00	137.00	27,608.00
40 BHP	SN@8.00	Ea	29,900.00	371.00	137.00	30,408.00
46 BHP	SN@8.00	Ea	34,100.00	371.00	137.00	34,608.00
52 BHP	SN@8.00	Ea	39,900.00	371.00	137.00	40,408.00
61 BHP	SN@8.00	Ea	44,000.00	371.00	137.00	44,508.00
70 BHP	SN@8.00	Ea	47,100.00	371.00	137.00	47,608.00
80 BHP	SN@8.00	Ea	48,500.00	371.00	137.00	49,008.00
100 BHP	SN@8.00	Ea	54,800.00	371.00	137.00	55,308.00
125 BHP	SN@16.0	Ea	65,100.00	742.00	273.00	66,115.00
150 BHP	SN@16.0	Ea	70,000.00	742.00	273.00	71,015.00
Add for LEED-certified boiler	—	%	4.0	—	—	—
Add for LEED inspection	—	Ea	—	—	—	2,060.00
Add for central monitor/ recorder	—	Ea	3,030.00	—	—	3,030.00
Add for recording thermostat	—	Ea	424.00	—	—	424.00

Description	Craft@Hrs	Unit	Material $	Labor $	Equipment $	Total $

Packaged watertube industrial high pressure boilers. Costs shown are based on boiler horsepower rating (BHP). Set in place only. Make additional allowances for associated appurtenances and pipe connections.

480 BHP	SN@32.0	Ea	396,000.00	1,480.00	546.00	398,026.00
526 BHP	SN@32.0	Ea	405,000.00	1,480.00	546.00	407,026.00
572 BHP	SN@32.0	Ea	428,000.00	1,480.00	546.00	430,026.00
618 BHP	SN@32.0	Ea	436,000.00	1,480.00	546.00	438,026.00
664 BHP	SN@32.0	Ea	455,000.00	1,480.00	546.00	457,026.00
710 BHP	SN@64.0	Ea	461,000.00	2,970.00	1,090.00	465,060.00
747 BHP	SN@64.0	Ea	469,000.00	2,970.00	1,090.00	473,060.00
792 BHP	SN@64.0	Ea	483,000.00	2,970.00	1,090.00	487,060.00
838 BHP	SN@64.0	Ea	498,000.00	2,970.00	1,090.00	502,060.00
884 BHP	SN@64.0	Ea	513,000.00	2,970.00	1,090.00	517,060.00
930 BHP	SN@64.0	Ea	519,000.00	2,970.00	1,090.00	523,060.00
1,000 BHP	SN@64.0	Ea	529,000.00	2,970.00	1,090.00	533,060.00
Add for LEED-certified boiler	—	%	4.0	—	—	—
Add for LEED inspection	—	Ea	—	—	—	2,060.00
Add for central monitor/ recorder	—	Ea	5,210.00	—	—	5,210.00
Add for recording thermostat	—	Ea	424.00	—	—	424.00

Gas fired hot water steel boilers, 12 PSIG Mid-efficiency (82%) atmospheric draft, standing pilot light, c/w primary loop circulating pump. Labor includes setting in place and connecting pump and controls. Make additional allowances for distribution piping connections, safety devices, flue, gas connections, wiring, etc.

50,000 Btu	P1@4.00	Ea	1,550.00	178.00	68.30	1,796.30
70,000 Btu	P1@4.25	Ea	1,770.00	189.00	72.20	2,031.20
105,000 Btu	P1@4.50	Ea	1,950.00	200.00	76.80	2,226.80
140,000 Btu	P1@4.75	Ea	2,240.00	211.00	81.10	2,532.10
175,000 Btu	P1@5.00	Ea	2,490.00	222.00	85.40	2,797.40
210,000 Btu	P1@5.50	Ea	3,160.00	244.00	93.90	3,497.90
245,000 Btu	P1@7.00	Ea	3,520.00	311.00	120.00	3,951.00
Electronic spark ignition Add	—	Ea	332.00	—	—	332.00
Add for LEED-certified boiler	—	%	15.0	—	—	—
Add for LEED inspection	—	Ea	—	—	—	2,060.00
Add for central monitor/ recorder	—	Ea	1,330.00	—	—	1,330.00
Add for recording thermostat	—	Ea	182.00	—	—	182.00

Description	Craft@Hrs	Unit	Material $	Labor $	Equipment $	Total $

Gas fired hot water cast iron boilers, 12 PSIG Mid-efficiency (82%) direct vent, forced draft electronic spark ignition, c/w primary loop circulating pump. Labor includes setting in place and connecting pump and controls. Make additional allowances for distribution piping connections, safety devices, flue, gas connections, wiring, etc.

Description	Craft@Hrs	Unit	Material $	Labor $	Equipment $	Total $
50,000 Btu	P1@4.00	Ea	1,940.00	178.00	68.30	2,186.30
65,000 Btu	P1@4.25	Ea	2,170.00	189.00	72.60	2,431.60
100,000 Btu	P1@4.50	Ea	2,370.00	200.00	76.80	2,646.80
130,000 Btu	P1@4.75	Ea	2,630.00	211.00	81.10	2,922.10
170,000 Btu	P1@5.00	Ea	3,040.00	222.00	85.40	3,347.40
200,000 Btu	P1@5.50	Ea	3,340.00	244.00	93.90	3,677.90
235,000 Btu	P1@7.00	Ea	3,690.00	311.00	120.00	4,121.00
Add for LEED-certified boiler	—	%	15.0	—	—	—
Add for LEED inspection	—	Ea	—	—	—	2,060.00
Add for central monitor/ recorder	—	Ea	1,330.00	—	—	1,330.00
Add for recording thermostat	—	Ea	182.00	—	—	182.00

Description	Craft@Hrs	Unit	Material $	Labor $	Total $

Steam heating boilers for residential and light commercial applications. Parker or equal 15 PSIG Spiral watertube type, with ASTM SA-53 grade tubing, full American Society of Mechanical Engineers "M", "H" and "S" ratings and Underwriters Laboratories certifications. Specifically designed for "twinning" and multiple unit additions and retrofits. Packaged type units with ASME rated pressure relief valve, low water cutoff valve, main and check valves, pressure gauge, and pre-wired Fireye/McDonnell Miller combustion and steam regulator controls. Also includes pre-assembled FM/IRI approved natural gas combustion train w/Fisher or equal diaphragm regulator valve, check trim and vent line fittings or pre-assembled FM/IRI fuel oil combustion train with oil preheater, valved recirculation loop, filter check trim and fuel oil pump. Does not include sheet metal stack or vents, installation permits and approvals, or inspection costs. (BHP = boiler horse power)

Description	Craft@Hrs	Unit	Material $	Labor $	Total $
37,500 Btu (1 BHP)	P1@3.00	Ea	2,060.00	133.00	2,193.00
50,000 Btu (1.5 BHP)	P1@3.00	Ea	2,500.00	133.00	2,633.00
100,000 Btu (2.0 BHP)	P1@4.00	Ea	3,060.00	178.00	3,238.00
150,000 Btu (2.5 BHP)	P1@4.25	Ea	3,860.00	189.00	4,049.00
200,000 Btu (3.0 BHP)	P1@5.00	Ea	4,090.00	222.00	4,312.00
Add for LEED-certified boiler	—	%	15.0	—	—
Add for LEED inspection	—	Ea	—	—	2,060.00
Add for central monitor/ recorder	—	Ea	1,390.00	—	1,390.00
Add for recording thermostat	—	Ea	182.00	—	182.00

Description	Craft@Hrs	Unit	Material $	Labor $	Equipment $	Total $

Commercial hot water boiler connection. Includes pipe, fittings, water feed and pressure relief (drains), pipe insulation, valves, gauges, vents and thermometers to connect boiler to distribution system.

Description	Craft@Hrs	Unit	Material $	Labor $	Equipment $	Total $
2½" supply	SN@18.0	Ea	2,420.00	834.00	12.00	3,266.00
3" supply	SN@20.5	Ea	3,040.00	950.00	13.70	4,003.70
4" supply	SN@26.0	Ea	4,390.00	1,210.00	17.40	5,617.40
6" supply	SN@32.0	Ea	6,460.00	1,480.00	21.40	7,961.40
8" supply	SN@39.5	Ea	8,610.00	1,830.00	26.40	10,466.40
10" supply	SN@48.0	Ea	12,200.00	2,220.00	32.10	14,452.10
12" supply	SN@54.0	Ea	15,300.00	2,500.00	36.10	17,836.10

Commercial steam boiler connection. Includes pipe, fittings, water feed and pressure relief (drains), pipe insulation, valves, gauges, vents and thermometers to connect boiler to distribution system.

Description	Craft@Hrs	Unit	Material $	Labor $	Equipment $	Total $
2½" supply	SN@34.0	Ea	4,650.00	1,580.00	22.80	6,252.80
3" supply	SN@38.0	Ea	4,880.00	1,760.00	25.40	6,665.40
4" supply	SN@43.0	Ea	5,690.00	1,990.00	28.80	7,708.80
6" supply	SN@53.0	Ea	7,800.00	2,460.00	35.50	10,295.50
8" supply	SN@62.0	Ea	10,000.00	2,870.00	41.50	12,911.50
10" supply	SN@70.0	Ea	13,700.00	3,240.00	46.90	16,986.90
12" supply	SN@82.0	Ea	17,100.00	3,800.00	54.90	20,954.90

Additional costs for industrial, commercial and institutional packaged boilers. See more detailed installation costs and boiler trim costs in the sections that follow.

Description	Craft@Hrs	Unit	Material $	Labor $	Equipment $	Total $
Form and pour an interior slab	CF@.156	SF	9.74	6.81	—	16.55
Place vibration pad and anchors	CF@.100	SF	5.92	4.37	—	10.29
Boltdown of boiler	P1@3.00	Ea	—	133.00	—	133.00
Fuel train and feedwater lines	P1@.200	LF	12.40	8.89	—	21.29
Mount interior boiler drain	P1@.500	LF	9.67	22.20	—	31.87
Bore pipe hole in concrete wall	P1@.250	Ea	—	11.10	—	11.10
Bore stack vent in concrete wall	P1@.250	Ea	—	11.10	—	11.10
Mount and edge-seal stack	SN@.100	LF	9.96	4.64	—	14.60
Mount circulating pump	P1@.450	Ea	12.40	20.00	—	32.40
Expansion tank or condensate return lines	P1@.250	LF	4.01	11.10	—	15.11
Electrical wiring	BE@.150	LF	2.37	7.25	—	9.62
Combustion controls	BE@.200	LF	4.37	9.67	—	14.04
Install flue gas monitoring system	BE@.200	LF	4.32	9.67	—	13.99

Description	Craft@Hrs	Unit	Material $	Labor $	Equipment $	Total $

Additional costs for steam heating boilers for residential and light commercial applications. See more detailed installation costs and boiler trim costs in the sections that follow.

Description	Craft@Hrs	Unit	Material $	Labor $	Equipment $	Total $
Form and pour a 4' x 4' interior slab	CF@2.50	Ea	155.00	109.00	—	264.00
Place 4' x 4' x 1/2" vibration pads	CF@.750	Ea	45.20	32.70	—	77.90
Boltdown of boiler	P1@1.00	Ea	—	44.50	—	44.50
Fuel train and feedwater lines	P1@2.50	Ea	—	111.00	—	111.00
Mount interior boiler drain	P1@.500	LF	9.09	22.20	—	31.29
Bore piping hole in basement wall	P1@.250	Ea	—	11.10	—	11.10
Bore stack vent in concrete wall	P1@.250	Ea	—	11.10	—	11.10
Mount and edge-seal stack	P1@.500	Ea	47.20	22.20	—	69.40
Mount circulating pump	P1@.450	Ea	—	20.00	—	20.00
Install expansion tank						
2.1 gallon	P1@.300	Ea	42.20	13.30	—	55.50
4.5 gallon	P1@.300	Ea	70.90	13.30	—	84.20

Costs for Combustion Trains for Boilers. Factory Mutual, Industrial Risk Insurers, and Underwriters' Laboratories certified, U.S. Green Building Council rated, LFUE (Levelized Fuel Utilization Efficiency) rated at 90% or better. Includes the burner, burner head, burner mount flange, associated refractory and gasketing, main fuel valve, draft damper and fan (for a power burner), or burner face plate (for an atmospheric or induced-draft burner). Note: BHP multiplied by 33.5 is the approximate required 1,000 Btu per hour input (MBH input). Cost of fuel line relief vent piping and trip valve to outside of the building are not required in all jurisdictions and are not assumed in these costs.

Description	Craft@Hrs	Unit	Material $	Labor $	Equipment $	Total $

Combustion train for Scotch marine firetube boilers. Costs shown are based on boiler horsepower rating (BHP).

20 BHP	SN@8.00	Ea	3,180.00	371.00	5.36	3,556.36
30 BHP	SN@8.00	Ea	4,300.00	371.00	5.36	4,676.36
40 BHP	SN@8.00	Ea	5,900.00	371.00	5.36	6,276.36
50 BHP	SN@8.00	Ea	7,310.00	371.00	5.36	7,686.36
60 BHP	SN@8.00	Ea	8,130.00	371.00	5.36	8,506.36
70 BHP	SN@8.00	Ea	8,900.00	371.00	5.36	9,276.36
80 BHP	SN@8.00	Ea	9,070.00	371.00	5.36	9,446.36
100 BHP	SN@8.00	Ea	9,910.00	371.00	5.36	10,286.36
125 BHP	SN@8.00	Ea	10,300.00	371.00	5.36	10,676.36
150 BHP	SN@8.00	Ea	11,300.00	371.00	5.36	11,676.36
200 BHP	SN@8.00	Ea	11,800.00	371.00	5.36	12,176.36
250 BHP	SN@8.00	Ea	13,100.00	371.00	5.36	13,476.36
300 BHP	SN@8.00	Ea	14,800.00	371.00	137.00	15,308.00
350 BHP	SN@8.00	Ea	15,000.00	371.00	137.00	15,508.00
400 BHP	SN@8.00	Ea	15,300.00	371.00	137.00	15,808.00
500 BHP	SN@16.0	Ea	16,600.00	742.00	273.00	17,615.00
600 BHP	SN@16.0	Ea	18,100.00	742.00	273.00	19,115.00
750 BHP	SN@32.0	Ea	19,100.00	1,480.00	546.00	21,126.00
800 BHP	SN@32.0	Ea	19,900.00	1,480.00	546.00	21,926.00

Combustion train for package firebox boilers. Costs shown are based on boiler horsepower rating (BHP).

13.4 BHP	SN@8.00	Ea	1,630.00	371.00	137.00	2,138.00
16.4 BHP	SN@8.00	Ea	2,170.00	371.00	137.00	2,678.00
17.4 BHP	SN@8.00	Ea	2,570.00	371.00	137.00	3,078.00
22.4 BHP	SN@8.00	Ea	3,100.00	371.00	137.00	3,608.00
28.4 BHP	SN@8.00	Ea	3,270.00	371.00	137.00	3,778.00
34.4 BHP	SN@8.00	Ea	3,610.00	371.00	137.00	4,118.00
40.3 BHP	SN@8.00	Ea	3,930.00	371.00	137.00	4,438.00
46.3 BHP	SN@8.00	Ea	4,320.00	371.00	137.00	4,828.00
52.3 BHP	SN@8.00	Ea	4,820.00	371.00	137.00	5,328.00
61.1 BHP	SN@8.00	Ea	5,240.00	371.00	137.00	5,748.00
70.0 BHP	SN@8.00	Ea	5,360.00	371.00	137.00	5,868.00
80.0 BHP	SN@8.00	Ea	5,580.00	371.00	137.00	6,088.00
100.0 BHP	SN@8.00	Ea	5,900.00	371.00	137.00	6,408.00
125.0 BHP	SN@8.00	Ea	6,310.00	371.00	137.00	6,818.00
150.0 BHP	SN@8.00	Ea	6,840.00	371.00	137.00	7,348.00

Description	Craft@Hrs	Unit	Material $	Labor $	Equipment $	Total $

Natural gas fuel train piping for firetube and watertube boilers. Meets requirements of Underwriters' Laboratories. Shipped pre-assembled. Add the cost of venting and piping to a gas meter. Costs shown are based on boiler horsepower rating (BHP).

20 BHP	SN@8.00	Ea	831.00	371.00	137.00	1,339.00
30 BHP	SN@8.00	Ea	1,130.00	371.00	137.00	1,638.00
40 BHP	SN@8.00	Ea	1,740.00	371.00	137.00	2,248.00
50 BHP	SN@8.00	Ea	2,060.00	371.00	137.00	2,568.00
60 BHP	SN@8.00	Ea	2,260.00	371.00	137.00	2,768.00
70 BHP	SN@8.00	Ea	2,400.00	371.00	137.00	2,908.00
80 BHP	SN@8.00	Ea	2,610.00	371.00	137.00	3,118.00
100 BHP	SN@8.00	Ea	3,020.00	371.00	137.00	3,528.00
125 BHP	SN@8.00	Ea	3,590.00	371.00	137.00	4,098.00
150 BHP	SN@8.00	Ea	4,260.00	371.00	137.00	4,768.00
200 BHP	SN@8.00	Ea	5,110.00	371.00	137.00	5,618.00
250 BHP	SN@8.00	Ea	6,310.00	371.00	137.00	6,818.00
300 BHP	SN@8.00	Ea	6,650.00	371.00	137.00	7,158.00
350 BHP	SN@8.00	Ea	7,120.00	371.00	137.00	7,628.00
400 BHP	SN@8.00	Ea	7,660.00	371.00	137.00	8,168.00
500 BHP	SN@16.0	Ea	8,610.00	742.00	273.00	9,625.00
600 BHP	SN@16.0	Ea	9,910.00	742.00	273.00	10,925.00
750 BHP	SN@16.0	Ea	10,700.00	742.00	273.00	11,715.00
800 BHP	SN@16.0	Ea	11,100.00	742.00	273.00	12,115.00

Natural gas fuel train piping for package firebox boilers. Costs shown are based on boiler horsepower rating (BHP).

13 BHP	SN@8.00	Ea	1,100.00	371.00	137.00	1,608.00
16 BHP	SN@8.00	Ea	1,280.00	371.00	137.00	1,788.00
19 BHP	SN@8.00	Ea	1,370.00	371.00	137.00	1,878.00
22 BHP	SN@8.00	Ea	1,500.00	371.00	137.00	2,008.00
28 BHP	SN@8.00	Ea	1,580.00	371.00	137.00	2,088.00
34 BHP	SN@8.00	Ea	1,740.00	371.00	137.00	2,248.00
40 BHP	SN@8.00	Ea	2,060.00	371.00	137.00	2,568.00
45 BHP	SN@8.00	Ea	2,170.00	371.00	137.00	2,678.00
50 BHP	SN@8.00	Ea	2,610.00	371.00	137.00	3,118.00
60 BHP	SN@8.00	Ea	2,960.00	371.00	137.00	3,468.00
70 BHP	SN@8.00	Ea	3,320.00	371.00	137.00	3,828.00
80 BHP	SN@8.00	Ea	3,490.00	371.00	137.00	3,998.00
100 BHP	SN@8.00	Ea	4,170.00	371.00	137.00	4,678.00
125 BHP	SN@8.00	Ea	4,310.00	371.00	137.00	4,818.00
150 BHP	SN@8.00	Ea	4,730.00	371.00	137.00	5,238.00
Add for code approval:						
Factory Mutual	—	%	7.0	—	—	—
Industrial Risk	—	%	10.0	—	—	—

Oil Fuel Train Piping. Costs shown include recirculating pump, electric oil preheater, regulating valve, check valves, isolating pump stop cocks, oil line pressure relief valve and pressure gauges. Add the cost of concrete pump pad, electrical connections, relief line piping back to the oil tank and tank connecting piping. Add for number 4-6 fuel oil as shown.

Description	Craft@Hrs	Unit	Material $	Labor $	Equipment $	Total $
Oil fuel train piping for boilers. Costs shown are based on boiler horsepower rating (BHP) using number 2 fuel oil.						
20 to 30 BHP	SN@4.00	Ea	1,770.00	185.00	68.30	2,023.30
40 to 60 BHP	SN@4.00	Ea	1,990.00	185.00	68.30	2,243.30
70 to 80 BHP	SN@4.00	Ea	2,160.00	185.00	68.30	2,413.30
100 BHP	SN@8.00	Ea	2,770.00	371.00	137.00	3,278.00
125 to 150 BHP	SN@8.00	Ea	3,590.00	371.00	137.00	4,098.00
200 BHP	SN@8.00	Ea	4,740.00	371.00	137.00	5,248.00
250 BHP	SN@8.00	Ea	5,610.00	371.00	137.00	6,118.00
300 BHP	SN@16.0	Ea	6,000.00	742.00	273.00	7,015.00
350 BHP	SN@16.0	Ea	6,700.00	742.00	273.00	7,715.00
400 BHP	SN@16.0	Ea	8,050.00	742.00	273.00	9,065.00
500 BHP	SN@32.0	Ea	8,140.00	1,480.00	546.00	10,166.00
600 BHP	SN@32.0	Ea	8,540.00	1,480.00	546.00	10,566.00
750 to 800 BHP	SN@32.0	Ea	9,330.00	1,480.00	546.00	11,356.00
900 to 1,000 BHP	SN@32.0	Ea	10,900.00	1,480.00	546.00	12,926.00
Add for #4-6 oil	—	%	10.0	—	—	—
Add for tie-in with:						
dual-fuel burner	—	%	5.0	—	—	—

Dual-fuel boiler burners. For a new installation or retrofit for Scotch marine firetube, firebox or package watertube boilers. Rated at the best available control technology (BACT) for natural gas. Complies with SCAQMD rules 1146.2, 222 and 219 and the Clean Air Act section 307(d). Includes control of CO_2 emissions as well as NOx, SOx, VMs. Cleaver Brooks or equal. Factory Mutual, Industrial Risk Insurers, and Underwriters' Laboratories certified, U.S. Green Building Council rated, LFUE (Levelized Fuel Utilization Efficiency) rated at 90% or better. Includes the burner, burner head, burner mount flange, associated refractory and gasketing, main natural gas and oil fuel valves, draft damper and fan (for a power burner), or burner face plate (for an atmospheric or induced-draft burner) and the natural gas fuel train. Add the cost of an oil fuel train to complete dual-fuel bid. The control panel includes the control circuit, burner mounted control panel, motor starter(s), panel signal lights and full modulation with manual potentiometer, flame safeguard controls, flue gas recirculator air inlet butterfly valve and relay actuator, combustion air proving switch and high fire air interlock switch. Note: BHP multiplied by 33.5 is the approximate required 1,000 Btu per hour input (MBH input). Add the installation costs below and the cost of natural gas fuel line relief vent piping and a trip valve to outside of the building if required.

Description	Craft@Hrs	Unit	Material $	Labor $	Equipment $	Total $
Dual-fuel boiler burners. Costs shown are based on boiler horsepower rating (BHP).						
20 BHP	SN@8.00	Ea	5,660.00	371.00	—	6,031.00
30 BHP	SN@8.00	Ea	6,580.00	371.00	—	6,951.00
40 BHP	SN@8.00	Ea	8,120.00	371.00	—	8,491.00
50 BHP	SN@8.00	Ea	9,280.00	371.00	—	9,651.00
60 BHP	SN@8.00	Ea	10,500.00	371.00	—	10,871.00
70 BHP	SN@8.00	Ea	11,600.00	371.00	—	11,971.00
80 BHP	SN@8.00	Ea	13,200.00	371.00	—	13,571.00
100 BHP	SN@8.00	Ea	15,600.00	371.00	—	15,971.00
125 BHP	SN@8.00	Ea	17,500.00	371.00	—	17,871.00
150 BHP	SN@8.00	Ea	18,400.00	371.00	—	18,771.00
200 BHP	SN@8.00	Ea	23,300.00	371.00	—	23,671.00
250 BHP	SN@8.00	Ea	29,800.00	371.00	—	30,171.00
300 BHP	SN@8.00	Ea	35,300.00	371.00	—	35,671.00
350 BHP	SN@8.00	Ea	39,200.00	371.00	—	39,571.00
400 BHP	SN@8.00	Ea	41,600.00	371.00	—	41,971.00
500 BHP	SN@16.0	Ea	42,700.00	742.00	—	43,442.00
600 BHP	SN@16.0	Ea	47,900.00	742.00	—	48,642.00
750 BHP	SN@32.0	Ea	51,500.00	1,480.00	—	52,980.00
800 BHP	SN@32.0	Ea	52,200.00	1,480.00	—	53,680.00

Description	Craft@Hrs	Unit	Material $	Labor $	Equipment $	Total $
Additional costs for installing dual-fuel boiler burners						
Form and pour 4' x 4' x 2" interior slab	CF@2.50	Ea	155.00	109.00	—	264.00
Bore fuel line hole through wall or pad	P1@.250	Ea	—	11.10	—	11.10
Bore flue gas hole through wall or pad	P1@.250	Ea	—	11.10	—	11.10
Place burner and control panel enclosure	ER@8.00	Ea	—	379.00	—	379.00
Mount, level and grout burner, duct and control panel	P1@.450	Ea	487.00	20.00	—	507.00
Piping for fuel lines and flue gas recirculator	P1@.500	LF	9.67	22.20	—	31.87
Electrical wiring	BE@.300	LF	4.01	14.50	—	18.51
Wire control panel	BE@.300	LF	1.86	14.50	—	16.36
Ground fault protection wiring	BE@.300	LF	5.87	14.50	—	20.37
Commission and test	BE@8.00	Ea	—	387.00	—	387.00

Costs for Electrical Service for Boilers, Subcontract. Costs shown include typical main power tie-in and fusing, circuit breaker panel, and main bus. Assumes controls, fan, blower switchgear and control console are pre-wired electrical service. These costs include the electrical subcontractor's overhead and profit.

Description	Craft@Hrs	Unit	Material $	Labor $	Total $
Electrical service for firetube and watertube boilers. Costs shown are based on boiler horsepower rating (BHP).					
20 to 100 BHP	—	Ea	—	—	1,560.00
125 to 200 BHP	—	Ea	—	—	1,880.00
250 to 350 BHP	—	Ea	—	—	2,130.00
400 to 500 BHP	—	Ea	—	—	2,450.00
500 to 600 BHP	—	Ea	—	—	3,530.00
750 to 800 BHP	—	Ea	—	—	6,370.00
Electrical service for package firebox boilers. Costs shown are based on boiler horsepower rating (BHP).					
13 to 70 BHP	—	Ea	—	—	1,510.00
80 to 125 BHP	—	Ea	—	—	1,540.00
150 BHP	—	Ea	—	—	1,900.00

Commercial Boiler Components and Accessories

Costs for Refractory for Boilers, Subcontract. For package Scotch marine firetube, firebox and watertube boilers. Most new boilers are shipped from the factory with the refractory brick installed. If refractory brick is installed on the job site to repair shipping damage or to create firedoor seals, use the figures below. These costs include the refractory subcontractor's overhead and profit. Costs shown are per 100 pounds of refractory used.

Description	Craft@Hrs	Unit	Material $	Labor $	Equipment $	Total $
Repairs and seals	—	100 lb	—	—	—	255.00

Costs for Boiler Feedwater Pumps. Costs shown are for the pump only. Add the cost of piping, pump actuating valves and concrete pump base.

Description	Craft@Hrs	Unit	Material $	Labor $	Equipment $	Total $
Feedwater pumps for firetube and watertube boilers. Costs shown are based on boiler horsepower rating (BHP).						
20 to 40 BHP	SN@8.00	Ea	11,800.00	371.00	137.00	12,308.00
50 to 80 BHP	SN@8.00	Ea	14,200.00	371.00	137.00	14,708.00
100 BHP	SN@8.00	Ea	15,800.00	371.00	137.00	16,308.00
125 to 200 BHP	SN@16.0	Ea	16,200.00	742.00	273.00	17,215.00
250 to 400 BHP	SN@32.0	Ea	18,100.00	1,480.00	546.00	20,126.00
500 BHP	SN@32.0	Ea	19,100.00	1,480.00	546.00	21,126.00
600 BHP	SN@32.0	Ea	19,600.00	1,480.00	546.00	21,626.00
750 BHP	SN@32.0	Ea	20,200.00	1,480.00	546.00	22,226.00
800 BHP	SN@32.0	Ea	27,300.00	1,480.00	546.00	29,326.00
Feedwater pumps for package firebox boilers. Costs shown are based on boiler horsepower rating (BHP).						
13 to 35 BHP	SN@8.00	Ea	6,340.00	371.00	137.00	6,848.00
40 to 45 BHP	SN@8.00	Ea	7,450.00	371.00	137.00	7,958.00
50 to 60 BHP	SN@8.00	Ea	9,330.00	371.00	137.00	9,838.00
70 to 80 BHP	SN@8.00	Ea	11,000.00	371.00	137.00	11,508.00
100 BHP	SN@8.00	Ea	14,200.00	371.00	137.00	14,708.00
125 BHP	SN@16.0	Ea	14,200.00	742.00	273.00	15,215.00
150 BHP	SN@16.0	Ea	15,800.00	742.00	273.00	16,815.00

Costs for Natural Gas Combustion Controls for Boilers. Instrumentation Society of America, Underwriters' Laboratories, Factory Mutual, and Industrial Risk Insurers certified. Fireye or equal. Includes automatic recycling or continuous pilot, spark igniter and coil (for electric ignition), timer to control the purge cycle, pilot, main burner (hi-lo-off) or main burner (proportional control), flame safeguard and sensor and combustion control panel.

Description	Craft@Hrs	Unit	Material $	Labor $	Equipment $	Total $
Natural gas combustion controls for firetube and watertube boilers. Costs shown are based on boiler horsepower rating (BHP).						
20 to 80 BHP	SN@4.00	Ea	4,820.00	185.00	68.30	5,073.30
100 BHP	SN@4.00	Ea	5,360.00	185.00	68.30	5,613.30
125 to 200 BHP	SN@8.00	Ea	5,360.00	371.00	137.00	5,868.00
250 BHP	SN@16.0	Ea	5,360.00	742.00	273.00	6,375.00
300 BHP	SN@16.0	Ea	6,430.00	742.00	273.00	7,445.00
350 or 400 BHP	SN@16.0	Ea	7,530.00	742.00	273.00	8,545.00
500 BHP	SN@32.0	Ea	7,530.00	1,480.00	546.00	9,556.00
600 BHP	SN@32.0	Ea	8,990.00	1,480.00	546.00	11,016.00
750 BHP	SN@64.0	Ea	10,500.00	2,970.00	1,090.00	14,560.00
800 BHP	SN@64.0	Ea	12,100.00	2,970.00	1,090.00	16,160.00

Description	Craft@Hrs	Unit	Material $	Labor $	Equipment $	Total $

Natural gas combustion controls for firebox boilers. Costs shown are based on boiler horsepower rating (BHP).

13.4-80 BHP	SN@4.00	Ea	4,630.00	185.00	68.30	4,883.30
100.0 BHP	SN@4.00	Ea	5,360.00	185.00	68.30	5,613.30
126.9 BHP	SN@8.00	Ea	5,360.00	371.00	137.00	5,868.00
150.0 BHP	SN@8.00	Ea	9,990.00	371.00	137.00	10,498.00
Add if burner controls are for:						
#2 oil-fired	—	%	—	—	—	10.0
Heavy oil-fired	—	%	—	—	—	20.0
Dual fuel burners	—	%	—	—	—	70.0

Costs for Water Softening Systems for Use with Boilers. Costs shown include the metering valve, tank, pump, and zeolite media. Add the cost of piping to connect the water softening system to the main feedwater tank and pump.

Water softening system for firetube and watertube boilers. Costs shown are based on boiler horsepower rating (BHP).

20 to 80 BHP	SN@4.00	Ea	7,720.00	185.00	68.30	7,973.30
100 BHP	SN@4.00	Ea	9,060.00	185.00	68.30	9,313.30
125 BHP	SN@8.00	Ea	1,050.00	371.00	137.00	1,558.00
150 or 200 BHP	SN@8.00	Ea	13,400.00	371.00	137.00	13,908.00
250 BHP	SN@8.00	Ea	15,800.00	371.00	137.00	16,308.00
300 or 350 BHP	SN@16.0	Ea	20,200.00	742.00	273.00	21,215.00
400 or 500 BHP	SN@16.0	Ea	25,300.00	742.00	273.00	26,315.00
600 BHP	SN@32.0	Ea	30,600.00	1,480.00	546.00	32,626.00
750 or 800 BHP	SN@64.0	Ea	35,600.00	2,970.00	1,090.00	39,660.00

Water softening system for package firebox boilers. Costs shown are based on boiler horsepower rating (BHP).

13.4 to 80 BHP	SN@4.00	Ea	9,080.00	185.00	68.30	9,333.30
100.0 BHP	SN@4.00	Ea	9,080.00	185.00	68.30	9,333.30
126.9 BHP	SN@8.00	Ea	10,700.00	371.00	137.00	11,208.00
150.0 BHP	SN@8.00	Ea	12,300.00	371.00	137.00	12,808.00

Costs for Start-up, Shakedown, and Calibration of Boilers. Costs shown include minor changes and adjustments to meet requirements of the boiler inspector, adjustment of combustion efficiency at the burner by a control technician, pump and metering calibration testing and steam or hot water metering on start-up.

Start-up, shakedown & calibration of firetube and watertube boilers. Costs shown are based on boiler horsepower rating (BHP.) If a factory technician performs the start-up, add this cost at a minimum of $800 (8 hours at $100.00 per hour).

13.4 to 100 BHP	S2@4.00	Ea	—	174.00	—	174.00
125 to 200 BHP	S2@8.00	Ea	—	348.00	—	348.00
250 to 400 BHP	S2@16.0	Ea	—	696.00	—	696.00
500 to 600 BHP	S2@24.0	Ea	—	1,040.00	—	1,040.00
750 to 800 BHP	S2@30.0	Ea	—	1,310.00	—	1,310.00

Costs for Stack Economizer Units for Boilers. Costs shown include piping connection to existing boiler, economizer fin tube coils, feedwater pump, check valves, thermostatic regulating valve, stack dampers, pressure relief valve, stack mount flanges and insulating enclosure, but no architectural modifications.

Description	Craft@Hrs	Unit	Material $	Labor $	Equipment $	Total $
Stack waste heat recovery system for boilers. Costs shown are based on boiler horsepower rating (BHP).						
500 BHP	SN@40.0	Ea	44,300.00	1,850.00	683.00	46,833.00
750 BHP	SN@48.0	Ea	55,100.00	2,220.00	820.00	58,140.00
800 BHP	SN@48.0	Ea	59,400.00	2,220.00	820.00	62,440.00
1,000 BHP	SN@56.0	Ea	74,200.00	2,600.00	956.00	77,756.00
Pollution control modules for gas-fired industrial package watertube boilers. NOx, SOx, VM (volatile materials) and particulates emissions reduction module. Free-standing unit. Verantis, Croll-Reynolds or equal. Designed to meet or exceed 2010 California Code, Tier IV, EPA and European ISO standards. Either vertical counter-flow or packed-bed type. For multi-dwelling, process and industrial applications. By boiler rating in millions of Btu's per hour (MMBtu/hr). See installation and efficiency optimizing control costs below. Oil-fired boilers will add about 15% to the installed cost, including a second stage reduction module, second stage pump, sulfur extraction, venturi particulate separator and sludge tank. To calculate the annual lease rate for a complete installation, divide the total installed cost by five. Then add 2% per year to cover vendor-supplied maintenance.						
1.0 MMBtu/Hr	—	Ea	41,100.00	—	—	41,100.00
1.7 MMBtu/Hr	—	Ea	73,100.00	—	—	73,100.00
3.4 MMBtu/Hr	—	Ea	130,000.00	—	—	130,000.00
4.1 MMBtu/Hr	—	Ea	135,000.00	—	—	135,000.00
5.0 MMBtu/Hr	—	Ea	146,000.00	—	—	146,000.00
6.7 MMBtu/Hr	—	Ea	157,000.00	—	—	157,000.00
13.4 MMBtu/Hr	—	Ea	163,000.00	—	—	163,000.00
20.1 MMBtu/Hr	—	Ea	181,000.00	—	—	181,000.00
26.8 MMBtu/Hr	—	Ea	192,000.00	—	—	192,000.00
33.5 MMBtu/Hr	—	Ea	209,000.00	—	—	209,000.00
Installation costs for pollution control modules for gas-fired industrial package watertube boilers.						
Form and pour foundation	ER@.040	SF	2.49	1.90	—	4.39
Position main housing, fans and pumps	ER@4.00	Ea	—	190.00	—	190.00
Tie-down and leveling	ER@4.00	Ea	598.00	190.00	—	788.00
Mount piping system frame supports	ER@6.00	Ea	1.81	284.00	—	285.81
Rig and mount plenum and ducts	SN@5.00	Ea	—	232.00	—	232.00
Rig and mount tank for emission reduction, fluid and housing inlet metering valve	ER@4.00	Ea	—	190.00	—	190.00

Description	Craft@Hrs	Unit	Material $	Labor $	Equipment $	Total $

Installation costs for pollution control modules for gas-fired industrial package watertube boilers – (continued).

Description	Craft@Hrs	Unit	Material $	Labor $	Equipment $	Total $
Weld systems piping	ER@8.00	Ea	2.52	379.00	—	381.52
Wire electrical system	BE@8.00	Ea	2.15	387.00	—	389.15
Wire controls system	BE@16.0	Ea	2.28	774.00	—	776.28
Level and balance rotating equipment	P1@8.00	Ea	241.00	356.00	—	597.00
Calibration and test	ER@2.00	Ea	—	94.80	—	94.80
Witnessed run trials (add the permit fee)	P1@4.00	Ea	—	178.00	—	178.00

Efficiency-optimizing centralized boiler burner controls. Retrofit or new construction. Preferred Instruments or equal. For use with emissions monitoring and digital reporting pollution control equipment installed for a gas-fired industrial boiler. Complies with the Clean Air Act section 307(d), California and Massachusetts boiler codes and NFPA 85. Includes a boiler-mounted control panel with chassis, servo motor controls for burner fuel line valves, LCD touchpad control panel, ultra-violet, infra-red or self-checking flame scanner, operator interface panel and interface to a remote PC (not included). Controls variable speed drives for burner forced-draft fans, induced draft fans and feedwater pumps.

Description	Craft@Hrs	Unit	Material $	Labor $	Equipment $	Total $
Mount automation module, less wiring	BE@.250	Ea	1,330.00	12.10	—	1,342.10
Mount flame safety sensors less wiring	BE@.250	Ea	211.00	12.10	—	223.10
Servomotor mounting less wiring	BE@.250	Ea	562.00	12.10	—	574.10
Mount operator interface terminal with power supply less wiring	BE@.250	Ea	96.20	12.10	—	108.30
Operator interface terminal bridge, less wiring	BE@.250	Ea	4,570.00	12.10	—	4,582.10
Bore control wiring hole through basement wall	CF@.250	Ea	—	10.90	—	10.90
Twisted pair Cat 6, 1 GBPS wire 4 pair	BE@.008	LF	.58	.39	—	.97
Panel wiring to emissions monitoring array	BE@.250	LF	3.03	12.10	—	15.13
Sensor wiring to controller	BE@.300	LF	2.76	14.50	—	17.26
Electric servomotor wiring to controller	BE@.400	LF	2.52	19.30	—	21.82
Mount and wire PC module	BE@2.00	LF	24.20	96.70	—	120.90
Ground fault protection wiring	BE@1.25	LF	150.00	60.40	—	210.40
Startup, programming and system test	BE@3.00	Ea	—	145.00	—	145.00
Commissioning and final test	BE@8.00	Ea	—	387.00	—	387.00

Costs for Condensate Receiver and Pumping Units for Boilers. Costs shown include a stainless steel factory assembled tank and pump with motor, float switch, condensate inlet, feedwater tank, connection, level gauge, thermometer, solenoid valve for direct boiler feed. No concrete mounting pad required.

Description	Craft@Hrs	Unit	Material $	Labor $	Equipment $	Total $
Condensate receiver and pumping unit for boilers. Unit sizing and costs are based on storage for one minute at the designed steam return rate. Set in place only. Make additional allowances for pipe and electrical connections.						
750 lbs per hour	SN@4.00	Ea	5,770.00	185.00	2.68	5,957.68
1,500 lbs per hour	SN@4.00	Ea	5,770.00	185.00	2.68	5,957.68
3,000 lbs per hour	SN@4.00	Ea	7,330.00	185.00	2.68	7,517.68
5,000 lbs per hour	SN@6.00	Ea	8,090.00	278.00	4.02	8,372.02
6,250 lbs per hour	SN@6.00	Ea	9,010.00	278.00	4.02	9,292.02

Costs for Deaerator/Condenser Units for Boilers. Pressurized jet spray type. Costs shown include an ASME storage receiver, stand, jet spray deaerator head, makeup control valve, makeup controller, pressure control valve, level controls, pressure and temperature controls, pumped condensate inlet, pumps and motor, steam inlet and system accessories. Add for a concrete mounting pad and piping to the condensate return line (inlet) and piping to the boiler feedwater system (outlet).

Description	Craft@Hrs	Unit	Material $	Labor $	Equipment $	Total $
Deaerator/condenser unit for boilers. Costs shown are based on boiler horsepower rating (BHP).						
100 BHP	SN@8.00	Ea	60,400.00	371.00	5.36	60,776.36
200 BHP	SN@12.0	Ea	62,700.00	556.00	8.03	63,264.03
300 BHP	SN@14.0	Ea	66,500.00	649.00	9.37	67,158.37
400 BHP	SN@16.0	Ea	75,400.00	742.00	10.70	76,152.70
600 BHP	SN@24.0	Ea	80,900.00	1,110.00	16.10	82,026.10
800 BHP	SN@36.0	Ea	88,400.00	1,670.00	24.10	90,094.10
1,000 BHP	SN@48.0	Ea	103,000.00	2,220.00	32.10	105,252.10
Rolairtrol type air separators. Priced by capacity in gallons per minute (GPM) and by pipe connection size. Includes hanging separator but no connection, (2" is a screwed connection, 2½" and up are flanged connections) using small tools.						
56 GPM, 2"	S2@2.50	Ea	1,080.00	109.00	1.67	1,190.67
90 GPM, 2½"	S2@2.50	Ea	1,180.00	109.00	1.67	1,290.67
170 GPM, 3"	S2@3.00	Ea	1,550.00	131.00	2.01	1,683.01
300 GPM, 4"	S2@3.50	Ea	2,330.00	152.00	2.34	2,484.34
500 GPM, 6"	S2@4.75	Ea	3,040.00	207.00	3.18	3,250.18
700 GPM, 6"	S2@4.75	Ea	3,920.00	207.00	3.18	4,130.18
1,300 GPM, 8"	SN@6.00	Ea	5,270.00	278.00	4.02	5,552.02
2,000 GPM, 10"	SN@6.00	Ea	7,380.00	278.00	4.02	7,662.02
2,750 GPM, 12"	SN@12.0	Ea	9,570.00	556.00	8.03	10,134.03

Description	Craft@Hrs	Unit	Material $	Labor $	Equipment $	Total $

Costs for Chemical Feed Duplex Pump, Packaged Units for Boilers. Costs shown include the chemical tanks, electronic metering, valve and stand. Add the cost of piping to the feedwater tank.

Chemical feed duplex pump for boilers. Costs shown are based on boiler horsepower rating (BHP).

20 to 100 BHP	SN@3.00	Ea	2,210.00	139.00	51.20	2,400.20
25 to 200 BHP	SN@3.00	Ea	2,900.00	139.00	51.20	3,090.20
200 to 400 BHP	SN@6.00	Ea	2,900.00	278.00	102.00	3,280.00
500 to 600 BHP	SN@6.00	Ea	6,590.00	278.00	102.00	6,970.00
700 to 800 BHP	SN@8.00	Ea	7,370.00	371.00	137.00	7,878.00
1,000 BHP	SN@10.0	Ea	9,960.00	464.00	171.00	10,595.00

Chemical feed simple in-line system for hot water heating system. Costs shown include 3 gallon pot feeder, two 3/4" isolation valves, 10' of ¾" pipe and two ¾" threadolets. Chemicals are drawn in through suction created by hot water system circulation pumps.

50 to 20 BHP	SN@3.00	Ea	514.00	139.00	51.20	704.20

Packaged Boiler Feedwater Systems for Boilers. Costs shown include duplex pumps with heater assembly, thermal lining, pressure gauges, makeup feeder valve, ASME feedwater tank and stand, single phase electric motor and level gauge. Add the costs for a concrete mounting pad and piping to the boilers.

Packaged boiler feedwater systems. Costs shown are based on boiler horsepower rating (BHP).

15 to 50 BHP	SN@5.00	Ea	15,500.00	232.00	85.40	15,817.40
60 to 80 BHP	SN@5.00	Ea	17,100.00	232.00	85.40	17,417.40
100 BHP	SN@6.00	Ea	17,500.00	278.00	102.00	17,880.00
125 to 150 BHP	SN@6.00	Ea	18,700.00	278.00	102.00	19,080.00
200 BHP	SN@8.00	Ea	19,100.00	371.00	137.00	19,608.00
250 BHP	SN@8.00	Ea	19,900.00	371.00	137.00	20,408.00
300 BHP	SN@8.00	Ea	21,200.00	371.00	137.00	21,708.00
400 BHP	SN@8.00	Ea	21,800.00	371.00	137.00	22,308.00
500 BHP	SN@8.00	Ea	26,600.00	371.00	137.00	27,108.00
600 BHP	SN@12.0	Ea	27,700.00	556.00	205.00	28,461.00
750 BHP	SN@12.0	Ea	27,600.00	556.00	205.00	28,361.00

Continuous Blowdown Heat Recovery System. Includes a flash receiver (50 PSI ASME rated tank), plate and frame heat exchanger, pneumatic level controller and valve, air filter regulator, water gauge, pressure gauge, backflush piping, pressure relief valve connected to mud drum and drain. Add for concrete pad mount, piping and electrical wiring. Add the cost of automatic controls from the following section.

Continuous blowdown heat recovery systems. Costs shown by pounds per hour (lbs per hr) of rated makeup capacity.

5,000 lbs per hr	SN@8.00	Ea	51,900.00	371.00	137.00	52,408.00
10,000 lbs per hr	SN@8.00	Ea	103,000.00	371.00	137.00	103,508.00
25,000 lbs per hr	SN@8.00	Ea	131,000.00	371.00	137.00	131,508.00
50,000 lbs per hr	SN@8.00	Ea	208,000.00	371.00	137.00	208,508.00

Commercial Boiler Components and Accessories

Boiler blowdown waste heat recovery system controls. Typical costs in rated pounds per hour for high-efficiency commercial or industrial controls for a manually-operated boiler blowdown waste heat recovery system. Wilson Equipment, Spirax Sarco, or equal. Meets EPA efficiency specifications and liquid waste control requirements. For use with package industrial watertube, firebox and firetube boilers up to 250 PSIG and up to 800 boiler horsepower. Includes timer, total dissolved solids sensor, digital control minipanel, valve, valve actuator and T-type security key for manual actuation. For new construction, add the appropriately sized blowdown waste heat recovery system shown above.

Description	Craft@Hrs	Unit	Material $	Labor $	Equipment $	Total $
5,000 lbs per hr	SN@8.00	Ea	13,300.00	371.00	—	13,671.00
10,000 lbs per hr	SN@8.00	Ea	18,200.00	371.00	—	18,571.00
25,000 lbs per hr	SN@8.00	Ea	31,800.00	371.00	—	32,171.00
50,000 lbs per hr	SN@8.00	Ea	38,500.00	371.00	—	38,871.00
Add for piping for blowdown valve	P1@.500	LF	9.79	22.20	—	31.99
Add for wiring the valve actuator	BE@.300	LF	3.82	14.50	—	18.32
Add for control panel wiring	BE@.300	LF	1.82	14.50	—	16.32
Add for ground fault protection wiring	BE@.300	LF	6.00	14.50	—	20.50
Add for commissioning and tests	BE@8.00	Ea	—	387.00	—	387.00

Boiler Stacks. Costs shown assume a 20-foot stack height and include rigging, breeching, connection of stack sections, guy wires where required, vendor-supplied stack supports and insulation. Add the cost of a concrete base.

Boiler stack. Costs shown assume a 20-foot stack height.

10" diameter	SN@12.0	Ea	3,700.00	556.00	205.00	4,461.00
12" diameter	SN@12.0	Ea	4,610.00	556.00	205.00	5,371.00
16" diameter	SN@16.0	Ea	5,510.00	742.00	273.00	6,525.00
20" diameter	SN@20.0	Ea	6,920.00	927.00	341.00	8,188.00
24" diameter	SN@24.0	Ea	9,150.00	1,110.00	410.00	10,670.00
30" diameter	SN@32.0	Ea	10,800.00	1,480.00	546.00	12,826.00
56" diameter	SN@40.0	Ea	21,400.00	1,850.00	683.00	23,933.00
Add for each 10' section (or portion thereof) beyond 20' high per 10' sect.	—	%	—	—	—	15.0

Costs for Tankless Indirect Water Heater. Provides domestic hot water from boiler coil insert. Costs shown include heat exchanger, coil, circulating pump, thermostatically controlled valve, bypass piping and air bleeder valve. Add the cost of an equipment base, electrical wiring, piping and an insulated hot water storage tank.

Tankless indirect water heater. Costs by gallons per hour (GPH).

240 GPH	SN@8.00	Ea	22,000.00	371.00	137.00	22,508.00
360 GPH	SN@8.00	Ea	27,500.00	371.00	137.00	28,008.00
480 GPH	SN@10.0	Ea	28,300.00	464.00	171.00	28,935.00
840 GPH	SN@12.0	Ea	64,100.00	556.00	205.00	64,861.00
920 GPH	SN@16.0	Ea	71,400.00	742.00	273.00	72,415.00
1,240 GPH	SN@20.0	Ea	80,100.00	927.00	341.00	81,368.00

Costs for Boiler Trim. Costs shown include the pressure relief valve, level gauge (for steam systems), level controller & low water cutoff, pressure and temperature gauge, feedwater valve & actuator, main valve, blowdown valve, cleanout manholes and gaskets.

Boiler trim. Costs shown are per boiler horsepower rating (BHP).

Description	Craft@Hrs	Unit	Material $	Labor $	Equipment $	Total $
20 to 100 BHP	SN@4.00	Ea	3,510.00	185.00	2.68	3,697.68
100 to 200 BHP	SN@6.00	Ea	3,950.00	278.00	4.02	4,232.02
200 to 400 BHP	SN@8.00	Ea	4,860.00	371.00	5.36	5,236.36
400 to 600 BHP	SN@8.00	Ea	7,440.00	371.00	5.36	7,816.36
600 to 800 BHP	SN@8.00	Ea	8,770.00	371.00	5.36	9,146.36
800 to 1,000 BHP	SN@10.0	Ea	8,780.00	464.00	6.69	9,250.69

Heat reclaimer condensate return & feedwater system, stainless steel non-siphon, including flash tank with level control and vacuum breaker, overflow relief drain, simplex centrifugal pump, pre-assembled piping to feedwater tank, structural steel stand and frame, check trim and pre-wired controls. Also provides potable hot water as part of the condensed steam return cycle. Capacity rating is in GPH (gallons per hour of condensate)

Description	Craft@Hrs	Unit	Material $	Labor $	Total $
33 gal. ½ HP pump	P1@3.00	Ea	1,090.00	133.00	1,223.00
46 gal. ¾ HP pump	P1@4.00	Ea	1,350.00	178.00	1,528.00
Vacuum breakers, rough brass, threaded, to be mounted on finned radiation tube inlet valve fitting					
½" vacuum breaker	P1@.250	Ea	51.10	11.10	62.20
Steam traps, float valve Sarco type or equal w/drain fitting, brazed					
¾" with check valve	P1@.250	Ea	48.60	11.10	59.70
1.0" with check valve	P1@.450	Ea	77.90	20.00	97.90
Baseboard fin tube radiation, per linear foot, 15 PSIG test thinwall copper tube with aluminum fins, braze fitted wall mounted					
¾" element with 8" cover	P1@.280	LF	12.70	12.40	25.10
1" element with 8" cover	P1@.350	LF	20.80	15.60	36.40
Add for corners, fillers and caps, average per foot	—	%	—	15.0	—
Solenoid valves, zone thermostatically-controlled steam electric, 15 PSIG rating, braze fitting. Thermostatic Honeywell-type or equal solenoid relay valves					
¾", brazed	P1@.250	Ea	16.10	11.10	27.20
1", brazed	P1@.250	Ea	16.10	11.10	27.20

Description	Craft@Hrs	Unit	Material $	Labor $	Equipment $	Total $

Additional costs for boiler pumps, condenser units, feedwater systems, heat recovery and oil fuel train piping.

Description	Craft@Hrs	Unit	Material $	Labor $	Equipment $	Total $
Form and pour concrete pad	CF@1.56	SF	9.64	68.10	—	77.74
Place vibration pads	CF@.750	Ea	46.20	32.70	—	78.90
Mount interior drain	P1@.500	LF	9.55	22.20	—	31.75
Bore hole through wall or pad	P1@.250	Ea	—	11.10	—	11.10
Piping to load (typical)	P1@.500	LF	9.59	22.20	—	31.79
Bore burner stack vent in exterior wall	P1@.250	Ea	—	11.10	—	11.10
Mounting and edge-sealing stack	P1@.500	Ea	46.80	22.20	—	69.00

Emissions sensing and recording equipment. For stack gas combustion monitoring of boilers. Monitors and records NOx, SOx, CO, H^2S, volatile materials and particulates emissions to meet 2011 EPA greenhouse gas regulations. CO_2 monitoring and recording is required on boilers producing 25,000 tons or more a year of carbon emissions. Signal Group Limited or equal. Includes a stack-mounted pulse type infra-red single-beam light source with solid-state detector. Sensitivity range is variable between .1 ppm and 1,000 ppm. Includes a hot-wire radiation source, rotating shutter with reference target for continuous unmanned recalibration, 240 x 64 pixel LCD display with remote panel-mounted solid-state digital analyzer and lockout type inspector-only access recording circuitry. Cost includes calibration and testing by a factory technician. Add the cost of the permit, inspection, test and documentation.

Description	Craft@Hrs	Unit	Material $	Labor $	Equipment $	Total $
Analyzer and recorder installation to remote panel	BE@3.00	Ea	30,600.00	145.00	—	30,745.00
Infrared sensor unit installation to stack	BE@2.00	Ea	9,590.00	96.70	—	9,686.70
Wiring of sensor to analyzer and recorder	BE@8.00	Ea	598.00	387.00	—	985.00
Circuit testing and grounding	BE@2.00	Ea	—	96.70	—	96.70

Description	Craft@Hrs	Unit	Material $	Labor $	Equipment $	Total $

Base-mounted, bronze fitted centrifugal pump, set in place only. Make additional allowances for pipe connection assembly.

½ HP	P1@1.50	Ea	2,000.00	66.70	—	2,066.70
¾ HP	P1@1.75	Ea	2,550.00	77.80	—	2,627.80
1 HP	P1@2.00	Ea	3,120.00	88.90	—	3,208.90
1½ HP	P1@2.40	Ea	3,780.00	107.00	—	3,887.00
2 HP	P1@2.70	Ea	4,650.00	120.00	—	4,770.00
3 HP	P1@3.00	Ea	5,540.00	133.00	—	5,673.00
5 HP	P1@3.50	Ea	6,320.00	156.00	—	6,476.00
7½ HP	P1@4.00	Ea	7,010.00	178.00	—	7,188.00
10 HP	P1@4.50	Ea	10,200.00	200.00	—	10,400.00
15 HP	ER@5.00	Ea	11,800.00	237.00	3.35	12,040.35
20 HP	ER@6.00	Ea	13,700.00	284.00	4.02	13,988.02
25 HP	ER@7.00	Ea	14,300.00	332.00	4.69	14,636.69
30 HP	ER@7.40	Ea	16,900.00	351.00	4.95	17,255.95
40 HP	ER@7.70	Ea	16,800.00	365.00	5.15	17,170.15
50 HP	ER@8.00	Ea	20,400.00	379.00	5.36	20,784.36

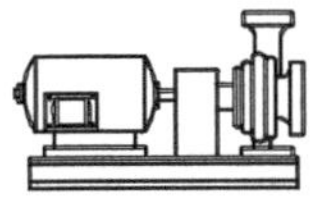

Base-mounted centrifugal pump connection assembly. Includes pipe, fittings, pipe insulation, valves, strainers, suction guides and gauges.

2½" supply	P1@12.0	Ea	2,970.00	533.00	—	3,503.00
3" supply	P1@13.5	Ea	3,440.00	600.00	—	4,040.00
4" supply	P1@16.0	Ea	4,720.00	711.00	—	5,431.00
6" supply	P1@22.0	Ea	7,810.00	978.00	—	8,788.00
8" supply	P1@28.0	Ea	11,300.00	1,240.00	—	12,540.00
10" supply	P1@35.0	Ea	18,700.00	1,560.00	—	20,260.00
12" supply	P1@42.0	Ea	24,900.00	1,870.00	—	26,770.00

Cast iron line-mounted centrifugal pump, including connection and supports.

1/6 HP	P1@1.25	Ea	496.00	55.60	—	551.60
1/4 HP	P1@1.40	Ea	634.00	62.20	—	696.20
1/3 HP	P1@1.50	Ea	783.00	66.70	—	849.70
1/2 HP	P1@1.70	Ea	1,000.00	75.60	—	1,075.60
3/4 HP	P1@1.80	Ea	1,350.00	80.00	—	1,430.00
1 HP	P1@2.00	Ea	1,610.00	88.90	—	1,698.90

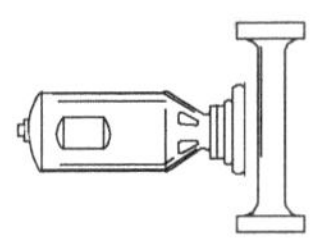

Description	Craft@Hrs	Unit	Material $	Labor $	Equipment $	Total $

2-pass shell and tube type heat exchanger, set in place only. Make additional allowances for pipe connection assembly.

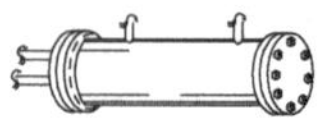

4" dia. x 30"	SN@2.00	Ea	1,900.00	92.70	1.34	1,994.04
4" dia. x 60"	SN@2.50	Ea	2,150.00	116.00	1.67	2,267.67
6" dia. x 30"	SN@3.25	Ea	2,630.00	151.00	2.18	2,783.18
6" dia. x 60"	SN@3.50	Ea	3,280.00	162.00	2.34	3,444.34
8" dia. x 30"	SN@4.25	Ea	4,070.00	197.00	2.85	4,269.85
8" dia. x 60"	SN@4.50	Ea	5,250.00	209.00	3.01	5,462.01
10" dia. x 30"	SN@5.50	Ea	6,030.00	255.00	3.68	6,288.68
10" dia. x 60"	SN@5.75	Ea	7,970.00	267.00	3.85	8,240.85
10" dia. x 90"	SN@6.00	Ea	9,930.00	278.00	4.02	10,212.02
12" dia. x 60"	SN@6.50	Ea	10,400.00	301.00	4.35	10,705.35
12" dia. x 90"	SN@7.00	Ea	13,400.00	324.00	4.69	13,728.69
12" dia. x 120"	SN@7.25	Ea	16,300.00	336.00	4.85	16,640.85
14" dia. x 60"	SN@7.50	Ea	15,400.00	348.00	5.02	15,753.02
14" dia. x 90"	SN@8.00	Ea	19,200.00	371.00	5.36	19,576.36
14" dia. x 120"	SN@8.25	Ea	23,100.00	382.00	5.52	23,487.52
16" dia. x 60"	SN@8.50	Ea	19,100.00	394.00	5.69	19,499.69
16" dia. x 90"	SN@9.00	Ea	23,800.00	417.00	6.02	24,223.02
16" dia. x 120"	SN@9.25	Ea	26,900.00	429.00	6.19	27,335.19

Heat exchanger connection assembly, hot water to hot water. Includes pipe, fittings, pipe insulation, valves, thermometers gauges and vents.

1½" supply	P1@32.0	Ea	4,430.00	1,420.00	—	5,850.00
2" supply	P1@38.0	Ea	5,470.00	1,690.00	—	7,160.00
2½" supply	P1@50.0	Ea	7,120.00	2,220.00	—	9,340.00
3" supply	P1@55.0	Ea	7,490.00	2,440.00	—	9,930.00
4" supply	P1@62.0	Ea	9,110.00	2,760.00	—	11,870.00
6" supply	P1@85.0	Ea	14,700.00	3,780.00	5,080.00	23,560.00
8" supply	P1@102.	Ea	20,900.00	4,530.00	6,100.00	31,530.00
10" supply	P1@130.	Ea	27,300.00	5,780.00	7,770.00	40,850.00

Heat exchanger connection assembly, steam to hot water. Includes pipe, fittings, pipe insulation, valves, thermometers gauges and vents.

1½" supply	P1@22.0	Ea	5,080.00	978.00	—	6,058.00
2" supply	P1@26.0	Ea	5,930.00	1,160.00	—	7,090.00
2½" supply	P1@41.0	Ea	7,360.00	1,820.00	—	9,180.00
3" supply	P1@44.0	Ea	8,530.00	1,960.00	—	10,490.00
4" supply	P1@54.5	Ea	9,720.00	2,420.00	—	12,140.00
6" supply	P1@67.0	Ea	15,400.00	2,980.00	4,000.00	22,380.00
8" supply	P1@80.0	Ea	21,800.00	3,560.00	4,780.00	30,140.00

Description	Craft@Hrs	Unit	Material $	Labor $	Equipment $	Total $

Ceiling-hung horizontal, 2 pipe hydronic fan coil unit with cabinet, set in place only. Make additional allowances for pipe connection assembly.

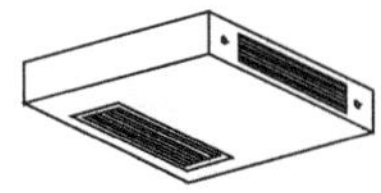

200 CFM	P1@2.00	Ea	1,040.00	88.90	—	1,128.90
400 CFM	P1@2.50	Ea	1,220.00	111.00	—	1,331.00
600 CFM	P1@3.00	Ea	1,450.00	133.00	—	1,583.00
800 CFM	P1@3.25	Ea	1,830.00	144.00	—	1,974.00
1,000 CFM	P1@3.50	Ea	2,290.00	156.00	—	2,446.00
Deduct for "less cabinet"—		%	-20.0	—	—	—
Add for "floor-mounted" —		%	10.0	—	—	—

Vertical, 4 pipe hydronic fan coil unit, set in place only. Make additional allowances for pipe connection assembly.

200 CFM	P1@2.15	Ea	1,180.00	95.60	—	1,275.60
400 CFM	P1@2.75	Ea	1,380.00	122.00	—	1,502.00
600 CFM	P1@3.25	Ea	1,630.00	144.00	—	1,774.00
800 CFM	P1@3.45	Ea	2,020.00	153.00	—	2,173.00
1,000 CFM	P1@3.60	Ea	2,530.00	160.00	—	2,690.00

2 pipe fan coil unit connection. Coil connection includes type L copper pipe and wrought fittings, pipe insulation, condensate drain, isolation valves and 2-way control valve.

½" supply	P1@2.50	Ea	179.00	111.00	—	290.00
¾" supply	P1@2.80	Ea	217.00	124.00	—	341.00
1" supply	P1@3.40	Ea	328.00	151.00	—	479.00

4 pipe fan coil unit connection. Coil connection includes type L copper pipe and wrought fittings, pipe insulation, condensate drain, isolation valves and 3-way control valve.

½" supply	P1@2.75	Ea	217.00	122.00	—	339.00
¾" supply	P1@2.95	Ea	251.00	131.00	—	382.00

Description	Craft@Hrs	Unit	Material $	Labor $	Equipment $	Total $

Duct-mounted flanged hot water reheat coil, set in place only. Make additional allowances for pipe connection assembly.

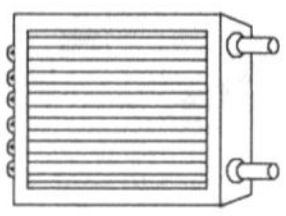

12" x 6"	SN@1.00	Ea	401.00	46.40	—	447.40
12" x 8"	SN@1.00	Ea	421.00	46.40	—	467.40
12" x 10"	SN@1.00	Ea	438.00	46.40	—	484.40
12" x 12"	SN@1.00	Ea	479.00	46.40	—	525.40
18" x 6"	SN@1.25	Ea	507.00	57.90	—	564.90
18" x 12"	SN@1.25	Ea	522.00	57.90	—	579.90
18" x 18"	SN@1.25	Ea	560.00	57.90	—	617.90
24" x 12"	SN@1.50	Ea	604.00	69.50	—	673.50
24" x 18"	SN@1.50	Ea	631.00	69.50	—	700.50
24" x 24"	SN@1.75	Ea	710.00	81.10	—	791.10

Duct-mounted hot water reheat coil connection assembly. Includes pipe, fittings, pipe insulation, valves and strainers.

½" steel	P1@2.25	Ea	217.00	100.00	—	317.00
¾" steel	P1@2.35	Ea	257.00	104.00	—	361.00
1" steel	P1@2.65	Ea	337.00	118.00	—	455.00
1¼" steel	P1@3.00	Ea	407.00	133.00	—	540.00
1½" steel	P1@3.25	Ea	434.00	144.00	—	578.00
2" steel	P1@3.75	Ea	484.00	167.00	—	651.00

Duct-mounted slip-in electric reheat coil, set in place only. Make additional allowances for electrical connection.

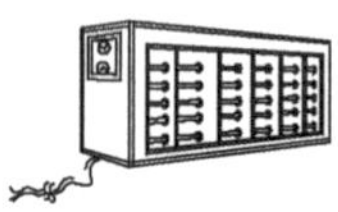

12" x 6"	SN@1.10	Ea	892.00	51.00	—	943.00
12" x 8"	SN@1.20	Ea	1,020.00	55.60	—	1,075.60
12" x 10"	SN@1.30	Ea	1,220.00	60.30	—	1,280.30
12" x 12"	SN@1.40	Ea	1,350.00	64.90	—	1,414.90
18" x 6"	SN@1.50	Ea	1,500.00	69.50	—	1,569.50
18" x 12"	SN@1.60	Ea	1,610.00	74.20	—	1,684.20
18" x 18"	SN@1.70	Ea	1,710.00	78.80	—	1,788.80
24" x 12"	SN@1.90	Ea	1,840.00	88.10	—	1,928.10
24" x 18"	SN@2.10	Ea	1,940.00	97.30	—	2,037.30
24" x 24"	SN@2.40	Ea	2,370.00	111.00	—	2,481.00

Description	Craft@Hrs	Unit	Material $	Labor $	Equipment $	Total $

Gas-fired unit heater, hang in place only. Make additional allowances for gas connection assembly.

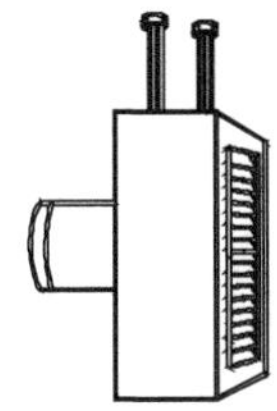

60mbh, 200cfm	P1@2.00	Ea	1,930.00	88.90	—	2,018.90
100mbh, 400cfm	P1@2.50	Ea	2,240.00	111.00	—	2,351.00
150mbh, 600cfm	P1@3.00	Ea	2,730.00	133.00	—	2,863.00
175mbh, 800cfm	P1@3.25	Ea	2,940.00	144.00	—	3,084.00
200mbh, 1,000cfm	P1@3.50	Ea	3,160.00	156.00	—	3,316.00
250mbh, 1,500cfm	P1@3.75	Ea	3,720.00	167.00	—	3,887.00
300mbh, 2,000cfm	P1@4.00	Ea	4,130.00	178.00	—	4,308.00

Hot water unit heater, hang in place only. Make additional allowances for pipe connection assembly.

8mbh, 400cfm	P1@2.00	Ea	508.00	88.90	4.69	601.59
12mbh, 400cfm	P1@2.00	Ea	654.00	88.90	4.69	747.59
16mbh, 400cfm	P1@2.00	Ea	690.00	88.90	4.69	783.59
30mbh, 700cfm	P1@3.00	Ea	923.00	133.00	7.03	1,063.03
43mbh, 1,040cfm	P1@3.50	Ea	1,140.00	156.00	8.20	1,304.20
57mbh, 1,370cfm	P1@4.00	Ea	1,210.00	178.00	9.37	1,397.37
68mbh, 1,715cfm	P1@4.00	Ea	1,930.00	178.00	9.37	2,117.37
105mbh, 2,360cfm	P1@5.00	Ea	2,230.00	222.00	11.70	2,463.70
123mbh, 2,900cfm	P1@6.00	Ea	2,860.00	267.00	14.10	3,141.10
140mbh, 3,530cfm	P1@6.00	Ea	3,310.00	267.00	14.10	3,591.10
156mbh, 4,250cfm	P1@6.00	Ea	3,470.00	267.00	14.10	3,751.10
210mbh, 4,450cfm	P1@8.00	Ea	6,160.00	356.00	18.70	6,534.70
123mbh, 5,200cfm	P1@6.00	Ea	5,080.00	267.00	14.10	5,361.10
257mbh, 5,350cfm	P1@8.00	Ea	10,400.00	356.00	18.70	10,774.70

Hot water unit heater connection. Includes Schedule 40 threaded and coupled supply and return piping with malleable iron fittings, pipe insulation, valves and vents Make additional allowances for electrical connection.

¾" supply	P1@2.50	Ea	303.00	111.00	—	414.00
1" supply	P1@3.00	Ea	355.00	133.00	—	488.00
1¼" supply	P1@3.30	Ea	422.00	147.00	—	569.00
1½" supply	P1@3.75	Ea	491.00	167.00	—	658.00
2" supply	P1@4.10	Ea	635.00	182.00	—	817.00

Gas-fired unit heater flue connection

2" B vent	SN@.090	LF	6.56	4.17	—	10.73
3" B vent	SN@.100	LF	8.14	4.64	—	12.78
4" B vent	SN@.110	LF	10.90	5.10	—	16.00
6" B vent	SN@.130	LF	12.30	6.03	—	18.33

Gas-fired unit heater fuel connection. Assembly includes 8' of black steel pipe, malleable fittings, plug valve and regulator.

¾" gas conn.	P1@1.30	Ea	110.00	57.80	—	167.80
1" gas conn.	P1@1.60	Ea	119.00	71.10	—	190.10
1¼" gas conn.	P1@2.10	Ea	212.00	93.30	—	305.30
1½" gas conn.	P1@2.50	Ea	245.00	111.00	—	356.00

Description	Craft@Hrs	Unit	Material $	Labor $	Equipment $	Total $

Centrifugal water-cooled chiller, set in place only. Unit comes factory assembled complete. Make additional allowances for cooling tower pipe connections.

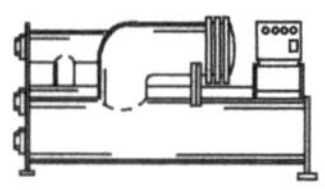

Description	Craft@Hrs	Unit	Material $	Labor $	Equipment $	Total $
100 tons	SN@20.0	Ea	113,000.00	927.00	341.00	114,268.00
150 tons	SN@20.0	Ea	171,000.00	927.00	341.00	172,268.00
200 tons	SN@20.0	Ea	184,000.00	927.00	341.00	185,268.00
250 tons	SN@20.0	Ea	216,000.00	927.00	341.00	217,268.00
300 tons	SN@25.0	Ea	231,000.00	1,160.00	427.00	232,587.00
350 tons	SN@25.0	Ea	239,000.00	1,160.00	427.00	240,587.00
400 tons	SN@25.0	Ea	282,000.00	1,160.00	427.00	283,587.00
500 tons	SN@25.0	Ea	353,000.00	1,160.00	427.00	354,587.00

Reciprocating water-cooled chiller, set in place only. Unit comes factory assembled complete. Make additional allowances for cooling tower pipe connections.

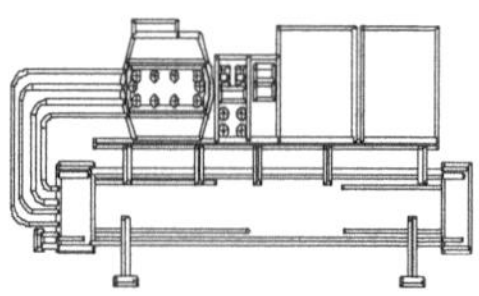

Description	Craft@Hrs	Unit	Material $	Labor $	Equipment $	Total $
20 tons	SN@12.0	Ea	34,600.00	556.00	205.00	35,361.00
25 tons	SN@12.0	Ea	39,600.00	556.00	205.00	40,361.00
30 tons	SN@12.0	Ea	43,100.00	556.00	205.00	43,861.00
40 tons	SN@12.0	Ea	51,800.00	556.00	205.00	52,561.00
50 tons	SN@16.0	Ea	73,200.00	742.00	273.00	74,215.00
60 tons	SN@16.0	Ea	94,100.00	742.00	273.00	95,115.00
75 tons	SN@16.0	Ea	101,000.00	742.00	273.00	102,015.00
100 tons	SN@20.0	Ea	108,000.00	927.00	341.00	109,268.00
125 tons	SN@20.0	Ea	131,000.00	927.00	341.00	132,268.00
150 tons	SN@20.0	Ea	149,000.00	927.00	341.00	150,268.00

Water-cooled chiller connection assembly. Includes condenser and chilled water pipe, fittings, pipe insulation, valves, thermometers and gauges.

Description	Craft@Hrs	Unit	Material $	Labor $	Equipment $	Total $
2½" chiller conn.	SN@36.0	Ea	3,480.00	1,670.00	—	5,150.00
3" chiller conn.	SN@38.0	Ea	3,830.00	1,760.00	—	5,590.00
4" chiller conn.	SN@51.0	Ea	5,100.00	2,360.00	—	7,460.00
6" chiller conn.	SN@65.0	Ea	8,550.00	3,010.00	1,110.00	12,670.00
8" chiller conn.	SN@77.0	Ea	13,100.00	3,570.00	1,310.00	17,980.00
10" chiller conn.	SN@100.	Ea	23,900.00	4,640.00	1,710.00	30,250.00
12" chiller conn.	SN@125.	Ea	27,600.00	5,790.00	2,130.00	35,520.00

Description	Craft@Hrs	Unit	Material $	Labor $	Equipment $	Total $

Air-cooled condensing unit, set in place only. Unit comes factory assembled complete. Make additional allowances for pipe and electrical connections.

3 tons	SN@2.00	Ea	4,680.00	92.70	34.10	4,806.80
5 tons	SN@3.00	Ea	7,960.00	139.00	51.20	8,150.20
7½ tons	SN@4.00	Ea	10,300.00	185.00	68.30	10,553.30
10 tons	SN@4.50	Ea	13,700.00	209.00	76.80	13,985.80
12½ tons	SN@5.00	Ea	16,100.00	232.00	85.40	16,417.40
15 tons	SN@5.50	Ea	18,700.00	255.00	93.90	19,048.90
20 tons	SN@6.00	Ea	24,600.00	278.00	102.00	24,980.00
25 tons	SN@7.00	Ea	31,000.00	324.00	120.00	31,444.00
30 tons	SN@8.00	Ea	37,100.00	371.00	137.00	37,608.00
40 tons	SN@9.00	Ea	48,400.00	417.00	154.00	48,971.00
50 tons	SN@10.0	Ea	60,900.00	464.00	171.00	61,535.00

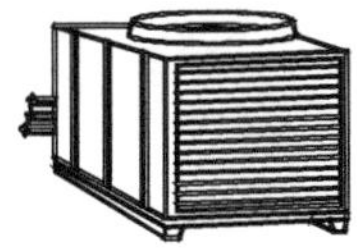

Galvanized steel cooling tower. Assemble and set in place only. Add for pipe connections.

10 tons	SN@16.0	Ea	4,730.00	742.00	273.00	5,745.00
15 tons	SN@16.0	Ea	5,360.00	742.00	273.00	6,375.00
20 tons	SN@18.0	Ea	8,420.00	834.00	307.00	9,561.00
25 tons	SN@20.0	Ea	10,400.00	927.00	341.00	11,668.00
30 tons	SN@25.0	Ea	11,900.00	1,160.00	427.00	13,487.00
40 tons	SN@30.0	Ea	15,100.00	1,390.00	512.00	17,002.00
50 tons	SN@35.0	Ea	16,600.00	1,620.00	598.00	18,818.00
60 tons	SN@40.0	Ea	17,600.00	1,850.00	683.00	20,133.00
80 tons	SN@45.0	Ea	19,000.00	2,090.00	768.00	21,858.00
100 tons	SN@50.0	Ea	22,600.00	2,320.00	854.00	25,774.00
125 tons	SN@60.0	Ea	27,500.00	2,780.00	1,020.00	31,300.00
150 tons	SN@65.0	Ea	31,700.00	3,010.00	1,110.00	35,820.00
175 tons	SN@70.0	Ea	35,800.00	3,240.00	1,200.00	40,240.00
200 tons	SN@70.0	Ea	39,500.00	3,240.00	1,200.00	43,940.00
300 tons	SN@75.0	Ea	56,800.00	3,480.00	1,280.00	61,560.00
400 tons	SN@80.0	Ea	71,900.00	3,710.00	1,370.00	76,980.00
500 tons	SN@80.0	Ea	86,100.00	3,710.00	1,370.00	91,180.00

Redwood or fir induced-draft cooling tower. Assemble and set in place only. Add for pipe connections.

100 tons	SN@75.0	Ea	14,500.00	3,480.00	1,280.00	19,260.00
200 tons	SN@80.0	Ea	24,500.00	3,710.00	1,370.00	29,580.00
300 tons	SN@85.0	Ea	35,200.00	3,940.00	1,450.00	40,590.00
400 tons	SN@90.0	Ea	40,200.00	4,170.00	1,540.00	45,910.00
500 tons	SN@90.0	Ea	46,900.00	4,170.00	1,540.00	52,610.00

Description	Craft@Hrs	Unit	Material $	Labor $	Equipment $	Total $

Redwood or fir forced-draft cooling tower. Assemble and set in place only. Add for pipe connections.

Description	Craft@Hrs	Unit	Material $	Labor $	Equipment $	Total $
100 tons	SN@75.0	Ea	23,400.00	3,480.00	1,280.00	28,160.00
200 tons	SN@80.0	Ea	44,900.00	3,710.00	1,370.00	49,980.00
300 tons	SN@85.0	Ea	52,300.00	3,940.00	1,450.00	57,690.00
400 tons	SN@90.0	Ea	69,300.00	4,170.00	1,540.00	75,010.00
500 tons	SN@90.0	Ea	84,200.00	4,170.00	1,540.00	89,910.00

Cooling tower connection assembly. Includes condenser water pipe, fittings, valves, thermometers and gauges.

Description	Craft@Hrs	Unit	Material $	Labor $	Equipment $	Total $
2½" connection	SN@42.0	Ea	6,410.00	1,950.00	—	8,360.00
3" connection	SN@47.0	Ea	7,270.00	2,180.00	—	9,450.00
4" connection	SN@62.0	Ea	10,500.00	2,870.00	—	13,370.00
6" connection	SN@96.0	Ea	16,500.00	4,450.00	64.30	21,014.30
8" connection	SN@118.	Ea	23,600.00	5,470.00	79.00	29,149.00
10" connection	SN@150.	Ea	32,700.00	6,950.00	100.00	39,750.00
12" connection	SN@215.	Ea	39,800.00	9,970.00	144.00	49,914.00
14" connection	SN@260.	Ea	45,600.00	12,100.00	174.00	57,874.00

Schedule 40 carbon steel (ASTM A-53 and A-120) pipe with 150 pound carbon steel welding fittings is commonly used for heating hot water, chilled water, glycol and low pressure steam systems where operating pressures don't exceed 250 PSIG.

Schedule 40 (ASTM A-53B SML) steel pipe with carbon steel welding fittings can also be used for refrigerant piping systems.

The cost estimates in this section are based on the conditions, limitations and wage rates described in the section "How to Use This Book" beginning on page 5.

Equipment cost, where shown, is $4.57 per hour for a 2-ton chain hoist and $6.49 per hour for 200 amp gas welder, trailer mounted.

Description	Craft@Hrs	Unit	Material $	Labor $	Equipment $	Total $

Schedule 40 carbon steel butt-welded horizontal pipe assembly. Horizontally hung in a building. Assembly includes fittings and hanger assemblies. Based on a reducing tee and a 45-degree elbow every 16 feet for ½" pipe. A reducing tee and a 45-degree elbow every 50 feet for 12" pipe. Hangers spaced to meet plumbing code. *(Distance between fittings in the pipe assembly increases as pipe diameter increases.)* Use these figures for preliminary estimates.

Description	Craft@Hrs	Unit	Material $	Labor $	Equipment $	Total $
½"	P1@.144	LF	2.24	6.40	.47	9.11
¾"	P1@.170	LF	2.42	7.56	.55	10.53
1"	P1@.206	LF	2.97	9.16	.67	12.80
1¼"	P1@.208	LF	3.95	9.25	.67	13.87
1½"	P1@.238	LF	4.23	10.60	.77	15.60
2"	P1@.286	LF	5.59	12.70	.93	19.22
2½"	P1@.344	LF	10.30	15.30	1.12	26.72
3"	P1@.390	LF	12.80	17.30	1.26	31.36
4"	P1@.507	LF	17.90	22.50	1.64	42.04
6"	ER@.712	LF	28.60	33.80	1.14	63.54
8"	ER@.847	LF	49.50	40.20	1.35	91.05
10"	ER@1.02	LF	72.20	48.40	1.63	122.23
12"	ER@1.25	LF	92.40	59.30	1.99	153.69

Schedule 40 carbon steel butt-welded vertical pipe assembly. Riser assembly, including fittings and riser clamps. Based on a reducing tee, and a riser clamp on every floor. Use these figures for preliminary estimates.

Description	Craft@Hrs	Unit	Material $	Labor $	Equipment $	Total $
½"	P1@.104	LF	1.70	4.62	.34	6.66
¾"	P1@.128	LF	1.90	5.69	.42	8.01
1"	P1@.162	LF	2.44	7.20	.53	10.17
1¼"	P1@.184	LF	4.23	8.18	.60	13.01
1½"	P1@.204	LF	4.73	9.07	.66	14.46
2"	P1@.247	LF	6.30	11.00	.80	18.10
2½"	P1@.304	LF	10.70	13.50	.99	25.19
3"	P1@.351	LF	13.00	15.60	1.14	29.74
4"	P1@.505	LF	17.10	22.40	1.64	41.14
6"	ER@.737	LF	30.70	34.90	1.18	66.78
8"	ER@.878	LF	55.50	41.60	1.40	98.50
10"	ER@1.07	LF	84.50	50.70	1.71	136.91
12"	ER@1.32	LF	111.00	62.60	2.11	175.71

Description	Craft@Hrs	Unit	Material $	Labor $	Equipment $	Total $

Schedule 40 carbon steel pipe, plain end, butt-welded joints, pipe only

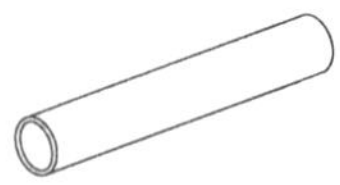

½"	P1@.070	LF	1.03	3.11	.23	4.37
¾"	P1@.080	LF	1.17	3.56	.26	4.99
1"	P1@.100	LF	1.72	4.45	.32	6.49
1¼"	P1@.110	LF	2.24	4.89	.36	7.49
1½"	P1@.120	LF	2.68	5.33	.39	8.40
2"	P1@.140	LF	3.64	6.22	.45	10.31
2½"	P1@.170	LF	6.49	7.56	.55	14.60
3"	P1@.190	LF	8.52	8.45	.62	17.59
4"	P1@.300	LF	12.10	13.30	.97	26.37
6"	ER@.420	LF	21.00	19.90	.67	41.57
8"	ER@.510	LF	27.30	24.20	.81	52.31
10"	ER@.600	LF	38.90	28.40	.96	68.26
12"	ER@.760	LF	46.80	36.00	1.21	84.01

Schedule 40 carbon steel 45-degree ell, butt-welded joints

½"	P1@.220	Ea	3.01	9.78	.71	13.50
¾"	P1@.330	Ea	3.01	14.70	1.07	18.78
1"	P1@.440	Ea	3.01	19.60	1.43	24.04
1¼"	P1@.560	Ea	3.18	24.90	1.82	29.90
1½"	P1@.710	Ea	3.18	31.60	2.30	37.08
2"	P1@.890	Ea	3.29	39.60	2.89	45.78
2½"	P1@1.11	Ea	5.93	49.30	3.60	58.83
3"	P1@1.33	Ea	7.17	59.10	4.31	70.58
4"	P1@1.78	Ea	10.60	79.10	5.77	95.47
6"	ER@2.67	Ea	27.30	127.00	4.26	158.56
8"	ER@3.20	Ea	64.60	152.00	5.11	221.71
10"	ER@4.00	Ea	98.90	190.00	6.38	295.28
12"	ER@4.80	Ea	145.00	228.00	7.66	380.66

Schedule 40 carbon steel 90-degree ell, butt-welded joints

½"	P1@.220	Ea	2.94	9.78	.71	13.43
¾"	P1@.330	Ea	2.94	14.70	1.07	18.71
1"	P1@.440	Ea	2.94	19.60	1.43	23.97
1¼"	P1@.560	Ea	3.13	24.90	1.82	29.85
1½"	P1@.670	Ea	3.13	29.80	2.30	35.23
2"	P1@.890	Ea	3.43	39.60	2.89	45.92
2½"	P1@1.11	Ea	6.80	49.30	3.60	59.70
3"	P1@1.33	Ea	9.13	59.10	4.31	72.54
4"	P1@1.78	Ea	14.70	79.10	5.77	99.57
6"	ER@2.67	Ea	41.70	127.00	4.26	172.96
8"	ER@3.20	Ea	94.80	152.00	5.11	251.91
10"	ER@4.00	Ea	170.00	190.00	6.38	366.38
12"	ER@4.80	Ea	251.00	228.00	7.66	486.66

Description	Craft@Hrs	Unit	Material $	Labor $	Equipment $	Total $

Schedule 40 carbon steel tee, butt-welded joints

Description	Craft@Hrs	Unit	Material $	Labor $	Equipment $	Total $
½"	P1@.330	Ea	9.32	14.70	1.07	25.09
¾"	P1@.500	Ea	9.32	22.20	1.62	33.14
1"	P1@.670	Ea	9.32	29.80	2.17	41.29
1¼"	P1@.830	Ea	10.40	36.90	2.69	49.99
1½"	P1@1.00	Ea	10.90	44.50	3.24	58.64
2"	P1@1.33	Ea	12.00	59.10	4.31	75.41
2½"	P1@1.67	Ea	20.80	74.20	5.42	100.42
3"	P1@2.00	Ea	23.20	88.90	6.49	118.59
4"	P1@2.67	Ea	31.70	119.00	8.66	159.36
6"	ER@4.00	Ea	58.00	190.00	6.38	254.38
8"	ER@4.80	Ea	135.00	228.00	7.66	370.66
10"	ER@6.00	Ea	232.00	284.00	9.58	525.58
12"	ER@7.20	Ea	347.00	341.00	11.50	699.50

Schedule 40 carbon steel reducing tee, butt-welded joints

Description	Craft@Hrs	Unit	Material $	Labor $	Equipment $	Total $
1½" x ½"	P1@.940	Ea	14.10	41.80	3.05	58.95
1½" x ¾"	P1@.940	Ea	14.10	41.80	3.05	58.95
1½" x 1"	P1@.940	Ea	14.10	41.80	3.05	58.95
1½" x 1¼"	P1@.940	Ea	14.10	41.80	3.05	58.95
2" x ¾"	P1@1.22	Ea	19.10	54.20	3.96	77.26
2" x 1"	P1@1.22	Ea	19.10	54.20	3.96	77.26
2" x 1¼"	P1@1.22	Ea	19.10	54.20	3.96	77.26
2" x 1½"	P1@1.22	Ea	19.10	54.20	3.96	77.26
2½" x 1"	P1@1.56	Ea	29.40	69.30	5.06	103.76
2½" x 1¼"	P1@1.56	Ea	29.40	69.30	5.06	103.76
2½" x 1½"	P1@1.56	Ea	29.40	69.30	5.06	103.76
2½" x 2"	P1@1.56	Ea	29.40	69.30	5.06	103.76
3" x 1¼"	P1@1.89	Ea	30.30	84.00	6.13	120.43
3" x 2"	P1@1.89	Ea	30.30	84.00	6.13	120.43
3" x 2½"	P1@1.89	Ea	30.30	84.00	6.13	120.43
4" x 1½"	P1@2.44	Ea	32.20	108.00	7.91	148.11
4" x 2"	P1@2.44	Ea	32.20	108.00	7.91	148.11
4" x 2½"	P1@2.44	Ea	32.20	108.00	7.91	148.11
4" x 3"	P1@2.44	Ea	32.20	108.00	7.91	148.11
6" x 2½"	ER@3.78	Ea	72.90	179.00	6.03	257.93
6" x 3"	ER@3.78	Ea	72.90	179.00	6.03	257.93
6" x 4"	ER@3.78	Ea	72.90	179.00	6.03	257.93
8" x 3"	ER@4.40	Ea	180.00	209.00	7.02	396.02
8" x 4"	ER@4.40	Ea	180.00	209.00	7.02	396.02
8" x 6"	ER@4.40	Ea	180.00	209.00	7.02	396.02
10" x 4"	ER@5.60	Ea	305.00	265.00	8.94	578.94
10" x 6"	ER@5.60	Ea	305.00	265.00	8.94	578.94
10" x 8"	ER@5.60	Ea	305.00	265.00	8.94	578.94
12" x 6"	ER@6.80	Ea	452.00	322.00	10.90	784.90
12" x 8"	ER@6.80	Ea	452.00	322.00	10.90	784.90
12" x 10"	ER@6.80	Ea	452.00	322.00	10.90	784.90

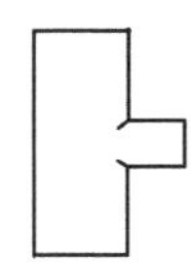

Description	Craft@Hrs	Unit	Material $	Labor $	Equipment $	Total $

Schedule 40 carbon steel concentric reducer, butt-welded joints

¾"	P1@.295	Ea	6.30	13.10	.96	20.36
1"	P1@.395	Ea	7.53	17.60	1.28	26.41
1¼"	P1@.500	Ea	7.96	22.20	1.62	31.78
1½"	P1@.600	Ea	8.12	26.70	1.95	36.77
2"	P1@.800	Ea	8.26	35.60	2.59	46.45
2½"	P1@1.00	Ea	11.90	44.50	3.24	59.64
3"	P1@1.20	Ea	13.10	53.30	3.89	70.29
4"	P1@1.60	Ea	18.60	71.10	5.19	94.89
6"	ER@2.40	Ea	34.30	114.00	3.83	152.13
8"	ER@2.88	Ea	65.30	137.00	4.60	206.90
10"	ER@3.60	Ea	100.00	171.00	5.75	276.75
12"	ER@4.30	Ea	154.00	204.00	6.86	364.86

Schedule 40 carbon steel eccentric reducer, butt-welded joints

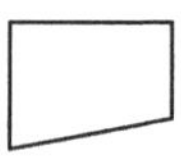

¾"	P1@.295	Ea	7.73	13.10	.96	21.79
1"	P1@.395	Ea	9.57	17.60	1.28	28.45
1¼"	P1@.500	Ea	11.10	22.20	1.62	34.92
1½"	P1@.600	Ea	11.70	26.70	1.95	40.35
2"	P1@.800	Ea	13.90	35.60	2.59	52.09
2½"	P1@1.00	Ea	19.90	44.50	3.24	67.64
3"	P1@1.20	Ea	17.50	53.30	3.89	74.69
4"	P1@1.60	Ea	24.90	71.10	5.19	101.19
6"	ER@2.40	Ea	49.50	114.00	3.83	167.33
8"	ER@2.88	Ea	91.20	137.00	4.60	232.80
10"	ER@3.60	Ea	157.00	171.00	5.75	333.75
12"	ER@4.30	Ea	240.00	204.00	6.86	450.86

Schedule 40 carbon steel cap, butt-welded joints

½"	P1@.150	Ea	7.14	6.67	.49	14.30
¾"	P1@.230	Ea	7.65	10.20	.75	18.60
1"	P1@.300	Ea	8.16	13.30	.97	22.43
1¼"	P1@.390	Ea	9.02	17.30	1.26	27.58
1½"	P1@.470	Ea	9.56	20.90	1.52	31.98
2"	P1@.620	Ea	10.20	27.60	2.01	39.81
2½"	P1@.770	Ea	14.00	34.20	2.50	50.70
3"	P1@.930	Ea	15.10	41.30	3.02	59.42
4"	P1@1.25	Ea	19.40	55.60	4.05	79.05
6"	ER@1.87	Ea	22.40	88.70	2.98	114.08
8"	ER@2.24	Ea	40.60	106.00	3.57	150.17
10"	ER@2.80	Ea	68.50	133.00	4.47	205.97
12"	ER@3.36	Ea	85.90	159.00	5.36	250.26

Description	Craft@Hrs	Unit	Material $	Labor $	Equipment $	Total $

Schedule 40 carbon steel union, butt-welded joints

Description	Craft@Hrs	Unit	Material $	Labor $	Equipment $	Total $
½"	P1@.220	Ea	5.15	9.78	.71	15.64
¾"	P1@.330	Ea	6.64	14.70	1.07	22.41
1"	P1@.440	Ea	9.18	19.60	1.43	30.21
1¼"	P1@.560	Ea	19.40	24.90	1.82	46.12
1½"	P1@.670	Ea	21.10	29.80	2.17	53.07
2"	P1@.890	Ea	28.40	39.60	2.89	70.89
2½"	P1@1.11	Ea	94.80	49.30	3.60	147.70
3"	P1@1.33	Ea	112.00	59.10	4.31	175.41

Carbon steel weldolet, Schedule 40

Description	Craft@Hrs	Unit	Material $	Labor $	Equipment $	Total $
½"	P1@.330	Ea	7.16	14.70	1.07	22.93
¾"	P1@.500	Ea	7.91	22.20	1.62	31.73
1"	P1@.670	Ea	8.81	29.80	2.17	40.78
1¼"	P1@.830	Ea	13.40	36.90	2.69	52.99
1½"	P1@1.00	Ea	13.40	44.50	3.24	61.14
2"	P1@1.33	Ea	14.60	59.10	4.31	78.01
2½"	P1@1.67	Ea	49.90	74.20	5.42	129.52
3"	P1@2.00	Ea	53.80	88.90	6.49	149.19
4"	P1@2.67	Ea	70.00	119.00	8.66	197.66
6"	P1@4.00	Ea	284.00	178.00	13.00	475.00
8"	P1@4.80	Ea	543.00	213.00	15.60	771.60
10"	P1@6.00	Ea	746.00	267.00	19.50	1,032.50
12"	P1@7.20	Ea	912.00	320.00	23.30	1,255.30

Carbon steel threadolet, Schedule 40

Description	Craft@Hrs	Unit	Material $	Labor $	Equipment $	Total $
¾"	P1@.330	Ea	4.77	14.70	1.07	20.54
1"	P1@.440	Ea	5.65	19.60	1.43	26.68
1¼"	P1@.560	Ea	9.76	24.90	1.82	36.48
1½"	P1@.670	Ea	11.00	29.80	2.17	42.97
2"	P1@.890	Ea	12.70	39.60	2.89	55.19
2½"	P1@1.11	Ea	32.00	49.30	3.60	84.90
3"	P1@1.33	Ea	35.00	59.10	4.31	98.41
4"	P1@1.78	Ea	48.00	79.10	5.77	132.87

150# forged steel companion flange, weld neck

Description	Craft@Hrs	Unit	Material $	Labor $	Equipment $	Total $
2½"	P1@.610	Ea	12.90	27.10	1.98	41.98
3"	P1@.730	Ea	12.00	32.40	2.37	46.77
4"	P1@.980	Ea	16.60	43.60	3.18	63.38
6"	ER@1.47	Ea	26.80	69.70	2.35	98.85
8"	ER@1.77	Ea	54.00	83.90	2.82	140.72
10"	ER@2.20	Ea	88.90	104.00	3.51	196.41
12"	ER@2.64	Ea	129.00	125.00	4.21	258.21

Class 125 bronze body gate valve, threaded ends

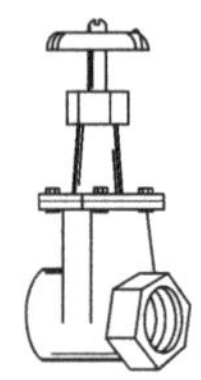

Description	Craft@Hrs	Unit	Material $	Labor $	Equipment $	Total $
½"	P1@.210	Ea	22.60	9.33	—	31.93
¾"	P1@.250	Ea	28.30	11.10	—	39.40
1"	P1@.300	Ea	39.70	13.30	—	53.00
1¼"	P1@.400	Ea	49.90	17.80	—	67.70
1½"	P1@.450	Ea	67.40	20.00	—	87.40
2"	P1@.500	Ea	113.00	22.20	—	135.20
2½"	P1@.750	Ea	186.00	33.30	—	219.30
3"	P1@.950	Ea	267.00	42.20	—	309.20

Class 125 iron body gate valve, flanged ends

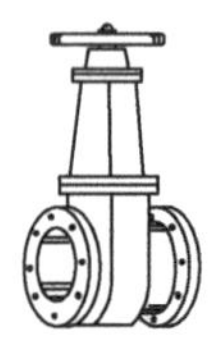

Description	Craft@Hrs	Unit	Material $	Labor $	Equipment $	Total $
2"	P1@.500	Ea	431.00	22.20	—	453.20
2½"	P1@.600	Ea	581.00	26.70	—	607.70
3"	P1@.750	Ea	630.00	33.30	—	663.30
4"	P1@1.35	Ea	925.00	60.00	—	985.00
6"	ER@2.50	Ea	1,770.00	119.00	1.67	1,890.67
8"	ER@3.00	Ea	2,040.00	142.00	2.01	2,184.01
10"	ER@4.00	Ea	3,340.00	190.00	2.68	3,532.68
12"	ER@4.50	Ea	4,620.00	213.00	3.01	4,836.01

Class 125 bronze body globe valve, threaded ends

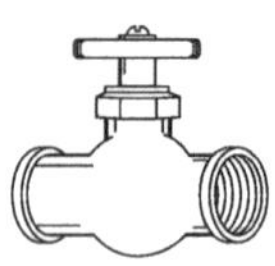

Description	Craft@Hrs	Unit	Material $	Labor $	Equipment $	Total $
½"	P1@.210	Ea	41.90	9.33	—	51.23
¾"	P1@.250	Ea	55.70	11.10	—	66.80
1"	P1@.300	Ea	96.00	13.30	—	109.30
1¼"	P1@.400	Ea	135.00	17.80	—	152.80
1½"	P1@.450	Ea	183.00	20.00	—	203.00
2"	P1@.500	Ea	296.00	22.20	—	318.20

Class 125 iron body globe valve, flanged ends

Description	Craft@Hrs	Unit	Material $	Labor $	Equipment $	Total $
2½"	P1@.600	Ea	488.00	26.70	—	514.70
3"	P1@.750	Ea	586.00	33.30	—	619.30
4"	P1@1.35	Ea	790.00	60.00	—	850.00
6"	ER@2.50	Ea	1,460.00	119.00	1.67	1,580.67
8"	ER@3.00	Ea	2,210.00	142.00	2.01	2,354.01
10"	ER@4.00	Ea	3,380.00	190.00	2.68	3,572.68

200 PSIG iron body butterfly valve, lug-type, lever operated

Description	Craft@Hrs	Unit	Material $	Labor $	Equipment $	Total $
2"	P1@.450	Ea	194.00	20.00	—	214.00
2½"	P1@.450	Ea	201.00	20.00	—	221.00
3"	P1@.550	Ea	212.00	24.40	—	236.40
4"	P1@.550	Ea	263.00	24.40	—	287.40
6"	ER@.800	Ea	443.00	37.90	.54	481.44
8"	ER@.800	Ea	605.00	37.90	.54	643.44
10"	ER@.900	Ea	841.00	42.70	.60	884.30
12"	ER@1.00	Ea	1,100.00	47.40	.67	1,148.07

Description	Craft@Hrs	Unit	Material $	Labor $	Equipment $	Total $

200 PSIG iron body butterfly valve, wafer-type, lever operated

2"	P1@.450	Ea	177.00	20.00	—	197.00
2½"	P1@.450	Ea	180.00	20.00	—	200.00
3"	P1@.550	Ea	194.00	24.40	—	218.40
4"	P1@.550	Ea	235.00	24.40	—	259.40
6"	ER@.800	Ea	410.00	37.90	.54	448.44
8"	ER@.800	Ea	568.00	37.90	.54	606.44
10"	ER@.900	Ea	794.00	42.70	.60	837.30
12"	ER@1.00	Ea	1,040.00	47.40	.67	1,088.07

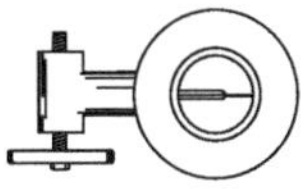

Class 125 bronze body swing check valve, threaded

½"	P1@.210	Ea	25.00	9.33	—	34.33
¾"	P1@.250	Ea	35.70	11.10	—	46.80
1"	P1@.300	Ea	46.70	13.30	—	60.00
1¼"	P1@.400	Ea	66.60	17.80	—	84.40
1½"	P1@.450	Ea	93.90	20.00	—	113.90
2"	P1@.500	Ea	155.00	22.20	—	177.20

Class 125 iron body swing check valve, flanged joints

2"	P1@.500	Ea	219.00	22.20	—	241.20
2½"	P1@.600	Ea	277.00	26.70	—	303.70
3"	P1@.750	Ea	345.00	33.30	—	378.30
4"	P1@1.35	Ea	506.00	60.00	—	566.00
6"	ER@2.50	Ea	983.00	119.00	1.67	1,103.67
8"	ER@3.00	Ea	1,770.00	142.00	2.01	1,914.01
10"	ER@4.00	Ea	2,880.00	190.00	2.68	3,072.68
12"	ER@4.50	Ea	3,950.00	213.00	3.01	4,166.01

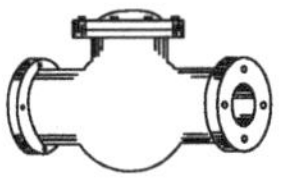

Class 125 iron body silent check valve, flanged joints

2"	P1@.500	Ea	160.00	22.20	—	182.20
2½"	P1@.600	Ea	178.00	26.70	—	204.70
3"	P1@.750	Ea	204.00	33.30	—	237.30
4"	P1@1.35	Ea	273.00	60.00	—	333.00
6"	ER@2.50	Ea	488.00	119.00	1.67	608.67
8"	ER@3.00	Ea	881.00	142.00	2.01	1,025.01
10"	ER@4.00	Ea	1,360.00	190.00	2.68	1,552.68

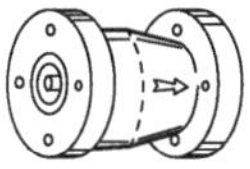

Class 125 bronze body strainer, threaded ends

½"	P1@.210	Ea	38.50	9.33	—	47.83
¾"	P1@.250	Ea	50.50	11.10	—	61.60
1"	P1@.300	Ea	61.90	13.30	—	75.20
1¼"	P1@.400	Ea	86.60	17.80	—	104.40
1½"	P1@.450	Ea	130.00	20.00	—	150.00
2"	P1@.500	Ea	226.00	22.20	—	248.20

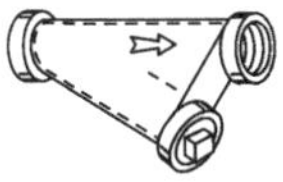

Description	Craft@Hrs	Unit	Material $	Labor $	Equipment $	Total $

Class 125 iron body strainer, flanged

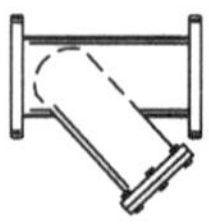

2"	P1@.500	Ea	173.00	22.20	—	195.20
2½"	P1@.600	Ea	192.00	26.70	—	218.70
3"	P1@.750	Ea	224.00	33.30	—	257.30
4"	P1@1.35	Ea	378.00	60.00	—	438.00
6"	ER@2.50	Ea	729.00	119.00	1.67	849.67
8"	ER@3.00	Ea	1,230.00	142.00	2.01	1,374.01

Installation of 2-way control valve, threaded joints

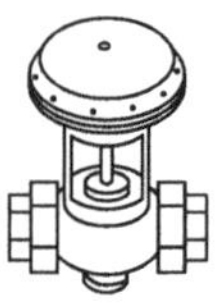

½"	P1@.210	Ea	—	9.33	—	9.33
¾"	P1@.250	Ea	—	11.10	—	11.10
1"	P1@.300	Ea	—	13.30	—	13.30
1¼"	P1@.400	Ea	—	17.80	—	17.80
1½"	P1@.450	Ea	—	20.00	—	20.00
2"	P1@.500	Ea	—	22.20	—	22.20
2½"	P1@.830	Ea	—	36.90	—	36.90
3"	P1@.990	Ea	—	44.00	—	44.00

Installation of 2-way control valve, flanged joints

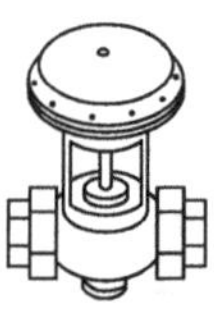

2"	P1@.500	Ea	—	22.20	—	22.20
2½"	P1@.600	Ea	—	26.70	—	26.70
3"	P1@.750	Ea	—	33.30	—	33.30
4"	P1@1.35	Ea	—	60.00	—	60.00
6"	P1@2.50	Ea	—	111.00	—	111.00
8"	P1@3.00	Ea	—	133.00	—	133.00
10"	P1@4.00	Ea	—	178.00	—	178.00
12"	P1@4.50	Ea	—	200.00	—	200.00

Installation of 3-way control valve, threaded joints

½"	P1@.260	Ea	—	11.60	—	11.60
¾"	P1@.365	Ea	—	16.20	—	16.20
1"	P1@.475	Ea	—	21.10	—	21.10
1¼"	P1@.575	Ea	—	25.60	—	25.60
1½"	P1@.680	Ea	—	30.20	—	30.20
2"	P1@.910	Ea	—	40.40	—	40.40
2½"	P1@1.12	Ea	—	49.80	—	49.80
3"	P1@1.33	Ea	—	59.10	—	59.10

Installation of 3-way control valve, flanged joints

2"	P1@.910	Ea	—	40.40	—	40.40
2½"	P1@1.12	Ea	—	49.80	—	49.80
3"	P1@1.33	Ea	—	59.10	—	59.10
4"	P1@2.00	Ea	—	88.90	—	88.90
6"	P1@3.70	Ea	—	164.00	—	164.00
8"	P1@4.40	Ea	—	196.00	—	196.00
10"	P1@5.90	Ea	—	262.00	—	262.00
12"	P1@6.50	Ea	—	289.00	—	289.00

Description	Craft@Hrs	Unit	Material $	Labor $	Equipment $	Total $

Bolt and gasket sets

Description	Craft@Hrs	Unit	Material $	Labor $	Equipment $	Total $
2"	P1@.500	Ea	4.64	22.20	—	26.84
2½"	P1@.650	Ea	5.42	28.90	—	34.32
3"	P1@.750	Ea	9.19	33.30	—	42.49
4"	P1@1.00	Ea	15.40	44.50	—	59.90
6"	P1@1.20	Ea	25.10	53.30	—	78.40
8"	P1@1.25	Ea	28.10	55.60	—	83.70
10"	P1@1.70	Ea	48.30	75.60	—	123.90
12"	P1@2.20	Ea	54.50	97.80	—	152.30

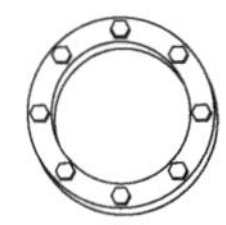

Thermometer with well

Description	Craft@Hrs	Unit	Material $	Labor $	Equipment $	Total $
7"	P1@.250	Ea	186.00	11.10	—	197.10
9"	P1@.250	Ea	192.00	11.10	—	203.10

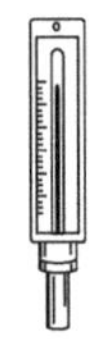

Dial-type pressure gauge

Description	Craft@Hrs	Unit	Material $	Labor $	Equipment $	Total $
2½"	P1@.200	Ea	38.50	8.89	—	47.39
3½"	P1@.200	Ea	51.20	8.89	—	60.09

Pressure/temperature tap

Description	Craft@Hrs	Unit	Material $	Labor $	Equipment $	Total $
Tap	P1@.150	Ea	17.70	6.67	—	24.37

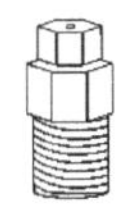

Hanger with swivel assembly

Description	Craft@Hrs	Unit	Material $	Labor $	Equipment $	Total $
2"	P1@.300	Ea	5.99	13.30	—	19.29
2½"	P1@.350	Ea	8.15	15.60	—	23.75
3"	P1@.350	Ea	10.10	15.60	—	25.70
4"	P1@.350	Ea	11.10	15.60	—	26.70
5"	P1@.450	Ea	19.10	20.00	—	39.10
6"	P1@.450	Ea	24.90	20.00	—	44.90
8"	P1@.450	Ea	33.10	20.00	—	53.10
10"	P1@.550	Ea	45.40	24.40	—	69.80
12"	P1@.600	Ea	60.20	26.70	—	86.90

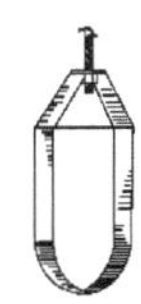

Riser clamp

Description	Craft@Hrs	Unit	Material $	Labor $	Equipment $	Total $
2"	P1@.115	Ea	6.62	5.11	—	11.73
2½"	P1@.120	Ea	6.93	5.33	—	12.26
3"	P1@.120	Ea	7.53	5.33	—	12.86
4"	P1@.125	Ea	9.61	5.56	—	15.17
5"	P1@.180	Ea	13.30	8.00	—	21.30
6"	P1@.200	Ea	16.00	8.89	—	24.89
8"	P1@.200	Ea	26.00	8.89	—	34.89
10"	P1@.250	Ea	38.60	11.10	—	49.70
12"	P1@.250	Ea	45.80	11.10	—	56.90

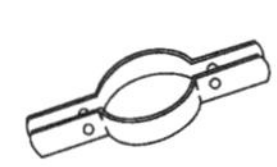

Carbon Steel, Schedule 40 with 150# M.I. Fittings & Threaded Joints

Schedule 40 carbon steel (ASTM A-53 and A-120) pipe with 150 pound malleable iron threaded fittings is commonly used for heating hot water, chilled water, glycol and low pressure steam systems where operating pressures do not exceed 125 PSIG and temperatures are below 250 degrees F., for pipe sizes 2" and smaller. For pipe sizes 2½" and larger, wrought steel fittings with either welded or grooved joints should be used.

Schedule 40 (ASTM A-53B SML) steel pipe with carbon steel welding fittings can also be used for refrigerant piping systems.

This section has been arranged to save the estimator's time by including all normally-used system components such as pipe, fittings, valves, hanger assemblies, riser clamps and miscellaneous items under one heading. Additional items can be found under "Plumbing and Piping Specialties." The cost estimates in this section are based on the conditions, limitations and wage rates described in the section "How to Use This Book" beginning on page 5.

Description	Craft@Hrs	Unit	Material $	Labor $	Equipment $	Total $

Schedule 40 carbon steel threaded horizontal pipe assembly. Horizontally hung in a building. Assembly includes fittings and hanger assemblies. Based on a reducing tee and a 45-degree elbow every 16 feet for ½" pipe. A reducing tee and a 45-degree elbow every 30 feet for 4" pipe. Hangers spaced to meet plumbing code. *(Distance between fittings in the pipe assembly increases as pipe diameter increases.)* Use these figures for preliminary estimates.

Description	Craft@Hrs	Unit	Material $	Labor $	Equipment $	Total $
½"	P1@.119	LF	2.01	5.29	—	7.30
¾"	P1@.132	LF	2.35	5.87	—	8.22
1"	P1@.144	LF	3.24	6.40	—	9.64
1¼"	P1@.160	LF	4.23	7.11	—	11.34
1½"	P1@.184	LF	4.70	8.18	—	12.88
2"	P1@.200	LF	7.12	8.89	—	16.01
2½"	P1@.236	LF	14.00	10.50	—	24.50
3"	P1@.271	LF	19.30	12.00	—	31.30
4"	P1@.398	LF	28.70	17.70	—	46.40

Schedule 40 carbon steel threaded vertical pipe assembly. Riser assembly, including fittings and riser clamps. Based on a reducing tee, and a riser clamp on every floor. Use these figures for preliminary estimates.

Description	Craft@Hrs	Unit	Material $	Labor $	Equipment $	Total $
½"	P1@.082	LF	1.53	3.64	—	5.17
¾"	P1@.096	LF	1.94	4.27	—	6.21
1"	P1@.121	LF	2.76	5.38	—	8.14
1¼"	P1@.138	LF	3.95	6.13	—	10.08
1½"	P1@.156	LF	4.49	6.93	—	11.42
2"	P1@.173	LF	6.70	7.69	—	14.39
2½"	P1@.200	LF	13.70	8.89	—	22.59
3"	P1@.236	LF	20.10	10.50	—	30.60
4"	P1@.354	LF	26.40	15.70	—	42.10

Carbon Steel, Schedule 40 with 150# M.I. Fittings & Threaded Joints

Schedule 40 carbon steel pipe, threaded and coupled, pipe only

Description	Craft@Hrs	Unit	Material $	Labor $	Equipment $	Total $
½"	P1@.060	LF	1.44	2.67	—	4.11
¾"	P1@.070	LF	1.60	3.11	—	4.71
1"	P1@.090	LF	2.43	4.00	—	6.43
1¼"	P1@.100	LF	3.15	4.45	—	7.60
1½"	P1@.110	LF	3.76	4.89	—	8.65
2"	P1@.120	LF	5.56	5.33	—	10.89
2½"	P1@.130	LF	8.54	5.78	—	14.32
3"	P1@.150	LF	11.30	6.67	—	17.97
4"	P1@.260	LF	16.20	11.60	—	27.80

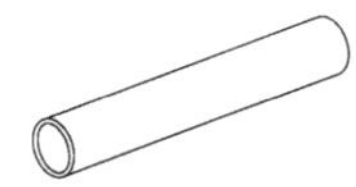

150# malleable iron 45-degree ell, threaded joints

Description	Craft@Hrs	Unit	Material $	Labor $	Equipment $	Total $
½"	P1@.120	Ea	2.02	5.33	—	7.35
¾"	P1@.130	Ea	2.46	5.78	—	8.24
1"	P1@.180	Ea	3.13	8.00	—	11.13
1¼"	P1@.240	Ea	5.55	10.70	—	16.25
1½"	P1@.300	Ea	6.86	13.30	—	20.16
2"	P1@.380	Ea	10.20	16.90	—	27.10
2½"	P1@.510	Ea	29.60	22.70	—	52.30
3"	P1@.640	Ea	38.40	28.40	—	66.80
4"	P1@.850	Ea	74.90	37.80	—	112.70

150# malleable iron 90-degree ell, threaded joints

Description	Craft@Hrs	Unit	Material $	Labor $	Equipment $	Total $
½"	P1@.120	Ea	1.25	5.33	—	6.58
¾"	P1@.130	Ea	1.53	5.78	—	7.31
1"	P1@.180	Ea	2.62	8.00	—	10.62
1¼"	P1@.240	Ea	4.28	10.70	—	14.98
1½"	P1@.300	Ea	5.67	13.30	—	18.97
2"	P1@.380	Ea	9.66	16.90	—	26.56
2½"	P1@.510	Ea	21.10	22.70	—	43.80
3"	P1@.640	Ea	31.40	28.40	—	59.80
4"	P1@.850	Ea	73.30	37.80	—	111.10

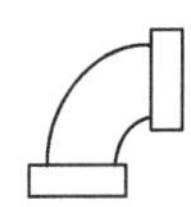

150# malleable iron tee, threaded joints

Description	Craft@Hrs	Unit	Material $	Labor $	Equipment $	Total $
½"	P1@.180	Ea	1.66	8.00	—	9.66
¾"	P1@.190	Ea	2.41	8.45	—	10.86
1"	P1@.230	Ea	4.08	10.20	—	14.28
1¼"	P1@.310	Ea	6.69	13.80	—	20.49
1½"	P1@.390	Ea	8.21	17.30	—	25.51
2"	P1@.490	Ea	13.90	21.80	—	35.70
2½"	P1@.660	Ea	29.50	29.30	—	58.80
3"	P1@.830	Ea	43.10	36.90	—	80.00
4"	P1@1.10	Ea	103.00	48.90	—	151.90

Description	Craft@Hrs	Unit	Material $	Labor $	Equipment $	Total $

150# malleable iron reducing tee, threaded joints

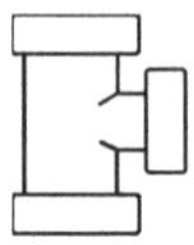

Description	Craft@Hrs	Unit	Material $	Labor $	Equipment $	Total $
¾" x ½"	P1@.180	Ea	3.56	8.00	—	11.56
1" x ½"	P1@.220	Ea	4.53	9.78	—	14.31
1" x ¾"	P1@.220	Ea	4.53	9.78	—	14.31
1¼" x ½"	P1@.290	Ea	8.21	12.90	—	21.11
1¼" x ¾"	P1@.290	Ea	7.23	12.90	—	20.13
1¼" x 1"	P1@.290	Ea	7.83	12.90	—	20.73
1½" x ½"	P1@.370	Ea	9.63	16.40	—	26.03
1½" x ¾"	P1@.370	Ea	9.09	16.40	—	25.49
1½" x 1"	P1@.370	Ea	9.09	16.40	—	25.49
1½" x 1¼"	P1@.370	Ea	11.20	16.40	—	27.60
2" x 1"	P1@.460	Ea	14.30	20.40	—	34.70
2" x 1¼"	P1@.460	Ea	15.50	20.40	—	35.90
2" x 1½"	P1@.460	Ea	15.50	20.40	—	35.90
4" x 1½"	P1@1.05	Ea	107.00	46.70	—	153.70
4" x 2"	P1@1.05	Ea	107.00	46.70	—	153.70
4" x 3"	P1@1.05	Ea	107.00	46.70	—	153.70

150# malleable iron reducer, threaded joints

Description	Craft@Hrs	Unit	Material $	Labor $	Equipment $	Total $
¾" x ½"	P1@.130	Ea	2.26	5.78	—	8.04
1" x ½"	P1@.160	Ea	3.53	7.11	—	10.64
1" x ¾"	P1@.160	Ea	3.53	7.11	—	10.64
1¼" x ¾"	P1@.210	Ea	3.53	9.33	—	12.86
1¼" x 1"	P1@.210	Ea	4.37	9.33	—	13.70
1½" x 1"	P1@.270	Ea	6.42	12.00	—	18.42
1½" x 1¼"	P1@.270	Ea	5.55	12.00	—	17.55
2" x 1¼"	P1@.340	Ea	9.18	15.10	—	24.28
2" x 1½"	P1@.340	Ea	8.06	15.10	—	23.16
3" x 2"	P1@.530	Ea	24.00	23.60	—	47.60
3" x 2½"	P1@.530	Ea	28.00	23.60	—	51.60
4" x 3"	P1@.750	Ea	28.00	33.30	—	61.30

150# malleable iron cross, threaded joints

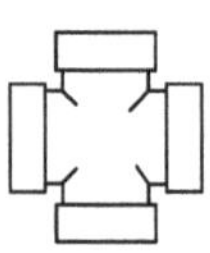

Description	Craft@Hrs	Unit	Material $	Labor $	Equipment $	Total $
½"	P1@.280	Ea	6.07	12.40	—	18.47
¾"	P1@.320	Ea	7.41	14.20	—	21.61
1"	P1@.360	Ea	9.09	16.00	—	25.09
1¼"	P1@.480	Ea	14.60	21.30	—	35.90
1½"	P1@.600	Ea	18.10	26.70	—	44.80
2"	P1@.700	Ea	30.00	31.10	—	61.10
2½"	P1@1.02	Ea	55.40	45.30	—	100.70
3"	P1@1.28	Ea	76.70	56.90	—	133.60

Description	Craft@Hrs	Unit	Material $	Labor $	Equipment $	Total $

150# malleable iron cap, threaded joint

Description	Craft@Hrs	Unit	Material $	Labor $	Equipment $	Total $
½"	P1@.090	Ea	1.26	4.00	—	5.26
¾"	P1@.100	Ea	1.68	4.45	—	6.13
1"	P1@.140	Ea	2.08	6.22	—	8.30
1¼"	P1@.180	Ea	2.69	8.00	—	10.69
1½"	P1@.230	Ea	3.66	10.20	—	13.86
2"	P1@.290	Ea	5.37	12.90	—	18.27
2½"	P1@.380	Ea	12.00	16.90	—	28.90
3"	P1@.480	Ea	17.70	21.30	—	39.00
4"	P1@.640	Ea	30.20	28.40	—	58.60

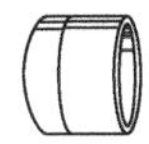

150# malleable iron plug, threaded joint

Description	Craft@Hrs	Unit	Material $	Labor $	Equipment $	Total $
½"	P1@.090	Ea	.99	4.00	—	4.99
¾"	P1@.100	Ea	1.07	4.45	—	5.52
1"	P1@.140	Ea	1.07	6.22	—	7.29
1¼"	P1@.180	Ea	1.70	8.00	—	9.70
1½"	P1@.230	Ea	2.44	10.20	—	12.64
2"	P1@.290	Ea	2.84	12.90	—	15.74
2½"	P1@.380	Ea	6.88	16.90	—	23.78
3"	P1@.480	Ea	8.00	21.30	—	29.30
4"	P1@.640	Ea	14.60	28.40	—	43.00

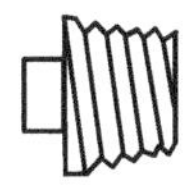

150# malleable iron union, threaded joints

Description	Craft@Hrs	Unit	Material $	Labor $	Equipment $	Total $
½"	P1@.140	Ea	5.51	6.22	—	11.73
¾"	P1@.150	Ea	6.32	6.67	—	12.99
1"	P1@.210	Ea	8.22	9.33	—	17.55
1¼"	P1@.280	Ea	11.80	12.40	—	24.20
1½"	P1@.360	Ea	14.60	16.00	—	30.60
2"	P1@.450	Ea	17.00	20.00	—	37.00
2½"	P1@.610	Ea	50.60	27.10	—	77.70
3"	P1@.760	Ea	60.90	33.80	—	94.70

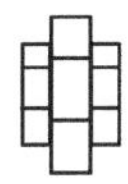

150# malleable iron coupling, threaded joints

Description	Craft@Hrs	Unit	Material $	Labor $	Equipment $	Total $
½"	P1@.120	Ea	1.68	5.33	—	7.01
¾"	P1@.130	Ea	2.00	5.78	—	7.78
1"	P1@.180	Ea	3.00	8.00	—	11.00
1¼"	P1@.240	Ea	3.88	10.70	—	14.58
1½"	P1@.300	Ea	5.17	13.30	—	18.47
2"	P1@.380	Ea	7.52	16.90	—	24.42

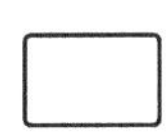

Description	Craft@Hrs	Unit	Material $	Labor $	Equipment $	Total $

Steel pipe nipples — Close nipple, standard right-hand thread

Description	Craft@Hrs	Unit	Material $	Labor $	Equipment $	Total $
½"	P1@.025	Ea	.65	1.11	—	1.76
¾"	P1@.025	Ea	.78	1.11	—	1.89
1"	P1@.025	Ea	1.20	1.11	—	2.31
1¼"	P1@.025	Ea	1.47	1.11	—	2.58
1½"	P1@.025	Ea	1.76	1.11	—	2.87
2"	P1@.025	Ea	2.27	1.11	—	3.38
2½"	P1@.030	Ea	8.82	1.33	—	10.15
3"	P1@.030	Ea	9.56	1.33	—	10.89
4"	P1@.040	Ea	15.20	1.78	—	16.98
6"	P1@.050	Ea	77.00	2.22	—	79.22
8"	P1@.065	Ea	135.00	2.89	—	137.89

Steel pipe nipples — 1½" long, standard right-hand thread

Description	Craft@Hrs	Unit	Material $	Labor $	Equipment $	Total $
½"	P1@.025	Ea	.69	1.11	—	1.80
¾"	P1@.025	Ea	.79	1.11	—	1.90

Steel pipe nipples — 2" long, standard right-hand thread

Description	Craft@Hrs	Unit	Material $	Labor $	Equipment $	Total $
½"	P1@.025	Ea	.69	1.11	—	1.80
¾"	P1@.025	Ea	.79	1.11	—	1.90
1"	P1@.025	Ea	1.20	1.11	—	2.31
1¼"	P1@.025	Ea	1.47	1.11	—	2.58
1½"	P1@.025	Ea	1.76	1.11	—	2.87

Steel pipe nipples — 2½" long, standard right-hand thread

Description	Craft@Hrs	Unit	Material $	Labor $	Equipment $	Total $
½"	P1@.025	Ea	.70	1.11	—	1.81
¾"	P1@.025	Ea	.83	1.11	—	1.94
1"	P1@.025	Ea	1.36	1.11	—	2.47
1¼"	P1@.025	Ea	1.74	1.11	—	2.85
1½"	P1@.025	Ea	2.16	1.11	—	3.27
2"	P1@.025	Ea	2.55	1.11	—	3.66

Steel pipe nipples — 3" long, standard right-hand thread

Description	Craft@Hrs	Unit	Material $	Labor $	Equipment $	Total $
½"	P1@.025	Ea	.82	1.11	—	1.93
¾"	P1@.025	Ea	.97	1.11	—	2.08
1"	P1@.025	Ea	1.41	1.11	—	2.52
1¼"	P1@.025	Ea	1.84	1.11	—	2.95
1½"	P1@.025	Ea	2.20	1.11	—	3.31
2"	P1@.030	Ea	2.73	1.33	—	4.06
2½"	P1@.030	Ea	9.17	1.33	—	10.50
3"	P1@.035	Ea	10.30	1.56	—	11.86

Description	Craft@Hrs	Unit	Material $	Labor $	Equipment $	Total $
Steel pipe nipples — 3½" long, standard right-hand thread						
½"	P1@.025	Ea	.94	1.11	—	2.05
¾"	P1@.025	Ea	1.12	1.11	—	2.23
1"	P1@.025	Ea	1.62	1.11	—	2.73
1¼"	P1@.025	Ea	2.07	1.11	—	3.18
1½"	P1@.030	Ea	2.61	1.33	—	3.94
2"	P1@.030	Ea	3.22	1.33	—	4.55
2½"	P1@.035	Ea	12.20	1.56	—	13.76
3"	P1@.035	Ea	14.00	1.56	—	15.56
4"	P1@.040	Ea	19.30	1.78	—	21.08
Steel pipe nipples — 4" long, standard right-hand thread						
½"	P1@.025	Ea	1.00	1.11	—	2.11
¾"	P1@.025	Ea	1.20	1.11	—	2.31
1"	P1@.025	Ea	1.72	1.11	—	2.83
1¼"	P1@.030	Ea	2.13	1.33	—	3.46
1½"	P1@.030	Ea	2.69	1.33	—	4.02
2"	P1@.030	Ea	3.39	1.33	—	4.72
2½"	P1@.035	Ea	10.60	1.56	—	12.16
3"	P1@.040	Ea	11.80	1.78	—	13.58
4"	P1@.040	Ea	17.20	1.78	—	18.98
Steel pipe nipples — 4½" long, standard right-hand thread						
½"	P1@.025	Ea	1.20	1.11	—	2.31
¾"	P1@.025	Ea	1.35	1.11	—	2.46
1"	P1@.025	Ea	1.84	1.11	—	2.95
1¼"	P1@.030	Ea	2.89	1.33	—	4.22
1½"	P1@.030	Ea	3.50	1.33	—	4.83
2"	P1@.030	Ea	3.76	1.33	—	5.09
2½"	P1@.035	Ea	13.50	1.56	—	15.06
3"	P1@.040	Ea	16.90	1.78	—	18.68
4"	P1@.050	Ea	21.20	2.22	—	23.42
6"	P1@.070	Ea	92.50	3.11	—	95.61
Steel pipe nipples — 5" long, standard right-hand thread						
½"	P1@.025	Ea	1.22	1.11	—	2.33
¾"	P1@.025	Ea	1.35	1.11	—	2.46
1"	P1@.030	Ea	2.00	1.33	—	3.33
1¼"	P1@.030	Ea	2.59	1.33	—	3.92
1½"	P1@.030	Ea	3.10	1.33	—	4.43
2"	P1@.035	Ea	3.90	1.56	—	5.46
2½"	P1@.040	Ea	14.30	1.78	—	16.08
3"	P1@.045	Ea	14.80	2.00	—	16.80
4"	P1@.050	Ea	23.20	2.22	—	25.42
6"	P1@.075	Ea	93.80	3.33	—	97.13
8"	P1@.100	Ea	157.00	4.45	—	161.45

Carbon Steel, Schedule 40 with 150# M.I. Fittings & Threaded Joints

Description	Craft@Hrs	Unit	Material $	Labor $	Equipment $	Total $
Steel pipe nipples — 6" long, standard right-hand thread						
½"	P1@.025	Ea	1.35	1.11	—	2.46
¾"	P1@.025	Ea	1.64	1.11	—	2.75
1"	P1@.030	Ea	2.20	1.33	—	3.53
1¼"	P1@.030	Ea	2.74	1.33	—	4.07
1½"	P1@.030	Ea	3.30	1.33	—	4.63
2"	P1@.035	Ea	4.45	1.56	—	6.01
2½"	P1@.040	Ea	12.80	1.78	—	14.58
3"	P1@.045	Ea	14.30	2.00	—	16.30
4"	P1@.060	Ea	21.20	2.67	—	23.87
6"	P1@.085	Ea	105.00	3.78	—	108.78
8"	P1@.120	Ea	168.00	5.33	—	173.33
Steel pipe nipples — 8" long, standard right-hand thread						
2½"	P1@.050	Ea	16.80	2.22	—	19.02
3"	P1@.055	Ea	20.60	2.44	—	23.04
4"	P1@.070	Ea	33.00	3.11	—	36.11
6"	P1@.100	Ea	110.00	4.45	—	114.45
8"	P1@.150	Ea	188.00	6.67	—	194.67
Steel pipe nipples — 10" long, standard right-hand thread						
2½"	P1@.055	Ea	20.30	2.44	—	22.74
3"	P1@.060	Ea	24.50	2.67	—	27.17
4"	P1@.080	Ea	39.90	3.56	—	43.46
6"	P1@.115	Ea	123.00	5.11	—	128.11
8"	P1@.165	Ea	211.00	7.33	—	218.33
Steel pipe nipples — 12" long, standard right-hand thread						
2½"	P1@.060	Ea	19.60	2.67	—	22.27
3"	P1@.065	Ea	24.10	2.89	—	26.99
4"	P1@.080	Ea	39.10	3.56	—	42.66
6"	P1@.130	Ea	140.00	5.78	—	145.78
8"	P1@.190	Ea	232.00	8.45	—	240.45
Class 125 bronze body gate valve, threaded ends						
½"	P1@.210	Ea	22.60	9.33	—	31.93
¾"	P1@.250	Ea	28.30	11.10	—	39.40
1"	P1@.300	Ea	39.70	13.30	—	53.00
1¼"	P1@.400	Ea	49.90	17.80	—	67.70
1½"	P1@.450	Ea	67.40	20.00	—	87.40
2"	P1@.500	Ea	113.00	22.20	—	135.20
2½"	P1@.750	Ea	186.00	33.30	—	219.30
3"	P1@.950	Ea	267.00	42.20	—	309.20

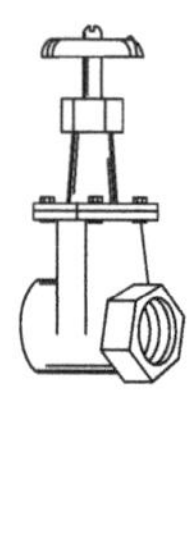

Description	Craft@Hrs	Unit	Material $	Labor $	Equipment $	Total $
Class 125 iron body gate valve, flanged ends						
2"	P1@.500	Ea	431.00	22.20	—	453.20
2½"	P1@.600	Ea	581.00	26.70	—	607.70
3"	P1@.750	Ea	630.00	33.30	—	663.30
4"	P1@1.35	Ea	925.00	60.00	—	985.00
Class 125 bronze body globe valve, threaded ends						
½"	P1@.210	Ea	41.90	9.33	—	51.23
¾"	P1@.250	Ea	55.70	11.10	—	66.80
1"	P1@.300	Ea	96.00	13.30	—	109.30
1¼"	P1@.400	Ea	135.00	17.80	—	152.80
1½"	P1@.450	Ea	183.00	20.00	—	203.00
2"	P1@.500	Ea	296.00	22.20	—	318.20
Class 125 iron body globe valve, flanged ends						
2"	P1@.500	Ea	488.00	22.20	—	510.20
2½"	P1@.600	Ea	586.00	26.70	—	612.70
3"	P1@.750	Ea	790.00	33.30	—	823.30
4"	P1@1.35	Ea	823.00	60.00	—	883.00
200 PSIG iron body butterfly valve, lug-type, lever operated						
2"	P1@.450	Ea	179.00	20.00	—	199.00
2½"	P1@.450	Ea	186.00	20.00	—	206.00
3"	P1@.550	Ea	196.00	24.40	—	220.40
4"	P1@.550	Ea	244.00	24.40	—	268.40
200 PSIG iron body butterfly valve, wafer-type, lever operated						
2"	P1@.450	Ea	164.00	20.00	—	184.00
2½"	P1@.450	Ea	167.00	20.00	—	187.00
3"	P1@.550	Ea	179.00	24.40	—	203.40
4"	P1@.550	Ea	217.00	24.40	—	241.40
Class 125 bronze body 2-piece ball valve, threaded ends						
½"	P1@.210	Ea	11.60	9.33	—	20.93
¾"	P1@.250	Ea	15.40	11.10	—	26.50
1"	P1@.300	Ea	26.70	13.30	—	40.00
1¼"	P1@.400	Ea	44.60	17.80	—	62.40
1½"	P1@.450	Ea	60.30	20.00	—	80.30
2"	P1@.500	Ea	76.40	22.20	—	98.60
3"	P1@.625	Ea	524.00	27.80	—	551.80
4"	P1@.690	Ea	684.00	30.70	—	714.70

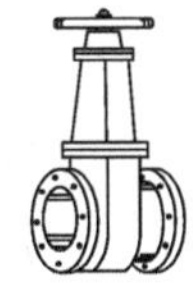
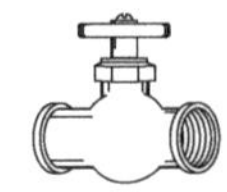
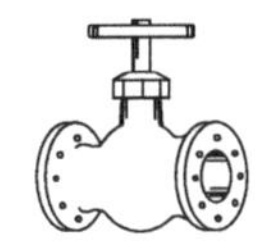
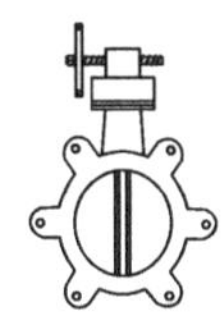
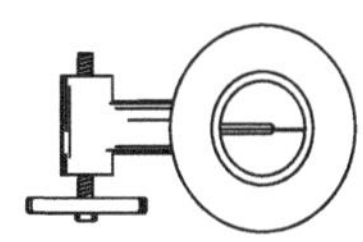
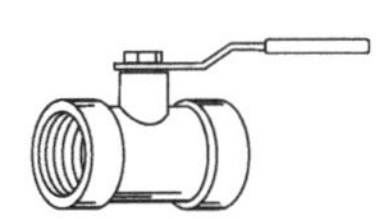

Description	Craft@Hrs	Unit	Material $	Labor $	Equipment $	Total $

Class 125 bronze body swing check valve, threaded

½"	P1@.210	Ea	23.90	9.33	—	33.23
¾"	P1@.250	Ea	34.10	11.10	—	45.20
1"	P1@.300	Ea	44.50	13.30	—	57.80
1¼"	P1@.400	Ea	63.60	17.80	—	81.40
1½"	P1@.450	Ea	89.60	20.00	—	109.60
2"	P1@.500	Ea	148.00	22.20	—	170.20

Class 125 iron body swing check valve, flanged ends

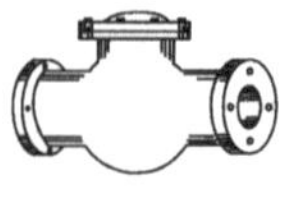

2"	P1@.500	Ea	209.00	22.20	—	231.20
2½"	P1@.600	Ea	264.00	26.70	—	290.70
3"	P1@.750	Ea	329.00	33.30	—	362.30
4"	P1@1.35	Ea	483.00	60.00	—	543.00

Class 125 iron body silent check valve, wafer-type

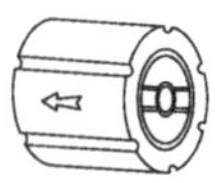

2"	P1@.500	Ea	153.00	22.20	—	175.20
2½"	P1@.600	Ea	170.00	26.70	—	196.70
3"	P1@.750	Ea	195.00	33.30	—	228.30
4"	P1@1.35	Ea	260.00	60.00	—	320.00

Class 125 bronze body strainer, threaded ends

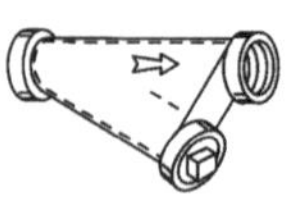

½"	P1@.210	Ea	37.40	9.33	—	46.73
¾"	P1@.250	Ea	49.10	11.10	—	60.20
1"	P1@.300	Ea	60.20	13.30	—	73.50
1¼"	P1@.400	Ea	84.20	17.80	—	102.00
1½"	P1@.450	Ea	126.00	20.00	—	146.00
2"	P1@.500	Ea	220.00	22.20	—	242.20

Class 125 iron body strainer, flanged ends

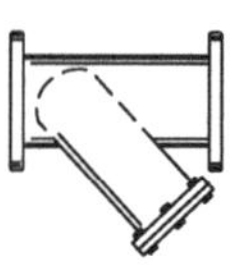

2"	P1@.500	Ea	168.00	22.20	—	190.20
2½"	P1@.600	Ea	187.00	26.70	—	213.70
3"	P1@.750	Ea	218.00	33.30	—	251.30
4"	P1@1.35	Ea	367.00	60.00	—	427.00

Installation of 2-way control valve, threaded joints

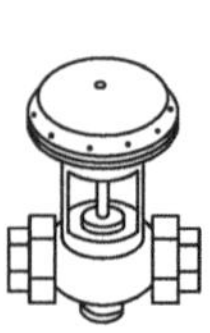

½"	P1@.210	Ea	—	9.33	—	9.33
¾"	P1@.275	Ea	—	12.20	—	12.20
1"	P1@.350	Ea	—	15.60	—	15.60
1¼"	P1@.430	Ea	—	19.10	—	19.10
1½"	P1@.505	Ea	—	22.40	—	22.40
2"	P1@.675	Ea	—	30.00	—	30.00
2½"	P1@.830	Ea	—	36.90	—	36.90
3"	P1@.990	Ea	—	44.00	—	44.00

Installation of 3-way control valve, threaded joints

Description	Craft@Hrs	Unit	Material $	Labor $	Equipment $	Total $
½"	P1@.260	Ea	—	11.60	—	11.60
¾"	P1@.365	Ea	—	16.20	—	16.20
1"	P1@.475	Ea	—	21.10	—	21.10
1¼"	P1@.575	Ea	—	25.60	—	25.60
1½"	P1@.680	Ea	—	30.20	—	30.20
2"	P1@.910	Ea	—	40.40	—	40.40
2½"	P1@1.12	Ea	—	49.80	—	49.80
3"	P1@1.33	Ea	—	59.10	—	59.10

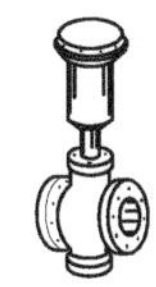

Companion flange, threaded

Description	Craft@Hrs	Unit	Material $	Labor $	Equipment $	Total $
2"	P1@.290	Ea	46.10	12.90	—	59.00
2½"	P1@.380	Ea	70.90	16.90	—	87.80
3"	P1@.460	Ea	76.60	20.40	—	97.00
4"	P1@.600	Ea	105.00	26.70	—	131.70

Bolt and gasket sets

Description	Craft@Hrs	Unit	Material $	Labor $	Equipment $	Total $
2"	P1@.500	Ea	4.64	22.20	—	26.84
2½"	P1@.650	Ea	5.42	28.90	—	34.32
3"	P1@.750	Ea	9.19	33.30	—	42.49
4"	P1@1.00	Ea	15.40	44.50	—	59.90

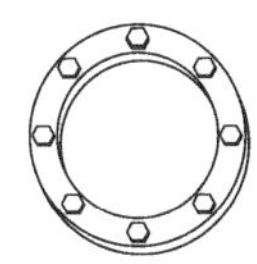

Thermometer with well

Description	Craft@Hrs	Unit	Material $	Labor $	Equipment $	Total $
7"	P1@.250	Ea	186.00	11.10	—	197.10
9"	P1@.250	Ea	192.00	11.10	—	203.10

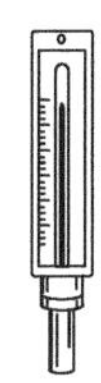

Dial-type pressure gauge

Description	Craft@Hrs	Unit	Material $	Labor $	Equipment $	Total $
2½"	P1@.200	Ea	38.50	8.89	—	47.39
3½"	P1@.200	Ea	51.20	8.89	—	60.09

Pressure/temperature tap

Description	Craft@Hrs	Unit	Material $	Labor $	Equipment $	Total $
Tap	P1@.150	Ea	17.70	6.67	—	24.37

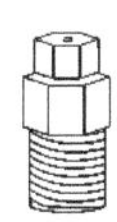

Hanger with swivel assembly

Description	Craft@Hrs	Unit	Material $	Labor $	Equipment $	Total $
½"	P1@.250	Ea	4.50	11.10	—	15.60
¾"	P1@.250	Ea	4.98	11.10	—	16.08
1"	P1@.250	Ea	5.20	11.10	—	16.30
1¼"	P1@.300	Ea	5.34	13.30	—	18.64
1½"	P1@.300	Ea	5.74	13.30	—	19.04
2"	P1@.300	Ea	5.99	13.30	—	19.29
2½"	P1@.350	Ea	8.15	15.60	—	23.75
3"	P1@.350	Ea	10.10	15.60	—	25.70
4"	P1@.350	Ea	11.10	15.60	—	26.70

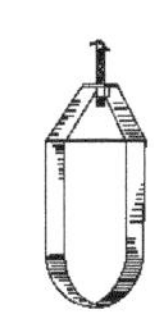

Description	Craft@Hrs	Unit	Material $	Labor $	Equipment $	Total $
Riser clamp						
½"	P1@.100	Ea	3.30	4.45	—	7.75
¾"	P1@.100	Ea	4.87	4.45	—	9.32
1"	P1@.100	Ea	4.91	4.45	—	9.36
1¼"	P1@.105	Ea	5.90	4.67	—	10.57
1½"	P1@.110	Ea	6.22	4.89	—	11.11
2"	P1@.115	Ea	6.62	5.11	—	11.73
2½"	P1@.120	Ea	6.93	5.33	—	12.26
3"	P1@.120	Ea	7.53	5.33	—	12.86
4"	P1@.125	Ea	9.61	5.56	—	15.17

Schedule 5 carbon steel (ASTM A-53, A135 and A-795) pipe with pressfit fittings. The pressfit jointing method offers economy, speed and reliability for joining ¾" to 2" diameter pipe for fire protection, heating and cooling water systems. *(Use roll-grooved pipe and fittings for 2½" and larger pipe diameters.)*

The pressfit system for carbon steel is BOCA approved. Listed by SBCCI PST and ESI and is UL/ULC listed and FM approved for 175 PSI (1,200 kPa) on fire protection services and is rated to 300 PSI (2,065 kPa) for heating water systems.

The pressfit system requires no special preparation of the pipe ends before assembly. Pipe should be square cut (+/–0.030") and de-burred. Pressfit system products are designed only for use on approved Schedule 5 carbon steel pipe.

O-ring selection as specified by the mechanical design engineer. Type O O-rings are rated for application temperatures of +20 degrees F to +300 degrees F. Type O O-rings are recommended for many oxidizing acids, petroleum oils, halogenated hydrocarbons, lubricants, hydraulic fluids, organic liquids and air with hydrocarbons within the specified temperature range. Not recommended for steam services.

Standard type E O-rings are rated for application temperatures of –30 degrees F to +230 degrees F. Type E O-rings are recommended for hot water service within the specified temperature range plus a variety of dilute acids, oil free air and many chemical services.

Type T O-rings are rated for application temperatures of –20 degrees F to +180 degrees F. Type T O-rings are recommended for petroleum products, air with oil vapors, vegetable or mineral oils within the specified temperature range. Not recommended for hot water services over 150 degrees F., or steam, or hot dry air over 140 degrees F.

A specialized hand-held electric pressfit tool is required to make the pipe/fitting connections. Equipment cost, where shown, is $11.97 (to 1-1/2" dia.) to $28.22 (2" and greater dia.) per hour for this electric pressfit tool.

Description	Craft@Hrs	Unit	Material $	Labor $	Equipment $	Total $
Schedule 5 carbon steel pipe, plain end, pressfit joints, pipe only. No hangers or fittings.						
¾"	P1@.030	Ea	2.33	1.33	.18	3.84
1"	P1@.040	LF	3.18	1.78	.24	5.20
1¼"	P1@.040	LF	4.05	1.78	.24	6.07
1½"	P1@.050	LF	4.84	2.22	.30	7.36
2"	P1@.050	LF	5.80	2.22	.71	8.73
45-degree pressfit ell, with Type O O-ring (P x P)						
¾"	P1@.170	Ea	33.10	7.56	1.02	41.68
1"	P1@.180	Ea	39.70	8.00	1.08	48.78
1¼"	P1@.190	Ea	54.60	8.45	1.14	64.19
1½"	P1@.200	Ea	69.60	8.89	1.20	79.69
2"	P1@.200	Ea	94.30	8.89	2.82	106.01

Description	Craft@Hrs	Unit	Material $	Labor $	Equipment $	Total $

90-degree standard radius pressfit ell, with Type O O-ring (P x P)

¾"	P1@.170	Ea	34.70	7.56	1.02	43.28
1"	P1@.180	Ea	39.10	8.00	1.08	48.18
1¼"	P1@.190	Ea	54.60	8.45	1.14	64.19
1½"	P1@.200	Ea	66.80	8.89	1.20	76.89
2"	P1@.200	Ea	92.40	8.89	2.82	104.11

90-degree short radius pressfit ell, with Type O O-ring (P x P)

¾"	P1@.170	Ea	39.70	7.56	1.02	48.28
1"	P1@.180	Ea	48.00	8.00	1.08	57.08
1¼"	P1@.190	Ea	73.00	8.45	1.14	82.59
1½"	P1@.200	Ea	90.20	8.89	1.20	100.29
2"	P1@.200	Ea	118.00	8.89	2.82	129.71

90-degree standard radius reducing pressfit x FIP ell, with Type O O-ring (P x F)

1" x ½"	P1@.185	Ea	44.80	8.22	1.11	54.13
1" x ¾"	P1@.185	Ea	45.30	8.22	1.11	54.63
1¼" x ½"	P1@.195	Ea	59.70	8.67	1.17	69.54
1¼" x ½"	P1@.195	Ea	61.40	8.67	1.17	71.24
1½" x ½"	P1@.215	Ea	75.00	9.56	1.29	85.85
1½" x ¾"	P1@.215	Ea	75.60	9.56	1.29	86.45

90-degree short radius reducing pressfit x FIP ell, with Type O O-ring (P x F)

1" x ¾"	P1@.185	Ea	40.70	8.22	1.11	50.03
¾" x 1"	P1@.185	Ea	41.30	8.22	1.11	50.63
1¼" x ¾"	P1@.195	Ea	57.20	8.67	1.17	67.04
1½" x ¾"	P1@.215	Ea	72.60	9.56	1.29	83.45

Standard coupling, pressfit, with Type O O-ring (P x P)

¾"	P1@.170	Ea	25.90	7.56	1.02	34.48
1"	P1@.180	Ea	33.70	8.00	1.08	42.78
1¼"	P1@.190	Ea	46.60	8.45	1.14	56.19
1½"	P1@.200	Ea	57.30	8.89	1.20	67.39
2"	P1@.200	Ea	76.80	8.89	2.82	88.51

Description	Craft@Hrs	Unit	Material $	Labor $	Equipment $	Total $

Slip coupling, pressfit, with Type O O-ring (P x P)

¾"	P1@.170	Ea	34.90	7.56	1.02	43.48
1"	P1@.180	Ea	44.80	8.00	1.08	53.88
1¼"	P1@.190	Ea	59.70	8.45	1.14	69.29
1½"	P1@.200	Ea	73.90	8.89	1.20	83.99
2"	P1@.200	Ea	98.20	8.89	2.82	109.91

Tee, pressfit, with Type O O-ring (P x P)

¾"	P1@.190	Ea	39.60	8.45	1.14	49.19
1"	P1@.220	Ea	46.80	9.78	1.32	57.90
1¼"	P1@.240	Ea	61.80	10.70	1.44	73.94
1½"	P1@.255	Ea	75.40	11.30	1.53	88.23
2"	P1@.270	Ea	103.00	12.00	3.81	118.81

Reducing tee, pressfit, with Type O O-ring (P x P x P)

1" x ¾"	P1@.220	Ea	31.40	9.78	1.32	42.50
1¼" x ¾"	P1@.240	Ea	63.50	10.70	1.44	75.64
1¼" x 1"	P1@.240	Ea	75.60	10.70	1.44	87.74
1½" x ¾"	P1@.255	Ea	73.90	11.30	1.53	86.73
1½" x 1"	P1@.255	Ea	85.50	11.30	1.53	98.33
2" x ¾"	P1@.270	Ea	97.60	12.00	3.81	113.41
2" x 1"	P1@.270	Ea	112.00	12.00	1.62	125.62
2" x 1½"	P1@.270	Ea	118.00	12.00	1.62	131.62

Male threaded pressfit adapter with Type O O-ring (P x M)

¾" x ½"	P1@.180	Ea	24.10	8.00	1.08	33.18
¾" x ¾"	P1@.190	Ea	25.90	8.45	1.14	35.49
¾" x 1"	P1@.190	Ea	28.60	8.45	1.14	38.19
1" x ¾"	P1@.190	Ea	33.60	8.45	1.14	43.19
1" x 1"	P1@.190	Ea	35.90	8.45	1.14	45.49
1¼" x 1¼"	P1@.200	Ea	49.50	8.89	1.20	59.59
1½" x 1½"	P1@.215	Ea	62.00	9.56	1.29	72.85
2" x 2"	P1@.225	Ea	80.30	10.00	3.17	93.47

Female threaded pressfit adapter with Type O O-ring (P x F)

¾" x ½"	P1@.180	Ea	24.10	8.00	1.08	33.18
¾" x ¾"	P1@.190	Ea	25.90	8.45	1.14	35.49
1" x ½"	P1@.190	Ea	26.80	8.45	1.14	36.39
1" x ¾"	P1@.190	Ea	35.90	8.45	1.14	45.49
1" x 1"	P1@.200	Ea	36.80	8.89	1.20	46.89

Carbon Steel, Schedule 80 with 300# Fittings & Butt-Welded Joints

Schedule 80 carbon steel (ASTM A-53 and A-120) pipe with 300 pound carbon steel welding fittings is commonly used for steam condensate and re-circulating water systems where operating pressures do not exceed 700 PSIG.

The cost estimates in this section are based on the conditions, limitations and wage rates described in the section "How to Use This Book" beginning on page 5.

Equipment cost, where shown, is $4.57 per hour for a 2-ton chain hoist and $6.49 per hour for 200 amp gas welder, trailer mounted.

Description	Craft@Hrs	Unit	Material $	Labor $	Equipment $	Total $
Schedule 80 carbon steel butt-welded horizontal pipe assembly. Horizontally hung in a building. Assembly includes fittings and hanger assemblies. Based on a reducing tee and a 45-degree elbow every 16 feet for ½" pipe. A reducing tee and a 45-degree elbow every 50 feet for 12" pipe. Hangers spaced to meet plumbing code. *(Distance between fittings in the pipe assembly increases as pipe diameter increases.)* Use these figures for preliminary estimates.						
½"	P1@.183	LF	5.32	8.13	.59	14.04
¾"	P1@.203	LF	5.95	9.02	.66	15.63
1"	P1@.241	LF	6.80	10.70	.78	18.28
1¼"	P1@.246	LF	7.64	10.90	.80	19.34
1½"	P1@.271	LF	8.27	12.00	.88	21.15
2"	P1@.350	LF	10.80	15.60	1.13	27.53
2½"	P1@.437	LF	16.00	19.40	1.42	36.82
3"	P1@.532	LF	20.90	23.60	1.73	46.23
4"	P1@.601	LF	30.20	26.70	1.95	58.85
6"	ER@.833	LF	42.20	39.50	1.33	83.03
8"	ER@.994	LF	65.30	47.10	1.59	113.99
10"	ER@1.20	LF	111.00	56.90	1.92	169.82
12"	ER@1.43	LF	130.00	67.80	2.28	200.08
Schedule 80 carbon steel butt-welded vertical pipe assembly. Riser assembly, including fittings and riser clamps. Based on a reducing tee, and a riser clamp on every floor. Use these figures for preliminary estimates.						
½"	P1@.153	LF	5.25	6.80	.50	12.55
¾"	P1@.163	LF	5.89	7.25	.53	13.67
1"	P1@.196	LF	6.74	8.71	.64	16.09
1¼"	P1@.229	LF	8.40	10.20	.74	19.34
1½"	P1@.240	LF	8.97	10.70	.78	20.45
2"	P1@.310	LF	11.20	13.80	1.01	26.01
2½"	P1@.395	LF	16.30	17.60	1.28	35.18
3"	P1@.516	LF	21.00	22.90	1.67	45.57
4"	P1@.610	LF	30.20	27.10	1.98	59.28
6"	ER@.877	LF	44.60	41.60	1.40	87.60
8"	ER@1.04	LF	71.20	49.30	1.66	122.16
10"	ER@1.28	LF	119.00	60.70	2.04	181.74
12"	ER@1.58	LF	147.00	74.90	2.52	224.42

Description	Craft@Hrs	Unit	Material $	Labor $	Equipment $	Total $

Schedule 80 carbon steel pipe, plain end, butt-welded

½"	P1@.080	LF	2.13	3.56	.26	5.95
¾"	P1@.090	LF	2.64	4.00	.29	6.93
1"	P1@.110	LF	3.63	4.89	.36	8.88
1¼"	P1@.120	LF	4.83	5.33	.39	10.55
1½"	P1@.130	LF	5.80	5.78	.42	12.00
2"	P1@.170	LF	8.06	7.56	.55	16.17
2½"	P1@.220	LF	12.40	9.78	.71	22.89
3"	P1@.280	LF	16.30	12.40	.91	29.61
4"	P1@.340	LF	24.10	15.10	1.10	40.30
6"	ER@.460	LF	36.40	21.80	.73	58.93
8"	ER@.560	LF	54.60	26.50	.89	81.99
10"	ER@.660	LF	94.80	31.30	1.05	127.15
12"	ER@.840	LF	109.00	39.80	1.34	150.14

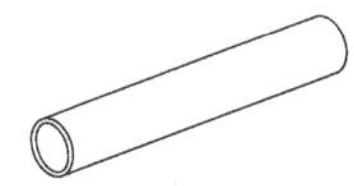

Schedule 80 carbon steel 45-degree ell, butt-welded

½"	P1@.300	Ea	7.71	13.30	.97	21.98
¾"	P1@.440	Ea	7.71	19.60	1.43	28.74
1"	P1@.590	Ea	7.71	26.20	1.91	35.82
1¼"	P1@.740	Ea	8.49	32.90	2.40	43.79
1½"	P1@.890	Ea	8.83	39.60	2.89	51.32
2"	P1@1.18	Ea	9.48	52.50	3.83	65.81
2½"	P1@1.49	Ea	11.30	66.20	4.83	82.33
3"	P1@1.77	Ea	14.30	78.70	5.74	98.74
4"	P1@2.37	Ea	21.70	105.00	7.69	134.39
6"	ER@3.55	Ea	60.60	168.00	5.67	234.27
8"	ER@4.26	Ea	98.20	202.00	6.80	307.00
10"	ER@5.32	Ea	157.00	252.00	8.49	417.49
12"	ER@6.38	Ea	236.00	302.00	10.20	548.20

Schedule 80 carbon steel 90-degree ell, butt-welded

½"	P1@.300	Ea	7.79	13.30	.97	22.06
¾"	P1@.440	Ea	7.79	19.60	1.43	28.82
1"	P1@.590	Ea	7.79	26.20	1.91	35.90
1¼"	P1@.740	Ea	8.95	32.90	2.40	44.25
1½"	P1@.890	Ea	9.59	39.60	2.89	52.08
2"	P1@1.18	Ea	14.40	52.50	3.83	70.73
2½"	P1@1.49	Ea	15.10	66.20	4.83	86.13
3"	P1@1.77	Ea	18.80	78.70	5.74	103.24
4"	P1@2.37	Ea	44.70	105.00	7.69	157.39
6"	ER@3.55	Ea	85.10	168.00	5.67	258.77
8"	ER@4.26	Ea	156.00	202.00	6.80	364.80
10"	ER@5.32	Ea	284.00	252.00	8.49	544.49
12"	ER@6.38	Ea	400.00	302.00	10.20	712.20

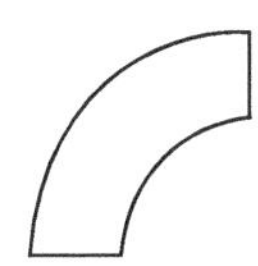

Schedule 80 carbon steel tee, butt-welded

Description	Craft@Hrs	Unit	Material $	Labor $	Equipment $	Total $
½"	P1@.440	Ea	26.10	19.60	1.43	47.13
¾"	P1@.670	Ea	26.10	29.80	2.17	58.07
1"	P1@.890	Ea	26.10	39.60	2.89	68.59
1¼"	P1@1.11	Ea	26.10	49.30	3.60	79.00
1½"	P1@1.33	Ea	29.50	59.10	4.31	92.91
2"	P1@1.77	Ea	36.10	78.70	5.74	120.54
2½"	P1@2.22	Ea	60.90	98.70	7.20	166.80
3"	P1@2.66	Ea	66.80	118.00	8.63	193.43
4"	P1@3.55	Ea	70.50	158.00	11.50	240.00
6"	ER@5.32	Ea	113.00	252.00	8.49	373.49
8"	ER@6.38	Ea	228.00	302.00	10.20	540.20
10"	ER@7.98	Ea	367.00	378.00	12.70	757.70
12"	ER@9.58	Ea	530.00	454.00	15.30	999.30

Description	Craft@Hrs	Unit	Material $	Labor $	Equipment $	Total $

Schedule 80 carbon steel reducing tee, butt-welded

1" x ¾"	P1@.810	Ea	31.90	36.00	2.63	70.53
1¼" x ½"	P1@1.04	Ea	31.90	46.20	3.37	81.47
1¼" x ¾"	P1@1.04	Ea	31.90	46.20	3.37	81.47
1¼" x 1"	P1@1.04	Ea	31.90	46.20	3.37	81.47
1½" x ½"	P1@1.26	Ea	31.90	56.00	4.09	91.99
1½" x ¾"	P1@1.26	Ea	31.90	56.00	4.09	91.99
1½" x 1"	P1@1.26	Ea	31.90	56.00	4.09	91.99
1½" x 1¼"	P1@1.26	Ea	31.90	56.00	4.09	91.99
2" x ¾"	P1@1.63	Ea	34.60	72.50	5.29	112.39
2" x 1"	P1@1.63	Ea	34.60	72.50	5.29	112.39
2" x 1¼"	P1@1.63	Ea	34.60	72.50	5.29	112.39
2" x 1½"	P1@1.63	Ea	34.60	72.50	5.29	112.39
2½" x 1"	P1@2.07	Ea	48.00	92.00	6.71	146.71
2½" x 1¼"	P1@2.07	Ea	48.00	92.00	6.71	146.71
2½" x 1½"	P1@2.07	Ea	48.00	92.00	6.71	146.71
2½" x 2"	P1@2.07	Ea	48.00	92.00	6.71	146.71
3" x 1¼"	P1@2.51	Ea	48.00	112.00	8.14	168.14
3" x 2"	P1@2.51	Ea	48.00	112.00	8.14	168.14
3" x 2½"	P1@2.51	Ea	48.00	112.00	8.14	168.14
4" x 1½"	P1@3.25	Ea	75.30	144.00	10.50	229.80
4" x 2"	P1@3.25	Ea	75.30	144.00	10.50	229.80
4" x 2½"	P1@3.25	Ea	75.30	144.00	10.50	229.80
4" x 3"	P1@3.25	Ea	75.30	144.00	10.50	229.80
6" x 2½"	ER@5.03	Ea	139.00	238.00	8.03	385.03
6" x 3"	ER@5.03	Ea	139.00	238.00	8.03	385.03
6" x 4"	ER@5.03	Ea	139.00	238.00	8.03	385.03
8" x 3"	ER@5.85	Ea	247.00	277.00	9.34	533.34
8" x 4"	ER@5.85	Ea	247.00	277.00	9.34	533.34
8" x 6"	ER@5.85	Ea	247.00	277.00	9.34	533.34
10" x 4"	ER@7.45	Ea	393.00	353.00	11.90	757.90
10" x 6"	ER@7.45	Ea	393.00	353.00	11.90	757.90
10" x 8"	ER@7.45	Ea	393.00	353.00	11.90	757.90
12" x 6"	ER@9.04	Ea	722.00	429.00	14.40	1,165.40
12" x 8"	ER@9.04	Ea	722.00	429.00	14.40	1,165.40
12" x 10"	ER@9.04	Ea	722.00	429.00	14.40	1,165.40

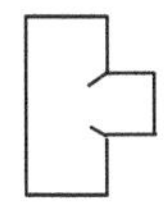

Schedule 80 carbon steel concentric reducer, butt-welded

¾"	P1@.360	Ea	15.50	16.00	1.17	32.67
1"	P1@.500	Ea	16.70	22.20	1.62	40.52
1¼"	P1@.665	Ea	14.40	29.60	2.16	46.16
1½"	P1@.800	Ea	15.30	35.60	2.59	53.49
2"	P1@1.00	Ea	16.30	44.50	3.24	64.04
2½"	P1@1.33	Ea	17.50	59.10	4.31	80.91
3"	P1@1.63	Ea	19.00	72.50	5.29	96.79
4"	P1@2.00	Ea	22.00	88.90	6.49	117.39
6"	ER@2.96	Ea	48.00	140.00	2.74	190.74
8"	ER@3.90	Ea	71.10	185.00	3.61	259.71
10"	ER@4.79	Ea	98.20	227.00	4.44	329.64
12"	ER@5.85	Ea	144.00	277.00	5.42	426.42

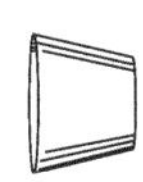

Schedule 80 carbon steel eccentric reducer, butt-welded

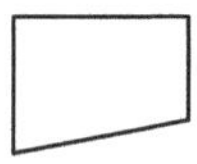

Description	Craft@Hrs	Unit	Material $	Labor $	Equipment $	Total $
¾"	P1@.360	Ea	17.20	16.00	1.17	34.37
1"	P1@.500	Ea	18.40	22.20	1.62	42.22
1¼"	P1@.665	Ea	19.00	29.60	2.16	50.76
1½"	P1@.800	Ea	20.30	35.60	2.59	58.49
2"	P1@1.00	Ea	21.50	44.50	3.24	69.24
2½"	P1@1.33	Ea	22.10	59.10	4.31	85.51
3"	P1@1.63	Ea	23.00	72.50	5.29	100.79
4"	P1@2.00	Ea	30.90	88.90	6.49	126.29
6"	ER@2.96	Ea	70.70	140.00	2.74	213.44
8"	ER@3.90	Ea	105.00	185.00	3.61	293.61
10"	ER@4.79	Ea	184.00	227.00	4.44	415.44
12"	ER@5.85	Ea	286.00	277.00	5.42	568.42

Schedule 80 carbon steel cap, butt-welded

Description	Craft@Hrs	Unit	Material $	Labor $	Equipment $	Total $
½"	P1@.180	Ea	7.31	8.00	.58	15.89
¾"	P1@.270	Ea	9.19	12.00	.88	22.07
1"	P1@.360	Ea	9.46	16.00	1.17	26.63
1¼"	P1@.460	Ea	9.74	20.40	1.49	31.63
1½"	P1@.560	Ea	10.50	24.90	1.82	37.22
2"	P1@.740	Ea	10.80	32.90	2.40	46.10
2½"	P1@.920	Ea	11.00	40.90	2.98	54.88
3"	P1@1.11	Ea	11.80	49.30	3.60	64.70
4"	P1@1.50	Ea	15.70	66.70	4.86	87.26
6"	ER@2.24	Ea	28.00	106.00	2.08	136.08
8"	ER@2.68	Ea	43.20	127.00	2.48	172.68
10"	ER@3.36	Ea	64.30	159.00	3.11	226.41
12"	ER@4.00	Ea	84.90	190.00	3.71	278.61

Schedule 80 carbon steel union, butt-welded

Description	Craft@Hrs	Unit	Material $	Labor $	Equipment $	Total $
½"	P1@.300	Ea	13.00	13.30	.97	27.27
¾"	P1@.440	Ea	15.50	19.60	1.43	36.53
1"	P1@.590	Ea	20.60	26.20	1.91	48.71
1¼"	P1@.740	Ea	34.60	32.90	2.40	69.90
1½"	P1@.890	Ea	37.50	39.60	2.89	79.99
2"	P1@1.18	Ea	50.10	52.50	3.83	106.43
2½"	P1@1.49	Ea	109.00	66.20	4.83	180.03
3"	P1@1.77	Ea	135.00	78.70	5.74	219.44

Carbon steel weldolet

Description	Craft@Hrs	Unit	Material $	Labor $	Equipment $	Total $
½"	P1@.440	Ea	12.80	19.60	1.43	33.83
¾"	P1@.670	Ea	13.80	29.80	2.17	45.77
1"	P1@.890	Ea	15.70	39.60	2.89	58.19
1¼"	P1@1.11	Ea	20.60	49.30	3.60	73.50
1½"	P1@1.33	Ea	20.80	59.10	4.31	84.21
2"	P1@1.77	Ea	22.00	78.70	5.74	106.44
2½"	P1@2.22	Ea	47.90	98.70	7.20	153.80
3"	P1@2.66	Ea	49.60	118.00	8.63	176.23
4"	P1@3.55	Ea	110.00	158.00	11.50	279.50

Carbon Steel, Schedule 80 with 300# Fittings & Butt-Welded Joints

Description	Craft@Hrs	Unit	Material $	Labor $	Equipment $	Total $
Carbon steel threadolet						
½"	P1@.300	Ea	10.40	13.30	.97	24.67
¾"	P1@.440	Ea	11.40	19.60	1.43	32.43
1"	P1@.590	Ea	14.10	26.20	1.91	42.21
1¼"	P1@.740	Ea	63.50	32.90	2.40	98.80
1½"	P1@.890	Ea	63.50	39.60	2.89	105.99
2"	P1@1.18	Ea	83.60	52.50	3.83	139.93
300# forged steel slip-on companion flange, welding-type						
2½"	P1@.810	Ea	35.30	36.00	2.63	73.93
3"	P1@.980	Ea	35.30	43.60	3.18	82.08
4"	P1@1.30	Ea	52.00	57.80	4.22	114.02
6"	ER@1.95	Ea	94.40	92.40	1.81	188.61
8"	ER@2.35	Ea	133.00	111.00	2.18	246.18
10"	ER@2.90	Ea	220.00	137.00	2.69	359.69
12"	ER@3.50	Ea	352.00	166.00	3.24	521.24
Class 300 bronze body gate valve, threaded joints						
½"	P1@.210	Ea	55.20	9.33	—	64.53
¾"	P1@.250	Ea	72.10	11.10	—	83.20
1"	P1@.300	Ea	87.30	13.30	—	100.60
1¼"	P1@.400	Ea	119.00	17.80	—	136.80
1½"	P1@.450	Ea	150.00	20.00	—	170.00
2"	P1@.500	Ea	221.00	22.20	—	243.20
2½"	P1@.750	Ea	452.00	33.30	—	485.30
3"	P1@.950	Ea	641.00	42.20	—	683.20
Class 250 iron body gate valve, flanged						
2½"	P1@.600	Ea	137.00	26.70	—	163.70
3"	P1@.750	Ea	992.00	33.30	—	1,025.30
4"	P1@1.35	Ea	1,460.00	60.00	—	1,520.00
6"	ER@2.50	Ea	1,950.00	119.00	1.67	2,070.67
8"	ER@3.00	Ea	6,690.00	142.00	2.01	6,834.01
10"	ER@4.00	Ea	10,900.00	190.00	2.68	11,092.68
12"	ER@4.50	Ea	17,800.00	213.00	3.01	18,016.01
Class 300 bronze body globe valve, threaded						
½"	P1@.210	Ea	63.30	9.33	—	72.63
¾"	P1@.250	Ea	80.30	11.10	—	91.40
1"	P1@.300	Ea	115.00	13.30	—	128.30
1¼"	P1@.400	Ea	167.00	17.80	—	184.80
1½"	P1@.450	Ea	221.00	20.00	—	241.00
2"	P1@.500	Ea	306.00	22.20	—	328.20

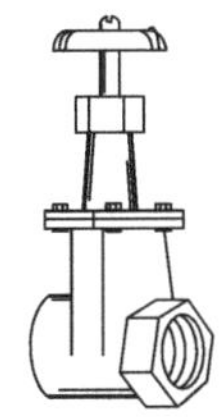

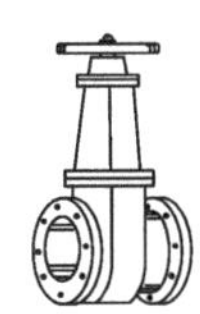

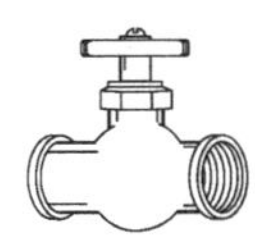

Description	Craft@Hrs	Unit	Material $	Labor $	Equipment $	Total $

Class 250 iron body globe valve, flanged

Description	Craft@Hrs	Unit	Material $	Labor $	Equipment $	Total $
2"	P1@.500	Ea	723.00	22.20	—	745.20
2½"	P1@.600	Ea	897.00	26.70	—	923.70
3"	P1@.750	Ea	1,070.00	33.30	—	1,103.30
4"	P1@1.35	Ea	1,480.00	60.00	—	1,540.00
6"	ER@2.50	Ea	2,910.00	119.00	1.67	3,030.67
8"	ER@3.00	Ea	6,440.00	142.00	2.01	6,584.01

200 PSIG iron body butterfly valve, lug-type, lever operated

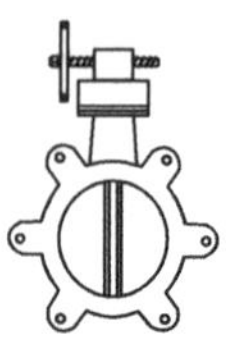

Description	Craft@Hrs	Unit	Material $	Labor $	Equipment $	Total $
2"	P1@.450	Ea	194.00	20.00	—	214.00
2½"	P1@.450	Ea	201.00	20.00	—	221.00
3"	P1@.550	Ea	212.00	24.40	—	236.40
4"	P1@.550	Ea	263.00	24.40	—	287.40
6"	ER@.800	Ea	443.00	37.90	.54	481.44
8"	ER@.800	Ea	605.00	37.90	.54	643.44
10"	ER@.900	Ea	841.00	42.70	.60	884.30
12"	ER@1.00	Ea	1,100.00	47.40	.67	1,148.07

200 PSIG iron body butterfly valve, wafer-type, lever operated

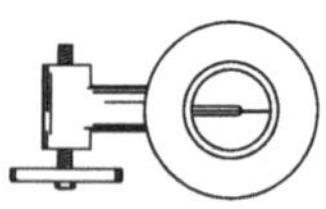

Description	Craft@Hrs	Unit	Material $	Labor $	Equipment $	Total $
2"	P1@.450	Ea	177.00	20.00	—	197.00
2½"	P1@.450	Ea	180.00	20.00	—	200.00
3"	P1@.550	Ea	194.00	24.40	—	218.40
4"	P1@.550	Ea	235.00	24.40	—	259.40
6"	ER@.800	Ea	410.00	37.90	.54	448.44
8"	ER@.800	Ea	568.00	37.90	.54	606.44
10"	ER@.900	Ea	794.00	42.70	.60	837.30
12"	ER@1.00	Ea	1,040.00	47.40	.67	1,088.07

Class 300 bronze body swing check valve, threaded

Description	Craft@Hrs	Unit	Material $	Labor $	Equipment $	Total $
½"	P1@.210	Ea	82.10	9.33	—	91.43
¾"	P1@.250	Ea	95.40	11.10	—	106.50
1"	P1@.300	Ea	148.00	13.30	—	161.30
1¼"	P1@.400	Ea	169.00	17.80	—	186.80
1½"	P1@.450	Ea	242.00	20.00	—	262.00
2"	P1@.500	Ea	338.00	22.20	—	360.20

Class 250 iron body swing check valve, flanged

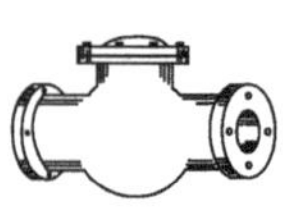

Description	Craft@Hrs	Unit	Material $	Labor $	Equipment $	Total $
2"	P1@.500	Ea	564.00	22.20	—	586.20
2½"	P1@.600	Ea	658.00	26.70	—	684.70
3"	P1@.750	Ea	817.00	33.30	—	850.30
4"	P1@1.35	Ea	1,000.00	60.00	—	1,060.00
6"	ER@2.50	Ea	1,940.00	119.00	1.67	2,060.67
8"	ER@3.00	Ea	3,630.00	142.00	2.01	3,774.01

Description	Craft@Hrs	Unit	Material $	Labor $	Equipment $	Total $

Class 250 iron body silent check valve, flanged

Description	Craft@Hrs	Unit	Material $	Labor $	Equipment $	Total $
2"	P1@.500	Ea	236.00	22.20	—	258.20
2½"	P1@.600	Ea	272.00	26.70	—	298.70
3"	P1@.750	Ea	302.00	33.30	—	335.30
4"	P1@1.35	Ea	410.00	60.00	—	470.00
6"	ER@2.50	Ea	721.00	119.00	1.67	841.67
8"	ER@3.00	Ea	1,310.00	142.00	2.01	1,454.01
10"	ER@4.00	Ea	2,080.00	190.00	2.68	2,272.68

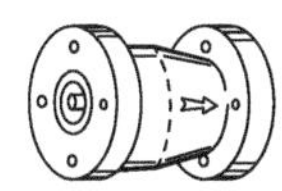

Class 250 bronze body strainer, threaded

Description	Craft@Hrs	Unit	Material $	Labor $	Equipment $	Total $
½"	P1@.210	Ea	48.40	9.33	—	57.73
¾"	P1@.250	Ea	65.30	11.10	—	76.40
1"	P1@.300	Ea	83.00	13.30	—	96.30
1¼"	P1@.400	Ea	118.00	17.80	—	135.80
1½"	P1@.450	Ea	154.00	20.00	—	174.00
2"	P1@.500	Ea	261.00	22.20	—	283.20

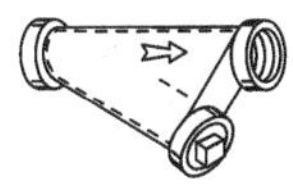

Class 250 iron body strainer, flanged

Description	Craft@Hrs	Unit	Material $	Labor $	Equipment $	Total $
2"	P1@.500	Ea	279.00	22.20	—	301.20
2½"	P1@.600	Ea	306.00	26.70	—	332.70
3"	P1@.750	Ea	356.00	33.30	—	389.30
4"	P1@1.35	Ea	633.00	60.00	—	693.00
6"	ER@2.50	Ea	1,010.00	119.00	1.67	1,130.67
8"	ER@3.00	Ea	1,850.00	142.00	2.01	1,994.01

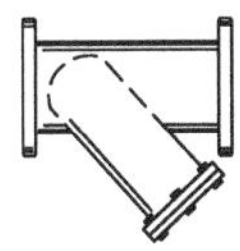

Installation of 2-way control valve, threaded joints

Description	Craft@Hrs	Unit	Material $	Labor $	Equipment $	Total $
½"	P1@.210	Ea	—	9.33	—	9.33
¾"	P1@.250	Ea	—	11.10	—	11.10
1"	P1@.300	Ea	—	13.30	—	13.30
1¼"	P1@.400	Ea	—	17.80	—	17.80
1½"	P1@.450	Ea	—	20.00	—	20.00
2"	P1@.500	Ea	—	22.20	—	22.20
2½"	P1@.830	Ea	—	36.90	—	36.90
3"	P1@.990	Ea	—	44.00	—	44.00

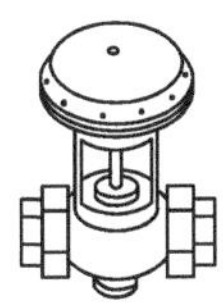

Installation of 2-way control valve, flanged joints

Description	Craft@Hrs	Unit	Material $	Labor $	Equipment $	Total $
2"	P1@.500	Ea	—	22.20	—	22.20
2½"	P1@.600	Ea	—	26.70	—	26.70
3"	P1@.750	Ea	—	33.30	—	33.30
4"	P1@1.35	Ea	—	60.00	—	60.00
6"	P1@2.50	Ea	—	111.00	—	111.00
8"	P1@3.00	Ea	—	133.00	—	133.00
10"	P1@4.00	Ea	—	178.00	—	178.00
12"	P1@4.50	Ea	—	200.00	—	200.00

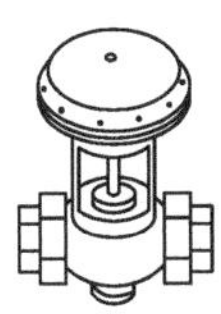

Installation of 3-way control valve, threaded joints

Description	Craft@Hrs	Unit	Material $	Labor $	Equipment $	Total $
½"	P1@.260	Ea	—	11.60	—	11.60
¾"	P1@.365	Ea	—	16.20	—	16.20
1"	P1@.475	Ea	—	21.10	—	21.10
1¼"	P1@.575	Ea	—	25.60	—	25.60
1½"	P1@.680	Ea	—	30.20	—	30.20
2"	P1@.910	Ea	—	40.40	—	40.40
2½"	P1@1.12	Ea	—	49.80	—	49.80
3"	P1@1.33	Ea	—	59.10	—	59.10

Installation of 3-way control valve, flanged joints

Description	Craft@Hrs	Unit	Material $	Labor $	Equipment $	Total $
2"	P1@.910	Ea	—	40.40	—	40.40
2½"	P1@1.12	Ea	—	49.80	—	49.80
3"	P1@1.33	Ea	—	59.10	—	59.10
4"	P1@2.00	Ea	—	88.90	—	88.90
6"	P1@3.70	Ea	—	164.00	—	164.00
8"	P1@4.40	Ea	—	196.00	—	196.00
10"	P1@5.90	Ea	—	262.00	—	262.00
12"	P1@6.50	Ea	—	289.00	—	289.00

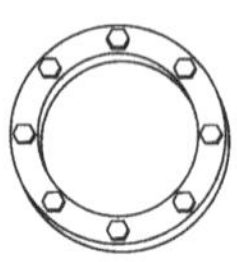

Bolt and gasket set

Description	Craft@Hrs	Unit	Material $	Labor $	Equipment $	Total $
2"	P1@.500	Ea	4.64	22.20	—	26.84
2½"	P1@.650	Ea	5.42	28.90	—	34.32
3"	P1@.750	Ea	9.19	33.30	—	42.49
4"	P1@1.00	Ea	15.40	44.50	—	59.90
6"	P1@1.20	Ea	25.30	53.30	—	78.60
8"	P1@1.25	Ea	28.40	55.60	—	84.00
10"	P1@1.70	Ea	48.90	75.60	—	124.50
12"	P1@2.20	Ea	55.10	97.80	—	152.90

Thermometer with well

Description	Craft@Hrs	Unit	Material $	Labor $	Equipment $	Total $
7"	P1@.250	Ea	186.00	11.10	—	197.10
9"	P1@.250	Ea	192.00	11.10	—	203.10

Dial-type pressure gauge

Description	Craft@Hrs	Unit	Material $	Labor $	Equipment $	Total $
2½"	P1@.200	Ea	38.50	8.89	—	47.39
3½"	P1@.200	Ea	51.20	8.89	—	60.09

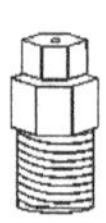

Pressure/temperature tap

Description	Craft@Hrs	Unit	Material $	Labor $	Equipment $	Total $
Tap	P1@.150	Ea	17.70	6.67	—	24.37

Description	Craft@Hrs	Unit	Material $	Labor $	Equipment $	Total $
Hanger with swivel assembly						
½"	P1@.250	Ea	4.50	11.10	—	15.60
¾"	P1@.250	Ea	4.98	11.10	—	16.08
1"	P1@.250	Ea	5.20	11.10	—	16.30
1½"	P1@.300	Ea	5.34	13.30	—	18.64
2"	P1@.300	Ea	5.74	13.30	—	19.04
2½"	P1@.350	Ea	5.99	15.60	—	21.59
3"	P1@.350	Ea	8.15	15.60	—	23.75
4"	P1@.350	Ea	10.10	15.60	—	25.70
5"	P1@.450	Ea	17.60	20.00	—	37.60
6"	P1@.450	Ea	19.60	20.00	—	39.60
8"	P1@.450	Ea	25.90	20.00	—	45.90
10"	P1@.550	Ea	35.50	24.40	—	59.90
12"	P1@.600	Ea	46.90	26.70	—	73.60

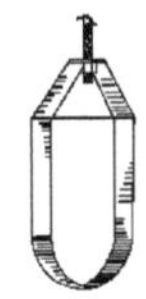

Description	Craft@Hrs	Unit	Material $	Labor $	Equipment $	Total $
Riser clamp						
½"	P1@.100	Ea	3.30	4.45	—	7.75
¾"	P1@.100	Ea	4.87	4.45	—	9.32
1"	P1@.100	Ea	4.91	4.45	—	9.36
1¼"	P1@.105	Ea	5.90	4.67	—	10.57
1½"	P1@.110	Ea	6.22	4.89	—	11.11
2"	P1@.115	Ea	6.62	5.11	—	11.73
2½"	P1@.120	Ea	6.93	5.33	—	12.26
3"	P1@.120	Ea	7.53	5.33	—	12.86
4"	P1@.125	Ea	9.61	5.56	—	15.17
5"	P1@.180	Ea	13.70	8.00	—	21.70
6"	P1@.200	Ea	16.50	8.89	—	25.39
8"	P1@.200	Ea	21.50	8.89	—	30.39
10"	P1@.250	Ea	32.10	11.10	—	43.20
12"	P1@.250	Ea	38.00	11.10	—	49.10

Carbon Steel, Schedule 80 with 300# M.I. Fittings & Threaded Joints

Schedule 80 carbon steel (ASTM A-53 and A-120) pipe with 300 pound malleable iron threaded fittings is commonly used for heating hot water, chilled water, steam and steam condensate systems where operating pressures do not exceed 250 PSIG for pipe sized 2" and smaller. For pipe sizes 2½" and larger wrought steel fittings either welded or grooved should be used.

The cost estimates in this section are based on the conditions, limitations and wage rates described in the section "How to Use This Book" beginning on page 5.

Equipment cost, where shown, is $4.57 per hour for a 2-ton chain hoist.

Description	Craft@Hrs	Unit	Material $	Labor $	Equipment $	Total $

Schedule 80 carbon steel threaded horizontal pipe assembly. Horizontally hung in a building. Assembly includes fittings and hanger assemblies. Based on a reducing tee and a 45-degree elbow every 16 feet for ½" pipe and a reducing tee and a 45-degree elbow every 30 feet for 4" pipe. Hangers spaced to meet plumbing code. *(Distance between fittings in the pipe assembly increases as pipe diameter increases.)* Use these figures for preliminary estimates.

Description	Craft@Hrs	Unit	Material $	Labor $	Equipment $	Total $
½"	P1@.121	LF	4.41	5.38	—	9.79
¾"	P1@.125	LF	5.04	5.56	—	10.60
1"	P1@.149	LF	6.32	6.62	—	12.94
1¼"	P1@.163	LF	7.26	7.25	—	14.51
1½"	P1@.186	LF	7.53	8.27	—	15.80
2"	P1@.210	LF	9.22	9.33	—	18.55
2½"	P1@.297	LF	12.20	13.20	—	25.40
3"	P1@.340	LF	26.10	15.10	—	41.20
4"	P1@.409	LF	39.50	18.20	—	57.70

Schedule 80 carbon steel threaded vertical pipe assembly. Riser assembly, including fittings and riser clamps. Based on a reducing tee, and a riser clamp on every floor. Use these figures for preliminary estimates.

Description	Craft@Hrs	Unit	Material $	Labor $	Equipment $	Total $
½"	P1@.094	LF	3.63	4.18	—	7.81
¾"	P1@.104	LF	4.14	4.62	—	8.76
1"	P1@.124	LF	5.44	5.51	—	10.95
1¼"	P1@.143	LF	6.36	6.36	—	12.72
1½"	P1@.161	LF	7.01	7.16	—	14.17
2"	P1@.179	LF	8.36	7.96	—	16.32
2½"	P1@.257	LF	11.80	11.40	—	23.20
3"	P1@.296	LF	35.20	13.20	—	48.40
4"	P1@.373	LF	58.40	16.60	—	75.00

Description	Craft@Hrs	Unit	Material $	Labor $	Equipment $	Total $

Schedule 80 carbon steel pipe, threaded and coupled joints

½"	P1@.070	LF	2.22	3.11	—	5.33
¾"	P1@.080	LF	2.80	3.56	—	6.36
1"	P1@.100	LF	3.80	4.45	—	8.25
1¼"	P1@.110	LF	5.01	4.89	—	9.90
1½"	P1@.120	LF	5.91	5.33	—	11.24
2"	P1@.130	LF	8.07	5.78	—	13.85
2½"	P1@.190	LF	12.60	8.45	—	21.05
3"	P1@.220	LF	16.80	9.78	—	26.58
4"	P1@.290	LF	24.10	12.90	—	37.00

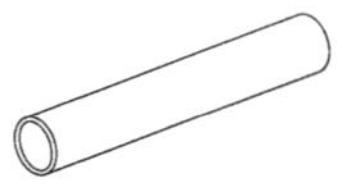

300# malleable iron 45-degree ell, threaded joints

½"	P1@.132	Ea	10.60	5.87	—	16.47
¾"	P1@.143	Ea	11.80	6.36	—	18.16
1"	P1@.198	Ea	13.10	8.80	—	21.90
1¼"	P1@.264	Ea	20.10	11.70	—	31.80
1½"	P1@.330	Ea	26.20	14.70	—	40.90
2"	P1@.418	Ea	39.10	18.60	—	57.70
2½"	P1@.561	Ea	61.70	24.90	—	86.60
3"	P1@.704	Ea	82.20	31.30	—	113.50
4"	P1@.935	Ea	158.00	41.60	—	199.60

300# malleable iron 90-degree ell, threaded joints

½"	P1@.132	Ea	8.13	5.87	—	14.00
¾"	P1@.143	Ea	9.42	6.36	—	15.78
1"	P1@.198	Ea	12.00	8.80	—	20.80
1¼"	P1@.264	Ea	16.80	11.70	—	28.50
1½"	P1@.330	Ea	20.10	14.70	—	34.80
2"	P1@.418	Ea	28.10	18.60	—	46.70
2½"	P1@.561	Ea	53.30	24.90	—	78.20
3"	P1@.704	Ea	72.90	31.30	—	104.20
4"	P1@.935	Ea	146.00	41.60	—	187.60

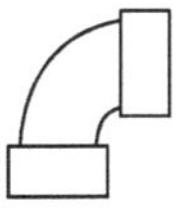

Description	Craft@Hrs	Unit	Material $	Labor $	Equipment $	Total $

300# malleable iron tee, threaded joints

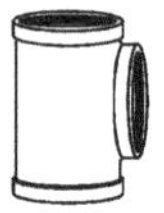

½"	P1@.198	Ea	11.40	8.80	—	20.20
¾"	P1@.209	Ea	12.60	9.29	—	21.89
1"	P1@.253	Ea	15.10	11.20	—	26.30
1¼"	P1@.341	Ea	20.10	15.20	—	35.30
1½"	P1@.429	Ea	24.80	19.10	—	43.90
2"	P1@.539	Ea	36.50	24.00	—	60.50
2½"	P1@.726	Ea	69.50	32.30	—	101.80
3"	P1@.913	Ea	116.00	40.60	—	156.60
4"	P1@1.21	Ea	198.00	53.80	—	251.80

300# malleable iron reducing tee, threaded joints

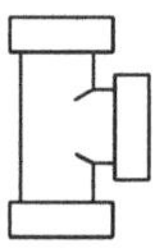

¾" x ½"	P1@.205	Ea	15.70	9.11	—	24.81
1" x ½"	P1@.205	Ea	19.80	9.11	—	28.91
1" x ¾"	P1@.205	Ea	19.80	9.11	—	28.91
1¼" x ½"	P1@.310	Ea	25.50	13.80	—	39.30
1¼" x ¾"	P1@.310	Ea	25.50	13.80	—	39.30
1¼" x 1"	P1@.310	Ea	25.50	13.80	—	39.30
1½" x ½"	P1@.400	Ea	31.50	17.80	—	49.30
1½" x ¾"	P1@.400	Ea	31.50	17.80	—	49.30
1½" x 1"	P1@.400	Ea	31.50	17.80	—	49.30
1½" x 1¼"	P1@.400	Ea	31.50	17.80	—	49.30
2" x 1"	P1@.500	Ea	46.70	22.20	—	68.90
2" x 1¼"	P1@.500	Ea	46.70	22.20	—	68.90
2" x 1½"	P1@.500	Ea	46.70	22.20	—	68.90
3" x 1¼"	P1@.830	Ea	243.00	36.90	—	279.90
3" x 1½"	P1@.830	Ea	243.00	36.90	—	279.90
3" x 2"	P1@.830	Ea	243.00	36.90	—	279.90

300# malleable iron reducer, threaded joints

¾" x ½"	P1@.135	Ea	9.34	6.00	—	15.34
1" x ½"	P1@.170	Ea	11.20	7.56	—	18.76
1" x ¾"	P1@.170	Ea	11.20	7.56	—	18.76
1¼" x ¾"	P1@.220	Ea	14.10	9.78	—	23.88
1¼" x 1"	P1@.220	Ea	14.10	9.78	—	23.88
1½" x 1"	P1@.285	Ea	19.40	12.70	—	32.10
1½" x 1¼"	P1@.285	Ea	19.40	12.70	—	32.10
2" x 1¼"	P1@.360	Ea	28.00	16.00	—	44.00
2" x 1½"	P1@.360	Ea	28.00	16.00	—	44.00
3" x 2"	P1@.550	Ea	67.90	24.40	—	92.30
3" x 2½"	P1@.550	Ea	67.90	24.40	—	92.30

Description	Craft@Hrs	Unit	Material $	Labor $	Equipment $	Total $
300# malleable iron cross, threaded joints						
½"	P1@.310	Ea	27.40	13.80	—	41.20
¾"	P1@.350	Ea	29.90	15.60	—	45.50
1"	P1@.400	Ea	34.00	17.80	—	51.80
1¼"	P1@.530	Ea	41.90	23.60	—	65.50
1½"	P1@.660	Ea	54.80	29.30	—	84.10
2"	P1@.770	Ea	78.70	34.20	—	112.90
300# malleable iron cap, threaded joint						
½"	P1@.100	Ea	5.48	4.45	—	9.93
¾"	P1@.110	Ea	7.29	4.89	—	12.18
1"	P1@.155	Ea	9.44	6.89	—	16.33
1¼"	P1@.200	Ea	12.80	8.89	—	21.69
1½"	P1@.250	Ea	15.20	11.10	—	26.30
2"	P1@.320	Ea	21.00	14.20	—	35.20
3"	P1@.530	Ea	43.80	23.60	—	67.40
4"	P1@.705	Ea	80.40	31.30	—	111.70
300# malleable iron plug, threaded joint						
½"	P1@.090	Ea	1.35	4.00	—	5.35
¾"	P1@.100	Ea	1.84	4.45	—	6.29
1"	P1@.140	Ea	2.20	6.22	—	8.42
1¼"	P1@.180	Ea	2.37	8.00	—	10.37
1½"	P1@.230	Ea	3.01	10.20	—	13.21
2"	P1@.290	Ea	4.67	12.90	—	17.57
2½"	P1@.380	Ea	6.60	16.90	—	23.50
3"	P1@.480	Ea	11.00	21.30	—	32.30
4"	P1@.640	Ea	31.50	28.40	—	59.90
300# malleable iron union, threaded joints						
½"	P1@.150	Ea	10.10	6.67	—	16.77
¾"	P1@.165	Ea	11.30	7.33	—	18.63
1"	P1@.230	Ea	14.90	10.20	—	25.10
1¼"	P1@.310	Ea	22.80	13.80	—	36.60
1½"	P1@.400	Ea	24.70	17.80	—	42.50
2"	P1@.500	Ea	31.50	22.20	—	53.70
2½"	P1@.670	Ea	69.50	29.80	—	99.30
3"	P1@.840	Ea	116.00	37.30	—	153.30
4"	P1@.960	Ea	327.00	42.70	—	369.70

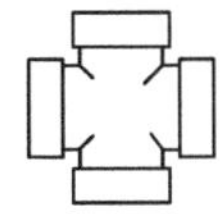

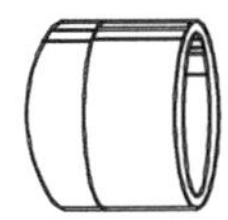

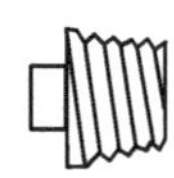

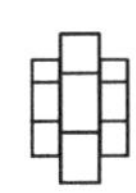

Description	Craft@Hrs	Unit	Material $	Labor $	Equipment $	Total $

300# malleable iron coupling, threaded joints

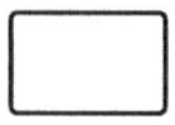

½"	P1@.132	Ea	6.11	5.87	—	11.98
¾"	P1@.143	Ea	7.15	6.36	—	13.51
1"	P1@.198	Ea	8.41	8.80	—	17.21
1¼"	P1@.264	Ea	10.10	11.70	—	21.80
1½"	P1@.330	Ea	15.10	14.70	—	29.80
2"	P1@.418	Ea	21.00	18.60	—	39.60
2½"	P1@.561	Ea	27.00	24.90	—	51.90
3"	P1@.704	Ea	43.80	31.30	—	75.10
4"	P1@.935	Ea	105.00	41.60	—	146.60

Class 300 bronze body gate valve, threaded joints

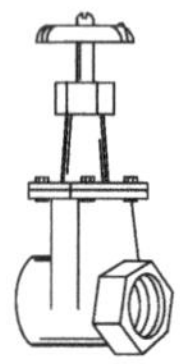

½"	P1@.210	Ea	55.20	9.33	—	64.53
¾"	P1@.250	Ea	72.10	11.10	—	83.20
1"	P1@.300	Ea	87.30	13.30	—	100.60
1¼"	P1@.400	Ea	119.00	17.80	—	136.80
1½"	P1@.450	Ea	150.00	20.00	—	170.00
2"	P1@.500	Ea	221.00	22.20	—	243.20

Class 250 iron body gate valve, flanged joints

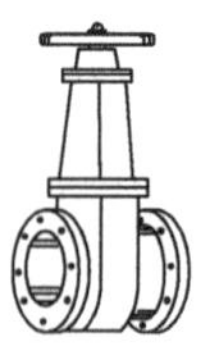

2"	P1@.500	Ea	137.00	22.20	—	159.20
2½"	P1@.600	Ea	992.00	26.70	—	1,018.70
3"	P1@.750	Ea	1,460.00	33.30	—	1,493.30
4"	P1@1.35	Ea	1,950.00	60.00	—	2,010.00

Class 300 bronze body globe valve, threaded joints

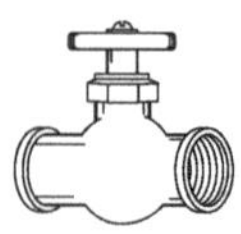

½"	P1@.210	Ea	63.30	9.33	—	72.63
¾"	P1@.250	Ea	80.30	11.10	—	91.40
1"	P1@.300	Ea	115.00	13.30	—	128.30
1¼"	P1@.400	Ea	167.00	17.80	—	184.80
1½"	P1@.450	Ea	221.00	20.00	—	241.00
2"	P1@.500	Ea	306.00	22.20	—	328.20

Class 250 iron body globe valve, flanged joints

2"	P1@.500	Ea	723.00	22.20	—	745.20
2½"	P1@.600	Ea	897.00	26.70	—	923.70
3"	P1@.750	Ea	1,070.00	33.30	—	1,103.30
4"	P1@1.35	Ea	1,480.00	60.00	—	1,540.00

200 PSIG iron body butterfly valve, lug-type, lever operated

2"	P1@.450	Ea	194.00	20.00	—	214.00
2½"	P1@.450	Ea	201.00	20.00	—	221.00
3"	P1@.550	Ea	212.00	24.40	—	236.40
4"	P1@.550	Ea	263.00	24.40	—	287.40

Description	Craft@Hrs	Unit	Material $	Labor $	Equipment $	Total $
200 PSIG iron body butterfly valve, wafer-type, lever operated						
2"	P1@.450	Ea	177.00	20.00	—	197.00
2½"	P1@.450	Ea	180.00	20.00	—	200.00
3"	P1@.550	Ea	194.00	24.40	—	218.40
4"	P1@.550	Ea	235.00	24.40	—	259.40

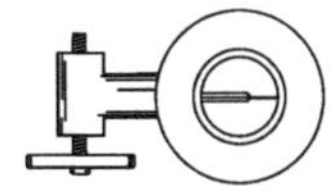

Description	Craft@Hrs	Unit	Material $	Labor $	Equipment $	Total $
Class 150 bronze body 2-piece ball valve, threaded joints						
½"	P1@.210	Ea	12.50	9.33	—	21.83
¾"	P1@.250	Ea	16.60	11.10	—	27.70
1"	P1@.300	Ea	28.90	13.30	—	42.20
1¼"	P1@.400	Ea	48.10	17.80	—	65.90
1½"	P1@.450	Ea	65.10	20.00	—	85.10
2"	P1@.500	Ea	82.50	22.20	—	104.70
3"	P1@.625	Ea	565.00	27.80	—	592.80
4"	P1@.690	Ea	739.00	30.70	—	769.70

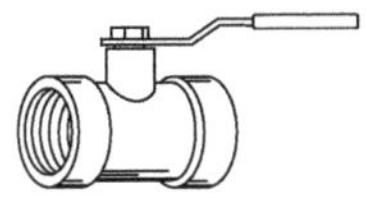

Description	Craft@Hrs	Unit	Material $	Labor $	Equipment $	Total $
Class 300 bronze body swing check valve, threaded joints						
½"	P1@.210	Ea	82.10	9.33	—	91.43
¾"	P1@.250	Ea	95.40	11.10	—	106.50
1"	P1@.300	Ea	148.00	13.30	—	161.30
1¼"	P1@.400	Ea	169.00	17.80	—	186.80
1½"	P1@.450	Ea	242.00	20.00	—	262.00
2"	P1@.500	Ea	338.00	22.20	—	360.20

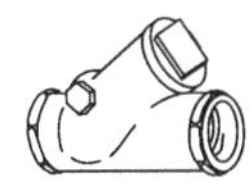

Description	Craft@Hrs	Unit	Material $	Labor $	Equipment $	Total $
Class 250 iron body swing check valve, flanged joints						
2"	P1@.500	Ea	564.00	22.20	—	586.20
2½"	P1@.600	Ea	658.00	26.70	—	684.70
3"	P1@.750	Ea	817.00	33.30	—	850.30
4"	P1@1.35	Ea	1,000.00	60.00	—	1,060.00

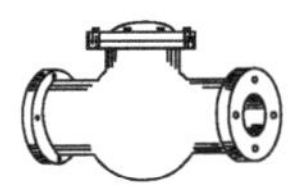

Description	Craft@Hrs	Unit	Material $	Labor $	Equipment $	Total $
Class 250 iron body silent check valve, wafer-type						
2"	P1@.500	Ea	236.00	22.20	—	258.20
2½"	P1@.600	Ea	272.00	26.70	—	298.70
3"	P1@.750	Ea	302.00	33.30	—	335.30
4"	P1@1.35	Ea	410.00	60.00	—	470.00

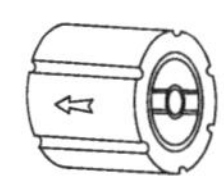

Description	Craft@Hrs	Unit	Material $	Labor $	Equipment $	Total $
Class 250 bronze body wye strainer, threaded joints						
½"	P1@.210	Ea	48.40	9.33	—	57.73
¾"	P1@.250	Ea	65.30	11.10	—	76.40
1"	P1@.300	Ea	83.00	13.30	—	96.30
1¼"	P1@.400	Ea	118.00	17.80	—	135.80
1½"	P1@.450	Ea	154.00	20.00	—	174.00
2"	P1@.500	Ea	261.00	22.20	—	283.20

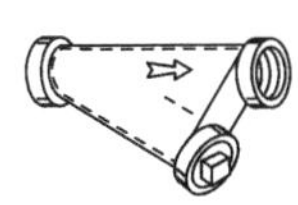

Class 250 iron body strainer, flanged ends

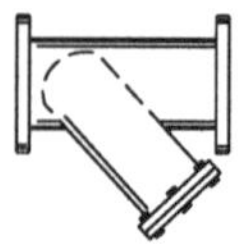

Description	Craft@Hrs	Unit	Material $	Labor $	Equipment $	Total $
2"	P1@.500	Ea	279.00	22.20	—	301.20
2½"	P1@.600	Ea	306.00	26.70	—	332.70
3"	P1@.750	Ea	356.00	33.30	—	389.30
4"	P1@1.35	Ea	633.00	60.00	—	693.00

Installation of 2-way control valve, threaded joints

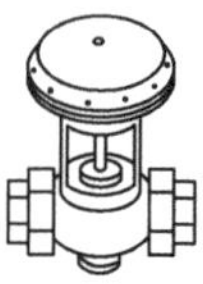

Description	Craft@Hrs	Unit	Material $	Labor $	Equipment $	Total $
½"	P1@.210	Ea	—	9.33	—	9.33
¾"	P1@.275	Ea	—	12.20	—	12.20
1"	P1@.350	Ea	—	15.60	—	15.60
1¼"	P1@.430	Ea	—	19.10	—	19.10
1½"	P1@.505	Ea	—	22.40	—	22.40
2"	P1@.675	Ea	—	30.00	—	30.00
2½"	P1@.830	Ea	—	36.90	—	36.90
3"	P1@.990	Ea	—	44.00	—	44.00

Installation of 3-way control valve, threaded joints

Description	Craft@Hrs	Unit	Material $	Labor $	Equipment $	Total $
½"	P1@.260	Ea	—	11.60	—	11.60
¾"	P1@.365	Ea	—	16.20	—	16.20
1"	P1@.475	Ea	—	21.10	—	21.10
1¼"	P1@.575	Ea	—	25.60	—	25.60
1½"	P1@.680	Ea	—	30.20	—	30.20
2"	P1@.910	Ea	—	40.40	—	40.40
2½"	P1@1.12	Ea	—	49.80	—	49.80
3"	P1@1.33	Ea	—	59.10	—	59.10

Threaded companion flange, 300# malleable iron

Description	Craft@Hrs	Unit	Material $	Labor $	Equipment $	Total $
2"	P1@.290	Ea	108.00	12.90	—	120.90
2½"	P1@.380	Ea	148.00	16.90	—	164.90
3"	P1@.460	Ea	141.00	20.40	—	161.40
4"	P1@.600	Ea	260.00	26.70	—	286.70

Bolt and gasket sets

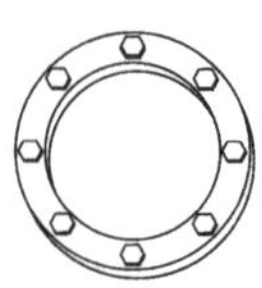

Description	Craft@Hrs	Unit	Material $	Labor $	Equipment $	Total $
2"	P1@.500	Ea	4.64	22.20	—	26.84
2½"	P1@.650	Ea	5.42	28.90	—	34.32
3"	P1@.750	Ea	9.19	33.30	—	42.49
4"	P1@1.00	Ea	15.40	44.50	—	59.90

Thermometer with well

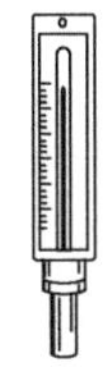

Description	Craft@Hrs	Unit	Material $	Labor $	Equipment $	Total $
7"	P1@.250	Ea	186.00	11.10	—	197.10
9"	P1@.250	Ea	192.00	11.10	—	203.10

Dial-type pressure gauge

Description	Craft@Hrs	Unit	Material $	Labor $	Equipment $	Total $
2½"	P1@.200	Ea	38.50	8.89	—	47.39
3½"	P1@.200	Ea	51.20	8.89	—	60.09

Pressure/temperature tap

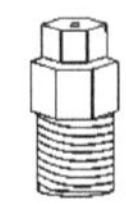

Description	Craft@Hrs	Unit	Material $	Labor $	Equipment $	Total $
Tap	P1@.150	Ea	17.70	6.67	—	24.37

Hanger with swivel assembly

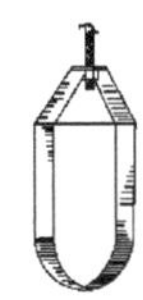

Description	Craft@Hrs	Unit	Material $	Labor $	Equipment $	Total $
½"	P1@.250	Ea	4.50	11.10	—	15.60
¾"	P1@.250	Ea	4.98	11.10	—	16.08
1"	P1@.250	Ea	5.20	11.10	—	16.30
1¼"	P1@.300	Ea	5.34	13.30	—	18.64
1½"	P1@.300	Ea	5.74	13.30	—	19.04
2"	P1@.300	Ea	5.99	13.30	—	19.29
2½"	P1@.350	Ea	8.15	15.60	—	23.75
3"	P1@.350	Ea	10.10	15.60	—	25.70
4"	P1@.350	Ea	11.10	15.60	—	26.70

Riser clamp

Description	Craft@Hrs	Unit	Material $	Labor $	Equipment $	Total $
½"	P1@.100	Ea	3.30	4.45	—	7.75
¾"	P1@.100	Ea	4.87	4.45	—	9.32
1"	P1@.100	Ea	4.91	4.45	—	9.36
1¼"	P1@.105	Ea	5.90	4.67	—	10.57
1½"	P1@.110	Ea	6.22	4.89	—	11.11
2"	P1@.115	Ea	6.62	5.11	—	11.73
2½"	P1@.120	Ea	6.93	5.33	—	12.26
3"	P1@.120	Ea	7.53	5.33	—	12.86
4"	P1@.125	Ea	9.61	5.56	—	15.17

Carbon Steel, Schedule 160 with 3,000-6,000# Fittings

Schedule 160 carbon steel (ASTM A-106) pipe with steel fittings is commonly used for steam, steam condensate and recirculating systems.

The cost estimates in this section are based on the conditions, limitations and wage rates described in the section "How to Use This Book" beginning on page 5.

Description	Craft@Hrs	Unit	Material $	Labor $	Equipment $	Total $

Schedule 160 carbon steel threaded horizontal pipe assembly. Horizontally hung in a building. Assembly includes fittings and hanger assemblies. Based on a reducing tee and a 45-degree elbow every 16 feet for ½" pipe. A reducing tee and a 45-degree elbow every 30 feet for 4" pipe. Hangers spaced to meet plumbing code. *(Distance between fittings in the pipe assembly increases as pipe diameter increases.)* Use these figures for preliminary estimates.

Description	Craft@Hrs	Unit	Material $	Labor $	Equipment $	Total $
½"	P1@.140	LF	6.44	6.22	—	12.66
¾"	P1@.145	LF	7.54	6.45	—	13.99
1"	P1@.172	LF	10.30	7.65	—	17.95
1¼"	P1@.184	LF	11.30	8.18	—	19.48
1½"	P1@.207	LF	13.10	9.20	—	22.30
2"	P1@.272	LF	19.80	12.10	—	31.90
2½"	P1@.388	LF	30.30	17.20	—	47.50
3"	P1@.434	LF	45.20	19.30	—	64.50
4"	P1@.530	LF	74.10	23.60	—	97.70

Schedule 160 carbon steel socket weld horizontal pipe assembly. Horizontally hung in a building. Assembly includes fittings and hanger assemblies. Based on a reducing tee and a 45-degree elbow every 16 feet for ½" pipe. A reducing tee and a 45-degree elbow every 30 feet for 4" pipe Hangers spaced to meet plumbing code. *(Distance between fittings in the pipe assembly increases as pipe diameter increases.)* Use these figures for preliminary estimates.

Description	Craft@Hrs	Unit	Material $	Labor $	Equipment $	Total $
½"	P1@.175	LF	6.26	7.78	.57	14.61
¾"	P1@.206	LF	7.24	9.16	.67	17.07
1"	P1@.247	LF	9.96	11.00	.80	21.76
1¼"	P1@.250	LF	11.10	11.10	.81	23.01
1½"	P1@.284	LF	12.00	12.60	.92	25.52
2"	P1@.375	LF	18.80	16.70	1.22	36.72
2½"	P1@.503	LF	28.30	22.40	1.63	52.33
3"	P1@.568	LF	44.00	25.20	1.84	71.04
4"	P1@.709	LF	72.60	31.50	2.30	106.40

Description	Craft@Hrs	Unit	Material $	Labor $	Equipment $	Total $

Schedule 160 carbon steel threaded vertical pipe assembly. Riser assembly, including fittings and riser clamps. Based on a reducing tee, and a riser clamp on every floor. Use these figures for preliminary estimates.

½"	P1@.114	LF	5.66	5.07	—	10.73
¾"	P1@.125	LF	6.76	5.56	—	12.32
1"	P1@.148	LF	9.36	6.58	—	15.94
1¼"	P1@.165	LF	11.40	7.33	—	18.73
1½"	P1@.183	LF	12.60	8.13	—	20.73
2"	P1@.242	LF	18.90	10.80	—	29.70
2½"	P1@.348	LF	28.30	15.50	—	43.80
3"	P1@.393	LF	43.50	17.50	—	61.00
4"	P1@.487	LF	69.70	21.60	—	91.30

Schedule 160 carbon steel socket weld vertical pipe assembly. Riser assembly, including fittings and riser clamps. Based on a reducing tee, and a riser clamp on every floor. Use these figures for preliminary estimates.

½"	P1@.133	LF	5.51	5.91	.43	11.85
¾"	P1@.162	LF	6.51	7.20	.53	14.24
1"	P1@.199	LF	9.15	8.85	.65	18.65
1¼"	P1@.227	LF	10.60	10.10	.74	21.44
1½"	P1@.255	LF	11.80	11.30	.83	23.93
2"	P1@.341	LF	18.00	15.20	1.11	34.31
2½"	P1@.467	LF	26.20	20.80	1.51	48.51
3"	P1@.532	LF	44.00	23.60	1.73	69.33
4"	P1@.674	LF	68.70	30.00	2.19	100.89

Schedule 160 carbon steel plain end pipe

½"	P1@.090	LF	5.22	4.00	—	9.22
¾"	P1@.100	LF	6.06	4.45	—	10.51
1"	P1@.120	LF	8.62	5.33	—	13.95
1¼"	P1@.130	LF	10.10	5.78	—	15.88
1½"	P1@.140	LF	15.80	6.22	—	22.02
2"	P1@.190	LF	13.00	8.45	—	21.45
3"	P1@.310	LF	25.40	13.80	—	39.20
4"	P1@.380	LF	39.30	16.90	—	56.20

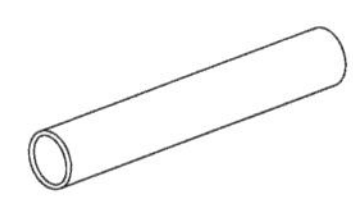

3,000# carbon steel 45-degree ell, threaded joints

Description	Craft@Hrs	Unit	Material $	Labor $	Equipment $	Total $
½"	P1@.132	Ea	10.20	5.87	—	16.07
¾"	P1@.143	Ea	11.50	6.36	—	17.86
1"	P1@.198	Ea	16.00	8.80	—	24.80
1¼"	P1@.264	Ea	22.10	11.70	—	33.80
1½"	P1@.330	Ea	33.40	14.70	—	48.10
2"	P1@.418	Ea	45.50	18.60	—	64.10
2½"	P1@.561	Ea	113.00	24.90	—	137.90
3"	P1@.704	Ea	183.00	31.30	—	214.30
4"	P1@.935	Ea	337.00	41.60	—	378.60

3,000# carbon steel 45-degree ell, socket-welded joints

Description	Craft@Hrs	Unit	Material $	Labor $	Equipment $	Total $
½"	P1@.300	Ea	8.23	13.30	.97	22.50
¾"	P1@.440	Ea	9.96	19.60	1.43	30.99
1"	P1@.590	Ea	13.00	26.20	1.91	41.11
1¼"	P1@.740	Ea	17.70	32.90	2.40	53.00
1½"	P1@.890	Ea	21.60	39.60	2.89	64.09
2"	P1@1.18	Ea	37.20	52.50	3.83	93.53
2½"	P1@1.49	Ea	96.10	66.20	4.83	167.13
3"	P1@1.77	Ea	162.00	78.70	5.74	246.44
4"	P1@2.37	Ea	320.00	105.00	7.69	432.69

6,000# carbon steel 45-degree ell, threaded joints

Description	Craft@Hrs	Unit	Material $	Labor $	Equipment $	Total $
½"	P1@.132	Ea	17.20	5.87	—	23.07
¾"	P1@.143	Ea	23.40	6.36	—	29.76
1"	P1@.198	Ea	26.20	8.80	—	35.00
1¼"	P1@.264	Ea	46.50	11.70	—	58.20
1½"	P1@.330	Ea	69.80	14.70	—	84.50
2"	P1@.418	Ea	97.40	18.60	—	116.00
2½"	P1@.561	Ea	370.00	24.90	—	394.90
3"	P1@.704	Ea	380.00	31.30	—	411.30

6,000# carbon steel 45-degree ell, socket-welded joints

Description	Craft@Hrs	Unit	Material $	Labor $	Equipment $	Total $
½"	P1@.300	Ea	13.30	13.30	.97	27.57
¾"	P1@.440	Ea	15.30	19.60	1.43	36.33
1"	P1@.590	Ea	20.90	26.20	1.91	49.01
1¼"	P1@.740	Ea	39.20	32.90	2.40	74.50
1½"	P1@.890	Ea	50.60	39.60	2.89	93.09
2"	P1@1.18	Ea	62.90	52.50	3.83	119.23
2½"	P1@1.49	Ea	249.00	66.20	4.83	320.03
3"	P1@1.77	Ea	294.00	78.70	5.74	378.44
4"	P1@2.37	Ea	352.00	105.00	7.69	464.69

3,000# carbon steel 90-degree ell, threaded joints

Description	Craft@Hrs	Unit	Material $	Labor $	Equipment $	Total $
½"	P1@.132	Ea	7.94	5.87	—	13.81
¾"	P1@.143	Ea	9.56	6.36	—	15.92
1"	P1@.198	Ea	13.80	8.80	—	22.60
1¼"	P1@.264	Ea	23.40	11.70	—	35.10
1½"	P1@.330	Ea	35.20	14.70	—	49.90
2"	P1@.418	Ea	42.20	18.60	—	60.80
2½"	P1@.561	Ea	105.00	24.90	—	129.90
3"	P1@.704	Ea	160.00	31.30	—	191.30
4"	P1@.935	Ea	352.00	41.60	—	393.60

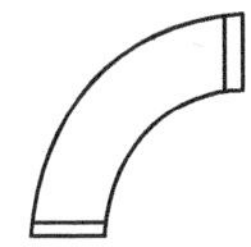

3,000# carbon steel 90-degree ell, socket-welded joints

Description	Craft@Hrs	Unit	Material $	Labor $	Equipment $	Total $
½"	P1@.300	Ea	7.87	13.30	.97	22.14
¾"	P1@.440	Ea	8.12	19.60	1.43	29.15
1"	P1@.590	Ea	10.50	26.20	1.91	38.61
1¼"	P1@.740	Ea	17.20	32.90	2.40	52.50
1½"	P1@.890	Ea	22.00	39.60	2.89	64.49
2"	P1@1.18	Ea	33.40	52.50	3.83	89.73
2½"	P1@1.49	Ea	80.20	66.20	4.83	151.23
3"	P1@1.77	Ea	132.00	78.70	5.74	216.44
4"	P1@2.37	Ea	389.00	105.00	7.69	501.69

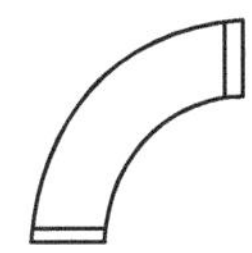

6,000# carbon steel 90-degree ell, threaded joints

Description	Craft@Hrs	Unit	Material $	Labor $	Equipment $	Total $
½"	P1@.132	Ea	12.80	5.87	—	18.67
¾"	P1@.143	Ea	16.30	6.36	—	22.66
1"	P1@.198	Ea	21.10	8.80	—	29.90
1¼"	P1@.264	Ea	35.60	11.70	—	47.30
1½"	P1@.330	Ea	55.00	14.70	—	69.70
2"	P1@.418	Ea	96.70	18.60	—	115.30
2½"	P1@.561	Ea	144.00	24.90	—	168.90
3"	P1@.704	Ea	326.00	31.30	—	357.30

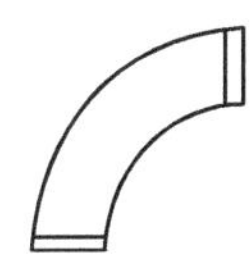

6,000# carbon steel 90-degree ell, socket-welded joints

Description	Craft@Hrs	Unit	Material $	Labor $	Equipment $	Total $
½"	P1@.300	Ea	10.90	13.30	.97	25.17
¾"	P1@.440	Ea	12.70	19.60	1.43	33.73
1"	P1@.590	Ea	17.70	26.20	1.91	45.81
1¼"	P1@.740	Ea	25.90	32.90	2.40	61.20
1½"	P1@.890	Ea	39.00	39.60	2.89	81.49
2"	P1@1.18	Ea	57.20	52.50	3.83	113.53
2½"	P1@1.49	Ea	119.00	66.20	4.83	190.03
3"	P1@1.77	Ea	210.00	78.70	5.74	294.44
4"	P1@2.37	Ea	463.00	105.00	7.69	575.69

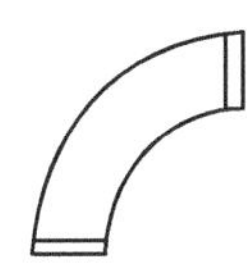

Carbon Steel, Schedule 160 with 3,000-6,000# Fittings

3,000# carbon steel tee, threaded joints

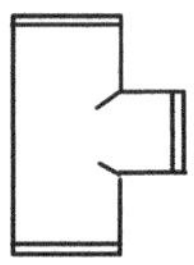

Description	Craft@Hrs	Unit	Material $	Labor $	Equipment $	Total $
½"	P1@.198	Ea	10.50	8.80	—	19.30
¾"	P1@.209	Ea	15.00	9.29	—	24.29
1"	P1@.253	Ea	19.10	11.20	—	30.30
1¼"	P1@.341	Ea	31.60	15.20	—	46.80
1½"	P1@.429	Ea	43.00	19.10	—	62.10
2"	P1@.539	Ea	57.30	24.00	—	81.30
2½"	P1@.726	Ea	138.00	32.30	—	170.30
3"	P1@.913	Ea	259.00	40.60	—	299.60
4"	P1@1.21	Ea	429.00	53.80	—	482.80

3,000# carbon steel tee, socket-welded joints

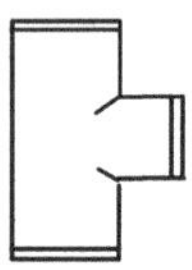

Description	Craft@Hrs	Unit	Material $	Labor $	Equipment $	Total $
½"	P1@.440	Ea	9.04	19.60	1.43	30.07
¾"	P1@.670	Ea	11.40	29.80	2.17	43.37
1"	P1@.890	Ea	15.80	39.60	2.89	58.29
1¼"	P1@1.11	Ea	21.10	49.30	3.60	74.00
1½"	P1@1.33	Ea	33.10	59.10	4.31	96.51
2"	P1@1.77	Ea	46.20	78.70	5.74	130.64
2½"	P1@2.22	Ea	111.00	98.70	7.20	216.90
3"	P1@2.66	Ea	262.00	118.00	8.63	388.63
4"	P1@3.55	Ea	421.00	158.00	11.50	590.50

6,000# carbon steel tee, threaded joints

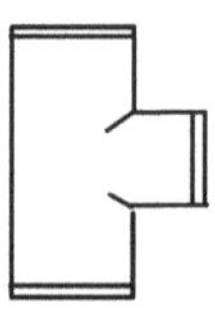

Description	Craft@Hrs	Unit	Material $	Labor $	Equipment $	Total $
½"	P1@.198	Ea	16.80	8.80	—	25.60
¾"	P1@.209	Ea	22.20	9.29	—	31.49
1"	P1@.253	Ea	37.20	11.20	—	48.40
1¼"	P1@.341	Ea	46.40	15.20	—	61.60
1½"	P1@.429	Ea	63.40	19.10	—	82.50
2"	P1@.539	Ea	124.00	24.00	—	148.00
2½"	P1@.726	Ea	444.00	32.30	—	476.30
3"	P1@.913	Ea	450.00	40.60	—	490.60

6,000# carbon steel tee, socket-welded joints

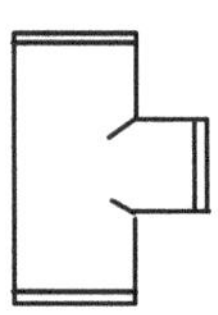

Description	Craft@Hrs	Unit	Material $	Labor $	Equipment $	Total $
½"	P1@.440	Ea	13.80	19.60	1.43	34.83
¾"	P1@.670	Ea	18.10	29.80	2.17	50.07
1"	P1@.890	Ea	22.70	39.60	2.89	65.19
1¼"	P1@1.11	Ea	38.90	49.30	3.60	91.80
1½"	P1@1.33	Ea	58.10	59.10	4.31	121.51
2"	P1@1.77	Ea	73.60	78.70	5.74	158.04
2½"	P1@2.22	Ea	156.00	98.70	7.20	261.90
3"	P1@2.66	Ea	261.00	118.00	8.63	387.63
4"	P1@3.55	Ea	496.00	158.00	11.50	665.50

Description	Craft@Hrs	Unit	Material $	Labor $	Equipment $	Total $
3,000# carbon steel union, threaded joints						
½"	P1@.150	Ea	11.50	6.67	—	18.17
¾"	P1@.165	Ea	13.80	7.33	—	21.13
1"	P1@.230	Ea	18.20	10.20	—	28.40
1¼"	P1@.310	Ea	30.60	13.80	—	44.40
1½"	P1@.400	Ea	33.40	17.80	—	51.20
2"	P1@.500	Ea	44.40	22.20	—	66.60
2½"	P1@.670	Ea	96.80	29.80	—	126.60
3"	P1@.840	Ea	120.00	37.30	—	157.30
3,000# carbon steel union, socket-welded joints						
½"	P1@.300	Ea	13.80	13.30	.97	28.07
¾"	P1@.440	Ea	15.80	19.60	1.43	36.83
1"	P1@.590	Ea	20.30	26.20	1.91	48.41
1¼"	P1@.740	Ea	27.90	32.90	2.40	63.20
1½"	P1@.890	Ea	33.80	39.60	2.89	76.29
2"	P1@1.18	Ea	47.80	52.50	3.83	104.13
2½"	P1@1.49	Ea	111.00	66.20	4.83	182.03
3"	P1@1.77	Ea	142.00	78.70	5.74	226.44
6,000# carbon steel union, threaded joints						
½"	P1@.150	Ea	27.00	6.67	—	33.67
¾"	P1@.165	Ea	30.60	7.33	—	37.93
1"	P1@.230	Ea	40.20	10.20	—	50.40
1¼"	P1@.310	Ea	65.90	13.80	—	79.70
1½"	P1@.400	Ea	81.30	17.80	—	99.10
2"	P1@.500	Ea	115.00	22.20	—	137.20
6,000# carbon steel union, socket-welded joints						
½"	P1@.300	Ea	27.90	13.30	.97	42.17
¾"	P1@.440	Ea	31.90	19.60	1.43	52.93
1"	P1@.590	Ea	46.50	26.20	1.91	74.61
1¼"	P1@.740	Ea	71.40	32.90	2.40	106.70
1½"	P1@.890	Ea	83.50	39.60	2.89	125.99
2"	P1@1.18	Ea	126.00	52.50	3.83	182.33
3,000# carbon steel reducer, socket-welded joints						
¾"	P1@.360	Ea	7.52	16.00	1.17	24.69
1"	P1@.500	Ea	9.14	22.20	1.62	32.96
1¼"	P1@.665	Ea	10.60	29.60	2.16	42.36
1½"	P1@.800	Ea	13.20	35.60	2.59	51.39
2"	P1@1.00	Ea	18.20	44.50	3.24	65.94
2½"	P1@1.33	Ea	125.00	59.10	4.31	188.41
3"	P1@1.63	Ea	158.00	72.50	5.29	235.79

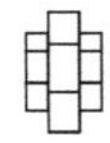
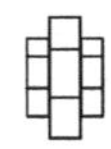
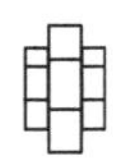
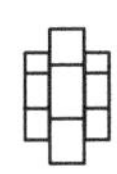
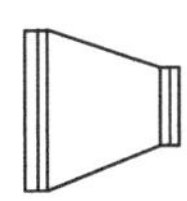

Description	Craft@Hrs	Unit	Material $	Labor $	Equipment $	Total $

6,000# carbon steel reducer, socket-welded joints

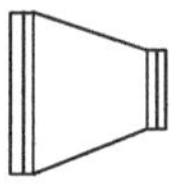

Description	Craft@Hrs	Unit	Material $	Labor $	Equipment $	Total $
¾"	P1@.360	Ea	28.80	16.00	1.17	45.97
1"	P1@.500	Ea	29.80	22.20	1.62	53.62
1¼"	P1@.665	Ea	63.90	29.60	2.16	95.66
1½"	P1@.800	Ea	63.90	35.60	2.59	102.09
2"	P1@1.00	Ea	85.00	44.50	3.24	132.74

3,000# carbon steel cap, socket-welded joint

Description	Craft@Hrs	Unit	Material $	Labor $	Equipment $	Total $
½"	P1@.180	Ea	4.49	8.00	.58	13.07
¾"	P1@.270	Ea	5.22	12.00	.88	18.10
1"	P1@.360	Ea	7.97	16.00	1.17	25.14
1¼"	P1@.460	Ea	9.36	20.40	1.49	31.25
1½"	P1@.560	Ea	12.70	24.90	1.82	39.42
2"	P1@.740	Ea	19.90	32.90	2.40	55.20
2½"	P1@.920	Ea	107.00	40.90	2.98	150.88
3"	P1@1.11	Ea	111.00	49.30	3.60	163.90
4"	P1@1.50	Ea	168.00	66.70	4.86	239.56

6,000# carbon steel cap, socket-welded joint

Description	Craft@Hrs	Unit	Material $	Labor $	Equipment $	Total $
½"	P1@.180	Ea	8.62	8.00	.58	17.20
¾"	P1@.270	Ea	12.50	12.00	.88	25.38
1"	P1@.360	Ea	13.20	16.00	1.17	30.37
1¼"	P1@.460	Ea	31.00	20.40	1.49	52.89
1½"	P1@.560	Ea	40.20	24.90	1.82	66.92
2"	P1@.740	Ea	59.80	32.90	2.40	95.10

3,000# carbon steel weldolet

Description	Craft@Hrs	Unit	Material $	Labor $	Equipment $	Total $
½"	P1@.580	Ea	15.90	25.80	1.88	43.58
¾"	P1@.890	Ea	16.30	39.60	2.89	58.79
1"	P1@1.18	Ea	17.90	52.50	3.83	74.23
1¼"	P1@1.48	Ea	22.20	65.80	4.80	92.80
1½"	P1@1.75	Ea	23.30	77.80	5.67	106.77
2"	P1@2.35	Ea	24.10	104.00	7.62	135.72
2½"	P1@2.90	Ea	54.40	129.00	9.40	192.80
3"	P1@3.50	Ea	56.00	156.00	11.30	223.30
4"	P1@4.75	Ea	69.30	211.00	15.40	295.70

6,000# carbon steel weldolet

Description	Craft@Hrs	Unit	Material $	Labor $	Equipment $	Total $
½"	P1@.690	Ea	69.30	30.70	2.24	102.24
¾"	P1@1.05	Ea	72.30	46.70	3.40	122.40
1"	P1@1.40	Ea	75.90	62.20	4.54	142.64
1¼"	P1@1.75	Ea	91.30	77.80	5.67	174.77
1½"	P1@2.10	Ea	102.00	93.30	6.81	202.11
2"	P1@2.80	Ea	118.00	124.00	9.08	251.08
2½"	P1@3.50	Ea	173.00	156.00	11.30	340.30
3"	P1@4.20	Ea	193.00	187.00	13.60	393.60
4"	P1@5.70	Ea	329.00	253.00	18.50	600.50

3,000# carbon steel threadolet

Description	Craft@Hrs	Unit	Material $	Labor $	Equipment $	Total $
½"	P1@.400	Ea	12.00	17.80	—	29.80
¾"	P1@.580	Ea	13.30	25.80	—	39.10
1"	P1@.790	Ea	16.30	35.10	—	51.40
1¼"	P1@.980	Ea	63.40	43.60	—	107.00
1½"	P1@1.20	Ea	73.90	53.30	—	127.20
2"	P1@1.60	Ea	97.70	71.10	—	168.80

6,000# carbon steel threadolet

Description	Craft@Hrs	Unit	Material $	Labor $	Equipment $	Total $
½"	P1@.480	Ea	18.40	21.30	—	39.70
¾"	P1@.690	Ea	20.10	30.70	—	50.80
1"	P1@.950	Ea	28.70	42.20	—	70.90
1¼"	P1@1.20	Ea	91.80	53.30	—	145.10
1½"	P1@1.45	Ea	107.00	64.50	—	171.50
2"	P1@1.90	Ea	175.00	84.50	—	259.50

Class 300 forged steel companion flange, weld neck

Description	Craft@Hrs	Unit	Material $	Labor $	Equipment $	Total $
½"	P1@.240	Ea	21.80	10.70	.78	33.28
¾"	P1@.320	Ea	21.80	14.20	1.04	37.04
1"	P1@.340	Ea	21.80	15.10	1.10	38.00
1¼"	P1@.425	Ea	28.20	18.90	1.38	48.48
1½"	P1@.560	Ea	28.20	24.90	1.82	54.92
2"	P1@.740	Ea	30.80	32.90	2.40	66.10
2½"	P1@.810	Ea	45.20	36.00	2.63	83.83
3"	P1@.980	Ea	46.30	43.60	3.18	93.08
4"	P1@1.30	Ea	68.10	57.80	4.22	130.12

Class 300 forged steel companion flange, slip on

Description	Craft@Hrs	Unit	Material $	Labor $	Equipment $	Total $
½"	P1@.240	Ea	17.30	10.70	—	28.00
¾"	P1@.320	Ea	17.30	14.20	—	31.50
1"	P1@.340	Ea	17.30	15.10	—	32.40
1½"	P1@.425	Ea	25.90	18.90	—	44.80
2"	P1@.560	Ea	26.70	24.90	—	51.60
2½"	P1@.740	Ea	41.60	32.90	—	74.50
3"	P1@.810	Ea	41.60	36.00	—	77.60
4"	P1@.980	Ea	61.50	43.60	—	105.10

Class 300 cast steel gate valve, flanged

Description	Craft@Hrs	Unit	Material $	Labor $	Equipment $	Total $
2"	P1@.500	Ea	2,500.00	22.20	—	2,522.20
2½"	P1@.600	Ea	3,390.00	26.70	—	3,416.70
3"	P1@.750	Ea	3,490.00	33.30	—	3,523.30
4"	P1@1.35	Ea	4,730.00	60.00	—	4,790.00

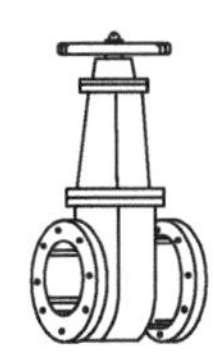

Description	Craft@Hrs	Unit	Material $	Labor $	Equipment $	Total $

Class 600 cast steel gate valve, flanged

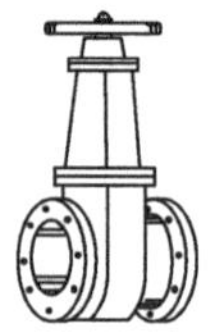

2"	P1@.500	Ea	4,160.00	22.20	—	4,182.20
2½"	P1@.600	Ea	6,000.00	26.70	—	6,026.70
3"	P1@.750	Ea	6,000.00	33.30	—	6,033.30
4"	P1@1.35	Ea	9,610.00	60.00	—	9,670.00

Class 300 cast steel body globe valve, flanged

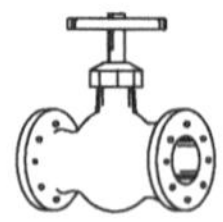

2"	P1@.500	Ea	1,710.00	22.20	—	1,732.20
2½"	P1@.600	Ea	2,450.00	26.70	—	2,476.70
3"	P1@.750	Ea	2,570.00	33.30	—	2,603.30
4"	P1@1.35	Ea	3,370.00	60.00	—	3,430.00

Class 600 cast steel body globe valve, flanged

2"	P1@.500	Ea	2,510.00	22.20	—	2,532.20
2½"	P1@.600	Ea	3,810.00	26.70	—	3,836.70
3"	P1@.750	Ea	4,000.00	33.30	—	4,033.30
4"	P1@1.35	Ea	6,150.00	60.00	—	6,210.00

Class 300 cast steel body swing check valve, flanged

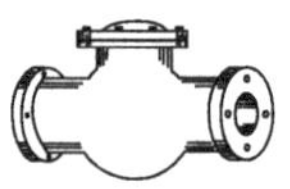

2"	P1@.500	Ea	1,700.00	22.20	—	1,722.20
2½"	P1@.600	Ea	2,560.00	26.70	—	2,586.70
3"	P1@.750	Ea	2,690.00	33.30	—	2,723.30
4"	P1@1.35	Ea	3,630.00	60.00	—	3,690.00

Class 600 cast steel body swing check valve, flanged

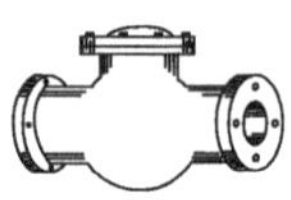

2"	P1@.500	Ea	2,610.00	22.20	—	2,632.20
2½"	P1@.600	Ea	4,120.00	26.70	—	4,146.70
3"	P1@.750	Ea	4,320.00	33.30	—	4,353.30
4"	P1@1.35	Ea	6,510.00	60.00	—	6,570.00

Installation of cast steel body 2-way control valve, threaded joints

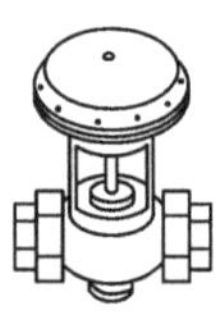

½"	P1@.210	Ea	—	9.33	—	9.33
¾"	P1@.250	Ea	—	11.10	—	11.10
1"	P1@.300	Ea	—	13.30	—	13.30
1¼"	P1@.400	Ea	—	17.80	—	17.80
1½"	P1@.450	Ea	—	20.00	—	20.00
2"	P1@.500	Ea	—	22.20	—	22.20
2½"	P1@.830	Ea	—	36.90	—	36.90
3"	P1@.990	Ea	—	44.00	—	44.00

Installation of cast steel body 2-way control valve, flanged joints

2"	P1@.500	Ea	—	22.20	—	22.20
2½"	P1@.600	Ea	—	26.70	—	26.70
3"	P1@.750	Ea	—	33.30	—	33.30
4"	P1@1.35	Ea	—	60.00	—	60.00

Installation of cast steel body 3-way control valve, threaded joints

Description	Craft@Hrs	Unit	Material $	Labor $	Equipment $	Total $
½"	P1@.260	Ea	—	11.60	—	11.60
¾"	P1@.365	Ea	—	16.20	—	16.20
1"	P1@.475	Ea	—	21.10	—	21.10
1¼"	P1@.575	Ea	—	25.60	—	25.60
1½"	P1@.680	Ea	—	30.20	—	30.20
2"	P1@.910	Ea	—	40.40	—	40.40
2½"	P1@1.12	Ea	—	49.80	—	49.80
3"	P1@1.33	Ea	—	59.10	—	59.10

Installation of cast steel body 3-way control valve, flanged joints

Description	Craft@Hrs	Unit	Material $	Labor $	Equipment $	Total $
2"	P1@.910	Ea	—	40.40	—	40.40
2½"	P1@1.12	Ea	—	49.80	—	49.80
3"	P1@1.33	Ea	—	59.10	—	59.10
4"	P1@2.00	Ea	—	88.90	—	88.90

300# bolt and gasket set, ring face

Description	Craft@Hrs	Unit	Material $	Labor $	Equipment $	Total $
2"	P1@.500	Ea	9.14	22.20	—	31.34
2½"	P1@.650	Ea	10.70	28.90	—	39.60
3"	P1@.750	Ea	18.40	33.30	—	51.70
4"	P1@1.00	Ea	29.90	44.50	—	74.40

300# bolt and gasket set, full face

Description	Craft@Hrs	Unit	Material $	Labor $	Equipment $	Total $
2"	P1@.500	Ea	18.40	22.20	—	40.60
2½"	P1@.650	Ea	21.40	28.90	—	50.30
3"	P1@.750	Ea	36.70	33.30	—	70.00
4"	P1@1.00	Ea	60.00	44.50	—	104.50

Thermometer with well

Description	Craft@Hrs	Unit	Material $	Labor $	Equipment $	Total $
7"	P1@.250	Ea	186.00	11.10	—	197.10
9"	P1@.250	Ea	192.00	11.10	—	203.10

Dial-type pressure gauge

Description	Craft@Hrs	Unit	Material $	Labor $	Equipment $	Total $
2½"	P1@.200	Ea	38.50	8.89	—	47.39
3½"	P1@.200	Ea	51.20	8.89	—	60.09

Pressure/temperature tap

Description	Craft@Hrs	Unit	Material $	Labor $	Equipment $	Total $
Tap	P1@.150	Ea	17.70	6.67	—	24.37

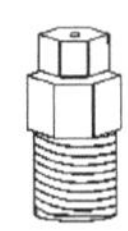

Carbon Steel, Schedule 160 with 3,000-6,000# Fittings

Description	Craft@Hrs	Unit	Material $	Labor $	Equipment $	Total $

Hanger with swivel assembly

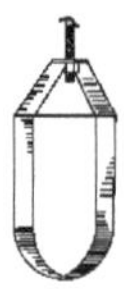

Description	Craft@Hrs	Unit	Material $	Labor $	Equipment $	Total $
½"	P1@.250	Ea	4.50	11.10	—	15.60
¾"	P1@.250	Ea	4.98	11.10	—	16.08
1"	P1@.250	Ea	5.20	11.10	—	16.30
1¼"	P1@.300	Ea	5.34	13.30	—	18.64
1½"	P1@.300	Ea	5.74	13.30	—	19.04
2"	P1@.300	Ea	5.99	13.30	—	19.29
2½"	P1@.350	Ea	8.15	15.60	—	23.75
3"	P1@.350	Ea	10.10	15.60	—	25.70
4"	P1@.350	Ea	11.10	15.60	—	26.70

Riser clamp

Description	Craft@Hrs	Unit	Material $	Labor $	Equipment $	Total $
½"	P1@.100	Ea	3.30	4.45	—	7.75
¾"	P1@.100	Ea	4.87	4.45	—	9.32
1¼"	P1@.105	Ea	4.91	4.67	—	9.58
1½"	P1@.110	Ea	5.90	4.89	—	10.79
2"	P1@.115	Ea	6.22	5.11	—	11.33
2½"	P1@.120	Ea	6.62	5.33	—	11.95
3"	P1@.120	Ea	6.93	5.33	—	12.26
4"	P1@.125	Ea	7.53	5.56	—	13.09

Schedule 40 carbon steel (ASTM A-53 and A-120) roll-grooved pipe with factory grooved malleable iron, ductile iron, or steel fittings is commonly used for heating hot water, chilled water and condenser water systems.

Consult the manufacturers for maximum operating temperature/pressure ratings for the various combinations of pipe and fittings and system applications.

The cost estimates in this section are based on the conditions, limitations and wage rates described in the section "How to Use This Book" beginning on page 5.

Equipment cost, where shown, is $4.57 per hour for a 2-ton chain hoist.

Schedule 40 carbon steel roll-grooved horizontal pipe assembly. Horizontally hung in a building. Assembly includes fittings, couplings and hanger assemblies. Based on a reducing tee and a 45-degree elbow every 16 feet for ½" pipe. A reducing tee and a 45-degree elbow every 50 feet for 12" pipe. Hangers spaced to meet plumbing code. *(Distance between fittings in the pipe assembly increases as pipe diameter increases.)* Use these figures for preliminary estimates.

Description	Craft@Hrs	Unit	Material $	Labor $	Equipment $	Total $
2"	P1@.183	LF	10.30	8.13	—	18.43
2½"	P1@.192	LF	13.60	8.53	—	22.13
3"	P1@.220	LF	15.30	9.78	—	25.08
4"	P1@.262	LF	22.20	11.60	—	33.80
6"	ER@.347	LF	41.00	16.50	.23	57.73
8"	ER@.414	LF	59.90	19.60	.28	79.78
10"	ER@.530	LF	89.00	25.10	.35	114.45
12"	ER@.617	LF	159.00	29.30	.41	188.71

Schedule 40 carbon steel roll-grooved vertical pipe assembly. Riser assembly, including fittings, couplings, and riser clamps. Based on a reducing tee, and a riser clamp on every floor. Use these figures for preliminary estimates.

Description	Craft@Hrs	Unit	Material $	Labor $	Equipment $	Total $
2"	P1@.120	LF	9.05	5.33	—	14.38
2½"	P1@.123	LF	12.80	5.47	—	18.27
3"	P1@.150	LF	13.40	6.67	—	20.07
4"	P1@.209	LF	20.10	9.29	—	29.39
6"	ER@.294	LF	39.10	13.90	.20	53.20
8"	ER@.395	LF	64.10	18.70	.26	83.06
10"	ER@.523	LF	89.00	24.80	.35	114.15
12"	ER@.608	LF	145.00	28.80	.41	174.21

Schedule 40 carbon steel pipe with roll-grooved joints

Description	Craft@Hrs	Unit	Material $	Labor $	Equipment $	Total $
2"	P1@.100	LF	4.09	4.45	—	8.54
2½"	P1@.115	LF	6.43	5.11	—	11.54
3"	P1@.135	LF	8.77	6.00	—	14.77
4"	P1@.190	LF	12.50	8.45	—	20.95
5"	ER@.240	LF	23.50	11.40	.16	35.06
6"	ER@.315	LF	35.70	14.90	.21	50.81
8"	ER@.370	LF	35.70	17.50	.25	53.45
10"	ER@.460	LF	50.00	21.80	.31	72.11
12"	ER@.550	LF	78.90	26.10	.37	105.37

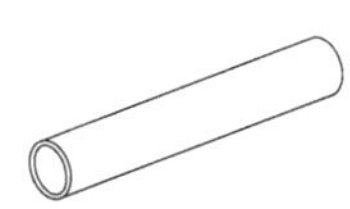

Carbon Steel, Schedule 40 with Roll-Grooved Joints

Description	Craft@Hrs	Unit	Material $	Labor $	Equipment $	Total $

Schedule 40 carbon steel 45-degree ell with roll-grooved joints

Description	Craft@Hrs	Unit	Material $	Labor $	Equipment $	Total $
2"	P1@.170	Ea	30.00	7.56	—	37.56
2½"	P1@.185	Ea	40.80	8.22	—	49.02
3"	P1@.200	Ea	53.80	8.89	—	62.69
4"	P1@.230	Ea	79.40	10.20	—	89.60
5"	ER@.250	Ea	194.00	11.90	.17	206.07
6"	ER@.350	Ea	226.00	16.60	.23	242.83
8"	ER@.500	Ea	472.00	23.70	.33	496.03
10"	ER@.750	Ea	637.00	35.60	.50	673.10
12"	ER@1.00	Ea	1,110.00	47.40	.67	1,158.07

Schedule 40 carbon steel 90-degree ell with roll-grooved joints

Description	Craft@Hrs	Unit	Material $	Labor $	Equipment $	Total $
2"	P1@.170	Ea	30.00	7.56	—	37.56
2½"	P1@.185	Ea	40.80	8.22	—	49.02
3"	P1@.200	Ea	53.80	8.89	—	62.69
4"	P1@.230	Ea	79.40	10.20	—	89.60
5"	ER@.250	Ea	194.00	11.90	.17	206.07
6"	ER@.350	Ea	226.00	16.60	.23	242.83
8"	ER@.500	Ea	472.00	23.70	.33	496.03
10"	ER@.750	Ea	733.00	35.60	.50	769.10
12"	ER@1.00	Ea	1,160.00	47.40	.67	1,208.07

Schedule 40 carbon steel tee with roll-grooved joints

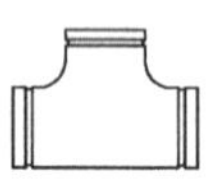

Description	Craft@Hrs	Unit	Material $	Labor $	Equipment $	Total $
2"	P1@.200	Ea	46.20	8.89	—	55.09
2½"	P1@.210	Ea	62.80	9.33	—	72.13
3"	P1@.230	Ea	87.60	10.20	—	97.80
4"	P1@.250	Ea	135.00	11.10	—	146.10
5"	ER@.300	Ea	317.00	14.20	.20	331.40
6"	ER@.500	Ea	368.00	23.70	.33	392.03
8"	ER@.750	Ea	807.00	35.60	.50	843.10
10"	ER@1.00	Ea	1,440.00	47.40	.67	1,488.07
12"	ER@1.50	Ea	1,980.00	71.10	1.00	2,052.10

Description	Craft@Hrs	Unit	Material $	Labor $	Equipment $	Total $
Schedule 40 carbon steel reducing tee with roll-grooved joints						
3" x 2"	P1@.230	Ea	119.00	10.20	—	129.20
4" x 2"	P1@.240	Ea	161.00	10.70	—	171.70
4" x 2½"	P1@.245	Ea	161.00	10.90	—	171.90
4" x 3"	P1@.250	Ea	161.00	11.10	—	172.10
5" x 2"	P1@.260	Ea	295.00	11.60	—	306.60
5" x 3"	P1@.280	Ea	324.00	12.40	—	336.40
5" x 4"	P1@.300	Ea	348.00	13.30	—	361.30
6" x 3"	ER@.400	Ea	383.00	19.00	.27	402.27
6" x 4"	ER@.450	Ea	383.00	21.30	.30	404.60
6" x 5"	ER@.500	Ea	383.00	23.70	.33	407.03
8" x 3"	ER@.600	Ea	553.00	28.40	.40	581.80
8" x 4"	ER@.650	Ea	807.00	30.80	.44	838.24
8" x 5"	ER@.700	Ea	807.00	33.20	.47	840.67
8" x 6"	ER@.750	Ea	807.00	35.60	.50	843.10
10" x 4"	ER@.800	Ea	860.00	37.90	.54	898.44
10" x 5"	ER@.900	Ea	876.00	42.70	.60	919.30
10" x 6"	ER@1.00	Ea	876.00	47.40	.67	924.07
10" x 8"	ER@1.10	Ea	876.00	52.20	.74	928.94
12" x 6"	ER@1.20	Ea	1,270.00	56.90	.80	1,327.70
12" x 8"	ER@1.35	Ea	1,320.00	64.00	.90	1,384.90
12" x 10"	ER@1.50	Ea	1,380.00	71.10	1.00	1,452.10
Schedule 40 carbon steel male adapter						
2"	P1@.170	Ea	19.60	7.56	—	27.16
2½"	P1@.185	Ea	23.20	8.22	—	31.42
3"	P1@.200	Ea	29.00	8.89	—	37.89
4"	P1@.230	Ea	48.40	10.20	—	58.60
5"	ER@.290	Ea	102.00	13.70	.19	115.89
6"	ER@.400	Ea	238.00	19.00	.27	257.27
Schedule 40 carbon steel female adapter						
2"	P1@.170	Ea	44.30	7.56	—	51.86
3"	P1@.200	Ea	68.50	8.89	—	77.39
4"	P1@.230	Ea	95.90	10.20	—	106.10
Schedule 40 carbon steel reducer, roll-grooved joints						
2"	P1@.160	Ea	39.90	6.24	—	46.14
3"	P1@.170	Ea	57.40	7.56	—	64.96
4"	P1@.210	Ea	69.90	9.33	—	79.23
5"	ER@.260	Ea	97.20	12.30	.17	109.67
6"	ER@.310	Ea	113.00	14.70	.21	127.91
8"	ER@.450	Ea	295.00	21.30	.30	316.60
10"	ER@.690	Ea	513.00	32.70	.46	546.16
12"	ER@.920	Ea	892.00	43.60	.62	936.22

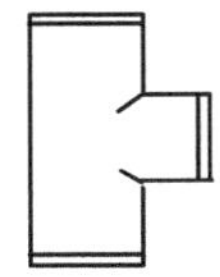

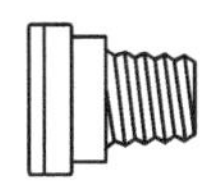

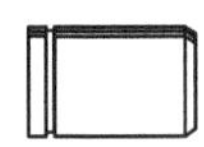

Carbon Steel, Schedule 40 with Roll-Grooved Joints

Description	Craft@Hrs	Unit	Material $	Labor $	Equipment $	Total $

Schedule 40 carbon steel cap, roll-grooved joint

Description	Craft@Hrs	Unit	Material $	Labor $	Equipment $	Total $
2"	P1@.110	Ea	21.40	4.89	—	26.29
2½"	P1@.120	Ea	33.80	5.33	—	39.13
3"	P1@.130	Ea	33.80	5.78	—	39.58
4"	P1@.140	Ea	36.40	6.22	—	42.62
5"	ER@.150	Ea	84.50	7.11	.10	91.71
6"	ER@.170	Ea	87.60	8.06	.11	95.77
8"	ER@.200	Ea	170.00	9.48	.13	179.61
10"	ER@.225	Ea	305.00	10.70	.15	315.85
12"	ER@.250	Ea	493.00	11.90	.17	505.07

Flange with gasket, grooved joint

Description	Craft@Hrs	Unit	Material $	Labor $	Equipment $	Total $
2"	P1@.280	Ea	64.30	12.40	—	76.70
2½"	P1@.340	Ea	76.00	15.10	—	91.10
3"	P1@.390	Ea	86.00	17.30	—	103.30
4"	P1@.560	Ea	124.00	24.90	—	148.90
6"	ER@.780	Ea	156.00	37.00	—	193.00
8"	ER@1.10	Ea	183.00	52.20	—	235.20
10"	ER@1.40	Ea	232.00	66.40	—	298.40
12"	ER@1.70	Ea	321.00	80.60	—	401.60

Roll-grooved coupling with gasket

Description	Craft@Hrs	Unit	Material $	Labor $	Equipment $	Total $
2"	P1@.300	Ea	33.30	13.30	—	46.60
2½"	P1@.350	Ea	40.10	15.60	—	55.70
3"	P1@.400	Ea	44.30	17.80	—	62.10
4"	P1@.500	Ea	64.30	22.20	—	86.50
5"	ER@.600	Ea	98.60	28.40	—	127.00
6"	ER@.700	Ea	115.00	33.20	—	148.20
8"	ER@.900	Ea	194.00	42.70	—	236.70
10"	ER@1.10	Ea	268.00	52.20	—	320.20
12"	ER@1.35	Ea	305.00	64.00	—	369.00

Class 125 iron body gate valve, flanged ends

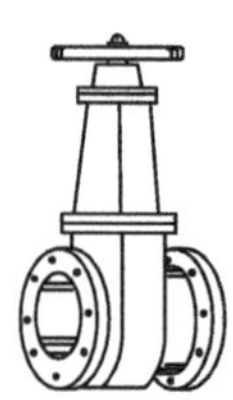

Description	Craft@Hrs	Unit	Material $	Labor $	Equipment $	Total $
2"	P1@.500	Ea	431.00	22.20	—	453.20
2½"	P1@.600	Ea	581.00	26.70	—	607.70
3"	P1@.750	Ea	630.00	33.30	—	663.30
4"	P1@1.35	Ea	925.00	60.00	—	985.00
6"	ER@2.50	Ea	1,770.00	119.00	1.67	1,890.67
8"	ER@3.00	Ea	3,140.00	142.00	2.01	3,284.01
10"	ER@4.00	Ea	5,110.00	190.00	2.68	5,302.68
12"	ER@4.50	Ea	7,100.00	213.00	3.01	7,316.01

Class 125 iron body globe valve, flanged joints

Description	Craft@Hrs	Unit	Material $	Labor $	Equipment $	Total $
2"	P1@.500	Ea	338.00	22.20	—	360.20
2½"	P1@.600	Ea	488.00	26.70	—	514.70
3"	P1@.750	Ea	586.00	33.30	—	619.30
4"	P1@1.35	Ea	790.00	60.00	—	850.00
6"	ER@2.50	Ea	1,490.00	119.00	1.67	1,610.67
8"	ER@3.00	Ea	2,270.00	142.00	2.01	2,414.01
10"	ER@4.00	Ea	3,460.00	190.00	2.68	3,652.68

200 PSIG iron body butterfly valve, lug-type, lever operated

Description	Craft@Hrs	Unit	Material $	Labor $	Equipment $	Total $
2"	P1@.450	Ea	179.00	20.00	—	199.00
2½"	P1@.450	Ea	186.00	20.00	—	206.00
3"	P1@.550	Ea	196.00	24.40	—	220.40
4"	P1@.550	Ea	244.00	24.40	—	268.40
6"	ER@.800	Ea	410.00	37.90	.54	448.44
8"	ER@.800	Ea	560.00	37.90	.54	598.44
10"	ER@.900	Ea	779.00	42.70	.60	822.30
12"	ER@1.00	Ea	1,020.00	47.40	.67	1,068.07

200 PSIG iron body butterfly valve, wafer-type, lever operated

Description	Craft@Hrs	Unit	Material $	Labor $	Equipment $	Total $
2"	P1@.450	Ea	165.00	20.00	—	185.00
2½"	P1@.450	Ea	169.00	20.00	—	189.00
3"	P1@.550	Ea	181.00	24.40	—	205.40
4"	P1@.550	Ea	219.00	24.40	—	243.40
6"	ER@.800	Ea	384.00	37.90	.54	422.44
8"	ER@.800	Ea	531.00	37.90	.54	569.44
10"	ER@.900	Ea	742.00	42.70	.60	785.30
12"	ER@1.00	Ea	975.00	47.40	.67	1,023.07

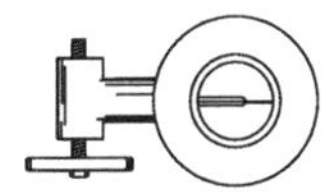

Class 125 iron body swing check valve, flanged ends

Description	Craft@Hrs	Unit	Material $	Labor $	Equipment $	Total $
2"	P1@.500	Ea	209.00	22.20	—	231.20
2½"	P1@.600	Ea	264.00	26.70	—	290.70
3"	P1@.750	Ea	329.00	33.30	—	362.30
4"	P1@1.35	Ea	483.00	60.00	—	543.00
6"	ER@2.50	Ea	937.00	119.00	1.67	1,057.67
8"	ER@3.00	Ea	1,690.00	142.00	2.01	1,834.01
10"	ER@4.00	Ea	2,740.00	190.00	2.68	2,932.68
12"	ER@4.50	Ea	3,760.00	213.00	3.01	3,976.01

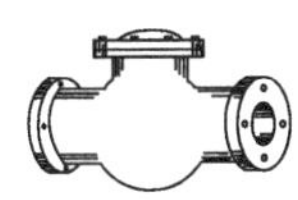

Class 125 iron body silent check valve, flanged joints

Description	Craft@Hrs	Unit	Material $	Labor $	Equipment $	Total $
2"	P1@.500	Ea	156.00	22.20	—	178.20
2½"	P1@.600	Ea	173.00	26.70	—	199.70
3"	P1@.750	Ea	199.00	33.30	—	232.30
4"	P1@1.35	Ea	265.00	60.00	—	325.00
5"	ER@2.00	Ea	384.00	94.80	1.34	480.14
6"	ER@2.50	Ea	475.00	119.00	1.67	595.67
8"	ER@3.00	Ea	856.00	142.00	2.01	1,000.01
10"	ER@4.00	Ea	1,320.00	190.00	2.68	1,512.68

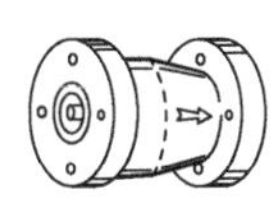

Class 125 iron body strainer, flanged

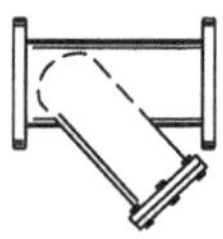

Description	Craft@Hrs	Unit	Material $	Labor $	Equipment $	Total $
2"	P1@.500	Ea	168.00	22.20	—	190.20
2½"	P1@.600	Ea	187.00	26.70	—	213.70
3"	P1@.750	Ea	218.00	33.30	—	251.30
4"	P1@1.35	Ea	367.00	60.00	—	427.00
6"	ER@2.50	Ea	709.00	119.00	1.67	829.67
8"	ER@3.00	Ea	883.00	142.00	2.01	1,027.01

Installation of 2-way control valve, flanged joints

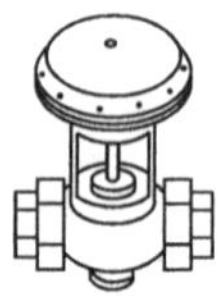

Description	Craft@Hrs	Unit	Material $	Labor $	Equipment $	Total $
2"	P1@.500	Ea	—	22.20	—	22.20
2½"	P1@.600	Ea	—	26.70	—	26.70
3"	P1@.750	Ea	—	33.30	—	33.30
4"	P1@1.35	Ea	—	60.00	—	60.00
6"	P1@2.50	Ea	—	111.00	—	111.00
8"	P1@3.00	Ea	—	133.00	—	133.00
10"	P1@4.00	Ea	—	178.00	—	178.00
12"	P1@4.50	Ea	—	200.00	—	200.00

Installation of 3-way control valve, flanged joints

Description	Craft@Hrs	Unit	Material $	Labor $	Equipment $	Total $
2"	P1@.910	Ea	—	40.40	—	40.40
2½"	P1@1.12	Ea	—	49.80	—	49.80
3"	P1@1.33	Ea	—	59.10	—	59.10
4"	P1@2.00	Ea	—	88.90	—	88.90
6"	P1@3.70	Ea	—	164.00	—	164.00
8"	P1@4.40	Ea	—	196.00	—	196.00
10"	P1@5.90	Ea	—	262.00	—	262.00
12"	P1@6.50	Ea	—	289.00	—	289.00

Bolt and gasket sets

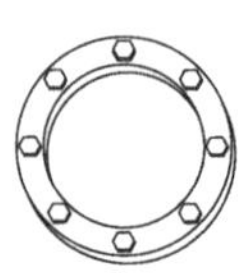

Description	Craft@Hrs	Unit	Material $	Labor $	Equipment $	Total $
2"	P1@.500	Ea	4.64	22.20	—	26.84
2½"	P1@.650	Ea	5.42	28.90	—	34.32
3"	P1@.750	Ea	9.19	33.30	—	42.49
4"	P1@1.00	Ea	15.40	44.50	—	59.90
6"	P1@1.20	Ea	25.30	53.30	—	78.60
8"	P1@1.25	Ea	28.40	55.60	—	84.00
10"	P1@1.70	Ea	49.20	75.60	—	124.80
12"	P1@2.20	Ea	55.10	97.80	—	152.90

Thermometer with well

Description	Craft@Hrs	Unit	Material $	Labor $	Equipment $	Total $
7"	P1@.250	Ea	186.00	11.10	—	197.10
9"	P1@.250	Ea	192.00	11.10	—	203.10

Dial-type pressure gauge

Description	Craft@Hrs	Unit	Material $	Labor $	Equipment $	Total $
2½"	P1@.200	Ea	38.50	8.89	—	47.39
3½"	P1@.200	Ea	51.20	8.89	—	60.09

Description	Craft@Hrs	Unit	Material $	Labor $	Equipment $	Total $

Pressure/temperature tap

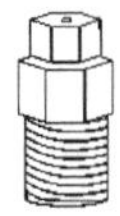

Tap	P1@.150	Ea	17.70	6.67	—	24.37

Hanger with swivel assembly

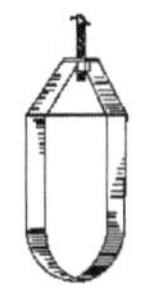

2"	P1@.300	Ea	5.99	13.30	—	19.29
2½"	P1@.350	Ea	8.15	15.60	—	23.75
3"	P1@.350	Ea	10.10	15.60	—	25.70
4"	P1@.350	Ea	11.10	15.60	—	26.70
5"	P1@.450	Ea	19.30	20.00	—	39.30
6"	P1@.450	Ea	25.10	20.00	—	45.10
8"	P1@.450	Ea	33.40	20.00	—	53.40
10"	P1@.550	Ea	46.00	24.40	—	70.40
12"	P1@.600	Ea	60.90	26.70	—	87.60

Riser clamp

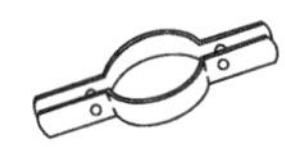

2"	P1@.115	Ea	3.30	5.11	—	8.41
2½"	P1@.120	Ea	4.87	5.33	—	10.20
3"	P1@.120	Ea	4.91	5.33	—	10.24
4"	P1@.125	Ea	5.90	5.56	—	11.46
5"	P1@.180	Ea	6.22	8.00	—	14.22
6"	P1@.200	Ea	6.62	8.89	—	15.51
8"	P1@.200	Ea	6.93	8.89	—	15.82
10"	P1@.250	Ea	7.53	11.10	—	18.63
12"	P1@.250	Ea	9.61	11.10	—	20.71

Carbon Steel, Schedule 10 with Roll-Grooved Joints

Schedule 10 carbon steel (ASTM A-53) roll-grooved pipe with factory grooved malleable iron, ductile iron, or steel fittings is commonly used for sprinkler systems, heating hot water, chilled water and condenser water systems.

Consult the manufacturers for maximum operating temperature/pressure ratings for the various combinations of pipe and fittings and system applications.

The cost estimates in this section are based on the conditions, limitations and wage rates described in the section "How to Use This Book" beginning on page 5.

Equipment cost, where shown, is $4.57 per hour for a 2-ton chain hoist.

Description	Craft@Hrs	Unit	Material $	Labor $	Equipment $	Total $

Schedule 10 carbon steel roll-grooved horizontal pipe assembly. Horizontally hung in a building. Assembly includes fittings, couplings and hanger assemblies. Based on a reducing tee and a 45-degree elbow every 16 feet for ½" pipe. A reducing tee and a 45-degree elbow every 50 feet for 12" pipe. Hangers spaced to meet plumbing code. *(Distance between fittings in the pipe assembly increases as pipe diameter increases.)* Use these figures for preliminary estimates.

Description	Craft@Hrs	Unit	Material $	Labor $	Equipment $	Total $
2"	P1@.178	LF	9.17	7.91	—	17.08
2½"	P1@.188	LF	11.40	8.36	—	19.76
3"	P1@.215	LF	13.30	9.56	—	22.86
4"	P1@.255	LF	18.90	11.30	—	30.20
6"	ER@.339	LF	35.70	16.10	.23	52.03
8"	ER@.404	LF	54.40	19.20	.27	73.87
10"	ER@.517	LF	81.50	24.50	.35	106.35
12"	ER@.602	LF	146.00	28.50	.40	174.90

Schedule 10 carbon steel roll-grooved vertical pipe assembly. Riser assembly, including fittings, couplings, and riser clamps. Based on a reducing tee, and a riser clamp on every floor. Use these figures for preliminary estimates.

Description	Craft@Hrs	Unit	Material $	Labor $	Equipment $	Total $
2"	P1@.117	LF	8.79	5.20	—	13.99
2½"	P1@.120	LF	11.90	5.33	—	17.23
3"	P1@.146	LF	12.20	6.49	—	18.69
4"	P1@.204	LF	18.80	9.07	—	27.87
6"	ER@.286	LF	37.00	13.60	.19	50.79
8"	ER@.385	LF	65.10	18.30	.26	83.66
10"	ER@.510	LF	90.40	24.20	.34	114.94
12"	ER@.593	LF	146.00	28.10	.40	174.50

Description	Craft@Hrs	Unit	Material $	Labor $	Equipment $	Total $

Schedule 10 carbon steel pipe, roll-grooved joints

Description	Craft@Hrs	Unit	Material $	Labor $	Equipment $	Total $
2"	P1@.080	LF	2.68	3.56	—	6.24
2½"	P1@.100	LF	3.65	4.45	—	8.10
3"	P1@.115	LF	4.43	5.11	—	9.54
4"	P1@.160	LF	5.89	7.11	—	13.00
5"	ER@.200	LF	10.10	9.48	.13	19.71
6"	ER@.260	LF	10.60	12.30	.17	23.07
8"	ER@.300	LF	14.60	14.20	.20	29.00
10"	ER@.380	LF	19.90	18.00	.25	38.15
12"	ER@.450	LF	28.20	21.30	.30	49.80

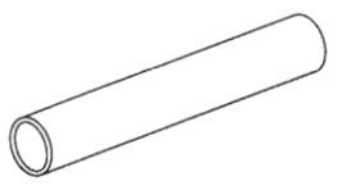

Schedule 10 carbon steel 45-degree ell, roll-grooved joints

Description	Craft@Hrs	Unit	Material $	Labor $	Equipment $	Total $
2"	P1@.170	Ea	30.00	7.56	—	37.56
2½"	P1@.185	Ea	40.80	8.22	—	49.02
3"	P1@.200	Ea	53.80	8.89	—	62.69
4"	P1@.230	Ea	79.40	10.20	—	89.60
5"	ER@.250	Ea	194.00	11.90	.17	206.07
6"	ER@.350	Ea	226.00	16.60	.23	242.83
8"	ER@.500	Ea	472.00	23.70	.33	496.03
10"	ER@.750	Ea	637.00	35.60	.50	673.10
12"	ER@1.00	Ea	1,110.00	47.40	.67	1,158.07

Schedule 10 carbon steel 90-degree ell, roll-grooved joints

Description	Craft@Hrs	Unit	Material $	Labor $	Equipment $	Total $
2"	P1@.170	Ea	30.00	7.56	—	37.56
2½"	P1@.185	Ea	40.80	8.22	—	49.02
3"	P1@.200	Ea	53.80	8.89	—	62.69
4"	P1@.230	Ea	79.40	10.20	—	89.60
5"	ER@.250	Ea	194.00	11.90	.17	206.07
6"	ER@.350	Ea	226.00	16.60	.23	242.83
8"	ER@.500	Ea	472.00	23.70	.33	496.03
10"	ER@.750	Ea	733.00	35.60	.50	769.10
12"	ER@1.00	Ea	1,160.00	47.40	.67	1,208.07

Schedule 10 carbon steel tee, roll-grooved joints

Description	Craft@Hrs	Unit	Material $	Labor $	Equipment $	Total $
2"	P1@.200	Ea	46.20	8.89	—	55.09
2½"	P1@.210	Ea	62.80	9.33	—	72.13
3"	P1@.230	Ea	87.60	10.20	—	97.80
4"	P1@.250	Ea	135.00	11.10	—	146.10
5"	ER@.300	Ea	317.00	14.20	.20	331.40
6"	ER@.500	Ea	368.00	23.70	.33	392.03
8"	ER@.750	Ea	807.00	35.60	.50	843.10
10"	ER@1.00	Ea	1,440.00	47.40	.67	1,488.07
12"	ER@1.50	Ea	1,980.00	71.10	1.00	2,052.10

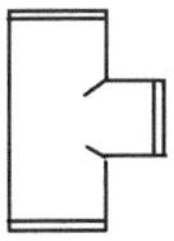

Schedule 10 carbon steel reducing tee, roll-grooved joints

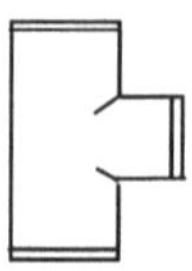

Description	Craft@Hrs	Unit	Material $	Labor $	Equipment $	Total $
3" x 2"	P1@.230	Ea	119.00	10.20	—	129.20
4" x 2"	P1@.240	Ea	161.00	10.70	—	171.70
4" x 2½"	P1@.245	Ea	161.00	10.90	—	171.90
4" x 3"	P1@.250	Ea	161.00	11.10	—	172.10
5" x 2"	P1@.260	Ea	295.00	11.60	—	306.60
5" x 3"	P1@.280	Ea	324.00	12.40	—	336.40
5" x 4"	P1@.300	Ea	348.00	13.30	—	361.30
6" x 3"	ER@.400	Ea	383.00	19.00	.27	402.27
6" x 4"	ER@.450	Ea	383.00	21.30	.30	404.60
6" x 5"	ER@.500	Ea	383.00	23.70	.33	407.03
8" x 3"	ER@.600	Ea	553.00	28.40	.40	581.80
8" x 4"	ER@.650	Ea	807.00	30.80	.44	838.24
8" x 5"	ER@.700	Ea	807.00	33.20	.47	840.67
8" x 6"	ER@.750	Ea	807.00	35.60	.50	843.10
10" x 4"	ER@.800	Ea	860.00	37.90	.54	898.44
10" x 5"	ER@.900	Ea	876.00	42.70	.60	919.30
10" x 6"	ER@1.00	Ea	876.00	47.40	.67	924.07
10" x 8"	ER@1.10	Ea	876.00	52.20	.74	928.94
12" x 6"	ER@1.20	Ea	1,270.00	56.90	.80	1,327.70
12" x 8"	ER@1.35	Ea	1,320.00	64.00	.90	1,384.90
12" x 10"	ER@1.50	Ea	1,380.00	71.10	1.00	1,452.10

Schedule 10 carbon steel male adapter

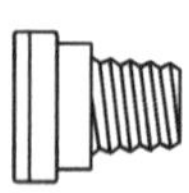

Description	Craft@Hrs	Unit	Material $	Labor $	Equipment $	Total $
2"	P1@.170	Ea	19.60	7.56	—	27.16
2½"	P1@.185	Ea	23.20	8.22	—	31.42
3"	P1@.200	Ea	29.00	8.89	—	37.89
4"	P1@.230	Ea	48.40	10.20	—	58.60
5"	ER@.290	Ea	102.00	13.70	—	115.70
6"	ER@.400	Ea	238.00	19.00	—	257.00

Schedule 10 carbon steel female adapter

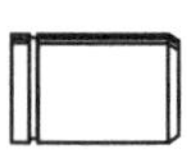

Description	Craft@Hrs	Unit	Material $	Labor $	Equipment $	Total $
2"	P1@.170	Ea	44.30	7.56	—	51.86
3"	P1@.200	Ea	68.50	8.89	—	77.39
4"	P1@.230	Ea	95.90	10.20	—	106.10

Schedule 10 carbon steel reducer, roll-grooved joints

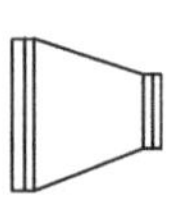

Description	Craft@Hrs	Unit	Material $	Labor $	Equipment $	Total $
2"	P1@.160	Ea	41.40	7.11	—	48.51
3"	P1@.170	Ea	57.40	7.56	—	64.96
4"	P1@.210	Ea	69.90	9.33	—	79.23
5"	ER@.260	Ea	97.20	12.30	—	109.50
6"	ER@.310	Ea	113.00	14.70	—	127.70
8"	ER@.450	Ea	295.00	21.30	—	316.30
10"	ER@.690	Ea	513.00	32.70	—	545.70
12"	ER@.920	Ea	892.00	43.60	—	935.60

Description	Craft@Hrs	Unit	Material $	Labor $	Equipment $	Total $
Schedule 10 carbon steel cap, roll-grooved joints						
2"	P1@.110	Ea	21.40	4.89	—	26.29
2½"	P1@.120	Ea	33.80	5.33	—	39.13
3"	P1@.130	Ea	33.80	5.78	—	39.58
4"	P1@.140	Ea	36.40	6.22	—	42.62
5"	ER@.150	Ea	84.50	7.11	—	91.61
6"	ER@.170	Ea	87.60	8.06	—	95.66
8"	ER@.200	Ea	170.00	9.48	—	179.48
10"	ER@.225	Ea	305.00	10.70	—	315.70
12"	ER@.250	Ea	493.00	11.90	—	504.90

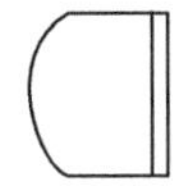

Description	Craft@Hrs	Unit	Material $	Labor $	Equipment $	Total $
Flange with gasket, grooved joint						
2"	P1@.280	Ea	64.30	12.40	—	76.70
2½"	P1@.340	Ea	76.00	15.10	—	91.10
3"	P1@.390	Ea	86.00	17.30	—	103.30
4"	P1@.560	Ea	124.00	24.90	—	148.90
6"	ER@.780	Ea	156.00	37.00	—	193.00
8"	ER@1.10	Ea	183.00	52.20	—	235.20
10"	ER@1.40	Ea	232.00	66.40	—	298.40
12"	ER@1.70	Ea	321.00	80.60	—	401.60

Description	Craft@Hrs	Unit	Material $	Labor $	Equipment $	Total $
Roll-grooved coupling with gasket						
2"	P1@.300	Ea	33.30	13.30	—	46.60
2½"	P1@.350	Ea	40.10	15.60	—	55.70
3"	P1@.400	Ea	44.30	17.80	—	62.10
4"	P1@.500	Ea	64.30	22.20	—	86.50
5"	ER@.600	Ea	98.60	28.40	—	127.00
6"	ER@.700	Ea	115.00	33.20	—	148.20
8"	ER@.900	Ea	194.00	42.70	—	236.70
10"	ER@1.10	Ea	268.00	52.20	—	320.20
12"	ER@1.35	Ea	305.00	64.00	—	369.00

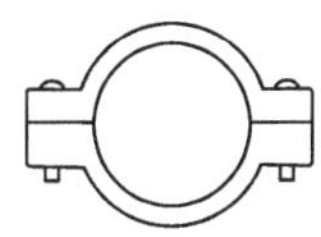

Description	Craft@Hrs	Unit	Material $	Labor $	Equipment $	Total $
Class 125 iron body gate valve, flanged joints						
2"	P1@.500	Ea	431.00	22.20	—	453.20
2½"	P1@.600	Ea	581.00	26.70	—	607.70
3"	P1@.750	Ea	630.00	33.30	—	663.30
4"	P1@1.35	Ea	925.00	60.00	—	985.00
5"	ER@2.00	Ea	2,470.00	94.80	1.34	2,566.14
6"	ER@2.50	Ea	2,470.00	119.00	1.67	2,590.67
8"	ER@3.00	Ea	4,030.00	142.00	2.01	4,174.01
10"	ER@4.00	Ea	6,700.00	190.00	2.68	6,892.68
12"	ER@4.50	Ea	9,140.00	213.00	3.01	9,356.01

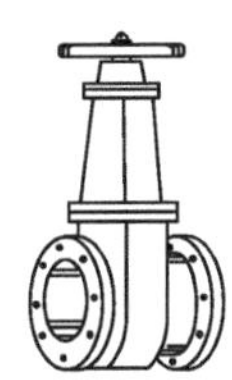

Description	Craft@Hrs	Unit	Material $	Labor $	Equipment $	Total $

Class 125 iron body globe valve, flanged joints

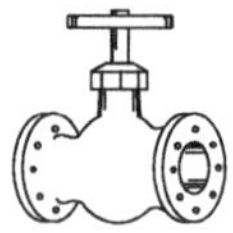

2"	P1@.500	Ea	488.00	22.20	—	510.20
2½"	P1@.600	Ea	586.00	26.70	—	612.70
3"	P1@.750	Ea	790.00	33.30	—	823.30
4"	P1@1.35	Ea	818.00	60.00	—	878.00
5"	ER@2.00	Ea	1,460.00	94.80	1.34	1,556.14
6"	ER@2.50	Ea	1,460.00	119.00	1.67	1,580.67
8"	ER@3.00	Ea	2,200.00	142.00	2.01	2,344.01
10"	ER@4.00	Ea	3,380.00	190.00	2.68	3,572.68

200 PSIG iron body butterfly valve, lug-type, lever operated

2"	P1@.450	Ea	179.00	20.00	—	199.00
2½"	P1@.450	Ea	186.00	20.00	—	206.00
3"	P1@.550	Ea	196.00	24.40	—	220.40
4"	P1@.550	Ea	244.00	24.40	—	268.40
5"	ER@.800	Ea	335.00	37.90	.54	373.44
6"	ER@.800	Ea	410.00	37.90	.54	448.44
8"	ER@.800	Ea	560.00	37.90	.54	598.44
10"	ER@.900	Ea	779.00	42.70	.60	822.30
12"	ER@1.00	Ea	1,020.00	47.40	.67	1,068.07

200 PSIG iron body butterfly valve, wafer-type, lever operated

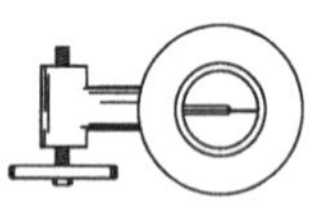

2"	P1@.450	Ea	164.00	20.00	—	184.00
2½"	P1@.450	Ea	167.00	20.00	—	187.00
3"	P1@.550	Ea	179.00	24.40	—	203.40
4"	P1@.550	Ea	217.00	24.40	—	241.40
5"	ER@.800	Ea	303.00	37.90	.54	341.44
6"	ER@.800	Ea	380.00	37.90	.54	418.44
8"	ER@.800	Ea	526.00	37.90	.54	564.44
10"	ER@.900	Ea	735.00	42.70	.60	778.30
12"	ER@1.00	Ea	966.00	47.40	.67	1,014.07

Class 125 iron body swing check valve, flanged joints

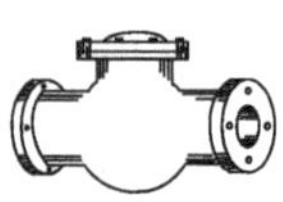

2"	P1@.500	Ea	213.00	22.20	—	235.20
2½"	P1@.600	Ea	269.00	26.70	—	295.70
3"	P1@.750	Ea	336.00	33.30	—	369.30
4"	P1@1.35	Ea	492.00	60.00	—	552.00
5"	ER@2.00	Ea	948.00	94.80	1.34	1,044.14
6"	ER@2.50	Ea	956.00	119.00	1.67	1,076.67
8"	ER@3.00	Ea	1,720.00	142.00	2.01	1,864.01
10"	ER@4.00	Ea	2,800.00	190.00	2.68	2,992.68
12"	ER@4.50	Ea	3,840.00	213.00	3.01	4,056.01

Description	Craft@Hrs	Unit	Material $	Labor $	Equipment $	Total $

Class 125 iron body silent check valve, flanged joints

Description	Craft@Hrs	Unit	Material $	Labor $	Equipment $	Total $
2"	P1@.500	Ea	156.00	22.20	—	178.20
2½"	P1@.600	Ea	173.00	26.70	—	199.70
3"	P1@.750	Ea	199.00	33.30	—	232.30
4"	P1@1.35	Ea	265.00	60.00	—	325.00
5"	ER@2.00	Ea	384.00	94.80	1.34	480.14
6"	ER@2.50	Ea	475.00	119.00	1.67	595.67
8"	ER@3.00	Ea	856.00	142.00	2.01	1,000.01
10"	ER@4.00	Ea	1,320.00	190.00	2.68	1,512.68

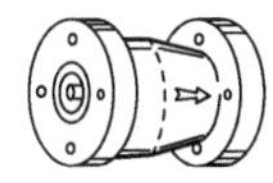

Class 125 iron body strainer, flanged

Description	Craft@Hrs	Unit	Material $	Labor $	Equipment $	Total $
2"	P1@.500	Ea	168.00	22.20	—	190.20
2½"	P1@.600	Ea	187.00	26.70	—	213.70
3"	P1@.750	Ea	218.00	33.30	—	251.30
4"	P1@1.35	Ea	367.00	60.00	—	427.00
5"	ER@2.00	Ea	708.00	94.80	1.34	804.14
6"	ER@2.50	Ea	708.00	119.00	1.67	828.67
8"	ER@3.00	Ea	1,190.00	142.00	2.01	1,334.01

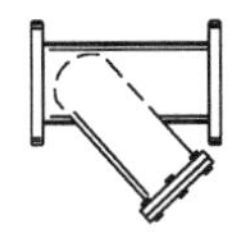

Installation of 2-way control valve, flanged joints

Description	Craft@Hrs	Unit	Material $	Labor $	Equipment $	Total $
2"	P1@.500	Ea	—	22.20	—	22.20
2½"	P1@.600	Ea	—	26.70	—	26.70
3"	P1@.750	Ea	—	33.30	—	33.30
4"	P1@1.35	Ea	—	60.00	—	60.00
6"	P1@2.50	Ea	—	111.00	—	111.00
8"	P1@3.00	Ea	—	133.00	—	133.00
10"	P1@4.00	Ea	—	178.00	—	178.00
12"	P1@4.50	Ea	—	200.00	—	200.00

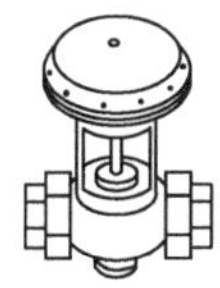

Installation of 3-way control valve, flanged joints

Description	Craft@Hrs	Unit	Material $	Labor $	Equipment $	Total $
2"	P1@.910	Ea	—	40.40	—	40.40
2½"	P1@1.12	Ea	—	49.80	—	49.80
3"	P1@1.33	Ea	—	59.10	—	59.10
4"	P1@2.00	Ea	—	88.90	—	88.90
6"	P1@3.70	Ea	—	164.00	—	164.00
8"	P1@4.40	Ea	—	196.00	—	196.00
10"	P1@5.90	Ea	—	262.00	—	262.00
12"	P1@6.50	Ea	—	289.00	—	289.00

Description	Craft@Hrs	Unit	Material $	Labor $	Equipment $	Total $

Bolt and gasket sets

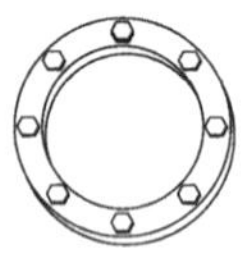

Description	Craft@Hrs	Unit	Material $	Labor $	Equipment $	Total $
2"	P1@.500	Ea	4.64	22.20	—	26.84
2½"	P1@.650	Ea	5.42	28.90	—	34.32
3"	P1@.750	Ea	9.19	33.30	—	42.49
4"	P1@1.00	Ea	15.40	44.50	—	59.90
5"	P1@1.10	Ea	25.30	48.90	—	74.20
6"	P1@1.20	Ea	25.30	53.30	—	78.60
8"	P1@1.25	Ea	28.40	55.60	—	84.00
10"	P1@1.70	Ea	48.90	75.60	—	124.50
12"	P1@2.20	Ea	55.10	97.80	—	152.90

Thermometer with well

Description	Craft@Hrs	Unit	Material $	Labor $	Equipment $	Total $
7"	P1@.250	Ea	186.00	11.10	—	197.10
9"	P1@.250	Ea	192.00	11.10	—	203.10

Dial-type pressure gauge

Description	Craft@Hrs	Unit	Material $	Labor $	Equipment $	Total $
2½"	P1@.200	Ea	38.50	8.89	—	47.39
3½"	P1@.200	Ea	51.20	8.89	—	60.09

Pressure/temperature tap

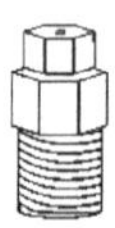

Description	Craft@Hrs	Unit	Material $	Labor $	Equipment $	Total $
Tap	P1@.150	Ea	17.70	6.67	—	24.37

Hanger with swivel assembly

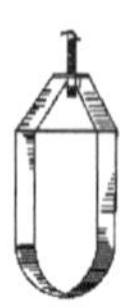

Description	Craft@Hrs	Unit	Material $	Labor $	Equipment $	Total $
2"	P1@.300	Ea	5.99	13.30	—	19.29
2½"	P1@.350	Ea	8.15	15.60	—	23.75
3"	P1@.350	Ea	10.10	15.60	—	25.70
4"	P1@.350	Ea	11.10	15.60	—	26.70
5"	P1@.450	Ea	20.20	20.00	—	40.20
6"	P1@.450	Ea	25.90	20.00	—	45.90
8"	P1@.450	Ea	34.60	20.00	—	54.60
10"	P1@.550	Ea	47.60	24.40	—	72.00
12"	P1@.600	Ea	63.20	26.70	—	89.90

Riser clamp

Description	Craft@Hrs	Unit	Material $	Labor $	Equipment $	Total $
2"	P1@.115	Ea	6.62	5.11	—	11.73
2½"	P1@.120	Ea	6.93	5.33	—	12.26
3"	P1@.120	Ea	7.53	5.33	—	12.86
4"	P1@.125	Ea	9.61	5.56	—	15.17
5"	P1@.180	Ea	13.70	8.00	—	21.70
6"	P1@.200	Ea	16.50	8.89	—	25.39
8"	P1@.200	Ea	26.90	8.89	—	35.79
10"	P1@.250	Ea	40.10	11.10	—	51.20
12"	P1@.250	Ea	47.50	11.10	—	58.60

Schedule 40 carbon steel (ASTM A-53 and A-120) cut-grooved pipe with factory-grooved malleable iron, ductile iron or steel fittings is commonly used for heating hot water, chilled water and condenser water systems.

Consult the manufacturers for maximum operating temperature/pressure ratings for various combinations of pipe and fittings.

This section has been arranged to save the estimator's time by including all normally-used system components such as pipe, fittings, valves, hanger assemblies, riser clamps and miscellaneous items under one heading. Additional items can be found under "Plumbing and Piping Specialties." The cost estimates in this section are based on the conditions, limitations and wage rates described in the section "How to Use This Book" beginning on page 5.

Equipment cost, where shown, is $4.57 per hour for a 2-ton chain hoist.

Description	Craft@Hrs	Unit	Material $	Labor $	Equipment $	Total $

Schedule 40 carbon steel pipe with cut-grooved joints

Description	Craft@Hrs	Unit	Material $	Labor $	Equipment $	Total $
2"	P1@.110	LF	4.56	4.89	—	9.45
2½"	P1@.130	LF	7.16	5.78	—	12.94
3"	P1@.150	LF	9.72	6.67	—	16.39
4"	P1@.210	LF	14.00	9.33	—	23.33
5"	ER@.270	LF	26.30	12.80	.18	39.28
6"	ER@.350	LF	39.70	16.60	.23	56.53
8"	ER@.410	LF	39.70	19.40	.27	59.37
10"	ER@.510	LF	55.70	24.20	.34	80.24
12"	ER@.610	LF	88.00	28.90	.41	117.31

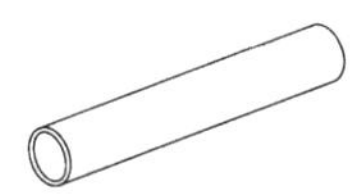

Schedule 40 carbon steel 45-degree ell with cut-grooved joints

Description	Craft@Hrs	Unit	Material $	Labor $	Equipment $	Total $
2"	P1@.170	Ea	30.00	7.56	—	37.56
2½"	P1@.185	Ea	41.00	8.22	—	49.22
3"	P1@.200	Ea	53.80	8.89	—	62.69
4"	P1@.230	Ea	79.40	10.20	—	89.60
5"	ER@.250	Ea	194.00	11.90	.17	206.07
6"	ER@.350	Ea	226.00	16.60	.23	242.83
8"	ER@.500	Ea	472.00	23.70	.33	496.03
10"	ER@.750	Ea	637.00	35.60	.50	673.10
12"	ER@1.00	Ea	1,110.00	47.40	.67	1,158.07

Schedule 40 carbon steel 90-degree ell with cut-grooved joints

Description	Craft@Hrs	Unit	Material $	Labor $	Equipment $	Total $
2"	P1@.170	Ea	30.00	7.56	—	37.56
2½"	P1@.185	Ea	41.00	8.22	—	49.22
3"	P1@.200	Ea	53.80	8.89	—	62.69
4"	P1@.230	Ea	79.40	10.20	—	89.60
5"	ER@.250	Ea	194.00	11.90	.17	206.07
6"	ER@.350	Ea	226.00	16.60	.23	242.83
8"	ER@.500	Ea	472.00	23.70	.33	496.03
10"	ER@.750	Ea	733.00	35.60	.50	769.10
12"	ER@1.00	Ea	1,980.00	47.40	.67	2,028.07

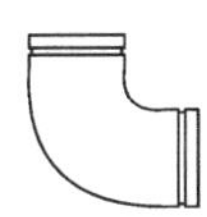

Description	Craft@Hrs	Unit	Material $	Labor $	Equipment $	Total $

Schedule 40 carbon steel tee with cut-grooved joints

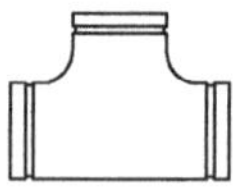

Description	Craft@Hrs	Unit	Material $	Labor $	Equipment $	Total $
2"	P1@.200	Ea	46.20	8.89	—	55.09
2½"	P1@.210	Ea	62.80	9.33	—	72.13
3"	P1@.230	Ea	87.60	10.20	—	97.80
4"	P1@.250	Ea	135.00	11.10	—	146.10
5"	ER@.300	Ea	317.00	14.20	.20	331.40
6"	ER@.500	Ea	368.00	23.70	.33	392.03
8"	ER@.750	Ea	807.00	35.60	.50	843.10
10"	ER@1.00	Ea	1,440.00	47.40	.67	1,488.07
12"	ER@1.50	Ea	1,980.00	71.10	1.00	2,052.10

Schedule 40 carbon steel reducing tee with cut-grooved joints

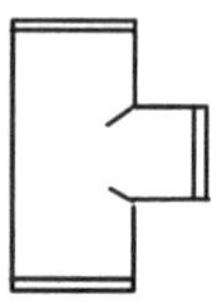

Description	Craft@Hrs	Unit	Material $	Labor $	Equipment $	Total $
3" x 2"	P1@.230	Ea	119.00	10.20	—	129.20
4" x 2"	P1@.240	Ea	161.00	10.70	—	171.70
4" x 2½"	P1@.245	Ea	161.00	10.90	—	171.90
4" x 3"	P1@.250	Ea	161.00	11.10	—	172.10
5" x 2"	P1@.260	Ea	295.00	11.60	—	306.60
5" x 3"	P1@.280	Ea	324.00	12.40	—	336.40
5" x 4"	P1@.300	Ea	348.00	13.30	—	361.30
6" x 3"	ER@.400	Ea	383.00	19.00	.27	402.27
6" x 4"	ER@.450	Ea	383.00	21.30	.30	404.60
6" x 5"	ER@.500	Ea	383.00	23.70	.33	407.03
8" x 3"	ER@.600	Ea	553.00	28.40	.40	581.80
8" x 4"	ER@.650	Ea	807.00	30.80	.44	838.24
8" x 5"	ER@.700	Ea	807.00	33.20	.47	840.67
8" x 6"	ER@.750	Ea	807.00	35.60	.50	843.10
10" x 4"	ER@.800	Ea	860.00	37.90	.54	898.44
10" x 5"	ER@.900	Ea	876.00	42.70	.60	919.30
10" x 6"	ER@1.00	Ea	876.00	47.40	.67	924.07
10" x 8"	ER@1.10	Ea	876.00	52.20	.74	928.94
12" x 6"	ER@1.20	Ea	1,270.00	56.90	.80	1,327.70
12" x 8"	ER@1.35	Ea	1,320.00	64.00	.90	1,384.90
12" x 10"	ER@1.50	Ea	1,380.00	71.10	1.00	1,452.10

Schedule 40 carbon steel male adapter

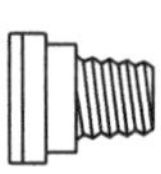

Description	Craft@Hrs	Unit	Material $	Labor $	Equipment $	Total $
2"	P1@.170	Ea	19.60	7.56	—	27.16
2½"	P1@.185	Ea	23.20	8.22	—	31.42
3"	P1@.200	Ea	29.00	8.89	—	37.89
4"	P1@.230	Ea	48.10	10.20	—	58.30
5"	ER@.290	Ea	102.00	13.70	—	115.70
6"	ER@.400	Ea	238.00	19.00	—	257.00

Schedule 40 carbon steel female adapter

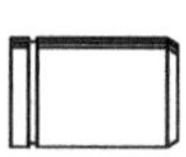

Description	Craft@Hrs	Unit	Material $	Labor $	Equipment $	Total $
2"	P1@.170	Ea	44.30	7.56	—	51.86
3"	P1@.200	Ea	68.50	8.89	—	77.39
4"	P1@.230	Ea	95.90	10.20	—	106.10

Description	Craft@Hrs	Unit	Material $	Labor $	Equipment $	Total $

Schedule 40 carbon steel reducer

2"	P1@.160	Ea	41.40	7.11	—	48.51
3"	P1@.170	Ea	57.40	7.56	—	64.96
4"	P1@.210	Ea	69.90	9.33	—	79.23
5"	ER@.260	Ea	97.20	12.30	—	109.50
6"	ER@.310	Ea	113.00	14.70	—	127.70
8"	ER@.450	Ea	295.00	21.30	—	316.30
10"	ER@.690	Ea	513.00	32.70	—	545.70
12"	ER@.920	Ea	892.00	43.60	—	935.60

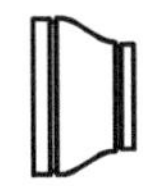

Schedule 40 carbon steel cap

2"	P1@.110	Ea	21.40	4.89	—	26.29
2½"	P1@.120	Ea	33.80	5.33	—	39.13
3"	P1@.130	Ea	33.80	5.78	—	39.58
4"	P1@.140	Ea	36.40	6.22	—	42.62
5"	ER@.150	Ea	84.50	7.11	—	91.61
6"	ER@.170	Ea	87.60	8.06	—	95.66
8"	ER@.200	Ea	170.00	9.48	—	179.48
10"	ER@.225	Ea	305.00	10.70	—	315.70
12"	ER@.250	Ea	493.00	11.90	—	504.90

Flange with gasket, grooved joint

2"	P1@.280	Ea	64.30	12.40	—	76.70
2½"	P1@.340	Ea	76.00	15.10	—	91.10
3"	P1@.390	Ea	86.00	17.30	—	103.30
4"	P1@.560	Ea	124.00	24.90	—	148.90
6"	ER@.780	Ea	156.00	37.00	—	193.00
8"	ER@1.10	Ea	183.00	52.20	—	235.20
10"	ER@1.40	Ea	232.00	66.40	—	298.40
12"	ER@1.70	Ea	321.00	80.60	—	401.60

Cut-grooved joint coupling with gasket

2"	P1@.300	Ea	33.30	13.30	—	46.60
2½"	P1@.350	Ea	40.10	15.60	—	55.70
3"	P1@.400	Ea	44.30	17.80	—	62.10
4"	P1@.500	Ea	64.30	22.20	—	86.50
5"	ER@.600	Ea	98.60	28.40	—	127.00
6"	ER@.700	Ea	115.00	33.20	—	148.20
8"	ER@.900	Ea	194.00	42.70	—	236.70
10"	ER@1.10	Ea	268.00	52.20	—	320.20
12"	ER@1.35	Ea	305.00	64.00	—	369.00

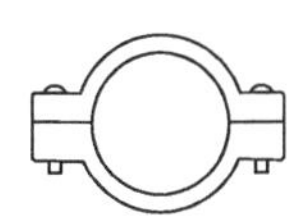

Class 125 iron body gate valve, flanged ends

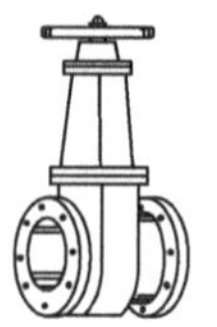

Description	Craft@Hrs	Unit	Material $	Labor $	Equipment $	Total $
2"	P1@.500	Ea	431.00	22.20	—	453.20
2½"	P1@.600	Ea	581.00	26.70	—	607.70
3"	P1@.750	Ea	630.00	33.30	—	663.30
4"	P1@1.35	Ea	925.00	60.00	—	985.00
6"	ER@2.50	Ea	2,670.00	119.00	1.67	2,790.67
8"	ER@3.00	Ea	4,330.00	142.00	2.01	4,474.01
10"	ER@4.00	Ea	7,150.00	190.00	2.68	7,342.68
12"	ER@4.50	Ea	9,790.00	213.00	3.01	10,006.01

Class 125 iron body globe valve, flanged joints

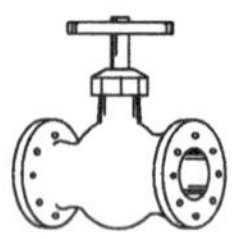

Description	Craft@Hrs	Unit	Material $	Labor $	Equipment $	Total $
2"	P1@.500	Ea	328.00	22.20	—	350.20
2½"	P1@.600	Ea	488.00	26.70	—	514.70
3"	P1@.750	Ea	586.00	33.30	—	619.30
4"	ER@1.35	Ea	790.00	64.00	—	854.00
6"	ER@2.50	Ea	1,460.00	119.00	1.67	1,580.67
8"	ER@3.00	Ea	2,210.00	142.00	2.01	2,354.01
10"	ER@4.00	Ea	3,380.00	190.00	2.68	3,572.68

200 PSIG iron body butterfly valve, lug-type, lever operated

Description	Craft@Hrs	Unit	Material $	Labor $	Equipment $	Total $
2"	P1@.450	Ea	179.00	20.00	—	199.00
2½"	P1@.450	Ea	186.00	20.00	—	206.00
3"	P1@.550	Ea	196.00	24.40	—	220.40
4"	P1@.550	Ea	244.00	24.40	—	268.40
6"	ER@.800	Ea	410.00	37.90	.54	448.44
8"	ER@.800	Ea	560.00	37.90	.54	598.44
10"	ER@.900	Ea	779.00	42.70	.60	822.30
12"	ER@1.00	Ea	1,020.00	47.40	.67	1,068.07

200 PSIG iron body butterfly valve, wafer-type, lever operated

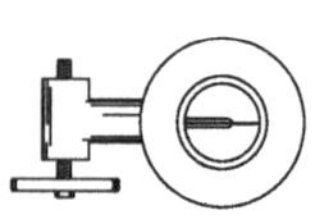

Description	Craft@Hrs	Unit	Material $	Labor $	Equipment $	Total $
2"	P1@.450	Ea	164.00	20.00	—	184.00
2½"	P1@.450	Ea	167.00	20.00	—	187.00
3"	P1@.550	Ea	179.00	24.40	—	203.40
4"	P1@.550	Ea	217.00	24.40	—	241.40
6"	ER@.800	Ea	380.00	37.90	.54	418.44
8"	ER@.800	Ea	526.00	37.90	.54	564.44
10"	ER@.900	Ea	735.00	42.70	.60	778.30
12"	ER@1.00	Ea	966.00	47.40	.67	1,014.07

Class 125 iron body swing check valve, flanged ends

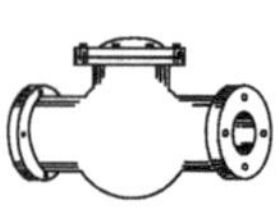

Description	Craft@Hrs	Unit	Material $	Labor $	Equipment $	Total $
2"	P1@.500	Ea	213.00	22.20	—	235.20
2½"	P1@.600	Ea	269.00	26.70	—	295.70
3"	P1@.750	Ea	336.00	33.30	—	369.30
4"	P1@1.35	Ea	492.00	60.00	—	552.00
6"	ER@2.50	Ea	956.00	119.00	1.67	1,076.67
8"	ER@3.00	Ea	1,720.00	142.00	2.01	1,864.01
10"	ER@4.00	Ea	2,800.00	190.00	2.68	2,992.68
12"	ER@4.50	Ea	3,840.00	213.00	3.01	4,056.01

Description	Craft@Hrs	Unit	Material $	Labor $	Equipment $	Total $

Class 125 iron body silent check valve, flanged joints

2"	P1@.500	Ea	156.00	22.20	—	178.20
2½"	P1@.600	Ea	173.00	26.70	—	199.70
3"	P1@.750	Ea	199.00	33.30	—	232.30
4"	P1@1.35	Ea	265.00	60.00	—	325.00
5"	ER@2.00	Ea	384.00	94.80	1.34	480.14
6"	ER@2.50	Ea	475.00	119.00	1.67	595.67
8"	ER@3.00	Ea	856.00	142.00	2.01	1,000.01
10"	ER@4.00	Ea	1,320.00	190.00	2.68	1,512.68

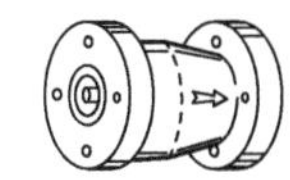

Class 125 iron body strainer, flanged

2"	P1@.500	Ea	168.00	22.20	—	190.20
2½"	P1@.600	Ea	187.00	26.70	—	213.70
3"	P1@.750	Ea	218.00	33.30	—	251.30
4"	P1@1.35	Ea	367.00	60.00	—	427.00
6"	ER@2.50	Ea	709.00	119.00	1.67	829.67
8"	ER@3.00	Ea	1,190.00	142.00	2.01	1,334.01

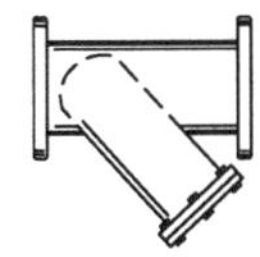

Installation of 2-way control valve, flanged joints

2"	P1@.500	Ea	—	22.20	—	22.20
2½"	P1@.600	Ea	—	26.70	—	26.70
3"	P1@.750	Ea	—	33.30	—	33.30
4"	P1@1.35	Ea	—	60.00	—	60.00
6"	P1@2.50	Ea	—	111.00	—	111.00
8"	P1@3.00	Ea	—	133.00	—	133.00
10"	P1@4.00	Ea	—	178.00	—	178.00
12"	P1@4.50	Ea	—	200.00	—	200.00

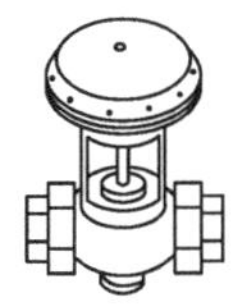

Installation of 3-way control valve, flanged joints

2"	P1@.910	Ea	—	40.40	—	40.40
2½"	P1@1.12	Ea	—	49.80	—	49.80
3"	P1@1.33	Ea	—	59.10	—	59.10
4"	P1@2.00	Ea	—	88.90	—	88.90
6"	P1@3.70	Ea	—	164.00	—	164.00
8"	P1@4.40	Ea	—	196.00	—	196.00
10"	P1@5.90	Ea	—	262.00	—	262.00
12"	P1@6.50	Ea	—	289.00	—	289.00

Bolt and gasket sets

2"	P1@.500	Ea	4.64	22.20	—	26.84
2½"	P1@.650	Ea	5.42	28.90	—	34.32
3"	P1@.750	Ea	9.19	33.30	—	42.49
4"	P1@1.00	Ea	15.40	44.50	—	59.90
6"	P1@1.20	Ea	25.30	53.30	—	78.60
8"	P1@1.25	Ea	28.40	55.60	—	84.00
10"	P1@1.70	Ea	48.90	75.60	—	124.50
12"	P1@2.20	Ea	55.10	97.80	—	152.90

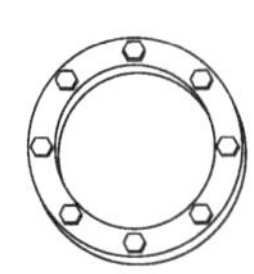

Description	Craft@Hrs	Unit	Material $	Labor $	Equipment $	Total $

Thermometer with well

7"	P1@.250	Ea	186.00	11.10	—	197.10
9"	P1@.250	Ea	192.00	11.10	—	203.10

Dial-type pressure gauge

2½"	P1@.200	Ea	38.50	8.89	—	47.39
3½"	P1@.200	Ea	51.20	8.89	—	60.09

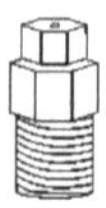

Pressure/temperature tap

Tap	P1@.150	Ea	17.70	6.67	—	24.37

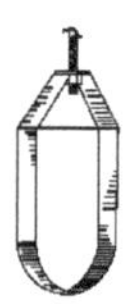

Hanger with swivel assembly

2"	P1@.300	Ea	5.99	13.30	—	19.29
2½"	P1@.350	Ea	8.15	15.60	—	23.75
3"	P1@.350	Ea	10.10	15.60	—	25.70
4"	P1@.350	Ea	11.10	15.60	—	26.70
5"	P1@.450	Ea	20.10	20.00	—	40.10
6"	P1@.450	Ea	25.80	20.00	—	45.80
8"	P1@.450	Ea	34.40	20.00	—	54.40
10"	P1@.550	Ea	47.40	24.40	—	71.80
12"	P1@.600	Ea	63.00	26.70	—	89.70

Riser clamp

2"	P1@.115	Ea	6.62	5.11	—	11.73
2½"	P1@.120	Ea	6.93	5.33	—	12.26
3"	P1@.120	Ea	7.53	5.33	—	12.86
4"	P1@.125	Ea	9.61	5.56	—	15.17
5"	P1@.180	Ea	13.60	8.00	—	21.60
6"	P1@.200	Ea	16.50	8.89	—	25.39
8"	P1@.200	Ea	26.80	8.89	—	35.69
10"	P1@.250	Ea	40.00	11.10	—	51.10
12"	P1@.250	Ea	47.30	11.10	—	58.40

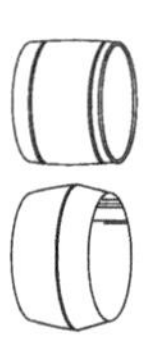

Galvanized steel pipe sleeves

2"	P1@.130	Ea	6.04	5.78	—	11.82
2½"	P1@.150	Ea	6.12	6.67	—	12.79
3"	P1@.180	Ea	6.27	8.00	—	14.27
4"	P1@.220	Ea	7.16	9.78	—	16.94
5"	P1@.250	Ea	8.86	11.10	—	19.96
6"	P1@.270	Ea	9.56	12.00	—	21.56
8"	P1@.270	Ea	11.00	12.00	—	23.00
10"	P1@.290	Ea	12.70	12.90	—	25.60
12"	P1@.310	Ea	17.50	13.80	—	31.30
14"	P1@.330	Ea	17.50	14.70	—	32.20

Description	Craft@Hrs	Unit	Material $	Labor $	Equipment $	Total $

Leadership in Energy & Environmental Design (LEED).

LEED offers an incentive to building owners to install energy-efficient and environmentally sensitive HVAC equipment. The LEED program awards points for application of the best available technology. The HVAC system can earn a maximum of 17 points toward LEED certification. A LEED-certified air conditioning system must comply with ASHRAE 90.1-2004 (efficiency) and ASHRAE 62.1-2004 (indoor air quality) standards. LEED certification also requires recording controls (Carrier ComfortView 3 or equal) to measure system performance. Zone sensor thermostats (Carrier Debonair or equal) are required to provide performance and efficiency data to the digital recorder. An indoor air quality CO_2 sensor, duct-mounted aspirator modules and a refrigerant such as Puron R-410A are also required.

Description	Craft@Hrs	Unit	Material $	Labor $	Equipment $	Total $
Add for a LEED-certified system with USGBC rating	—	%	20.0	—	—	—
Add for LEED registration and inspection	—	Ea	2,070.00	—	—	2,070.00
Add for LEED central digital performance monitor/recorder, ComfortView 3 or equal	—	Ea	2,600.00	—	—	2,600.00
Add for LEED-certified zone recording thermostats, Carrier DebonAir, USB remote-ready	—	Ea	385.00	—	—	385.00
Add for IAQ (Indoor Air Quality) CO_2 sensor and duct-mounted aspirator box	—	Ea	471.00	—	—	471.00

Residential A/C Cooling System.

With remote exterior condenser/compressor, field installed DX coil in existing supply air plenum. Includes electrical & control wiring, R-410A refrigerant piping & connections, programmable digital thermostat, start-up and testing. Costs based on a maximum distance between outdoor condenser and coil of 60' and include a contractor mark up of 25%. Use these figures for preliminary budget purposes. Costs are based on using 650 SF per ton as the cooling requirement with a thermal envelope of R20 or greater, i.e., a 1.5-ton A/C unit will serve a 975 SF home. Typical subcontract prices.

Description	Craft@Hrs	Unit	Material $	Labor $	Equipment $	Total $
1.5 tons	—	Ea	—	—	—	3,830.00
2 tons	—	Ea	—	—	—	4,200.00
2.5 tons	—	Ea	—	—	—	5,110.00
3 tons	—	Ea	—	—	—	6,970.00
5 tons	—	Ea	—	—	—	10,500.00
Add for high heat loads	—	%	—	—	—	12.0

Description	Craft@Hrs	Unit	Material $	Labor $	Equipment $	Total $

Residential forced-air heating, cooling & ventilation system. Including a remote exterior condenser/compressor, field installed R-410A refrigerant DX "A" coil in supply air plenum, a gas-fired high efficiency furnace, programmable digital thermostat, ductwork, grilles & registers, HRV (heat recovery ventilator), kitchen and bathroom exhaust fans, ventilation and exhaust ductwork and wall hoods, electrical & control wiring, gas, refrigerant & flue piping & associated connections, start-up, testing, and a contractor markup of 25%. Use these figures for preliminary budget purposes. Heating & cooling load calculations based on a thermal envelope of R20 or greater. Typical subcontract prices.

900 – 1,000 SF	—	Ea	—	—	—	19,400.00
1,100 – 1,400 SF	—	Ea	—	—	—	22,000.00
1,500 – 1,750 SF	—	Ea	—	—	—	27,300.00
1,800 – 2000 SF	—	Ea	—	—	—	29,100.00
2,200 – 2,500 SF	—	Ea	—	—	—	32,700.00
2,600 – 3,000 SF	—	Ea	—	—	—	39,500.00
Deduct for mid-efficient unit	—	Ea	—	—	—	-347.00
Deduct for electric furnace	—	Ea	—	—	—	-347.00

Description	Craft&Hrs	Unit	Material $	Labor $	Equipment $	Total $

Packaged self-contained rooftop DX air conditioning units. EnergySmart certified, ASHRAE, U.S. Green Building Council and Underwriters Laboratories approved. Includes cooling coils, compressor, heat rejection coils, regulator valves, refrigerant tank and remote digital single-zone control package. Set in place with a 13,000 lb. truck crane. Add installation costs from the section that follows.

2-T, 800 CFM	SN@5.00	Ea	4,770.00	232.00	54.90	5,056.90
2½-T, 1,000 CFM	SN@6.00	Ea	5,710.00	278.00	65.90	6,053.90
3-T, 1,200 CFM	SN@7.00	Ea	6,830.00	324.00	76.90	7,230.90
4-T, 1,600 CFM	SN@8.00	Ea	8,360.00	371.00	87.80	8,818.80
5-T, 2,000 CFM	SN@9.00	Ea	10,400.00	417.00	98.80	10,915.80
7½-T, 3,000 CFM	SN@10.0	Ea	14,500.00	464.00	110.00	15,074.00
10-T, 4,000 CFM	SN@12.0	Ea	18,100.00	556.00	132.00	18,788.00
12-T, 5,000 CFM	SN@14.0	Ea	21,500.00	649.00	154.00	22,303.00
15-T, 6,000 CFM	SN@15.0	Ea	26,400.00	695.00	165.00	27,260.00
20-T, 8,000 CFM	SN@16.0	Ea	35,000.00	742.00	176.00	35,918.00
25-T, 10,000 CFM	SN@17.0	Ea	42,600.00	788.00	187.00	43,575.00
30-T, 12,000 CFM	SN@19.0	Ea	48,700.00	881.00	209.00	49,790.00
40-T, 16,000 CFM	SN@22.0	Ea	60,800.00	1,020.00	242.00	62,062.00

Rooftop DX air conditioning unit, hot water coil. EnergySmart certified, ASHRAE, U.S. Green Building Council, and Underwriters Laboratories approved. Add for the hot water coil. Single zone controls. Add installation costs from the section that follows.

2-T, 800 CFM	SN@5.00	Ea	4,650.00	232.00	54.90	4,936.90
2½-T, 1,000 CFM	SN@6.00	Ea	5,590.00	278.00	65.90	5,933.90
3-T, 1,200 CFM	SN@7.00	Ea	6,690.00	324.00	76.90	7,090.90
4-T, 1,600 CFM	SN@8.00	Ea	8,150.00	371.00	87.80	8,608.80
5-T, 2,000 CFM	SN@9.00	Ea	10,200.00	417.00	98.80	10,715.80
7½-T, 3,000 CFM	SN@10.0	Ea	15,400.00	464.00	110.00	15,974.00
10-T, 4,000 CFM	SN@12.0	Ea	17,700.00	556.00	132.00	18,388.00
12-T, 5,000 CFM	SN@14.0	Ea	20,900.00	649.00	154.00	21,703.00
15-T, 6,000 CFM	SN@15.0	Ea	25,700.00	695.00	165.00	26,560.00
20-T, 8,000 CFM	SN@16.0	Ea	34,300.00	742.00	176.00	35,218.00
25-T, 10,000 CFM	SN@17.0	Ea	41,300.00	788.00	187.00	42,275.00
30-T, 12,000 CFM	SN@19.0	Ea	48,700.00	881.00	209.00	49,790.00
40-T, 16,000 CFM	SN@22.0	Ea	60,800.00	1,020.00	242.00	62,062.00

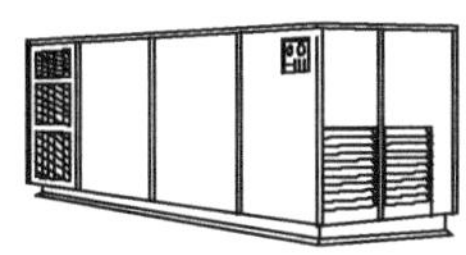

Description	Craft@Hrs	Unit	Material $	Labor $	Equipment $	Total $
Installation costs for packaged rooftop cooling Typical installation costs for packaged rooftop cooling. Costs will be higher for larger capacity units.						
Cut, frame and gasket downcomer hole in roof	S2@1.00	Ea	29.80	43.50	—	73.30
Mount duct hangers	S2@.250	Ea	4.50	10.90	—	15.40
Cut and mount sheet metal duct	S2@.350	LF	7.13	15.20	—	22.33
Apply and coat duct insulation	S2@.150	LF	3.06	6.53	—	9.59
Install piping for gas line	P1@.100	LF	5.39	4.45	—	9.84
Install piping for chilled or hot water/steam line	P1@.100	LF	4.09	4.45	—	8.54
Install electrical wiring	BE@.150	LF	2.19	7.25	—	9.44
Install HVAC controls	BE@.500	Ea	—	24.20	—	24.20
Commission and test	P1@4.00	Ea	—	178.00	—	178.00
Air balance and fine-tune	P1@4.00	Ea	—	178.00	—	178.00

Packaged air handler with chilled water and hot water/steam coil. Costs shown based on 400 CFM per ton cooling. Unit includes insulated single wall casing, fan section, cooling coil section, heating coil section, mixing plenum section, bag filter section, fan motor, variable pitch drive, vibration isolators and drain pan. Set in place only. Make additional allowances for coil connections, controls, motor starters and power wiring. (12,000 BTUs equals 1 ton cooling.) Use these costs for preliminary estimates.

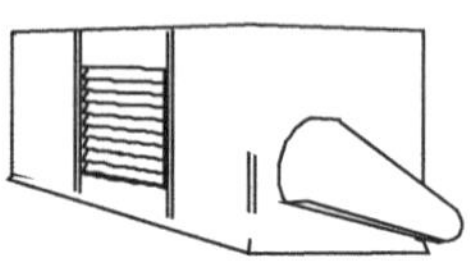

Description	Craft@Hrs	Unit	Material $	Labor $	Equipment $	Total $
3-T, 1,200 CFM	SN@4.00	Ea	3,850.00	185.00	43.90	4,078.90
4-T, 1,600 CFM	SN@5.50	Ea	5,300.00	255.00	60.40	5,615.40
5-T, 2,000 CFM	SN@7.00	Ea	6,300.00	324.00	76.90	6,700.90
7½-T, 3,000 CFM	SN@9.00	Ea	8,890.00	417.00	98.80	9,405.80
10-T, 4,000 CFM	SN@11.0	Ea	11,300.00	510.00	121.00	11,931.00
12-T, 5,000 CFM	SN@13.0	Ea	13,300.00	603.00	143.00	14,046.00
15-T, 6,000 CFM	SN@14.0	Ea	15,100.00	649.00	154.00	15,903.00
20-T, 8,000 CFM	SN@15.0	Ea	17,900.00	695.00	165.00	18,760.00
25-T, 10,000 CFM	SN@16.0	Ea	17,800.00	742.00	176.00	18,718.00
30-T, 12,000 CFM	SN@18.0	Ea	22,300.00	834.00	198.00	23,332.00
40-T, 16,000 CFM	SN@21.0	Ea	27,500.00	973.00	231.00	28,704.00

Air handling unit accessories and options. Use these costs for preliminary estimates.

Description	Craft@Hrs	Unit	Material $	Labor $	Equipment $	Total $
DDC controls per zone complete	SN@2.75	Ea	1,190.00	127.00	—	1,317.00
Electric controls per zone complete	SN@2.75	Ea	637.00	127.00	—	764.00
Pneumatic controls per zone complete	SN@2.75	Ea	599.00	127.00	—	726.00
Variable speed drive 5 HP	SN@4.00	Ea	1,880.00	185.00	—	2,065.00
Variable speed drive 7.5HP	SN@4.00	Ea	2,750.00	185.00	—	2,935.00
Variable speed drive 10 HP	SN@6.00	Ea	3,440.00	278.00	—	3,718.00
Variable speed drive 15 HP	SN@8.00	Ea	4,580.00	371.00	—	4,951.00
Variable speed drive 20 HP	SN@12.0	Ea	5,490.00	556.00	—	6,046.00
Variable speed drive 25 HP	SN@16.0	Ea	6,470.00	742.00	—	7,212.00
Variable speed drive 30 HP	SN@18.0	Ea	7,630.00	834.00	—	8,464.00
Variable speed drive 40 HP	SN@22.0	Ea	9,420.00	1,020.00	—	10,440.00
Variable speed drive 50 HP	SN@26.0	Ea	12,000.00	1,210.00	—	13,210.00
Variable inlet vanes 1,000-1,500 CFM	SN@4.00	Ea	247.00	185.00	2.68	434.68
Variable inlet vanes 1,600-2,500 CFM	SN@4.00	Ea	536.00	185.00	2.68	723.68
Variable inlet vanes 2,600-5,000 CFM	SN@6.00	Ea	1,500.00	278.00	4.02	1,782.02
Variable inlet vanes 6,000-10,000 CFM	SN@8.00	Ea	2,470.00	371.00	5.36	2,846.36
Variable inlet vanes 11,000-20,000 CFM	SN@10.0	Ea	4,090.00	464.00	6.69	4,560.69

Heat Recovery Ventilators — Commercial

Heat Recovery Ventilators provide a fresh air supply to tightly sealed building envelopes. An HRV extracts heat from the stale indoor air being exhausted and transfers the heat to the fresh air being drawn into the building through the HRV. Heat recovery ventilators are also excellent dehumidifiers.

HRV unit features include energy-efficient defrost cycle, cross-flow polypropylene heat exchanger, acoustically lined cabinet, outdoor air filter.

Allow 15-30 cfm per person or .04 to.05 cfm per square foot when sizing HRV unit. (ASHRAE62-19890) (cfm = cubic feet per minute)

Description	Craft@Hrs	Unit	Material $	Labor $	Equipment $	Total $

Commercial heat recovery ventilator. Set/hang in place only. Make additional allowances for duct, diffusers, controls, air balancing, electrical connections and condensate drain.

Description	Craft@Hrs	Unit	Material $	Labor $	Equipment $	Total $
HRV 700 cfm	SN@4.75	Ea	7,100.00	220.00	81.10	7,401.10
HRV 1,200 cfm	SN@5.25	Ea	7,660.00	243.00	89.60	7,992.60
HRV 2,500 cfm	SN@6.75	Ea	31,500.00	313.00	115.00	31,928.00

Swimming pool heat recovery ventilator. Set/hang in place only. Make additional allowances for duct, diffusers, controls, air balancing, electrical connections and condensate drain.

Description	Craft@Hrs	Unit	Material $	Labor $	Equipment $	Total $
HRV 700 cfm	SN@5.00	Ea	8,220.00	232.00	85.40	8,537.40
HRV 1,200 cfm	SN@5.75	Ea	9,330.00	267.00	98.20	9,695.20

Description	Craft@Hrs	Unit	Material $	Labor $	Equipment $	Total $

Conventional heat recovery ventilator. Hang in place only. Make additional allowances for duct, diffusers, controls, air balancing, electrical connections and condensate drain.

HRV 65–150 cfm	S2@2.45	Ea	1,310.00	107.00	—	1,417.00
HRV 115–200 cfm	S2@2.65	Ea	1,540.00	115.00	—	1,655.00

Compact heat recovery ventilator. Hang in place only. Make additional allowances for duct, diffusers, controls, air balancing, electrical connections and condensate drain.

HRV 65–127 cfm	S2@2.45	Ea	1,400.00	107.00	—	1,507.00
HRV 115–195 cfm	S2@2.65	Ea	1,520.00	115.00	—	1,635.00

High-efficiency heat recovery ventilator. Hang in place only. Make additional allowances for duct, diffusers, controls, air balancing, electrical connections and condensate drain.

HRV 65–127 cfm	S2@2.65	Ea	1,530.00	115.00	—	1,645.00
HRV 115–180cfm	S2@2.90	Ea	2,350.00	126.00	—	2,476.00
HRV 180–265cfm	S2@3.25	Ea	2,440.00	141.00	—	2,581.00

Heat recovery ventilator controls

HRV basic control	S2@1.00	Ea	118.00	43.50	—	161.50
HRV std. control	S2@1.00	Ea	190.00	43.50	—	233.50
HRV auto control	S2@1.00	Ea	257.00	43.50	—	300.50
60 minute timer	S2@.600	Ea	79.50	26.10	—	105.60
Interlock relay	S2@.600	Ea	94.50	26.10	—	120.60

Heat recovery ventilator accessories

6" diffusers	S2@.500	Ea	20.20	21.80	—	42.00
6" wall hoods	S2@1.50	Ea	34.90	65.30	—	100.20
6" tee fittings	S2@.350	Ea	4.67	15.20	—	19.87
6" flex duct	S2@.025	LF	1.11	1.09	—	2.20
6" flex insul duct	S2@.025	LF	2.12	1.09	—	3.21
HRV filters	S2@.450	Ea	28.50	19.60	—	48.10

Description	Craft@Hrs	Unit	Material $	Labor $	Equipment $	Total $
Installation costs for heat recovery ventilators						
Hang heat recovery ventilator, 150 CFM	S2@2.00	Ea	5.76	87.00	—	92.76
Hang heat recovery ventilator, 250 CMF	S2@2.25	LF	8.63	97.90	—	106.53
Heat recovery ventilator ducting, 150 CFM	S2@4.00	LF	384.00	174.00	—	558.00
Heat recovery ventilator ducting, 250 CFM	S2@4.75	LF	432.00	207.00	—	639.00
Heat recovery ventilator drain line, gravity	P1@1.50	LF	28.90	66.70	—	95.60
Heat recovery ventilator drain line, pumped	P1@2.00	LF	259.00	88.90	—	347.90
Heat recovery ventilator power wiring, 115 volt	BE@2.00	Ea	97.90	96.70	—	194.60
Heat recovery ventilator power wiring, 24 volt	BE@2.00	Ea	23.00	96.70	—	119.70
Commission and test	P1@1.50	Ea	—	66.70	—	66.70
Run air balance and fine tune system	P1@1.75	Ea	—	77.80	—	77.80

Engineering drawings of coil piping details have a bad reputation among HVAC contractors and estimators. They're notorious for what they leave out. They'll rarely show more than one coil bank, no matter how big the system. Furthermore, the drawings hardly ever call out sizes for either the piping or the control valves. Don't be taken in by the apparent simplicity of the system as shown in these drawings. It's likely to be only the tip of the iceberg. For example, unless the air handling capacity of the system is less than 16,000 CFM, the single coil bank shown won't be adequate. Add one or two more coil banks and you're looking at a lot more piping — and a more complex system that takes longer to install.

You probably haven't even decided which equipment supplier to use. This is hardly the time for you to start researching heating and cooling coils. Nevertheless, you need better, more complete and realistic data to come up with a competitive estimate.

I'll tell you how and where to track down the hard data that you need, and also pass along a few tips on estimating water coil piping. They'll help you avoid leaving something out of your estimates — a real pitfall for any beginner. It's those little (but essential) items, so easily overlooked, that are so deadly to a profit margin. Finally, check the next two pages of diagrams with tables. The data given there, combined with the data you collected earlier, forms the basis for informed guesswork.

The Hard Data

There are two things you absolutely must know to estimate water coil piping. First, the size of the branch piping to the coils; second, the CFM rating of the air handling units for the system.

To find the pipe sizes, look at either the floor plans or the details for the equipment room. They'll list the sizes of the branch run-out pipes. Once you know them, you can make a good guess at the right size for the control valves. (See the diagrams and tables on pages 295 and 296.) The only information you need is the CFM ratings of the air handling units and the branch piping sizes to the coils. If the system's capacity is over 16,000 CFM, you need two or three coil banks.

A Few Tips on Estimating Water Coil Piping

1) Coil connection sizes seldom match branch pipe sizes. Be sure to include the reducing fittings you'll need in any estimate.

2) Two-way and three-way control valves are both usually one pipe size smaller than the pipe where they're installed. That means you'll need either two or three reducing fittings per control valve. Be sure you include their cost in your estimates.

Using the Diagrams

In the following diagrams, for clarity, some items aren't included. These items are: balance valves, shut-off valves, reducers, strainers, gauges and gauge taps. Any details you need about these items for your estimate are in the engineer's coil piping details.

Water Coil Piping

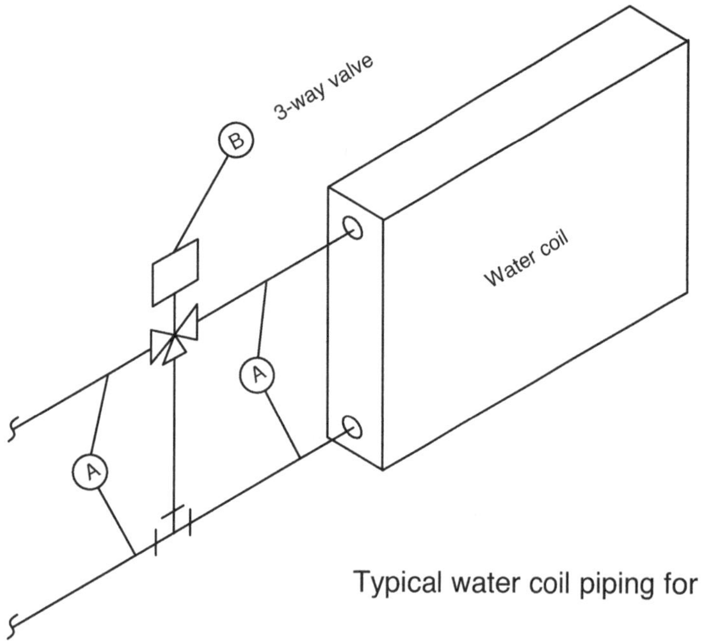

If (A) is:	Then (B) is:
1"	¾"
1¼"	1"
1½"	1¼"
2"	1½"
2½"	2"
3"	2½"

Typical water coil piping for A.H. units up to 16,000 CFM

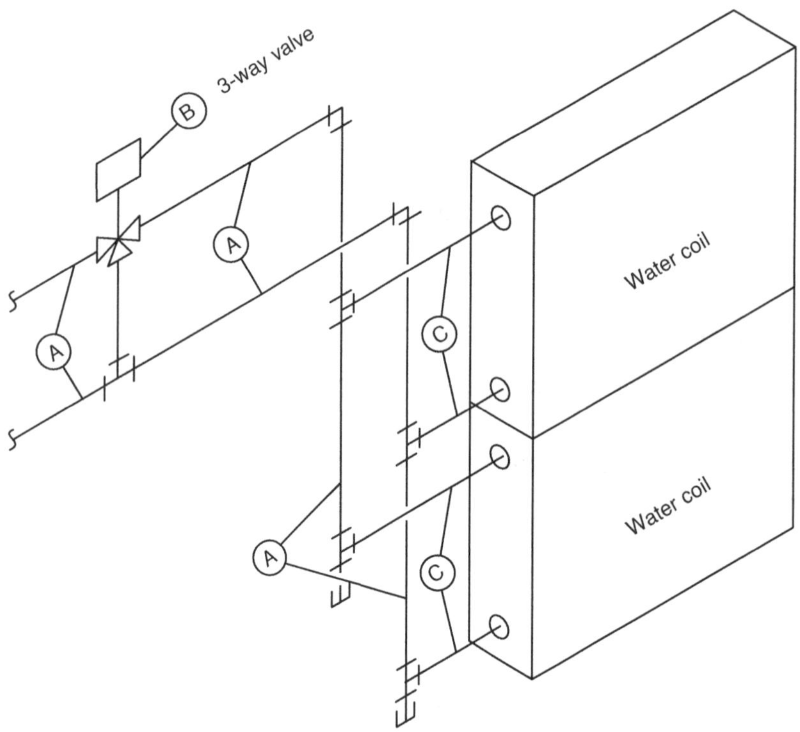

If (A) is:	Then (B) is:	And (C) is:
2½"	2"	2"
3"	2½"	2½"
4"	3"	3"

Typical water coil piping for A.H. units from 16,000 to 26,000 CFM

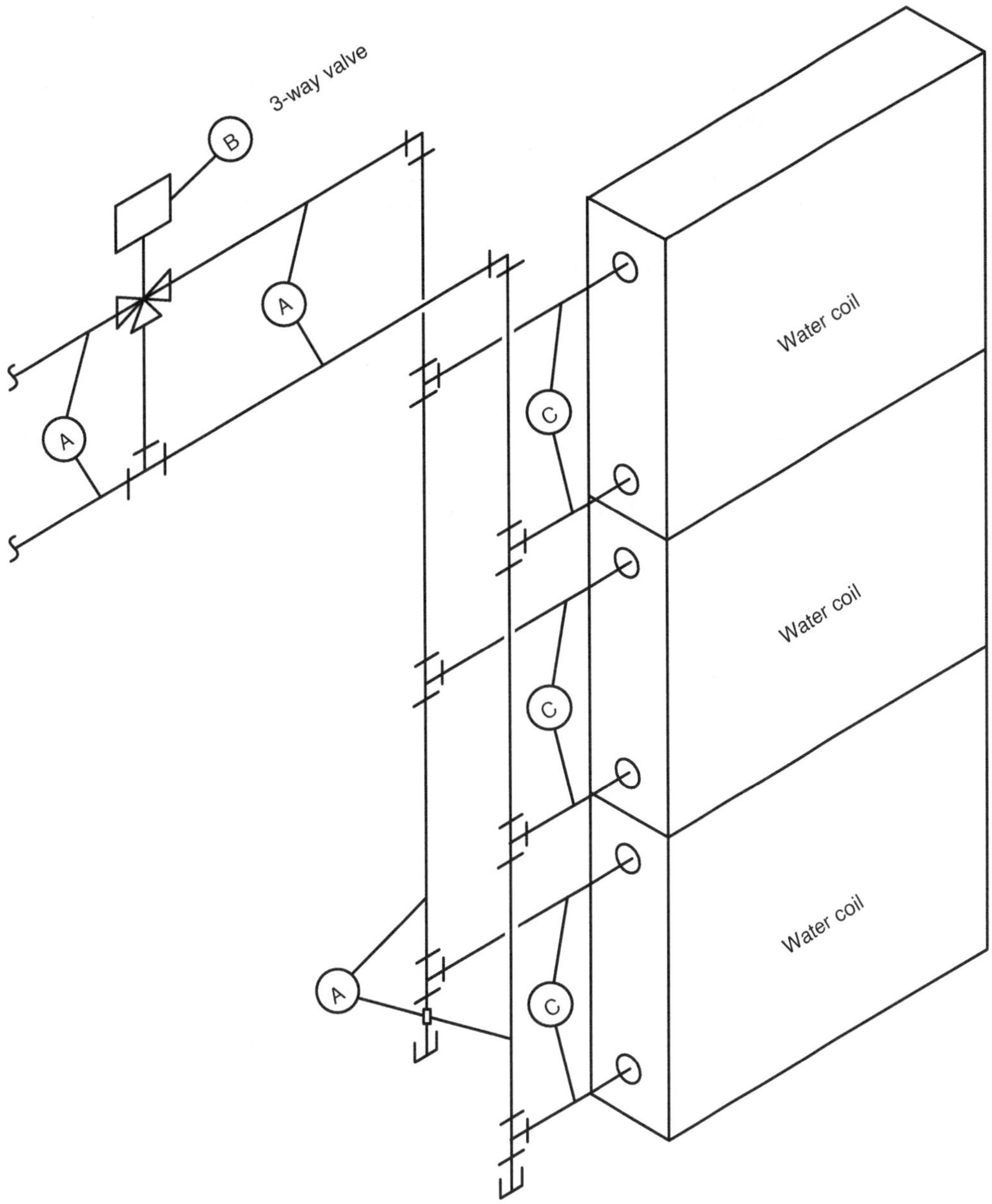

If (A) is:	Then (B) is:	And (C) is:
4"	3"	2½"
6"	4"	3"
8"	6"	4"
10"	8"	6"

Typical water coil piping for A.H. units over 26,000 CFM

Description	Craft@Hrs	Unit	Material $	Labor $	Equipment $	Total $

Air handling unit coil connection, one row coil bank, non-regulated flow. Connection assembly includes pipe, fittings, pipe insulation, valves, gauges, thermometers and vents.

1½" supply	SN@14.0	Ea	2,080.00	649.00	—	2,729.00
2" supply	SN@18.0	Ea	2,560.00	834.00	—	3,394.00
2½" supply	SN@28.0	Ea	3,420.00	1,300.00	—	4,720.00
3" supply	SN@31.5	Ea	3,750.00	1,460.00	—	5,210.00
4" supply	SN@34.0	Ea	4,020.00	1,580.00	—	5,600.00
6" supply	SN@41.0	Ea	5,820.00	1,900.00	27.40	7,747.40

Air handling unit coil connection, one row coil bank, 2-way control valve design. Connection assembly includes pipe and fittings, pipe insulation, valves, gauges, thermometers and vents.

1½" supply	SN@16.5	Ea	2,350.00	765.00	—	3,115.00
2" supply	SN@21.0	Ea	3,030.00	973.00	—	4,003.00
2½" supply	SN@33.6	Ea	3,690.00	1,560.00	—	5,250.00
3" supply	SN@36.0	Ea	3,960.00	1,670.00	—	5,630.00
4" supply	SN@43.8	Ea	4,810.00	2,030.00	—	6,840.00
6" supply	SN@54.0	Ea	7,700.00	2,500.00	36.10	10,236.10

Air handling unit coil connection, one row coil bank, 3-way control valve design. Connection assembly includes pipe and fittings, pipe insulation, valves, gauges, thermometers and vents.

1½" supply	SN@18.0	Ea	2,710.00	834.00	—	3,544.00
2" supply	SN@26.0	Ea	3,380.00	1,210.00	—	4,590.00
2½" supply	SN@41.5	Ea	4,250.00	1,920.00	—	6,170.00
3" supply	SN@43.8	Ea	4,670.00	2,030.00	—	6,700.00
4" supply	SN@54.6	Ea	6,020.00	2,530.00	—	8,550.00
6" supply	SN@61.4	Ea	8,580.00	2,850.00	41.10	11,471.10

Air handling unit coil connection, two row coil bank, non-regulated flow. Connection assembly includes pipe and fittings, pipe insulation, valves, gauges, thermometers and vents.

2½" supply	SN@48.0	Ea	5,140.00	2,220.00	—	7,360.00
3" supply	SN@51.5	Ea	6,250.00	2,390.00	—	8,640.00
4" supply	SN@63.6	Ea	8,270.00	2,950.00	—	11,220.00
6" supply	SN@79.2	Ea	12,100.00	3,670.00	53.00	15,823.00

Air handling unit coil connection, two row coil bank, 2-way control valve design. Connection assembly includes pipe and fittings, pipe insulation, valves, gauges, thermometers and vents.

2½" supply	SN@54.0	Ea	5,750.00	2,500.00	—	8,250.00
3" supply	SN@58.5	Ea	7,220.00	2,710.00	—	9,930.00
4" supply	SN@67.0	Ea	8,890.00	3,110.00	—	12,000.00
6" supply	SN@83.5	Ea	12,400.00	3,870.00	55.90	16,325.90

Description	Craft@Hrs	Unit	Material $	Labor $	Equipment $	Total $

Air handling unit coil connection, two row coil bank, 3-way control valve design. Connection assembly includes pipe and fittings, pipe insulation, valves, gauges, thermometers and vents.

2½" supply	SN@58.8	Ea	6,020.00	2,730.00	—	8,750.00
3" supply	SN@64.0	Ea	7,600.00	2,970.00	—	10,570.00
4" supply	SN@72.0	Ea	9,620.00	3,340.00	—	12,960.00
6" supply	SN@87.0	Ea	12,800.00	4,030.00	58.20	16,888.20

Air handling unit coil connection, three-row coil bank, non-regulated flow. Connection assembly includes pipe and fittings, pipe insulation, valves, gauges, thermometers and vents.

4" supply	SN@72.0	Ea	8,710.00	3,340.00	—	12,050.00
6" supply	SN@94.0	Ea	15,200.00	4,360.00	—	19,560.00
8" supply	SN@120.	Ea	20,300.00	5,560.00	80.30	25,940.30

Air handling unit coil connection, three-row coil bank, 2-way control valve design. Connection assembly includes pipe and fittings, pipe insulation, valves, gauges, thermometers and vents.

4" supply	SN@78.0	Ea	9,260.00	3,620.00	—	12,880.00
6" supply	SN@105.	Ea	16,700.00	4,870.00	—	21,570.00
8" supply	SN@132.	Ea	22,800.00	6,120.00	88.40	29,008.40

Air handling unit coil connection, three-row coil bank, 3-way control valve design. Connection assembly includes pipe and fittings, pipe insulation, valves, gauges, thermometers and vents.

4" supply	SN@82.0	Ea	10,700.00	3,800.00	—	14,500.00
6" supply	SN@112.	Ea	17,200.00	5,190.00	—	22,390.00
8" supply	SN@140.	Ea	25,600.00	6,490.00	93.70	32,183.70

Description	Craft@Hrs	Unit	Material $	Labor $	Equipment $	Total $

Residential gas-fired upflow furnace. Set in place only. Make additional allowances for gas and electrical connections. Costs shown per thousand BTU per hour (MBH) Mid-efficiency, 80% AFUE.

50 MBH input	SN@2.00	Ea	1,040.00	92.70	—	1,132.70
75 MBH input	SN@2.00	Ea	1,360.00	92.70	2.34	1,455.04
100 MBH input	SN@2.00	Ea	1,580.00	92.70	2.34	1,675.04
125 MBH input	SN@2.00	Ea	1,820.00	92.70	2.34	1,915.04

Residential combination gas-fired furnace and A/C unit. With remote exterior condenser/compressor. Set in place only. Make additional allowances for gas, electrical and DX connections. Costs shown per thousand BTU per hour (MBH) Mid-efficiency, 80% AFUE.

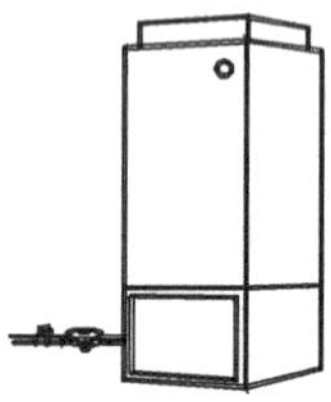

30 MBH, 1.5 tons	SN@2.50	Ea	3,200.00	116.00	—	3,316.00
36 MBH, 2 tons	SN@2.75	Ea	3,800.00	127.00	—	3,927.00
42 MBH, 2 tons	SN@3.00	Ea	4,270.00	139.00	—	4,409.00
48 MBH, 2.5 tons	SN@3.50	Ea	4,970.00	162.00	—	5,132.00
60 MBH, 3 tons	SN@4.00	Ea	5,940.00	185.00	—	6,125.00

High-efficiency, condensing gas furnace. Direct drive, atmospheric draft, electronic ignition. Set in place only. Make additional allowances for flue, gas and electrical connections. Costs shown per thousand BTU per hour (MBH) High efficiency 94% AFUE.

40 MBH input	S2@2.00	Ea	2,170.00	87.00	—	2,257.00
60 MBH input	S2@2.00	Ea	2,170.00	87.00	—	2,257.00
70 MBH input	S2@2.00	Ea	2,070.00	87.00	—	2,157.00
90 MBH input	S2@2.00	Ea	2,070.00	87.00	—	2,157.00
100 MBH input	S2@2.00	Ea	2,950.00	87.00	—	3,037.00
120 MBH input	S2@2.00	Ea	3,050.00	87.00	—	3,137.00
2 speed fan & limit control kit	S2@1.25	Ea	42.30	54.40	—	96.70
Propane conversion kit	S2@1.75	Ea	164.00	76.20	—	240.20

Description	Craft@Hrs	Unit	Material $	Labor $	Equipment $	Total $

Central dehumidification system. Upflow or counterflow, 400 CFM, 30 gallon per day baseload capacity at 0.5 water column pressure differential. With inlet filter, drain pan, drain line, heater plate, cooling draft exchanger, humidistat control sensor, control relay, actuator, cabinet, two 24" by 24" by 4' long duct transitions, gaskets, hangers and ground fault protection. Whirlpool UGD160UH or equal. Add the cost of control wiring and electrical wiring.

Central dehumidification system	SW@0.50	Ea	1,790.00	24.80	—	1,814.80
Cut, frame and gasket hole in wall, ceiling or roof	SW@1.00	Ea	34.60	49.60	—	84.20
Mount duct hangers	SW@0.25	Ea	4.36	12.40	—	16.76
Cut and mount sheet metal duct	SW@0.35	LF	6.72	17.40	—	24.12
Apply and coat duct insulation	SW@0.15	LF	2.90	7.45	—	10.35
Install electrical wiring	BE@0.15	LF	2.09	7.25	—	9.34
Install dehumidifier controls	BE@0.50	Ea	—	24.20	—	24.20
Test and balance	SW@1.50	Ea	—	74.50	—	74.50

Wall furnace, mid-efficiency, pilot light ignition. Gas-fired, atmospheric draft (radiant heat, no fan). Set in place only. Make additional allowances for gas and flue connections. Costs shown per thousand BTU per hour (MBH). Mid-efficiency 80% AFUE.

35 MBH input	S2@2.75	Ea	1,350.00	120.00	—	1,470.00
50 MBH input	S2@2.75	Ea	1,520.00	120.00	—	1,640.00
65 MBH input	S2@2.75	Ea	1,530.00	120.00	—	1,650.00
Fan blower kit	S2@1.75	Ea	218.00	76.20	—	294.20

Wall furnace, mid-efficiency, electronic ignition. Gas-fired, atmospheric draft (radiant heat, no fan). Set in place only. Make additional allowances for gas and flue connections. Costs shown per thousand BTU per hour (MBH). Mid-efficiency 80% AFUE.

35 MBH input	S2@2.75	Ea	1,690.00	120.00	—	1,810.00
65 MBH input	S2@2.75	Ea	1,890.00	120.00	—	2,010.00
Fan blower kit	S2@1.75	Ea	218.00	76.20	—	294.20

Wall furnace, direct vent (included). Gas-fired, pilot light ignition, mid-efficiency (radiant heat, no fan). Set in place only. Make additional allowances for vent, gas and electrical connections. Costs shown per thousand BTU per hour (MBH). Mid-efficiency 80% AFUE.

15 MBH input	S2@2.75	Ea	905.00	120.00	—	1,025.00
25 MBH input	S2@2.75	Ea	1,090.00	120.00	—	1,210.00
33 MBH input	S2@2.75	Ea	1,170.00	120.00	—	1,290.00
40 MBH input	S2@2.75	Ea	1,610.00	120.00	—	1,730.00
65 MBH input	S2@2.75	Ea	1,760.00	120.00	—	1,880.00
Direct vent kit	S2@2.50	Ea	153.00	109.00	—	262.00
Fan blower kit	S2@1.75	Ea	218.00	76.20	—	294.20

Description	Craft@Hrs	Unit	Material $	Labor $	Equipment $	Total $

Enthalpy energy recovery system. Steam-to-air or hot-water-to-air heat exchange. For boiler room downstream retrofits or new construction to qualify for EPA EnergyStar certification. Equipment cost is for a 40' rolling platform scissor lift. Add the cost of the energy recovery wheel ventilator.

Reinforced attic joists	P1@3.00	Ea	72.90	133.00	—	205.90
Vibration isolators	S2@.750	LS	47.00	32.60	—	79.60
Drill and mount coil service drain	P1@.150	LF	3.51	6.67	—	10.18
Install electrical wiring	BE@.150	LF	2.09	7.25	—	9.34
Mount unit heater or enthalpy wheel	P1@.500	Ea	—	22.20	—	22.20
Connect ductwork mains to manifold	SW@4.50	Ea	179.00	223.00	—	402.00
Coil piping and condensate drain	P1@4.00	Ea	9.83	178.00	—	187.83
Zone control thermostat, manual	BE@.750	Ea	148.00	36.30	—	184.30
Master heat controller with auto change-over	BE@.750	Ea	347.00	36.30	—	383.30
Wiring of controls	BE@3.00	Ea	1.74	145.00	—	146.74
Calibration test	BE@2.00	Ea	—	96.70	—	96.70

Enthalpy energy recovery wheel. Modine Slimvent, Reznor XAWS or equal. For both heating and cooling applications in commercial and institutional buildings. Complies with the ARI Standard 1060-2000. Provides recovery of up to 75% of heat energy and 80% of cooling energy.

800-1,000 CFM	SL@5.00	Ea	4,340.00	221.00	—	4,561.00
1,300-2,000 CFM	SL@5.00	Ea	5,450.00	221.00	—	5,671.00
2,400-3,000 CFM	SL@5.00	Ea	6,330.00	221.00	—	6,551.00
3,200-3,800 CFM	SL@5.00	Ea	8,330.00	221.00	—	8,551.00
4,000-4,800 CFM	SL@5.00	Ea	9,420.00	221.00	—	9,641.00
5,000-6,000 CFM	SL@5.00	Ea	10,500.00	221.00	—	10,721.00

Horizontal hot water unit heaters. Fan-driven, with integral check valves, bypass piping, drain line fitting and vacuum breaker, Modine HSB series or equal.

12.5 MBtu, 200 CFM	SL@2.00	Ea	510.00	88.20	4.69	602.89
17 MBtu, 300 CFM	SL@2.00	Ea	562.00	88.20	4.69	654.89
25 MBtu, 500 CFM	SL@2.50	Ea	644.00	110.00	5.86	759.86
30 MBtu, 700 CFM	SL@3.00	Ea	678.00	132.00	7.03	817.03
50 MBtu, 1,000 CFM	SL@3.50	Ea	788.00	154.00	8.20	950.20
60 MBtu, 1,300 CFM	SL@4.00	Ea	935.00	176.00	9.37	1,120.37

Description	Craft@Hrs	Unit	Material $	Labor $	Equipment $	Total $

Vertical hot water unit heaters. Fan-driven, with integral check valves, bypass piping, drain line fitting and vacuum breaker, Modine V500 series or equal.

Description	Craft@Hrs	Unit	Material $	Labor $	Equipment $	Total $
12.5 MBtu, 200 CFM	SL@2.00	Ea	571.00	88.20	4.69	663.89
17 MBtu, 300 CFM	SL@2.00	Ea	584.00	88.20	4.69	676.89
25 MBtu, 500 CFM	SL@2.50	Ea	757.00	110.00	5.86	872.86
30 MBtu, 700 CFM	SL@3.00	Ea	791.00	132.00	7.03	930.03
50 MBtu, 1,000 CFM	SL@3.50	Ea	952.00	154.00	8.20	1,114.20
60 MBtu, 1,300 CFM	SL@4.00	Ea	1,050.00	176.00	9.37	1,235.37

Horizontal steam coil unit heaters. Fan-driven, with integral check valves, bypass piping, drain line fitting and vacuum breaker, Modine HSB series or equal.

Description	Craft@Hrs	Unit	Material $	Labor $	Equipment $	Total $
12.5 MBtu, 200 CFM	SL@2.00	Ea	830.00	88.20	4.69	922.89
17 MBtu, 300 CFM	SL@2.00	Ea	959.00	88.20	4.69	1,051.89
25 MBtu, 500 CFM	SL@2.50	Ea	1,050.00	110.00	5.86	1,165.86
30 MBtu, 700 CFM	SL@3.00	Ea	1,150.00	132.00	7.03	1,289.03
50 MBtu, 1,000 CFM	SL@3.50	Ea	1,390.00	154.00	8.20	1,552.20
60 MBtu, 1,300 CFM	SL@4.00	Ea	1,740.00	176.00	9.37	1,925.37

Vertical steam coil unit heaters. Fan-driven, with integral check valves, bypass piping, drain line fitting and vacuum breaker, Modine V500 series or equal.

Description	Craft@Hrs	Unit	Material $	Labor $	Equipment $	Total $
12.5 MBtu, 200 CFM	SL@2.00	Ea	896.00	88.20	4.69	988.89
17 MBtu, 300 CFM	SL@2.00	Ea	982.00	88.20	4.69	1,074.89
25 MBtu, 500 CFM	SL@2.50	Ea	1,150.00	110.00	5.86	1,265.86
30 MBtu, 700 CFM	SL@3.00	Ea	1,530.00	132.00	7.03	1,669.03
50 MBtu, 1,000 CFM	SL@3.50	Ea	1,800.00	154.00	8.20	1,962.20
60 MBtu, 1,300 CFM	SL@4.00	Ea	1,910.00	176.00	9.37	2,095.37

Description	Craft@Hrs	Unit	Material $	Labor $	Equipment $	Total $

Unit heater, gas, direct-fired, mid-efficiency, propeller fan, atmospheric draft, single-stage, standing pilot, aluminized burner and heat exchanger. Hang in place only. Make additional allowances for flue, gas and electrical connections. Costs shown per thousand BTU per hour (MBH). Mid-efficiency 78% AFUE.

30 MBH input	S2@4.00	Ea	1,230.00	174.00	—	1,404.00
50 MBH input	S2@4.00	Ea	1,320.00	174.00	—	1,494.00
75 MBH input	S2@4.00	Ea	1,470.00	174.00	—	1,644.00
100 MBH input	S2@4.00	Ea	1,630.00	174.00	—	1,804.00
125 MBH input	S2@4.00	Ea	1,940.00	174.00	—	2,114.00
145 MBH input	S2@4.00	Ea	2,010.00	174.00	—	2,184.00
175 MBH input	S2@4.00	Ea	2,220.00	174.00	—	2,394.00
200 MBH input	S2@4.00	Ea	2,440.00	174.00	—	2,614.00
250 MBH input	S2@4.00	Ea	2,570.00	174.00	—	2,744.00
300 MBH input	S2@4.00	Ea	3,330.00	174.00	—	3,504.00
350 MBH input	S2@4.00	Ea	3,930.00	174.00	—	4,104.00
400 MBH input	S2@4.00	Ea	4,470.00	174.00	—	4,644.00

Unit heater, gas-fired, high-efficiency, propeller fan, power vented (included), direct-fired, single-stage, intermittent pilot ignition, aluminized burner and heat exchanger. Hang in place and install power vent. Make additional allowances for gas and electrical connections. Costs shown per thousand BTU per hour (MBH). High-efficiency 85% AFUE.

30 MBH input	S2@5.25	Ea	1,770.00	228.00	—	1,998.00
50 MBH input	S2@5.25	Ea	1,870.00	228.00	—	2,098.00
75 MBH input	S2@5.25	Ea	3,000.00	228.00	—	3,228.00
100 MBH input	S2@5.25	Ea	2,270.00	228.00	—	2,498.00
125 MBH input	S2@5.25	Ea	2,530.00	228.00	—	2,758.00
145 MBH input	S2@5.25	Ea	2,680.00	228.00	—	2,908.00
175 MBH input	S2@5.25	Ea	2,840.00	228.00	—	3,068.00
200 MBH input	S2@5.25	Ea	3,020.00	228.00	—	3,248.00
250 MBH input	S2@5.25	Ea	3,500.00	228.00	—	3,728.00
300 MBH input	S2@5.25	Ea	4,120.00	228.00	—	4,348.00
350 MBH input	S2@5.25	Ea	4,620.00	228.00	—	4,848.00
400 MBH input	S2@5.25	Ea	5,060.00	228.00	—	5,288.00

Unit heater, gas, indirect-fired, mid-efficiency, propeller fan, power vented (included), single-stage, intermittent pilot ignition, aluminized burner and heat exchanger, Hang in place and install power vent. Make additional allowances for gas and electrical connections. Costs shown per thousand BTU per hour (MBH). Mid-efficiency 78% AFUE.

130 MBH input	S2@5.25	Ea	3,220.00	228.00	—	3,448.00
150 MBH input	S2@5.25	Ea	3,500.00	228.00	—	3,728.00
170 MBH input	S2@5.25	Ea	3,620.00	228.00	—	3,848.00
225 MBH input	S2@5.25	Ea	4,060.00	228.00	—	4,288.00
280 MBH input	S2@5.25	Ea	4,820.00	228.00	—	5,048.00
340 MBH input	S2@5.25	Ea	5,450.00	228.00	—	5,678.00

Description	Craft@Hrs	Unit	Material $	Labor $	Equipment $	Total $

High-intensity infrared heater, gas-fired. Ceramic grid. Automatic spark ignition. Including reflector. Approved for unvented commercial indoor installations. Not for use in residential applications. Labor includes hang in place only. Make additional allowances for gas and 110 volt electrical connections. Refer to local gas code for proper installation requirements.

30,000 BTU	P1@1.20	Ea	657.00	53.30	2.81	713.11
60,000 BTU	P1@1.80	Ea	839.00	80.00	4.22	923.22
100,000 BTU	P1@2.50	Ea	1,990.00	111.00	5.86	2,106.86
130,000 BTU	P1@2.75	Ea	2,460.00	122.00	6.44	2,588.44
160,000 BTU	P1@2.75	Ea	2,810.00	122.00	6.44	2,938.44
Add for propane	—	Ea	163.00	—	—	163.00

High-intensity infrared heater, gas-fired. Ceramic grid. Millivolt standing pilot ignition. Approved for unvented commercial indoor installations. Not for use in residential applications. Labor includes hang in place only. Make additional allowances for gas connections. Millivolt standing pilot does not require electrical source.

30,000 BTU	P1@1.20	Ea	1,530.00	53.30	2.81	1,586.11
60,000 BTU	P1@1.80	Ea	1,990.00	80.00	4.22	2,074.22
100,000 BTU	P1@2.50	Ea	2,480.00	111.00	5.86	2,596.86
130,000 BTU	P1@2.75	Ea	2,960.00	122.00	6.44	3,088.44
160,000 BTU	P1@2.75	Ea	3,450.00	122.00	6.44	3,578.44
Add for propane	—	Ea	163.00	—	—	163.00

Infrared tube heater, burner unit only, gas-fired, negative pressure type. Labor includes hang in place only. Not for use in residential applications. Make additional allowances for burner tube, burner tube reflector, venting, gas piping and electrical requirements.

30,000 BTU	P1@1.20	Ea	1,710.00	53.30	2.81	1,766.11
60,000 BTU	P1@1.80	Ea	1,800.00	80.00	4.22	1,884.22
100,000 BTU	P1@2.50	Ea	2,050.00	111.00	5.86	2,166.86
150,000 BTU	P1@2.75	Ea	2,180.00	122.00	6.44	2,308.44
175,000 BTU	P1@2.75	Ea	2,540.00	122.00	6.44	2,668.44
200,000 BTU	P1@2.75	Ea	2,930.00	122.00	6.44	3,058.44
Add for propane	—	Ea	163.00	—	—	163.00

Infrared tube heater, 4" diameter, 10' lengths. 16 gauge burner tube and .024" mill finish aluminum reflector. Labor includes hang in place only. Not for use in residential applications. Make additional allowances for burner unit, venting, gas piping and electrical requirements.

Burner tubing	P1@.200	LF	7.78	8.89	—	16.67
Tubing U-bend	P1@.850	Ea	277.00	37.80	—	314.80
Reflector shield	P1@.200	LF	7.78	8.89	—	16.67
Vent terminal	P1@.600	Ea	122.00	26.70	—	148.70

Heat Pumps

In general, heat pumps are more effective in milder winter climates. In colder winter climates, secondary/supplemental electric heating elements (or equivalent heating source) are recommended.

A geothermal heat pump system's efficient operation is largely dependent upon proper design, layout, sizing and installation of the ground loop piping system for closed loop systems. In the case of well-to-well, the efficient operation of open loop systems is largely dependent upon the ability of the supply and return, or dump wells to meet the water volume demand of the heat pumps.

Rules of thumb when sizing loop & well systems:

Heating capacity, when considered or measured in tons cooling or refrigeration, is generally 3 to 4 times that of required cooling capacity for climates within 100 miles of the 49th parallel, i.e., if 2 tons of cooling is required, allow 6-8 tons of heating capacity (1 ton = 12,000 Btu).

Use 250 square feet per ton, e.g., a 2,000-square-foot home requires a 6-ton heat pump (heating capacity).

(Calculation based on R-20 wall and R-40 attic insulation, minimum dual pane argon-charged windows.)

Closed Vertical Loop:

Vertical Bore Hole Loop — ¾" or 1" diameter, socket fusioned loop = 200'-250' total bore hole depth per ton. For example, a 6-ton unit requires between 1,200'-1,500' (twelve to fifteen 5" bore holes, each 100' deep) with the associated 12-15 loop drops, spaced a minimum of 20' apart, connected to an adequately sized header, immersed in a bentonite slurry and applied with a tremie tube.

Closed Horizontal Loop:

Straight run — ¾" or 1" diameter manifolded loop = 400 horizontal feet per ton. For example, a 6-ton unit requires 2,400 horizontal feet (eight 300' runs, 150' out, 150' back, minimum 2' radius return bend, placed a minimum 10' apart, minimum 4' buried depth, or buried below potential winter frost line).

Slinky — ¾" or 1" diameter manifolded loop = 400 horizontal feet per ton. For example, a 6-ton unit requires 2,400 horizontal feet (eight 300' runs, 50' out, 50' back, coiled, placed a minimum 20' apart, minimum 4' buried depth, or buried below potential winter frost line. (Below 7' at 100 miles, or farther, north of the 49th parallel.)

Open Loop:

Well-to-Well — 2.5 gallons per minute per ton, each well (two required: supply & return). For example, a 6-ton unit requires 12-15 gallons per minute capacity per well.

Leadership in Energy & Environmental Design (LEED)

LEED offers an incentive to building owners to install energy-efficient and environmentally sensitive HVAC equipment. The LEED program awards points for application of the best available technology. The HVAC system can earn a maximum of 17 points toward LEED certification. A LEED-certified air conditioning system must comply with ASHRAE 90.1-2004 (efficiency) and ASHRAE 62.1-2004 (indoor air quality) standards. LEED certification also requires recording controls (Carrier ComfortView 3 or equal) to measure system performance. Zone sensor thermostats (Carrier Debonair or equal) are required to provide performance and efficiency data to the digital recorder. An indoor air quality CO2 sensor, duct-mounted aspirator modules and a refrigerant such as Puron R-410A are also required.

Description	Craft@Hrs	Unit	Material $	Labor $	Equipment $	Total $

Geothermal ground or water source heat pump (open or closed loop), w/active cooling. Glycol/water mixture or water-to-air thermal transfer. The refrigeration circuit operates in the heating and cooling mode as required (sometimes referred to as active heating & cooling or mechanical heating & cooling). This unit is suited for either open or closed loop systems. These units provide a higher degree of cooling capacity than the open loop, passive "free" cooling units. See table below. Optional de-superheater system provides domestic hot water temperature preheat or boost. Make additional allowances for the thermal source, i.e., horizontal ground loop, vertical bore hole ground loop, supply and return water wells, pond or lake source loop, etc. Make additional allowances for connections, ancillary components, start up and testing. Based on R-410A refrigerant.

Description	Craft@Hrs	Unit	Material $	Labor $	Equipment $	Total $
Heat pump, open or closed loop, ½ ton	P1@2.25	Ea	2,090.00	100.00	—	2,190.00
Heat pump, open or closed loop, ¾ ton	P1@2.25	Ea	2,140.00	100.00	—	2,240.00
Heat pump, open or closed loop, 1 ton	P1@2.25	Ea	2,240.00	100.00	—	2,340.00
Heat pump, open or closed loop, 1½ ton	P1@2.25	Ea	2,620.00	100.00	—	2,720.00
Heat pump, open or closed loop, 2 ton	P1@2.50	Ea	2,700.00	111.00	—	2,811.00
Heat pump, open or closed loop, 2½ ton	P1@2.50	Ea	3,350.00	111.00	—	3,461.00
Heat pump, open or closed loop, 3 ton	P1@3.00	Ea	3,490.00	133.00	—	3,623.00
Heat pump, open or closed loop, 4 ton	P1@3.50	Ea	3,970.00	156.00	—	4,126.00
Heat pump, open or closed loop, 5 ton	P1@4.00	Ea	7,840.00	178.00	—	8,018.00
Heat pump, open or closed loop, 7½ ton	P1@4.75	Ea	8,400.00	211.00	—	8,611.00
Optional de-superheater coil	—	Ea	952.00	—	—	952.00
Optional supplementary electric coil 5 KW	—	Ea	206.00	—	—	206.00
Optional supplementary electric coil 10 KW	—	Ea	243.00	—	—	243.00
Optional supplementary electric coil 15 KW	—	Ea	527.00	—	—	527.00
Optional supplementary electric coil 20 KW	—	Ea	585.00	—	—	585.00
Optional supplementary electric coil 25 KW	—	Ea	718.00	—	—	718.00
Water-to-water thermal transfer, additional	—	Ea	468.00	—	—	468.00

Description	Craft@Hrs	Unit	Material $	Labor $	Equipment $	Total $

Geothermal water source heat pump (open loop), w/free cooling. Water-to-air thermal transfer. The refrigeration circuit is only operated in the heating mode while passive "free" cooling is provided by pumping cold well water directly to a unit-mounted coil that bypasses the refrigeration (compressor) unit. This unit is suited to areas where the temperature of ground water supply is below 50°F. Optional de-superheater system provides domestic hot water temperature preheat or boost. Make additional allowances for the water source, i.e., supply and return water wells, pond or lake source, etc. Make additional allowances for connections, ancillary components, start up and testing.

Heat pump, open loop, free cooling, 2 ton	P1@2.50	Ea	3,690.00	111.00	—	3,801.00
Heat pump, open loop, free cooling, 3 ton	P1@3.00	Ea	4,520.00	133.00	—	4,653.00
Heat pump, open loop, free cooling, 4 ton	P1@3.50	Ea	5,080.00	156.00	—	5,236.00
Heat pump, open loop, free cooling, 5 ton	P1@4.00	Ea	8,940.00	178.00	—	9,118.00
Heat pump, open loop, free cooling, 7½ ton	P1@4.75	Ea	9,690.00	211.00	—	9,901.00
Optional de-superheater coil	—	Ea	952.00	—	—	952.00
Optional supplementary electric coil 5 KW	—	Ea	206.00	—	—	206.00
Optional supplementary electric coil 10 KW	—	Ea	243.00	—	—	243.00
Optional supplementary electric coil 15 KW	—	Ea	527.00	—	—	527.00
Optional supplementary electric coil 20 KW	—	Ea	585.00	—	—	585.00
Optional supplementary electric coil 25 KW	—	Ea	718.00	—	—	718.00
Water-to-water thermal transfer additional	—	Ea	468.00	—	—	468.00

Description	Craft@Hrs	Unit	Material $	Labor $	Equipment $	Total $

Geothermal water source heat pump (closed refrigerant loop). Water-to-air thermal transfer. Optional de-superheater system provides domestic hot water temperature preheat or boost. Make additional allowances for the heat transfer loop, i.e., refrigerant loop. Make additional allowances for connections, ancillary components, start up and testing.

Description	Craft@Hrs	Unit	Material $	Labor $	Equipment $	Total $
Heat pump, 2 ton	P1@2.50	Ea	2,210.00	111.00	—	2,321.00
Heat pump, 3 ton	P1@3.00	Ea	2,880.00	133.00	—	3,013.00
Heat pump, 4 ton	P1@3.50	Ea	3,330.00	156.00	—	3,486.00
Heat pump, 5 ton	P1@4.00	Ea	7,020.00	178.00	—	7,198.00
Heat pump, 7½ ton	P1@4.75	Ea	6,670.00	211.00	—	6,881.00
Optional de-superheater coil	—	Ea	914.00	—	—	914.00
Optional supplementary electric coil 5 KW	—	Ea	213.00	—	—	213.00
Optional supplementary electric coil 10 KW	—	Ea	251.00	—	—	251.00
Optional supplementary electric coil 15 KW	—	Ea	548.00	—	—	548.00
Optional supplementary electric coil 20 KW	—	Ea	608.00	—	—	608.00
Optional supplementary electric coil 25 KW	—	Ea	745.00	—	—	745.00

LEED certification for geothermal heat pumps

Description	Craft@Hrs	Unit	Material $	Labor $	Equipment $	Total $
Add for a LEED-certified heat pump with USGBC rating	—	%	20.0	—	—	—
Add for LEED registration and inspection	—	Ea	2,180.00	—	—	2,180.00
Add for LEED central digital performance monitor/recorder, ComfortView 3 or equal	—	Ea	2,730.00	—	—	2,730.00
Add for LEED-certified zone recording thermostats, Carrier DebonAir, USB remote-ready	—	Ea	406.00	—	—	406.00
Add for IAQ (Indoor Air Quality) CO_2 sensor and duct-mounted aspirator box	—	Ea	491.00	—	—	491.00

Description	Craft@Hrs	Unit	Material $	Labor $	Equipment $	Total $

Heat pump, residential air-to-air split system. Outdoor unit installed on pad or wall bracket. Indoor coil installed in supply air plenum upstream of supply air fan (air handler). Units rated at 13SEER (Seasonal Energy Efficiency Rating). System includes outdoor condenser/compressor unit, indoor evaporative coil, programmable thermostat, indoor/outdoor sensor. Condenser/compressor unit comes pre-charged from the factory with R410A refrigerant. 208/230 Volt, single phase. Rated in nominal tonnage. (12,000 BTU's equals 1 ton cooling).

Description	Craft@Hrs	Unit	Material $	Labor $	Equipment $	Total $
Setting outdoor unit on pad or wall bracket						
2.0-ton outdoor unit	P1@1.50	Ea	2,700.00	66.70	—	2,766.70
3.0-ton outdoor unit	P1@1.75	Ea	2,960.00	77.80	—	3,037.80
3.5-ton outdoor unit	P1@2.00	Ea	3,590.00	88.90	—	3,678.90
4.0-ton outdoor unit	P1@2.25	Ea	3,840.00	100.00	—	3,940.00
5.0-ton outdoor unit	P1@2.50	Ea	4,510.00	111.00	—	4,621.00
Mounting evaporative coil in air stream						
2.0-ton evap. coil	P1@1.15	Ea	29.30	51.10	—	80.40
3.0-ton evap. coil	P1@1.35	Ea	35.20	60.00	—	95.20
3.5-ton evap. coil	P1@1.50	Ea	41.00	66.70	—	107.70
4.0-ton evap. coil	P1@2.00	Ea	46.80	88.90	—	135.70
5.0-ton evap. coil	P1@2.25	Ea	52.80	100.00	—	152.80
Install and connect (braze) refrigerant line set – 50' (Liquid x Suction). Connect refrigerant line set. Condensor/compressor unit to evaporative coil.						
2.0-ton 3/8" - 5/8"	P1@2.00	Ea	258.00	88.90	—	346.90
3.0-ton 1/2" - 7/8"	P1@2.25	Ea	298.00	100.00	—	398.00
3.5-ton 5/8" - 1"	P1@3.75	Ea	451.00	167.00	—	618.00
4.0-ton 7/8" -1-1/8"	P1@4.50	Ea	539.00	200.00	—	739.00
5.0-ton 1" - 1-3/8"	P1@6.00	Ea	727.00	267.00	—	994.00
Control wiring						
24 volt	BE@3.00	Ea	32.20	145.00	—	177.20
Power wiring compressor						
208/230 Volt	BE@2.00	Ea	193.00	96.70	—	289.70
Evaporative coil drain	P1@1.00	Ea	22.30	44.50	—	66.80
Equipment pad, concrete slab,						
4' x 4' x 2"	CF@4.00	Ea	31.40	175.00	—	206.40
Equipment pad, patio stones,						
2' x 2' x 1–1/2" (4 ea)	CF@2.00	Ea	41.70	87.30	—	129.00

Description	Craft@Hrs	Unit	Material $	Labor $	Equipment $	Total $

Heat pump, split system, single-speed blower, single-phase scroll compressor – 208 volt, 13 SEER (Seasonal Energy Efficiency Rating)

Air-to-air thermal transfer. Material costs include outdoor stat / low ambient, short cycle timer, high-pressure cut out and digital heat/cool thermostat with manual change-over. Labor includes setting outdoor condenser, indoor blower and coil sections in place only. Make additional allowances for refrigerant piping / connections, electrical / temperature control wiring / connections and evaporator coil condensate drain.

Description	Craft@Hrs	Unit	Material $	Labor $	Equipment $	Total $
1.5 tons cooling	S2@4.00	Ea	3,560.00	174.00	—	3,734.00
2 tons cooling	S2@4.25	Ea	3,660.00	185.00	—	3,845.00
2.5 tons cooling	S2@4.50	Ea	4,260.00	196.00	—	4,456.00
3 tons cooling	S2@5.00	Ea	4,760.00	218.00	—	4,978.00
3.5 tons cooling	S2@5.75	Ea	5,220.00	250.00	—	5,470.00
4 tons cooling	S2@6.25	Ea	5,770.00	272.00	—	6,042.00
5 tons cooling	S2@7.50	Ea	6,690.00	326.00	—	7,016.00

Heat pump, split system, single-speed blower, three-phase scroll compressor – 230 volt, 13 SEER (Seasonal Energy Efficiency Rating)

Air-to-air thermal transfer. Material costs include outdoor stat / low ambient, short cycle timer, high-pressure cut out and digital heat/cool thermostat with manual change over. Labor includes setting outdoor condenser, indoor blower and coil sections in place only. Make additional allowances for refrigerant piping / connections, electrical / temperature control wiring / connections and evaporator coil condensate drain.

Description	Craft@Hrs	Unit	Material $	Labor $	Equipment $	Total $
3.5 tons cooling	S2@5.75	Ea	5,820.00	250.00	—	6,070.00
4 tons cooling	S2@6.25	Ea	6,350.00	272.00	—	6,622.00
5 tons cooling	S2@7.50	Ea	7,270.00	326.00	—	7,596.00

Heat pump, split system, variable-speed blower, single-phase scroll compressor – 208 volt, 13 SEER (Seasonal Energy Efficiency Rating)

Air-to-air thermal transfer. Material costs include outdoor stat / low ambient, short cycle timer, high-pressure cut out and digital heat/cool thermostat with manual change over. Labor includes setting outdoor condenser, indoor blower and coil sections in place only. Make additional allowances for refrigerant piping / connections, electrical / temperature control wiring / connections and evaporator coil condensate drain.

Description	Craft@Hrs	Unit	Material $	Labor $	Equipment $	Total $
1.5 tons cooling	S2@4.00	Ea	4,270.00	174.00	—	4,444.00
2 tons cooling	S2@4.25	Ea	4,340.00	185.00	—	4,525.00
2.5 tons cooling	S2@4.50	Ea	4,980.00	196.00	—	5,176.00
3 tons cooling	S2@5.00	Ea	5,510.00	218.00	—	5,728.00
3.5 tons cooling	S2@5.75	Ea	5,920.00	250.00	—	6,170.00
4 tons cooling	S2@6.25	Ea	6,690.00	272.00	—	6,962.00
5 tons cooling	S2@7.50	Ea	7,580.00	326.00	—	7,906.00

Description	Craft@Hrs	Unit	Material $	Labor $	Equipment $	Total $

Heat pump, split system, variable-speed blower, three-phase scroll compressor – 230 volt, 13 SEER (Seasonal Energy Efficiency Rating)

Air-to-air thermal transfer. Material costs include outdoor stat / low ambient, short cycle timer, high-pressure cut out and digital heat/cool thermostat with manual change over. Labor includes setting outdoor condenser, indoor blower and coil sections in place only. Make additional allowances for refrigerant piping / connections, electrical / temperature control wiring / connections and evaporator coil condensate drain.

3.5 tons cooling	S2@5.75	Ea	6,560.00	250.00	—	6,810.00
4 tons cooling	S2@6.25	Ea	7,250.00	272.00	—	7,522.00
5 tons cooling	S2@7.50	Ea	8,230.00	326.00	—	8,556.00

Heat pump, split system, single-speed blower, single-phase scroll compressor – 208 volt, 14 SEER (Seasonal Energy Efficiency Rating)

Air-to-air thermal transfer. Material costs include outdoor stat / low ambient, short cycle timer, high-pressure cut out and digital heat/cool thermostat with manual change over. Labor includes setting outdoor condenser, indoor blower and coil sections in place only. Make additional allowances for refrigerant piping / connections, electrical / temperature control wiring / connections and evaporator coil condensate drain.

2 tons cooling	S2@4.25	Ea	4,920.00	185.00	—	5,105.00
2.5 tons cooling	S2@4.50	Ea	5,570.00	196.00	—	5,766.00
3 tons cooling	S2@5.00	Ea	5,920.00	218.00	—	6,138.00
3.5 tons cooling	S2@5.75	Ea	6,760.00	250.00	—	7,010.00
4 tons cooling	S2@6.25	Ea	7,540.00	272.00	—	7,812.00
5 tons cooling	S2@7.50	Ea	8,250.00	326.00	—	8,576.00

Heat pump, split system, variable-speed blower, single-phase scroll compressor – 208 volt, 14 SEER (Seasonal Energy Efficiency Rating)

Air-to-air thermal transfer. Material costs include outdoor stat / low ambient, short cycle timer, high-pressure cut out and digital heat/cool thermostat with manual change over. Labor includes setting outdoor condenser, indoor blower and coil sections in place only. Make additional allowances for refrigerant piping / connections, electrical / temperature control wiring / connections and evaporator coil condensate drain.

2 tons cooling	S2@4.25	Ea	5,610.00	185.00	—	5,795.00
2.5 tons cooling	S2@4.50	Ea	6,140.00	196.00	—	6,336.00
3 tons cooling	S2@5.00	Ea	6,570.00	218.00	—	6,788.00
3.5 tons cooling	S2@5.75	Ea	7,400.00	250.00	—	7,650.00
4 tons cooling	S2@6.25	Ea	8,030.00	272.00	—	8,302.00
5 tons cooling	S2@7.50	Ea	11,800.00	326.00	—	12,126.00

Description	Craft@Hrs	Unit	Material $	Labor $	Equipment $	Total $

Heat pump system accessories

Fossil fuel kit	S2@2.50	Ea	236.00	109.00	—	345.00
High pressure cut out kit	S2@.500	Ea	82.00	21.80	—	103.80
Outdoor stat/low ambient sensors	S2@.500	Ea	77.30	21.80	—	99.10
Short cycle protection	S2@.500	Ea	61.00	21.80	—	82.80
Digital heat/cool thermostat w/manual change-over	S2@.750	Ea	179.00	32.60	—	211.60
Digital, programmable heat/cool thermostat w/auto change-over	S2@.750	Ea	560.00	32.60	—	592.60

Heat pump, supplemental electric heating coil

Set in place only. Make additional allowances for electrical connections.

5 KW, 240 Volt	S2@1.50	Ea	202.00	65.30	—	267.30
7.5 KW, 240 Volt	S2@1.75	Ea	243.00	76.20	—	319.20
10 KW, 240 Volt	S2@2.00	Ea	283.00	87.00	—	370.00
12.5 KW, 240 V	S2@2.25	Ea	469.00	97.90	—	566.90
15 KW, 240 Volt	S2@2.75	Ea	594.00	120.00	—	714.00
20 KW, 240 Volt	S2@3.50	Ea	665.00	152.00	—	817.00
25 KW, 240 Volt	S2@4.25	Ea	801.00	185.00	—	986.00

LEED certification for geothermal heat pumps

Add for a LEED-certified heat pump with USGBC rating	—	%	20.0	—	—	—
Add for LEED registration and inspection	—	Ea	2,180.00	—	—	2,180.00
Add for LEED central digital performance monitor/recorder, ComfortView 3 or equal	—	Ea	2,730.00	—	—	2,730.00
Add for LEED-certified zone recording thermostats, Carrier DebonAir, USB remote-ready	—	Ea	406.00	—	—	406.00
Add for IAQ (Indoor Air Quality) CO_2 sensor and duct-mounted aspirator box	—	Ea	491.00	—	—	491.00

Water pump, submersible. Domestic water, geothermal open loop, well, lake or pond applications. Stainless steel jacket with multiple-stage impellers and integral foot valve.

Description	Craft@Hrs	Unit	Material $	Labor $	Equipment $	Total $
Submersible pump, 120/240 volt, ½ HP	P1@2.00	Ea	798.00	88.90	—	886.90
Submersible pump, 240 volt, ¾ HP	P1@2.00	Ea	994.00	88.90	—	1,082.90
Submersible pump, 240 volt, 1 HP	P1@2.00	Ea	1,400.00	88.90	—	1,488.90
Submersible pump, 240 volt, 1½ HP	P1@2.00	Ea	1,830.00	88.90	—	1,918.90
Submersible pump, 240 volt, 2 HP	P1@2.00	Ea	2,130.00	88.90	—	2,218.90
Submersible pump, 240 volt, 3 HP	P1@2.00	Ea	2,400.00	88.90	—	2,488.90
Submersible pump, 240 volt, 5 HP	P1@2.00	Ea	2,800.00	88.90	—	2,888.90
Pump controller – constant pressure, VSD	P1@1.50	Ea	1,350.00	66.70	—	1,416.70
Pump control box, 120 volt	P1@.600	Ea	79.80	26.70	—	106.50
Pump control box, 240 volt	P1@.600	Ea	97.90	26.70	—	124.60
Pump stand, PVC	P1@1.00	Ea	214.00	44.50	—	258.50
Pump stand, galvanized steel	P1@1.35	Ea	290.00	60.00	—	350.00

Water pump, jet. Domestic water well, lake and pond applications. 110 volt, cast iron body, single-stage impeller, including pressure switch & pressure gauge. Use for shallow depths to 25' maximum lift. Deep well applications to 120' maximum lift.

Description	Craft@Hrs	Unit	Material $	Labor $	Equipment $	Total $
Jet pump, shallow depth ½ HP	P1@2.00	Ea	371.00	88.90	—	459.90
Jet pump, shallow depth ¾ HP	P1@2.00	Ea	429.00	88.90	—	517.90
Jet pump, deep well ½ HP	P1@2.00	Ea	503.00	88.90	—	591.90
Jet pump, deep well ¾ HP	P1@2.00	Ea	584.00	88.90	—	672.90
Optional pump-mounted pressure tank 5 gallon	P1@.500	Ea	170.00	22.20	—	192.20

Description	Craft@Hrs	Unit	Material $	Labor $	Equipment $	Total $

Water pump system accessories. Pressure tanks are factory pre-charged. VSD = variable speed drive.

Description	Craft@Hrs	Unit	Material $	Labor $	Equipment $	Total $
Pressure tank, bladder type, 5 gallon	P1@1.00	Ea	153.00	44.50	—	197.50
Pressure tank, bladder type, 10 gallon	P1@2.00	Ea	175.00	88.90	—	263.90
Pressure tank, bladder type, 20 gallon	P1@2.00	Ea	208.00	88.90	—	296.90
Pressure tank, bladder type, 30 gallon	P1@2.00	Ea	297.00	88.90	—	385.90
Pressure tank, bladder type, 40 gallon	P1@2.00	Ea	394.00	88.90	—	482.90
Brass tank tee, ¾" or 1" w/ union	P1@.300	Ea	96.30	13.30	—	109.60
Brass tank tee, ¾" or 1" w/o union	P1@.300	Ea	77.50	13.30	—	90.80
Constant pressure VSD pump controller	P1@1.50	Ea	742.00	66.70	—	808.70
Pressure switch, 110/220 volt	P1@.350	Ea	44.80	15.60	—	60.40
Pressure gauge, 0-100 psi, ¼"	P1@.200	Ea	11.20	8.89	—	20.09
Sediment faucet (hose bibb), ½"	P1@1.50	Ea	9.09	66.70	—	75.79
Double injector deep well venturi, 1", brass	P1@.600	Ea	340.00	26.70	—	366.70
Double injector deep well venturi, 1¼", brass	P1@.650	Ea	390.00	28.90	—	418.90
Double injector deep well venturi, 1½", brass	P1@.700	Ea	440.00	31.10	—	471.10
Double injector deep well venturi, 1", plastic	P1@.550	Ea	80.80	24.40	—	105.20
Double injector deep well venturi, 1¼", plastic	P1@.600	Ea	99.70	26.70	—	126.40
Double injector deep well venturi, 1½", plastic	P1@.650	Ea	108.00	28.90	—	136.90
Foot valve, 1", brass	P1@.400	Ea	106.00	17.80	—	123.80
Foot valve, 1¼", brass	P1@.400	Ea	110.00	17.80	—	127.80
Foot valve, 1½", brass	P1@.400	Ea	117.00	17.80	—	134.80
Foot valve, 1", plastic	P1@.400	Ea	27.80	17.80	—	45.60
Foot valve, 1¼", plastic	P1@.400	Ea	34.00	17.80	—	51.80
Foot valve, 1½", plastic	P1@.400	Ea	37.70	17.80	—	55.50
Barbed adapter, FIP/MIP, 1" brass	P1@.230	Ea	17.40	10.20	—	27.60
Barbed adapter, FIP/MIP, 1¼" brass	P1@.250	Ea	20.00	11.10	—	31.10

Description	Craft@Hrs	Unit	Material $	Labor $	Equipment $	Total $
Barbed adapter, FIP/MIP, 1½" brass	P1@.265	Ea	24.50	11.80	—	36.30
Barbed adapter, FIP/MIP, 1" plastic	P1@.215	Ea	13.50	9.56	—	23.06
Barbed adapter, FIP/MIP, 1¼" plastic	P1@.220	Ea	14.80	9.78	—	24.58
Barbed adapter, FIP/MIP, 1½" plastic	P1@.235	Ea	19.80	10.40	—	30.20
Pitless adapter, 1" x 5" brass	P1@1.25	Ea	258.00	55.60	—	313.60
Pitless adapter, 1¼" x 5" brass	P1@1.25	Ea	277.00	55.60	—	332.60
Pitless adapter, 1½" x 5" brass	P1@1.25	Ea	311.00	55.60	—	366.60
Pitless adapter, 2" x 6" brass	P1@1.80	Ea	472.00	80.00	—	552.00
Poly pipe, 75 psi, 1"	P1@.020	LF	.35	.89	—	1.24
Poly pipe, 75 psi, 1¼"	P1@.021	LF	.43	.93	—	1.36
Poly pipe, 75 psi, 1½"	P1@.022	LF	.45	.98	—	1.43
Poly pipe, 100 psi, 1"	P1@.021	LF	.35	.93	—	1.28
Poly pipe, 100 psi, 1¼"	P1@.022	LF	.43	.98	—	1.41
Poly pipe, 100 psi, 1½"	P1@.023	LF	.45	1.02	—	1.47
Electrical wire (110 volt), #12/4	P1@.015	LF	.40	.67	—	1.07
Electrical wire (220 volt), #10/4	P1@.015	LF	.42	.67	—	1.09
Heat shrink kit	—	Ea	5.36	—	—	5.36

Geothermal water wells

A geothermal water well is constructed and operates exactly like a water well used for drinking water. In many residential applications, the supply well of the geothermal system also functions as the domestic water supply well.

The cost of a drilled well depends on many variables, such as costs to get the drilling rig and crew to the site, the type of geologic formation the well drilling rig must drill through, the depth the driller must achieve to find water in that particular location and the distance from an existing available water source to provide the water required to drill (to refill the water truck if necessary).

Well drillers charge by the foot and usually have a 100' minimum charge-out fee. They can only speculate to what depth they must drill to find water. Water being found at a particular depth nearby doesn't always translate to finding water at another given location at the same depth. It's important to remember that drilling a relatively small hole in the ground and hoping to hit what is described as a vein of water as part of an underground water table is always a gamble. As such, it is, without proper research, difficult to determine a realistic budget for well drilling. It's always prudent to call the water authority or a well driller in the area for the most accurate information on the water and geological characteristics of a given location.

Use the following to begin to formulate a budget exclusive of costs associated with factors listed above. Because of the inherent uncertainties of the process, these prices are given only as typical subcontract costs.

Description	Craft@Hrs	Unit	Material $	Labor $	Equipment $	Total $
Drilled well, 5" diameter, PVC casing, set into bedrock. Based on a 2-man crew, 1 rotary drill rig and a 1,500 gallon water truck. Minimum 100' charge in most cases. Typical subcontract costs.						
Drilled through till, clay, sand	—	LF	—	—	—	41.20
Drilled through till and boulder	—	LF	—	—	—	53.20
Drilled through limestone soft bedrock	—	LF	—	—	—	59.90
Drilled through granite hard bedrock	—	LF	—	—	—	120.00
Add for galvanized steel casing	—	LF	—	—	—	17.70
Drilled well, 6" diameter, PVC casing, set into bedrock. Based on a 2-man crew, 1 rotary drill rig and a 1,500 gallon water truck. Minimum 100' charge in most cases. Typical subcontract costs.						
Drilled through till, clay, sand	—	LF	—	—	—	45.80
Drilled through till and boulder	—	LF	—	—	—	58.30
Drilled through limestone soft bedrock	—	LF	—	—	—	63.20
Drilled through granite hard bedrock	—	LF	—	—	—	129.00
Add for galvanized steel casing	—	LF	—	—	—	20.20

Description	Craft@Hrs	Unit	Material $	Labor $	Equipment $	Total $

Drilled well, 8" diameter, PVC casing, set into bedrock. Based on a 2-man crew, 1-rotary drill rig and a 1,500 gallon water truck. Minimum 100' charge in most cases. Typical subcontract costs.

Description	Craft@Hrs	Unit	Material $	Labor $	Equipment $	Total $
Drilled through till, clay, sand	—	LF	—	—	—	58.30
Drilled through till and boulder	—	LF	—	—	—	66.50
Drilled through limestone soft bedrock	—	LF	—	—	—	74.90
Drilled through granite hard bedrock	—	LF	—	—	—	151.00
Add for galvanized steel casing	—	LF	—	—	—	40.30

Drilled well, 10" diameter, PVC casing, set into bedrock. Based on a 2-man crew, 1 rotary drill rig and a 1,500 gallon water truck. Minimum 100' charge in most cases. Typical subcontract costs.

Description	Craft@Hrs	Unit	Material $	Labor $	Equipment $	Total $
Drilled through till, clay, sand	—	LF	—	—	—	69.80
Drilled through till and boulder	—	LF	—	—	—	78.00
Drilled through limestone soft bedrock	—	LF	—	—	—	86.40
Drilled through granite hard bedrock	—	LF	—	—	—	200.00
Add for galvanized steel casing	—	LF	—	—	—	51.30

Geothermal bore hole (vertical loop), 3" diameter, drilled to a maximum of 30'. Includes installation of a polypropylene fluid loop using a 180-degree return bend with socket fusion joints. Bore hole filled with bentonite slurry through a tremie tube and mud pump. Includes socket fusioned header and supply & return mains to heat pump. Based on a minimum of 1,000 vertical feet (2,000' of pipe) Based on a 2-man crew, 1 skid steer-mounted rotary drill rig and a 50 gallon water tank. Typical subcontract costs.

Description	Craft@Hrs	Unit	Material $	Labor $	Equipment $	Total $
¾" vertical loop	—	LF	—	—	—	14.10
1" vertical loop	—	LF	—	—	—	15.40
Add for thermally-enhanced Bentonite	—	LF	—	—	—	4.01

Geothermal horizontal loop, directionally bored. Includes installation of a polypropylene fluid loop. Based on a minimum of 2,000 horizontal feet (2,000' of pipe). Based on a 3-man crew, 1 directional boring/horizontal drill machine. Typical subcontract costs.

Description	Craft@Hrs	Unit	Material $	Labor $	Equipment $	Total $
¾" straight horizontal loop	—	LF	—	—	—	8.55
1" straight horizontal loop	—	LF	—	—	—	9.32

Description	Craft@Hrs	Unit	Material $	Labor $	Equipment $	Total $

Geothermal horizontal loop, slinky type. Includes installation of a polypropylene fluid loop. Based on a minimum of 2,000 horizontal feet (2,000' of pipe). Based on a 3-man crew, 1 excavator backhoe. Typical subcontract costs.

Description	Craft@Hrs	Unit	Material $	Labor $	Equipment $	Total $
¾" slinky loop	—	LF	—	—	—	6.48
1" slinky loop	—	LF	—	—	—	7.31
¾" slinky pond loop	—	LF	—	—	—	5.06
1" slinky pond loop	—	LF	—	—	—	5.91

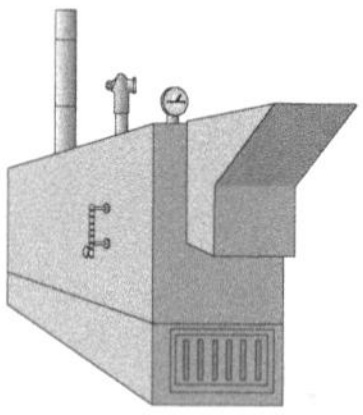

Biomass-fired hot water boilers for central heating systems. For residential, commercial and light industrial applications. 15 PSI, EPA, ASME and UL rated, with wood-pellet and/or corn fuel gravity-feed hopper, ash hood cleanout, secondary natural gas or #2 oil burner, platinum stack transition emissions catalyst, forced draft fan, factory-installed boiler trim, feedwater pump and valves, and tempering valve. Add the cost of the following if required: a concrete pad, optional stainless steel 2 GPM potable hot water coil insert, expansion tank and circulating pump, stack and stack liner, CO_2 fire extinguisher system and electrical connection. Add the cost of post-installation inspection by both the fire marshal and the mechanical inspector to validate federal, state and local energy tax credits or energy tax credit offsets. See pollution control stack costs below. Equipment cost is for an appliance dolly, 3,000 lb. come-along, and ½ ton chain hoist. Listed by boiler horsepower (BHP) rating and input capacity in thousands of BTU per hour (MBH).

Description	Craft@Hrs	Unit	Material $	Labor $	Equipment $	Total $
3 BHP, 102 MBH	SN@24.0	Ea	12,900.00	1,110.00	69.50	14,079.50
3 BHP, 104 MBH	SN@24.0	Ea	11,800.00	1,110.00	69.50	12,979.50
6 BHP, 200 MBH	SN@28.0	Ea	15,200.00	1,300.00	81.10	16,581.10
14 BHP, 500 MBH	SN@30.0	Ea	18,100.00	1,390.00	86.90	19,576.90
27 BHP, 900 MBH	SN@44.0	Ea	26,200.00	2,040.00	127.00	28,367.00

LEED certification requires 88% boiler efficiency, recording controls, compliance with SCAQMD 1146.2 emission standards and zone thermostats.

Add for LEED-certified boiler with USGBC rating	—	%	15.0	—	—	—
Add for LEED registration and inspection	—	Ea	2,150.00	—	—	2,150.00
Add for central performance monitor/ recorder	—	Ea	2,850.00	—	—	2,850.00
Add for zone recording thermostat	—	Ea	422.00	—	—	422.00

Retrofit pollution control stack for biomass hot water boilers. Stack-mounted platinum catalytic screen for biomass-fired hot water boilers for central radiant heating systems. E.P.A. and California Code standards compliant. UL rated. Includes stack mounting transition and extensions, baghouse structural steel frame, stack transition cleanout and inspection stack access bolted ports and an integral CO_2 fire extinguisher system. Add the cost of a concrete pad, electrical connection, post-installation calibration, inspection by both the fire marshal and the mechanical inspector to validate federal, state and local emission reduction tax credits and associated energy tax credit offsets. Equipment cost is for a 2-ton capacity hydraulic truck crane. Listed by boiler horsepower (BHP) rating and input capacity in thousands of BTU per hour (MBH). Use these figures to estimate the cost of adding a pollution control stack to an existing biomass boiler to meet environmental regulations.

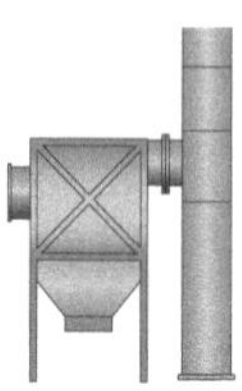

3 BHP, 102 MBH	SN@20.0	Ea	1,550.00	927.00	57.90	2,534.90
3 BHP, 104 MBH	SN@20.0	Ea	1,590.00	927.00	57.90	2,574.90
6 BHP, 200 MBH	SN@20.0	Ea	2,380.00	927.00	57.90	3,364.90
14 BHP, 500 MBH	SN@40.0	Ea	2,920.00	1,850.00	116.00	4,886.00
27 BHP, 900 MBH	SN@40.0	Ea	3,960.00	1,850.00	116.00	5,926.00

Description	Craft@Hrs	Unit	Material $	Labor $	Equipment $	Total $

Biomass-fired water or steam boilers for multi-dwelling, process and industrial applications. 15 PSI, EPA, ASME and UL rated. Includes wood pellet and/or corn fuel auger feed hopper, ash auger cleanout, secondary natural gas or #2 oil burner, platinum stack transition emissions catalyst, forced draft fan, factory-installed boiler trim, feedwater pump and valves, and tempering valve. Add the following costs if required: a concrete pad, optional stainless steel 10 GPM potable hot water coil insert, circulating pump, expansion tank, shell and tube heat exchanger, stack and stack liner, CO_2 fire extinguisher system, pollution control baghouse and filter, electrical connection. Add the cost of post-installation inspection by both the fire marshal and the mechanical inspector to validate federal, state and local energy tax credits or energy tax credit offsets. See pollution control stack costs below. Equipment cost is for a 5-ton capacity forklift. Listed by boiler horsepower (BHP) rating and input capacity in thousands of BTU per hour (MBH).

30 BHP, 1,000 MBH	SN@168	Ea	93,700.00	7,790.00	1,300.00	102,790.00
50 BHP, 1,700 MBH	SN@200	Ea	172,000.00	9,270.00	1,540.00	182,810.00
100 BHP, 3,400 MBH	SN@400	Ea	267,000.00	18,500.00	3,090.00	288,590.00
125 BHP, 4,100 MBH	SN@400	Ea	338,000.00	18,500.00	3,090.00	359,590.00
150 BHP, 5,000 MBH	SN@500	Ea	328,000.00	23,200.00	3,860.00	355,060.00
200 BHP, 6,700 MBH	SN@600	Ea	451,000.00	27,800.00	4,630.00	483,430.00

LEED certification requires 88% boiler efficiency, recording controls, compliance with SCAQMD 1146.2 emission standards and zone thermostats.

Add for LEED-certified boiler with USGBC rating	—	%	6.0	—	—	—
Add for LEED registration and inspection	—	Ea	2,150.00	—	—	2,150.00
Add for central performance monitor/ recorder	—	Ea	2,850.00	—	—	2,850.00
Add for zone recording thermostat	—	Ea	422.00	—	—	422.00

Biomass boiler stack tie-in to baghouse filter. Ionically bonded silicon-lined carbon steel. Costs assume a 20-foot vertical stack height, connection via T-shaped connector with cleanout to the baghouse inlet plenum. Includes rigging, breeching, connecting stack sections from back "turnaround" plenum of boiler to baghouse inlet plenum, vendor-supplied structural steel stack support brackets and aluminized external coating. Equipment cost assumes rental of a 2-ton hydraulic truck crane. Biomass boilers, stoves and heaters smaller than 30 BHP can use a conventional boiler stack (without a baghouse) and require only a platinum catalytic screen. Costs shown assume a 20-foot stack height and 10' horizontal run from T-connector. Cost per 10' length, including a bolt and gasket set on each end. Add 15% for each 10' section (or portion thereof) beyond 20' vertically or horizontally.

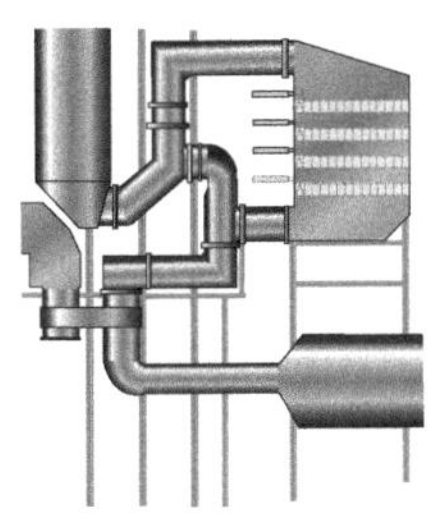

10" diameter	SN@12.0	Ea	3,310.00	556.00	205.00	4,071.00
12" diameter	SN@12.0	Ea	3,700.00	556.00	205.00	4,461.00
16" diameter	SN@16.0	Ea	3,960.00	742.00	273.00	4,975.00
20" diameter	SN@20.0	Ea	5,020.00	927.00	341.00	6,288.00
24" diameter	SN@24.0	Ea	5,550.00	1,110.00	410.00	7,070.00
30" diameter	SN@32.0	Ea	8,210.00	1,480.00	546.00	10,236.00
56" diameter	SN@40.0	Ea	14,000.00	1,850.00	683.00	16,533.00

Description	Craft@Hrs	Unit	Material $	Labor $	Equipment $	Total $

Retrofit pollution control stack for biomass water or steam boiler. Stack-mounted platinum screen catalytic reactor with air filtration fabric baghouse. For use with biomass-fired hot water or low-pressure steam boilers for central radiant heating or process systems. Meets EPA, NFPA, California Code and UL standards for multi-dwelling, process and industrial applications. Cost includes equipment stack transitions and fittings, transition access bolted inspection and cleanout ports, structural steel frame equipment support and ladders, bottom ash cleanout for baghouse, alarmed stack gas composition monitor and recorder, urea tank and urea pump & regulating valve and an integral CO_2 fire extinguisher system. Costs do not include the following: concrete pad, electrical tie-in, calibration and commissioning, post-installation inspection by the fire marshal and the mechanical inspector to validate federal, state and local energy tax credits or energy tax credit offsets. Equipment cost is for a 2-ton capacity hydraulic truck crane. Listed by boiler horsepower (BHP) rating and input capacity in thousands of BTU per hour (MBH). Use these figures to estimate the cost of adding a pollution control stack to an existing biomass boiler to meet environmental regulations.

30 BHP, 1,000 MBH	SN@40.0	Ea	46,100.00	1,850.00	683.00	48,633.00
50 BHP, 1,700 MBH	SN@80.0	Ea	81,800.00	3,710.00	1,370.00	86,880.00
100 BHP, 3,400 MBH	SN@80.0	Ea	144,000.00	3,710.00	1,370.00	149,080.00
125 BHP, 4,100 MBH	SN@120	Ea	153,000.00	5,560.00	2,050.00	160,610.00
150 BHP, 5,000 MBH	SN@120	Ea	163,000.00	5,560.00	2,050.00	170,610.00
200 BHP, 6,700 MBH	SN@160	Ea	173,000.00	7,420.00	2,730.00	183,150.00

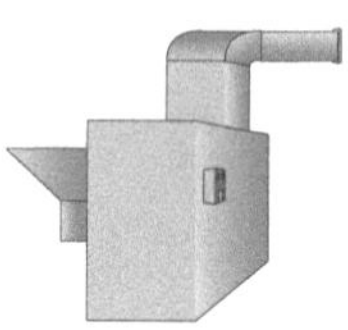

Biomass-fired central air space heaters for residential, multi-dwelling, and light commercial applications. 15 PSI, EPA, ASHRAE and UL rated. Includes a wood pellet and/or corn fuel gravity-feed hopper, ash hood cleanout, secondary natural gas or #2 oil burner, platinum stack transition emissions catalyst, forced draft fan, ducted hot-air circulation plenum and integral three- speed circulation duct blower. Add the cost of a concrete pad, optional stainless steel 2 GPM potable hot water coil insert, expansion tank and circulating pump, stack and stack liner, CO_2 fire extinguisher system and electrical connection. These costs do not include post-installation inspection by both the fire marshal and the mechanical inspector to validate federal, state and local energy tax credits or energy tax credit offsets. Equipment cost is for a 2-ton capacity forklift. Listed by input capacity in thousands of BTU per hour (MBH).

90 MBH	SN@16.0	Ea	3,820.00	742.00	101.00	4,663.00
120 MBH	SN@16.0	Ea	4,950.00	742.00	101.00	5,793.00
170 MBH	SN@20.0	Ea	8,660.00	927.00	126.00	9,713.00
500 MBH	SN@30.0	Ea	13,400.00	1,390.00	189.00	14,979.00
950 MBH	SN@44.0	Ea	16,000.00	2,040.00	278.00	18,318.00

Description	Craft@Hrs	Unit	Material $	Labor $	Equipment $	Total $

Water mist fire extinguishing systems for biomass boilers. 20 minute full saturation cycle, with automatic reset to enable instantaneous reactivation for re-ignition. NFPA rated. Includes high pressure spray heads, piping, pumps, valves, alarm horns, emergency halogen lighting, calibration, training, insurance inspection and safety procedure documentation. Based on remote sensor relays for the fuel hopper, fuel conveyor, combustion chamber and baghouse filter.

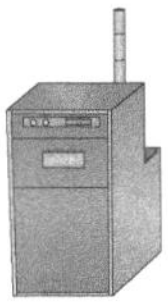

30 BHP boiler	ST@60.0	Ea	16,600.00	3,050.00	—	19,650.00
50 BHP boiler	ST@60.0	Ea	20,800.00	3,050.00	—	23,850.00
100 BHP boiler	ST@80.0	Ea	30,400.00	4,070.00	—	34,470.00
125 BHP boiler	ST@80.0	Ea	32,900.00	4,070.00	—	36,970.00
150 BHP boiler	ST@120	Ea	37,900.00	6,100.00	—	44,000.00
200 BHP boiler	ST@120	Ea	44,100.00	6,100.00	—	50,200.00
Add for CO_2 based extinguishing system	—	%	—	—	—	20.0
Add for Halon-based extinguishing system	—	%	—	—	—	30.0

Pulse-type boilers for central radiant heating. 15 PSI, EPA, ASME and UL rated. Modular design for multiple installations. Natural gas only. With forced draft fan, factory-installed boiler trim, feedwater pump and valves, and tempering valve. For residential, commercial and light industrial radiant heating and potable water applications. Add the cost of a concrete pad, circulating pump, stack and stack liner, CO_2 fire extinguisher system, electrical connection, and post-installation inspection by both the fire marshal and the mechanical inspector to validate federal, state and local energy tax credits or energy tax credit offsets. Equipment cost is for an appliance dolly, come-along and chain hoist. Listed by boiler horsepower (BHP) rating and input capacity in thousands of BTU per hour (MBH).

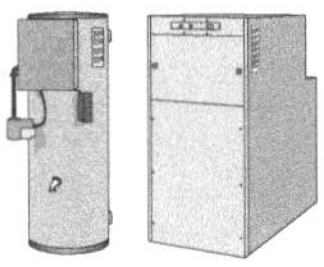

3 BHP, 90 MBH	P1@24.0	Ea	5,300.00	1,070.00	243.00	6,613.00
3 BHP, 150 MBH	P1@24.0	Ea	7,670.00	1,070.00	243.00	8,983.00
6 BHP, 299 MBH	P1@28.0	Ea	10,500.00	1,240.00	284.00	12,024.00

LEED certification requires 88% boiler efficiency, recording controls, compliance with SCAQMD 1146.2 emission standards and zone thermostats.

Add for LEED-certified boiler with USGBC rating	—	%	15.0	—	—	—
Add for LEED registration and inspection	—	Ea	2,150.00	—	—	2,150.00
Add for central performance monitor/ recorder	—	Ea	1,320.00	—	—	1,320.00
Add for zone recording thermostat	—	Ea	183.00	—	—	183.00

Description	Craft@Hrs	Unit	Material $	Labor $	Equipment $	Total $

Passive flat-panel solar water heater. Ambient pressure, modular design for multiple installation, EPA, DOE and UL rated, with piping, tempering valve, circulating pump, controls, storage tank and heating coil. Cost does not include electrical connection, roofing modification (bracing) to accommodate the additional weight of water-filled panels, insulation or padding between roof and panels. Add the cost of post-installation inspection by the mechanical inspector to validate federal, state and local energy tax credits or energy tax credit offsets. For larger arrays (laundries, institutional facilities, food processing plants), develop an estimate based on required daily capacity and multiply these costs by the number of panels required. Equipment cost is for a 1-ton capacity hydraulic crane truck.

Description	Craft@Hrs	Unit	Material $	Labor $	Equipment $	Total $
One 32.3 SF solar panel, 80 gallons per day	P1@16.0	Ea	3,970.00	711.00	956.00	5,637.00
Two 32.3 SF solar panels, 120 gallons per day	P1@20.0	Ea	5,560.00	889.00	1,200.00	7,649.00

Description	Craft@Hrs	Unit	Material $	Labor $	Equipment $	Total $

Centrifugal air-foil wheel blower, with motor, drive and isolation, set in place only.

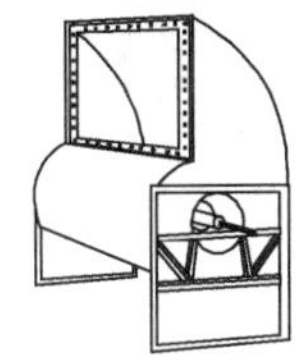

Description	Craft@Hrs	Unit	Material $	Labor $	Equipment $	Total $
1/3 HP, 12" dia.	SN@2.50	Ea	1,850.00	116.00	—	1,966.00
1/2 HP, 15" dia.	SN@4.50	Ea	3,210.00	209.00	—	3,419.00
1 HP, 18" dia.	SN@7.00	Ea	3,710.00	324.00	—	4,034.00
2 HP, 18" dia.	SN@7.50	Ea	4,390.00	348.00	—	4,738.00
3 HP, 27" dia.	SN@9.00	Ea	6,940.00	417.00	6.02	7,363.02
5 HP, 30" dia.	SN@11.0	Ea	8,190.00	510.00	7.36	8,707.36
7½ HP, 36" dia.	SN@12.0	Ea	9,920.00	556.00	8.03	10,484.03
10 HP, 40" dia.	SN@14.0	Ea	13,800.00	649.00	9.37	14,458.37
15 HP, 44" dia.	SN@17.0	Ea	18,500.00	788.00	11.40	19,299.40
20 HP, 60" dia.	SN@20.0	Ea	20,900.00	927.00	13.40	21,840.40

Utility centrifugal blower, with motor, drive and isolation, set in place only.

Description	Craft@Hrs	Unit	Material $	Labor $	Equipment $	Total $
1/3 HP, 10" dia.	SN@2.50	Ea	1,190.00	116.00	—	1,306.00
1/2 HP, 12" dia.	SN@4.50	Ea	1,560.00	209.00	—	1,769.00
1 HP, 15" dia.	SN@6.50	Ea	2,550.00	301.00	—	2,851.00
2 HP, 18" dia.	SN@7.50	Ea	2,900.00	348.00	—	3,248.00
3 HP, 24" dia.	SN@9.00	Ea	4,180.00	417.00	6.02	4,603.02
5 HP, 27" dia.	SN@11.0	Ea	4,800.00	510.00	7.36	5,317.36
7½ HP, 36" dia.	SN@12.0	Ea	7,720.00	556.00	8.03	8,284.03
10 HP, 40" dia.	SN@14.0	Ea	9,310.00	649.00	9.37	9,968.37
15 HP, 44" dia.	SN@17.0	Ea	10,100.00	788.00	11.40	10,899.40

Vane-axial fan, set in place only.

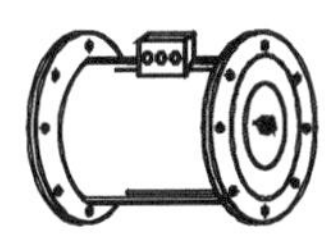

Description	Craft@Hrs	Unit	Material $	Labor $	Equipment $	Total $
30 HP, 36" dia.	SN@16.0	Ea	24,500.00	742.00	273.00	25,515.00
50 HP, 42" dia.	SN@24.0	Ea	27,200.00	1,110.00	410.00	28,720.00
100 HP, 48" dia.	SN@32.0	Ea	33,900.00	1,480.00	546.00	35,926.00
200 HP, 54" dia.	SN@40.0	Ea	48,000.00	1,850.00	683.00	50,533.00

Description	Craft@Hrs	Unit	Material $	Labor $	Equipment $	Total $
Tube-axial fan, set in place only.						
15 HP, 24" dia.	SN@10.0	Ea	7,230.00	464.00	171.00	7,865.00
25 HP, 30" dia.	SN@13.0	Ea	11,200.00	603.00	222.00	12,025.00
30 HP, 36" dia.	SN@16.0	Ea	15,000.00	742.00	273.00	16,015.00
75 HP, 40" dia.	SN@22.0	Ea	22,300.00	1,020.00	376.00	23,696.00
100 HP, 48" dia.	SN@32.0	Ea	34,100.00	1,480.00	546.00	36,126.00

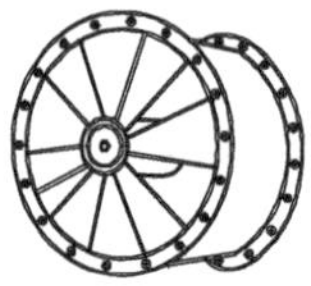

Roof exhaust fan, ASHRAE/American National Standards Institute specification 62.2-2007 and Air Conditioning Contractors of America Standard ACCA 5QI-2007, "HVAC Indoor Air Quality Installation Specification 2007." U.S. Green Building Council, Underwriters laboratories and CSA approved. Cost includes fan and bearing, shaft mount and power transmission assembly or direct drive, enclosure, mount plates, electric motor, 110 or 408, 3-phase, 60 Hz, switch or relay assembly, motor fusing and ground fault protection, fan blade guard. Set in place only. Add installation costs from the figures below.

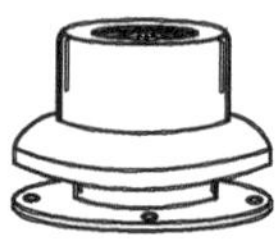

Description	Craft@Hrs	Unit	Material $	Labor $	Equipment $	Total $
1/4 HP, 14" dia.	SN@2.50	Ea	1,910.00	116.00	—	2,026.00
1/3 HP, 22" dia.	SN@2.90	Ea	2,350.00	134.00	—	2,484.00
1/2 HP, 24" dia.	SN@3.50	Ea	2,680.00	162.00	—	2,842.00
1 HP, 30" dia.	SN@4.50	Ea	3,710.00	209.00	—	3,919.00
2 HP, 36" dia.	SN@5.00	Ea	5,310.00	232.00	85.40	5,627.40
3 HP, 40" dia.	SN@6.00	Ea	7,940.00	278.00	102.00	8,320.00
5 HP, 48" dia.	SN@7.00	Ea	11,000.00	324.00	120.00	11,444.00
Cut, frame and gasket downcomer hole in roof	S2@1.00	Ea	37.80	43.50	—	81.30
Mount duct hangers	S2@.250	Ea	4.55	10.90	—	15.45
Cut and mount sheet metal duct	S2@.350	LF	7.00	15.20	—	22.20
Apply and coat duct insulation	S2@.150	LF	3.01	6.53	—	9.54
Install electrical wiring	BE@.150	LF	2.24	7.25	—	9.49
Install HVAC controls	BE@.500	Ea	—	24.20	—	24.20
Commission and test	P1@2.00	Ea	—	88.90	—	88.90
Run air balance and fine-tune system	P1@1.00	Ea	—	44.50	—	44.50

Description	Craft@Hrs	Unit	Material $	Labor $	Equipment $	Total $

Ceiling-mounted commercial ventilator. Low-profile metal grille. Galvanized steel housing. Designed for continuous operation. Acoustically insulated. Can be installed horizontally, vertically or in-line. Labor includes layout and fasten in place only. Make additional allowances for vent duct and electrical requirements.

100 CFM	P1@1.05	Ea	193.00	46.70	—	239.70
150 CFM	P1@1.15	Ea	244.00	51.10	—	295.10
200 CFM	P1@1.35	Ea	307.00	60.00	—	367.00
250 CFM	P1@1.45	Ea	386.00	64.50	—	450.50
300 CFM	P1@1.65	Ea	461.00	73.30	—	534.30
400 CFM	P1@2.15	Ea	593.00	95.60	—	688.60
500 CFM	P1@2.50	Ea	721.00	111.00	—	832.00
700 CFM	P1@2.75	Ea	1,000.00	122.00	—	1,122.00
900 CFM	P1@2.85	Ea	1,110.00	127.00	170.00	1,407.00
1,500 CFM	P1@3.50	Ea	1,540.00	156.00	209.00	1,905.00
2,000 CFM	P1@3.85	Ea	1,690.00	171.00	230.00	2,091.00
3,500 CFM	P1@4.25	Ea	1,930.00	189.00	254.00	2,373.00

Powered attic ventilator. Thermostatically controlled. 1/10 Hp, 120V/6-Hz motor. 8" x 24" metal housing. Labor includes layout and fasten in place only. Make additional allowances for electrical and roofing requirements.

1,200 CFM	P1@2.00	Ea	244.00	88.90	—	332.90

Exterior wall/roof exhaust fan. Fully enclosed. Thermally protected. Lifetime lubricated 120V/6-Hz motor. Weather-resistant aluminum housing. Balance blower wheel. Neoprene vibration isolation. Includes backdraft damper, flashing plate and critter screen. 10" discharge outlet. Labor includes layout and fasten in place only. Make additional allowances for vent duct, electrical and roofing requirements.

600 CFM	P1@2.50	Ea	721.00	111.00	—	832.00
900 CFM	P1@2.85	Ea	1,120.00	127.00	—	1,247.00

Room ventilator fan. Interior thru-wall room-to-room air transfer application. High efficiency. Manually switched operation with built-in variable speed control. 8" and 10" discharge outlets. Built in damper. Includes intake and discharge grilles. Labor includes layout and fasten in place only. Make additional allowances for electrical requirements.

180 CFM	P1@1.25	Ea	149.00	55.60	—	204.60
270 CFM	P1@1.50	Ea	225.00	66.70	—	291.70
380 CFM	P1@1.75	Ea	307.00	77.80	—	384.80

Ceiling exhaust fan, with automatic motion response control. Adjustable (5-60 minutes) time interval operation. Manual override feature. 4" discharge outlet. Pre-wired outlet box. Balanced blower wheel. Permanently lubricated motor. Neoprene vibration isolators. Quiet operation. Suitable for washroom exhaust applications. Labor includes layout and fasten in place only. Make additional allowances for vent duct and electrical requirements.

90 CFM	P1@1.00	Ea	231.00	44.50	—	275.50
130 CFM	P1@1.15	Ea	358.00	51.10	—	409.10

Description	Craft@Hrs	Unit	Material $	Labor $	Equipment $	Total $

Ceiling exhaust fan, with automatic humidity response control. Adjustable time interval operation (5-60 minutes). Manual override feature. 4" discharge outlet. Pre-wired outlet box. Balanced blower wheel. Permanently lubricated motor. Neoprene vibration isolators. Quiet operation. Suitable for washroom exhaust applications. Labor includes layout and fasten in place only. Make additional allowances for vent duct and electrical requirements.

90 CFM	P1@1.00	Ea	272.00	44.50	—	316.50
130 CFM	P1@1.15	Ea	399.00	51.10	—	450.10

Ceiling exhaust fan, manually-switched operation. 4" discharge outlet. Pre-wired outlet box. 4 pole motor. Suitable for washroom exhaust applications. Labor includes layout and fasten in place only. Make additional allowances for vent duct and electrical requirements.

50 CFM	P1@1.00	Ea	51.70	44.50	—	96.20
70 CFM	P1@1.00	Ea	77.20	44.50	—	121.70
90 CFM	P1@1.00	Ea	90.00	44.50	—	134.50
110 CFM	P1@1.05	Ea	108.00	46.70	—	154.70

Ceiling exhaust fan, high efficiency. Manually switched operation. 8" and 10" discharge outlets. Labor includes layout and fasten in place only. Suitable for kitchens, laundry rooms, rec rooms and workshops. Make additional allowances for vent duct and electrical requirements.

160 CFM	P1@1.25	Ea	97.60	55.60	—	153.20
180 CFM	P1@1.25	Ea	141.00	55.60	—	196.60
270 CFM	P1@1.50	Ea	206.00	66.70	—	272.70
350 CFM	P1@1.75	Ea	259.00	77.80	—	336.80

Discharge hood ducting kit. Including discharge hood, 4" x 10' of flexible duct, duct clamps, 3" x 4" transition coupling.

Wall hood, 4"	P1@2.00	Ea	25.10	88.90	—	114.00
Roof hood, 4"	P1@2.25	Ea	49.90	100.00	—	149.90
Wall hood, 6"	P1@2.55	Ea	112.00	113.00	—	225.00

Fan controls. Switch only.

1 hour timer	BE@.155	Ea	79.30	7.49	—	86.79
Humidistat	BE@.155	Ea	54.70	7.49	—	62.19
Speed controller	BE@.155	Ea	74.60	7.49	—	82.09

Description	Craft@Hrs	Unit	Material $	Labor $	Equipment $	Total $

Air admittance valve. For ventless sanitary plumbing. Provides a water seal in traps without roof penetration and vent piping. Complies with ASHRAE/American National Standards Institute spec. 62.2-2007 and American Society for Sanitary Engineering specs 1050 and 1051.

Cut, frame and gasket for wall inset and vent valve access door	SW@1.00	Ea	39.60	49.60	—	89.20
Mount valve	P1@.500	Ea	—	22.20	—	22.20
Connect valve to sanitary S- or P-trap	P1@1.00	Ea	—	44.50	—	44.50

Bathroom fans and heaters. Comply with ASHRAE/American National Standards Institute spec 62.2-2007 and meet new indoor air quality code requirements for dehumidification, mold and mildew prevention and radon mitigation. Continental Fan, Tamarack, Broan or equal. Includes enclosure, mount plates, 110 or 408 volt electric motor, switch or relay assembly, motor fusing, ground fault protection. Add the following installation costs, if required.

Cut, frame and gasket exhaust hole in sidewall	SW@1.00	Ea	39.60	49.60	—	89.20
Cut and mount sheet metal duct	SW@.150	LF	7.36	7.45	—	14.81
Apply and coat duct insulation	S2@.050	LF	3.17	2.18	—	5.35
Install electrical wiring (typical)	BE@.150	LF	2.35	7.25	—	9.60
Install controls	BE@.500	Ea	—	24.20	—	24.20
Test and adjust	P1@.250	Ea	—	11.10	—	11.10

Bathroom exhaust fan with heater, commercial grade, Continental Fan, Broan or equal.

106 CFM fan, 4" duct	BE@.500	Ea	271.00	24.20	—	295.20
152 CFM fan, 4" duct	BE@.500	Ea	308.00	24.20	—	332.20
247 CFM fan, 6" duct	BE@.500	Ea	513.00	24.20	—	537.20
418 CFM fan, 8" duct	BE@.500	Ea	661.00	24.20	—	685.20

Infrared bulb heater and fan. 4-point adjustable mounting brackets span up to 24 inches. Uses 250 watt, 120 volt R-40 infrared bulbs (not included). Heater and fan units include 70 CFM, 3.5 sones ventilator fan. Damper and duct connector included. Plastic matte-white molded grille and compact housings. CFM (cubic feet of air per minute).

1-bulb, 70 CFM, 3.5 sones	BE@1.00	Ea	76.90	48.40	—	125.30
2-bulb heater, 7" duct	BE@1.10	Ea	96.70	53.20	—	149.90

Description	Craft@Hrs	Unit	Material $	Labor $	Equipment $	Total $

Heat-A-Lamp® bulb heater and fan. NuTone. Swivel-mount. Torsion spring holds plates firmly to ceiling. 250 watt radiant heat from one R-40 infrared heat lamp. Uses 4-inch duct. Adjustable socket for ceilings up to 1-inch thick. Adjustable hanger bars. Automatic reset for thermal protection. White polymeric finish. UL listed.

2.6 amp, 1 lamp, 12-1/2" x 10"	BE@1.50	Ea	79.40	72.50	—	151.90
5.0 amp, 2 lamp, 15-3/8" x 11"	BE@1.50	Ea	102.00	72.50	—	174.50

Surface-mount ceiling bath resistance heater and fan. 1,250 watt. 4,266 BTU. 120 volt. Chrome alloy wire element for instant heat. Built-in fan. Automatic overheat protection. Low-profile housing. Permanently lubricated motor. Mounts to standard 3¼-inch round or 4-inch octagonal ceiling electrical box. Satin-finish aluminum grille extends 2¾ inch from ceiling. 11" diameter x 2¾" deep.

10.7 amp	BE@.750	Ea	80.90	36.30	—	117.20

Designer Series bath heater and exhaust fan. 1,500 watt fan-forced heater. 120 watt light capacity, 7 watt night-light (bulbs not included). Permanently lubricated motor. Polymeric damper. Torsion-spring grille mount. Non-glare light-diffusing glass lens. Fits single gang opening. Suitable for use with insulation. 100 CFM, 1.2 sones.

4" duct, 100 watt lamp	BE@1.00	Ea	328.00	48.40	—	376.40

Ventilation exhausters

Includes all-weather enclosure, air balance control, software module and NFPA-certified fusible plug-type vertical duct riser gravity dampers, and efficiency optimization arrays. Add the cost of roof curb, control wiring, electrical wiring, permits and final inspection. For wall- or roof-mount in single- and multi-unit residences, commercial (i.e. hotels), institutional (i.e. hospitals and schools) buildings. Set in place only. Add installation costs from the section below. Costs assume rental of an appliance dolly, a 3,000 lb. come-a-long and a ½-ton chain hoist.

Downblast ventilation exhausters. Rooftop mounted, direct drive, centrifugal type. Suitable for either ducted or nonducted applications.

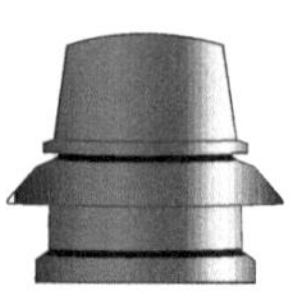

1/4 HP, 14" dia.	SN@2.50	Ea	1,940.00	116.00	—	2,056.00
1/3 HP, 22" dia.	SN@2.90	Ea	2,390.00	134.00	—	2,524.00
1/2 HP, 24" dia.	SN@3.50	Ea	2,730.00	162.00	—	2,892.00
1 HP, 30" dia.	SN@4.50	Ea	3,790.00	209.00	—	3,999.00
2 HP, 36" dia.	SN@5.00	Ea	5,410.00	232.00	14.50	5,656.50
3 HP, 40" dia.	SN@6.00	Ea	8,100.00	278.00	17.40	8,395.40
5 HP, 48" dia.	SN@7.00	Ea	11,200.00	324.00	20.30	11,544.30

Description	Craft@Hrs	Unit	Material $	Labor $	Equipment $	Total $

Upblast ventilation exhausters. Direct drive, centrifugal, for roof or wall mount.

1/4 HP, 14" dia.	SN@2.50	Ea	2,100.00	116.00	—	2,216.00
1/3 HP, 22" dia.	SN@2.90	Ea	2,540.00	134.00	—	2,674.00
1/2 HP, 24" dia.	SN@3.50	Ea	2,910.00	162.00	—	3,072.00
1 HP, 30" dia.	SN@4.50	Ea	4,050.00	209.00	—	4,259.00
2 HP, 36" dia.	SN@5.00	Ea	5,690.00	232.00	14.50	5,936.50
3 HP, 40" dia.	SN@6.00	Ea	8,370.00	278.00	17.40	8,665.40
5 HP, 48" dia.	SN@7.00	Ea	11,500.00	324.00	20.30	11,844.30

Energy-efficient exhauster arrays. Upblast belt-drive centrifugal UL762 roof exhausters. Low profile metal grille. Galvanized steel housing. Designed for continuous operation. Acoustically insulated. Can be installed horizontally, vertically or in-line. Labor includes layout and fasten in place only. Make additional allowances for vent duct and electrical requirements.

100 CFM	P1@1.05	Ea	299.00	46.70	—	345.70
150 CFM	P1@1.15	Ea	329.00	51.10	—	380.10
200 CFM	P1@1.35	Ea	358.00	60.00	—	418.00
250 CFM	P1@1.45	Ea	463.00	64.50	—	527.50
300 CFM	P1@1.65	Ea	524.00	73.30	—	597.30
400 CFM	P1@2.15	Ea	657.00	95.60	—	752.60
500 CFM	P1@2.50	Ea	762.00	111.00	—	873.00
700 CFM	P1@2.75	Ea	1,090.00	122.00	—	1,212.00
900 CFM	P1@2.85	Ea	1,190.00	127.00	8.25	1,325.25
1,500 CFM	P1@3.50	Ea	1,810.00	156.00	10.10	1,976.10
2,000 CFM	P1@3.85	Ea	2,010.00	171.00	11.10	2,192.10
3,500 CFM	P1@4.25	Ea	2,240.00	189.00	12.30	2,441.30

Installation costs for ventilation exhausters.

Cut, frame and gasket downcomer hole in sidewall or roof	S2@1.00	Ea	38.30	43.50	—	81.80
Mount duct hangers	S2@.250	Ea	4.55	10.90	—	15.45
Cut & mount sheet metal duct	S2@.350	LF	7.00	15.20	—	22.20
Apply and coat duct insulation	S2@.150	LF	3.01	6.53	—	9.54
Install electrical wiring	BE@.150	LF	2.24	7.25	—	9.49
Install exhauster controls	BE@.500	Ea	—	24.20	—	24.20
Commission and test	P1@4.00	Ea	—	178.00	—	178.00
Run air balance and fine-tune system	P1@4.00	Ea	—	178.00	—	178.00

Apparatus Housing

Unitary or factory-packaged air conditioning units are often too small to meet the heating and cooling needs of a large single-zone space. In such cases, engineers design what's called a "built-up" system. They're custom fitted to the client's needs and the space. A typical "built-up" system includes the following equipment:

- centrifugal blowers or axial-flow fans
- damper assemblies
- coil banks
- filter banks

Once the system is assembled, it's enclosed in a shop-fabricated housing or casing. The illustration on page 333 shows a standard apparatus housing. The walls and roof are shop-built panels. These panels consist of a layer of thermal insulation covered on one or both sides with sheets of 18 gauge galvanized steel. An interior framework of steel angles or channels supports the panels. Seal all mating surfaces to minimize air leakage, using gaskets, caulking or sealant.

Apparatus Housing

There are two types of panels used in building apparatus housings: "double-skin" panels and "single-skin" panels. Double-skin panels have a layer of thermal insulation material sandwiched between two sheets of 18 gauge galvanized steel. Single-skin panels also have a layer of insulation. However, only the outside surface is clad in 18 gauge steel sheeting. On the inside the panel insulation is exposed.

The costs of constructing and installing apparatus housings vary greatly. This is due to the wide range of possible conditions and needs that you face on each job. Some of the many factors that influence costs include the following:

- internal operating pressures
- external wind loads
- seismic requirements
- physical locations

That's why I recommend that you always prepare a detailed cost estimate. Then you can take all the particulars into account. However, for budget estimating purposes only, use the following cost data:

Single-skin panels	$12.30 per SF*
Double-skin panels	$20.90 per SF*
Access doors (20" x 60")	$428.00 each

*Total area of walls and roof, less doors.

When you're estimating material costs for apparatus housings, remember to add in an extra 25 percent to cover the following costs: waste, seams, gaskets, sealants and miscellaneous assembly hardware.

Labor for Shop-Fabrication and Field Assembly of Apparatus Housings. The costs listed are manhours per square foot (SF) of panel. To find the total SF of panel, add up areas for all the panels.

Description	Shop Labor (MH per SF)	Field Labor (MH per SF)
Single-skin panels	.040	.250
Double-skin panels	.085	.300
Add for access doors	.400	.240

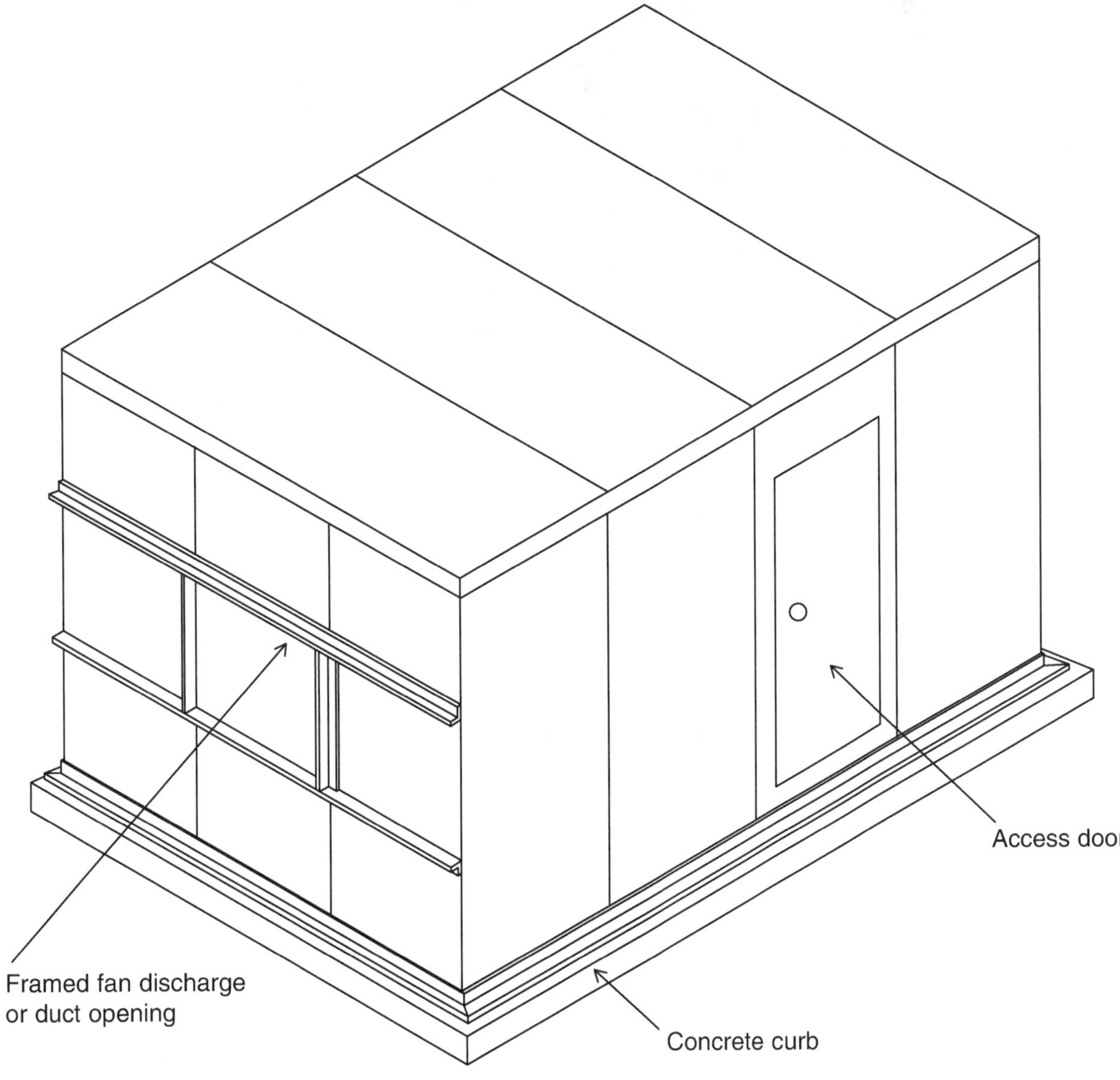

Figure 1 - Typical apparatus housing

Supply air side-wall register, fixed pattern, with damper. Steel construction.

Description	Craft@Hrs	Unit	Material $	Labor $	Equipment $	Total $
8" x 4"	S2@.250	Ea	12.20	10.90	—	23.10
8" x 6"	S2@.250	Ea	12.60	10.90	—	23.50
10" x 4"	S2@.250	Ea	13.70	10.90	—	24.60
10" x 6"	S2@.250	Ea	23.40	10.90	—	34.30
10" x 8"	S2@.250	Ea	22.40	10.90	—	33.30
12" x 6"	S2@.250	Ea	23.70	10.90	—	34.60
12" x 8"	S2@.250	Ea	19.20	10.90	—	30.10
14" x 6"	S2@.250	Ea	19.40	10.90	—	30.30
14" x 8"	S2@.250	Ea	22.10	10.90	—	33.00

Return air side-wall grille, fixed pattern, without damper. Steel construction.

Description	Craft@Hrs	Unit	Material $	Labor $	Equipment $	Total $
14" x 6"	S2@.250	Ea	10.50	10.90	—	21.40
14" x 8"	S2@.250	Ea	10.20	10.90	—	21.10
16" x 16"	S2@.250	Ea	30.80	10.90	—	41.70
20" x 10"	S2@.300	Ea	45.80	13.10	—	58.90
20" x 16"	S2@.300	Ea	50.20	13.10	—	63.30
20" x 20"	S2@.300	Ea	66.40	13.10	—	79.50
24" x 24"	S2@.350	Ea	86.40	15.20	—	101.60

Supply air double-deflection side-wall register, adjustable pattern. Aluminum construction.

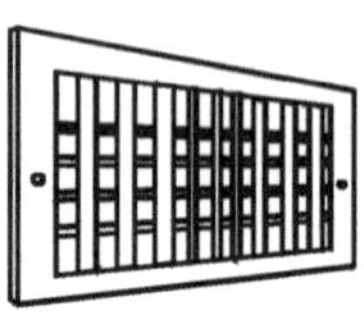

Description	Craft@Hrs	Unit	Material $	Labor $	Equipment $	Total $
8" x 4"	S2@.250	Ea	39.50	10.90	—	50.40
8" x 6"	S2@.250	Ea	44.00	10.90	—	54.90
10" x 4"	S2@.250	Ea	50.50	10.90	—	61.40
10" x 6"	S2@.250	Ea	51.90	10.90	—	62.80
10" x 8"	S2@.250	Ea	55.80	10.90	—	66.70
12" x 6"	S2@.250	Ea	57.70	10.90	—	68.60
12" x 8"	S2@.250	Ea	64.60	10.90	—	75.50
12" x 10"	S2@.250	Ea	73.30	10.90	—	84.20
14" x 6"	S2@.250	Ea	73.30	10.90	—	84.20
14" x 8"	S2@.250	Ea	74.10	10.90	—	85.00
14" x 10"	S2@.250	Ea	78.90	10.90	—	89.80
16" x 8"	S2@.250	Ea	78.90	10.90	—	89.80
16" x 10"	S2@.250	Ea	86.40	10.90	—	97.30
18" x 8"	S2@.300	Ea	90.00	13.10	—	103.10
18" x 10"	S2@.300	Ea	96.50	13.10	—	109.60
18" x 12"	S2@.300	Ea	102.00	13.10	—	115.10

Description	Craft@Hrs	Unit	Material $	Labor $	Equipment $	Total $

Return air single-deflection side-wall grille, adjustable pattern. Aluminum construction.

Description	Craft@Hrs	Unit	Material $	Labor $	Equipment $	Total $
8" x 4"	S2@.250	Ea	35.80	10.90	—	46.70
8" x 6"	S2@.250	Ea	39.50	10.90	—	50.40
10" x 4"	S2@.250	Ea	43.30	10.90	—	54.20
10" x 6"	S2@.250	Ea	46.80	10.90	—	57.70
10" x 8"	S2@.250	Ea	50.00	10.90	—	60.90
12" x 6"	S2@.250	Ea	51.60	10.90	—	62.50
12" x 8"	S2@.250	Ea	58.10	10.90	—	69.00
12" x 10"	S2@.250	Ea	65.50	10.90	—	76.40
14" x 6"	S2@.250	Ea	65.50	10.90	—	76.40
14" x 8"	S2@.250	Ea	67.20	10.90	—	78.10
14" x 10"	S2@.250	Ea	70.40	10.90	—	81.30
16" x 8"	S2@.250	Ea	70.40	10.90	—	81.30
16" x 10"	S2@.300	Ea	77.90	13.10	—	91.00
18" x 8"	S2@.300	Ea	81.10	13.10	—	94.20
18" x 10"	S2@.300	Ea	87.10	13.10	—	100.20
18" x 12"	S2@.300	Ea	91.80	13.10	—	104.90

4-way adjustable pattern ceiling diffuser, 24" x 24" with collar and volume damper. Steel construction.

Description	Craft@Hrs	Unit	Material $	Labor $	Equipment $	Total $
6" neck	S2@.350	Ea	101.00	15.20	—	116.20
8" neck	S2@.350	Ea	106.00	15.20	—	121.20
10" neck	S2@.350	Ea	115.00	15.20	—	130.20
12" neck	S2@.350	Ea	119.00	15.20	—	134.20
14" neck	S2@.400	Ea	143.00	17.40	—	160.40

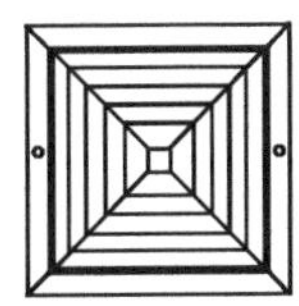

Fixed pattern ceiling diffuser, 24" x 24" with collar, without volume damper. Steel construction.

Description	Craft@Hrs	Unit	Material $	Labor $	Equipment $	Total $
6" neck	S2@.350	Ea	61.10	15.20	—	76.30
8" neck	S2@.350	Ea	65.80	15.20	—	81.00
10" neck	S2@.350	Ea	73.90	15.20	—	89.10
12" neck	S2@.350	Ea	78.30	15.20	—	93.50
14" neck	S2@.350	Ea	101.00	15.20	—	116.20

Perforated ceiling diffuser, 24" x 24" fixed pattern with collar, without volume damper. Steel construction.

Description	Craft@Hrs	Unit	Material $	Labor $	Equipment $	Total $
6" neck	S2@.350	Ea	81.20	15.20	—	96.40
8" neck	S2@.350	Ea	84.20	15.20	—	99.40
10" neck	S2@.350	Ea	90.20	15.20	—	105.40
12" neck	S2@.350	Ea	96.50	15.20	—	111.70
14" neck	S2@.400	Ea	108.00	17.40	—	125.40
16" neck	S2@.720	Ea	182.00	31.30	—	213.30

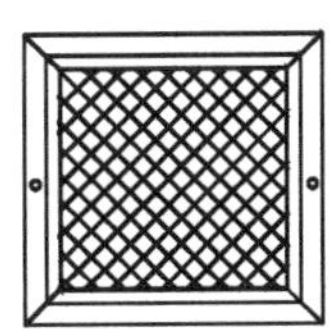

Description	Craft@Hrs	Unit	Material $	Labor $	Equipment $	Total $

Round 3-cone, 2-position ceiling diffuser with collar, without volume damper. Steel construction.

6"	S2@.350	Ea	83.90	15.20	—	99.10
8"	S2@.350	Ea	96.50	15.20	—	111.70
10"	S2@.350	Ea	117.00	15.20	—	132.20
12"	S2@.350	Ea	137.00	15.20	—	152.20
14"	S2@.350	Ea	183.00	15.20	—	198.20

Round adjustable ceiling diffuser with collar and volume damper. Steel construction.

6"	S2@.350	Ea	126.00	15.20	—	141.20
8"	S2@.350	Ea	150.00	15.20	—	165.20
10"	S2@.350	Ea	117.00	15.20	—	132.20
12"	S2@.350	Ea	183.00	15.20	—	198.20
14"	S2@.350	Ea	341.00	15.20	—	356.20
16"	S2@.450	Ea	399.00	19.60	—	418.60
18"	S2@.450	Ea	460.00	19.60	—	479.60
20"	S2@.550	Ea	545.00	23.90	—	568.90
24"	S2@.650	Ea	748.00	28.30	—	776.30

Description	Craft@Hrs	Unit	Material $	Labor $	Equipment $	Total $

Makeup air unit. Makeup air unit uses 100% direct-heated outdoor air for warehouse, factory and institutional applications to provide higher thermal efficiency and balanced distribution of system air while pre-heating air for use in boiler or other HVAC applications. ANSI standard Z83.4/CSA 3.7 certified. With weatherproof metal enclosure, burner, analog electric controls for burner and fan, thermostat, actuator, natural gas burner array, and either roof-mount supports or sidewall external mounts and transition and gaskets. Set in place only. Add installation costs.

Description	Craft@Hrs	Unit	Material $	Labor $	Equipment $	Total $
200 to 399 CFM	SN@2.25	Ea	752.00	104.00	—	856.00
400 to 599 CFM	SN@2.25	Ea	831.00	104.00	—	935.00
600 to 799 CFM	SN@2.75	Ea	955.00	127.00	47.00	1,129.00
800 to 999 CFM	SN@3.00	Ea	1,150.00	139.00	51.20	1,340.20
1,000 to 1,499 CFM	SN@3.50	Ea	1,520.00	162.00	59.80	1,741.80
1,500 to 1,999 CFM	SN@4.00	Ea	1,920.00	185.00	68.30	2,173.30
4,000 to 5,999 CFM	SN@6.00	Ea	4,870.00	278.00	102.00	5,250.00
7,500 to 10,000 CFM	SN@8.00	Ea	8,330.00	371.00	137.00	8,838.00
15,000 to 25,000 CFM	SN@10.0	Ea	11,800.00	464.00	171.00	12,435.00
30,000 to 40,000 CFM	SN@16.0	Ea	17,500.00	742.00	273.00	18,515.00
60,000 to 80,000 CFM	SN@24.0	Ea	30,600.00	1,110.00	410.00	32,120.00

Installation costs for makeup air units.

Description	Craft@Hrs	Unit	Material $	Labor $	Equipment $	Total $
Cut, frame and gasket downcomer/ sidewall hole	S2@1.00	Ea	35.20	43.50	—	78.70
Mount duct hangers	S2@.250	Ea	4.23	10.90	—	15.13
Cut and mount sheet metal duct	S2@.350	LF	6.54	15.20	—	21.74
Apply and coat duct insulation	S2@.150	LF	2.83	6.53	—	9.36
Install piping for gas line	P1@.100	LF	4.68	4.45	—	9.13
Install electrical wiring	BE@.150	LF	2.09	7.25	—	9.34
Install makeup air unit controls	BE@.500	Ea	—	24.20	—	24.20
Commission and test	P1@4.00	Ea	—	178.00	—	178.00
Run air balance and fine-tune system	P1@4.00	Ea	—	178.00	—	178.00

Variable-air volume units regulate supply air to ducted air conditioning systems. These systems are certified to ANSI standard Z83.4/CSA 3.7 and can be either roof-mounted or upper sidewall externally-mounted. Costs include a weatherproof metal enclosure, heater coil or electrical heater fixture, fan and fan controls, downcomer or sidewall entry transition and gaskets, and either roof-mount supports or sidewall external mounts. Includes set in place only. Add installation costs from the table that follows.

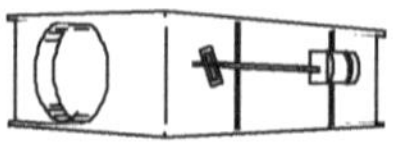

Description	Craft@Hrs	Unit	Material $	Labor $	Equipment $	Total $
Variable-air volume terminal unit with analog electric controls, thermostat, actuator and controller.						
200- 400 CFM	SN@2.00	Ea	678.00	92.70	34.10	804.80
400- 600 CFM	SN@2.00	Ea	756.00	92.70	34.10	882.80
600- 800 CFM	SN@2.50	Ea	795.00	116.00	42.70	953.70
800-1,000 CFM	SN@2.75	Ea	863.00	127.00	47.00	1,037.00
1,000-1,500 CFM	SN@3.00	Ea	1,250.00	139.00	51.20	1,440.20
1,500-2,000 CFM	SN@3.50	Ea	1,590.00	162.00	59.80	1,811.80
Variable-air volume terminal reheat unit with analog electric controls, thermostat, actuator, controller and a two-row hot water coil.						
200- 400 CFM	SN@2.25	Ea	782.00	104.00	38.40	924.40
400- 600 CFM	SN@2.25	Ea	863.00	104.00	38.40	1,005.40
600- 800 CFM	SN@2.75	Ea	994.00	127.00	47.00	1,168.00
800-1,000 CFM	SN@3.00	Ea	1,190.00	139.00	51.20	1,380.20
1,000-1,500 CFM	SN@3.50	Ea	1,590.00	162.00	59.80	1,811.80
1,500-2,000 CFM	SN@4.00	Ea	2,010.00	185.00	68.30	2,263.30
Installation costs for variable-air volume units.						
Cut, frame and gasket downcomer hole in roof	S2@1.00	Ea	34.40	43.50	—	77.90
Mount duct hangers	S2@.250	Ea	4.15	10.90	—	15.05
Cut and mount sheet metal duct	S2@.350	LF	6.41	15.20	—	21.61
Apply and coat duct insulation	S2@.150	LF	2.77	6.53	—	9.30
Install piping for chilled or hot water/steam	P1@.100	LF	3.46	4.45	—	7.91
Install electrical wiring	BE@.150	LF	2.04	7.25	—	9.29
Install VAV controls	BE@.500	Ea	—	24.20	—	24.20
Commission and test	P1@4.00	Ea	—	178.00	—	178.00
Run air balance and fine-tune system	P1@4.00	Ea	—	178.00	—	178.00

Air balance software and control modules for variable-air volume ducted zone control. For commercial, institutional and industrial applications. Johnson Controls or equal. The zone and duct-mounted sensor and response unit comprises a digital controller, actuator and differential pressure transducer. These components are wired together and securely housed inside a plenum-rated enclosure which mounts directly to the VAV "serviced zone" box. Complies with Underwriters Laboratories file E107041, CCN PAZX, UL916 Energy Management Equipment. Interfaces with Windows-based commercial desktop computers per the digital control manufacturer's specifications.

Description	Craft@Hrs	Unit	Material $	Labor $	Equipment $	Total $
Fully programmable 28 point controller, 16 universal inputs, 12 universal outputs, 12 HOA switches, a battery-backed real-time clock, plus preconfigured function blocks for a real-time clock. Stores schedules and trending information.						
Programmable controller	BE@.500	Ea	1,140.00	24.20	—	1,164.20
Configurable variable air volume pressure independent terminal box controller with wireless transceiver (868.3 MHz)						
Terminal box controller	BE@.500	Ea	3,420.00	24.20	—	3,444.20
Configurable controller, preconfigured LNS plug-in, and all input devices required for typical rooftop unit application						
Preconfigured controller	BE@.500	Ea	2,260.00	24.20	—	2,284.20
Variable-air volume controller system software for a Windows computer. Includes a software license and support for two concurrent client browser connections with no limit on the number of concurrent users and no data tag limit.						
System software	BE@18.0	Ea	7,450.00	870.00	—	8,320.00

Opposed-blade damper for rectangular ductwork

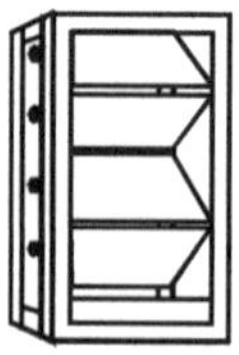

Description	Craft@Hrs	Unit	Material $	Labor $	Equipment $	Total $
12" x 6"	S2@.600	Ea	73.80	26.10	—	99.90
12" x 10"	S2@.650	Ea	87.30	28.30	—	115.60
12" x 12"	S2@.700	Ea	107.00	30.50	—	137.50
18" x 12"	S2@.800	Ea	122.00	34.80	—	156.80
18" x 18"	S2@.850	Ea	136.00	37.00	—	173.00
24" x 12"	S2@.850	Ea	145.00	37.00	—	182.00
24" x 18"	S2@.900	Ea	158.00	39.20	—	197.20
24" x 24"	S2@1.00	Ea	175.00	43.50	—	218.50
30" x 12"	S2@.900	Ea	164.00	39.20	—	203.20
30" x 18"	S2@1.00	Ea	183.00	43.50	—	226.50
30" x 24"	S2@1.10	Ea	209.00	47.90	—	256.90
30" x 30"	S2@1.20	Ea	223.00	52.20	—	275.20
36" x 18"	S2@1.10	Ea	215.00	47.90	—	262.90
36" x 36"	S2@1.40	Ea	295.00	60.90	—	355.90
42" x 24"	S2@1.30	Ea	300.00	56.60	—	356.60
42" x 42"	S2@1.50	Ea	410.00	65.30	—	475.30
48" x 24"	S2@1.40	Ea	312.00	60.90	—	372.90
48" x 48"	S2@1.60	Ea	432.00	69.60	—	501.60
54" x 24"	S2@1.45	Ea	395.00	63.10	—	458.10
54" x 54"	S2@1.70	Ea	493.00	74.00	—	567.00

Single-blade damper for rectangular ductwork

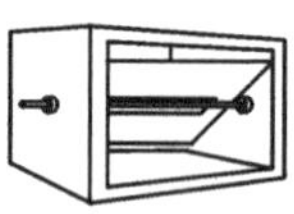

Description	Craft@Hrs	Unit	Material $	Labor $	Equipment $	Total $
12" x 6"	S2@.600	Ea	46.20	26.10	—	72.30
12" x 8"	S2@.625	Ea	50.60	27.20	—	77.80
12" x 10"	S2@.650	Ea	56.90	28.30	—	85.20
12" x 12"	S2@.700	Ea	66.70	30.50	—	97.20
18" x 12"	S2@.800	Ea	74.30	34.80	—	109.10
18" x 18"	S2@.850	Ea	84.30	37.00	—	121.30
24" x 12"	S2@.850	Ea	89.40	37.00	—	126.40
24" x 18"	S2@.900	Ea	93.00	39.20	—	132.20
24" x 24"	S2@1.00	Ea	102.00	43.50	—	145.50

Single-blade butterfly damper for round ductwork

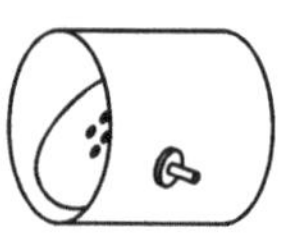

Description	Craft@Hrs	Unit	Material $	Labor $	Equipment $	Total $
6"	S2@.600	Ea	46.20	26.10	—	72.30
8"	S2@.550	Ea	43.60	23.90	—	67.50
10"	S2@.600	Ea	50.50	26.10	—	76.60
12"	S2@.650	Ea	57.10	28.30	—	85.40
14"	S2@.700	Ea	65.10	30.50	—	95.60
16"	S2@.765	Ea	72.20	33.30	—	105.50
18"	S2@.835	Ea	79.90	36.30	—	116.20
20"	S2@.900	Ea	89.90	39.20	—	129.10

Description	Craft@Hrs	Unit	Material $	Labor $	Equipment $	Total $

Curtain-type duct fire damper, 1½-hour rating, U.L. label

Description	Craft@Hrs	Unit	Material $	Labor $	Equipment $	Total $
12" x 6"	S2@1.00	Ea	84.00	43.50	—	127.50
12" x 8"	S2@1.00	Ea	90.90	43.50	—	134.40
12" x 10"	S2@1.05	Ea	99.80	45.70	—	145.50
12" x 12"	S2@1.10	Ea	107.00	47.90	—	154.90
18" x 12"	S2@1.20	Ea	150.00	52.20	—	202.20
18" x 18"	S2@1.30	Ea	163.00	56.60	—	219.60
24" x 12"	S2@1.40	Ea	180.00	60.90	—	240.90
24" x 18"	S2@1.50	Ea	206.00	65.30	—	271.30
24" x 24"	S2@1.60	Ea	244.00	69.60	—	313.60
30" x 12"	S2@1.50	Ea	210.00	65.30	—	275.30
30" x 18"	S2@1.80	Ea	232.00	78.30	—	310.30
30" x 24"	S2@2.00	Ea	310.00	87.00	—	397.00
30" x 30"	S2@2.30	Ea	333.00	100.00	—	433.00
36" x 18"	S2@2.00	Ea	266.00	87.00	—	353.00
36" x 36"	S2@2.40	Ea	396.00	104.00	—	500.00
42" x 24"	S2@2.40	Ea	344.00	104.00	—	448.00
42" x 42"	S2@2.90	Ea	498.00	126.00	—	624.00
48" x 24"	S2@2.90	Ea	374.00	126.00	—	500.00
48" x 48"	S2@3.50	Ea	618.00	152.00	—	770.00
54" x 24"	S2@3.00	Ea	404.00	131.00	—	535.00
54" x 54"	S2@3.80	Ea	719.00	165.00	—	884.00

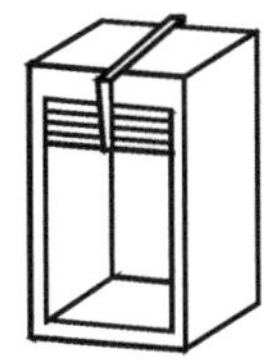

Correction factors:

(1) For 22 gauge break-away connection, multiply material cost by 1.30

(2) For blades out of air stream cap, multiply material cost by 1.25

Description	Craft@Hrs	Unit	Material $	Labor $	Equipment $	Total $

Fusible plug dampers for air duct risers. Vertical riser and header air duct fusible (melting) and heat-sensing automatic counterweight closure actuators and closure dampers. For large A/C systems in multi-story institutions such as schools, hospitals and office buildings. 1½-hour rating. The liner and insulation must meet the requirements of U.L. 181 and NFPA No. 90 A-96, as modified, "Standard for the Installation of Air Conditioning and Ventilation Systems."

Description	Craft@Hrs	Unit	Material $	Labor $	Equipment $	Total $
12" x 6"	S2@1.00	Ea	111.00	43.50	—	154.50
12" x 8"	S2@1.00	Ea	114.00	43.50	—	157.50
12" x 10"	S2@1.05	Ea	117.00	45.70	—	162.70
12" x 12"	S2@1.10	Ea	119.00	47.90	—	166.90
18" x 12"	S2@1.20	Ea	151.00	52.20	—	203.20
18" x 18"	S2@1.30	Ea	194.00	56.60	—	250.60
24" x 12"	S2@1.40	Ea	209.00	60.90	—	269.90
24" x 18"	S2@1.50	Ea	232.00	65.30	—	297.30
24" x 24"	S2@1.60	Ea	266.00	69.60	—	335.60
30" x 12"	S2@1.50	Ea	250.00	65.30	—	315.30
30" x 18"	S2@1.80	Ea	317.00	78.30	—	395.30
30" x 24"	S2@2.00	Ea	367.00	87.00	—	454.00
30" x 30"	S2@2.30	Ea	441.00	100.00	—	541.00
36" x 18"	S2@2.00	Ea	354.00	87.00	—	441.00
36" x 36"	S2@2.40	Ea	472.00	104.00	—	576.00
42" x 24"	S2@2.40	Ea	441.00	104.00	—	545.00
42" x 42"	S2@2.90	Ea	536.00	126.00	—	662.00
48" x 24"	S2@2.90	Ea	480.00	126.00	—	606.00
48" x 48"	S2@3.50	Ea	1,070.00	152.00	—	1,222.00
54" x 24"	S2@3.00	Ea	1,370.00	131.00	—	1,501.00
54" x 54"	S2@3.80	Ea	2,240.00	165.00	—	2,405.00
Cut, frame and gasket access hole in sidewall	S2@1.00	Ea	63.40	43.50	—	106.90
Mount duct hangers	S2@.250	Ea	4.12	10.90	—	15.02
Cut and mount sheet metal duct	S2@.350	LF	6.34	15.20	—	21.54
Apply and coat duct insulation	S2@.150	LF	2.73	6.53	—	9.26

Description	Craft@Hrs	Unit	Material $	Labor $	Equipment $	Total $

Single-skin 2" fixed-position turning vanes for duct

Description	Craft@Hrs	Unit	Material $	Labor $	Equipment $	Total $
12" x 6"	S2@.280	Ea	9.99	12.20	—	22.19
12" x 8"	S2@.285	Ea	12.40	12.40	—	24.80
12" x 10"	S2@.290	Ea	16.00	12.60	—	28.60
12" x 12"	S2@.300	Ea	19.70	13.10	—	32.80
18" x 12"	S2@.375	Ea	29.90	16.30	—	46.20
18" x 18"	S2@.450	Ea	36.30	19.60	—	55.90
24" x 12"	S2@.450	Ea	39.80	19.60	—	59.40
24" x 18"	S2@.500	Ea	43.30	21.80	—	65.10
24" x 24"	S2@.525	Ea	49.40	22.80	—	72.20
30" x 12"	S2@.550	Ea	49.80	23.90	—	73.70
30" x 18"	S2@.750	Ea	71.90	32.60	—	104.50
30" x 24"	S2@.800	Ea	95.50	34.80	—	130.30
30" x 30"	S2@.900	Ea	120.00	39.20	—	159.20
36" x 18"	S2@.750	Ea	85.20	32.60	—	117.80
36" x 36"	S2@.800	Ea	173.00	34.80	—	207.80
42" x 24"	S2@.900	Ea	134.00	39.20	—	173.20
42" x 42"	S2@1.00	Ea	236.00	43.50	—	279.50
48" x 24"	S2@1.10	Ea	153.00	47.90	—	200.90
48" x 48"	S2@1.50	Ea	306.00	65.30	—	371.30
54" x 24"	S2@1.15	Ea	171.00	50.00	—	221.00
54" x 54"	S2@1.25	Ea	390.00	54.40	—	444.40
60" x 24"	S2@1.30	Ea	192.00	56.60	—	248.60
60" x 36"	S2@1.40	Ea	290.00	60.90	—	350.90
60" x 60"	S2@1.50	Ea	479.00	65.30	—	544.30
72" x 36"	S2@1.65	Ea	351.00	71.80	—	422.80
72" x 54"	S2@2.40	Ea	516.00	104.00	—	620.00

Note: Labor includes cost to cut, assemble and install vanes and rails. For 4-inch air-foil vanes, multiply costs by 1.80 and labor costs by .90

Description	Craft@Hrs	Unit	Material $	Labor $	Equipment $	Total $

Spin-in galvanized steel collars without volume damper

Description	Craft@Hrs	Unit	Material $	Labor $	Equipment $	Total $
4"	S2@.200	Ea	7.45	8.70	—	16.15
5"	S2@.200	Ea	8.36	8.70	—	17.06
6"	S2@.200	Ea	9.22	8.70	—	17.92
7"	S2@.200	Ea	10.10	8.70	—	18.80
8"	S2@.250	Ea	11.70	10.90	—	22.60
9"	S2@.250	Ea	12.90	10.90	—	23.80
10"	S2@.300	Ea	13.80	13.10	—	26.90
12"	S2@.350	Ea	16.70	15.20	—	31.90
14"	S2@.400	Ea	19.80	17.40	—	37.20
16"	S2@.450	Ea	23.60	19.60	—	43.20
18"	S2@.500	Ea	29.60	21.80	—	51.40

Spin-in galvanized steel collar with volume damper

Description	Craft@Hrs	Unit	Material $	Labor $	Equipment $	Total $
4"	S2@.250	Ea	24.90	10.90	—	35.80
5"	S2@.250	Ea	25.30	10.90	—	36.20
6"	S2@.250	Ea	25.50	10.90	—	36.40
7"	S2@.250	Ea	26.20	10.90	—	37.10
8"	S2@.300	Ea	26.80	13.10	—	39.90
9"	S2@.300	Ea	27.10	13.10	—	40.20
10"	S2@.350	Ea	27.80	15.20	—	43.00
12"	S2@.400	Ea	29.80	17.40	—	47.20
14"	S2@.450	Ea	32.00	19.60	—	51.60
16"	S2@.500	Ea	33.90	21.80	—	55.70
18"	S2@.550	Ea	36.30	23.90	—	60.20

Duct-to-equipment flexible connection

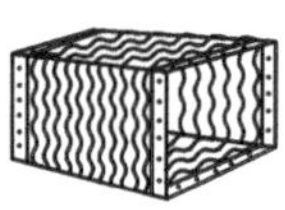

Description	Craft@Hrs	Unit	Material $	Labor $	Equipment $	Total $
18" x 12"	S2@.650	Ea	25.20	28.30	—	53.50
18" x 18"	S2@.850	Ea	30.50	37.00	—	67.50
24" x 12"	S2@.850	Ea	30.50	37.00	—	67.50
24" x 18"	S2@1.00	Ea	35.60	43.50	—	79.10
24" x 24"	S2@1.30	Ea	40.70	56.60	—	97.30
30" x 12"	S2@1.00	Ea	35.60	43.50	—	79.10
30" x 18"	S2@1.30	Ea	40.70	56.60	—	97.30
30" x 30"	S2@1.60	Ea	51.00	69.60	—	120.60
36" x 12"	S2@1.30	Ea	40.70	56.60	—	97.30
36" x 24"	S2@1.60	Ea	51.00	69.60	—	120.60
36" x 36"	S2@1.80	Ea	60.40	78.30	—	138.70
48" x 24"	S2@1.80	Ea	60.40	78.30	—	138.70
48" x 48"	S2@2.00	Ea	80.00	87.00	—	167.00

The costs for fabricating and installing galvanized steel ductwork are usually based on the total weights of the duct and fittings. Because of cost differences, fitting weights should be kept separate from straight duct weights.

Round spiral duct: Normally-used duct gauge/size relationships for low-pressure systems, up to 2 inches static pressure, are as follows:

Duct Diameter (inches)	U.S. Standard Gauge
Up to 12	26
13 through 24	24
26 through 36	22

Rectangular duct: Normally-used duct gauge/size relationships for low-pressure systems, up to 2 inches static pressure, are as follows:

Duct Size (inches)	U.S. Standard Gauge
Up to 12	26
13 through 30	24
31 through 54	22
55 through 72	20

The fitting weights in the following tables are in accordance with the above gauges.

Other material costs: All weights in the tables are net weights; they do not include bracing, cleats, scrap, hangers, end closures, sealants and miscellaneous hardware. Add 15% to calculated duct/fitting weights to cover these items if ductwork is purchased. Add 20% if ducting is manufactured in contractor's shop. The additional 5% covers the costs for scrap which is already included in an outside vendor's price.

Ductwork costs: Typical costs for purchased ductwork are as follows:

Straight duct, less than 1,000 pounds per order	$3.66/lb.
Straight duct, over 1,000 pounds per order	$3.32/lb.
Fittings, less than 1,000 pounds per order	$7.44/lb.
Fittings, over 1,000 pounds per order	$6.66/lb.
Lined ducts and fittings, add	$1.98/lb.*
Delivery costs	$0.16/lb.

**Duct sizes shown on drawings are always inside, or net, dimensions and must be increased in size when estimating lined ductwork.*

If ductwork is manufactured in the contractor's fabrication shop, the above costs can be reduced by about 20%, depending on the shop's rate of productivity.

Installation costs: An average crew can install approximately 25 pounds of duct and fittings per manhour under normal conditions. See "Applying Correction Factors" on page 6 for situations that do not conform to the definition of a standard labor unit.

EXAMPLE:

What is the cost to furnish and install 2,585 pounds of unlined duct and 560 pounds of unlined fittings? The ductwork will be purchased from an outside vendor. A 15% allowance for miscellaneous material is included in the weights.

Material:	
2,585 pounds of straight duct x $3.32/lb.	= $8,582.20
560 pounds of fittings x $7.44/lb.	= 4,166.40
Delivery cost: (2,585 + 560) x $.16/lb.	= 503.20
Total material cost	= $13,251.80*
Labor:	
$\frac{2{,}585 + 560}{25} = 125.8\ MH$; $125.8\ MH \times \$43.52$	= $5,474.82
Total installed cost: $13,251.80 + $5,474.82	= $18,726.62*

**Sales tax not included.*

Description	Craft@Hrs	Unit	Material $	Labor $	Equipment $	Total $

Installed ductwork per lb. (under 1,000 lbs.). Duct purchased from an independent fabrication shop, installed by this contractor's shop. Material price includes scrap, cleats, hangers, sealant, miscellaneous hardware, delivery and fabrication shop markups.

Straight duct	S2@.042	Lb	3.66	1.83	—	5.49
Duct fittings	S2@.042	Lb	7.44	1.83	—	9.27
Duct & fittings	S2@.042	Lb	4.92	1.83	—	6.75

Installed ductwork per lb. (over 1,000 lbs.). Duct purchased from an independent fabrication shop, installed by this contractor's shop. Material price includes scrap, cleats, hangers, sealant, miscellaneous hardware, delivery and subcontractor markups.

Straight duct	S2@.039	Lb	3.32	1.70	—	5.02
Duct fittings	S2@.039	Lb	6.66	1.70	—	8.36
Duct & fittings	S2@.039	Lb	4.45	1.70	—	6.15

Installed lined ductwork per lb. (under 1,000 lbs.). Duct purchased from an independent fabrication shop, installed by this contractor's shop. Material price includes scrap, cleats, hangers, sealant, miscellaneous hardware, delivery and subcontractor markups.

Straight duct	S2@.042	Lb	5.65	1.83	—	7.48
Duct fittings	S2@.042	Lb	9.36	1.83	—	11.19
Duct & fittings	S2@.042	Lb	6.89	1.83	—	8.72

Installed lined ductwork per lb. (over 1,000 lbs.). Duct purchased from an independent fabrication shop, installed by this contractor's shop. Material price includes scrap, cleats, hangers, sealant, miscellaneous hardware, delivery and subcontractor markups.

Straight duct	S2@.039	Lb	5.31	1.70	—	7.01
Duct fittings	S2@.039	Lb	8.67	1.70	—	10.37
Duct & fittings	S2@.039	Lb	6.40	1.70	—	8.10

Installed ductwork per lb. (under 1,000 lbs.). Duct fabricated and installed by this contractor's shop. Material price includes scrap, cleats, hangers, sealant, miscellaneous hardware and delivery.

Straight duct	S2@.051	Lb	2.26	2.22	—	4.48
Duct fittings	S2@.074	Lb	2.59	3.22	—	5.81
Duct & fittings	S2@.059	Lb	2.37	2.57	—	4.94

Installed ductwork per lb. (over 1,000 lbs.). Duct fabricated and installed by this contractor's shop. Material price includes scrap, cleats, hangers, sealant, miscellaneous hardware and delivery.

Straight duct	S2@.046	Lb	2.08	2.00	—	4.08
Duct fittings	S2@.067	Lb	2.39	2.92	—	5.31
Duct & fittings	S2@.053	Lb	2.16	2.31	—	4.47

Description	Craft@Hrs	Unit	Material $	Labor $	Equipment $	Total $

Installed lined ductwork per lb. (under 1,000 lbs.). Duct fabricated and installed by this contractor's shop. Material price includes scrap, cleats, hangers, sealant, miscellaneous hardware and delivery.

Description	Craft@Hrs	Unit	Material $	Labor $	Equipment $	Total $
Straight duct	S2@.062	Lb	3.34	2.70	—	6.04
Duct fittings	S2@.085	Lb	3.69	3.70	—	7.39
Duct & fittings	S2@.070	Lb	3.47	3.05	—	6.52

Installed lined ductwork per lb. (over 1,000 lbs.). Duct fabricated and installed by this contractor's shop. Material price includes scrap, cleats, hangers, sealant, miscellaneous hardware and delivery. Use for preliminary estimates.

Description	Craft@Hrs	Unit	Material $	Labor $	Equipment $	Total $
Straight duct	S2@.057	Lb	3.17	2.48	—	5.65
Duct fittings	S2@.078	Lb	3.53	3.39	—	6.92
Duct & fittings	S2@.064	Lb	6.13	2.79	—	8.92

Weights of Galvanized Steel Spiral Duct (pounds per LF)					
Diameter (inches)	U.S. Standard Gauge				
	26	24	22	20	18
3	.76	1.01	1.23	1.46	—
4	1.02	1.35	1.64	1.94	—
5	1.28	1.69	2.06	2.42	—
6	1.54	2.03	2.47	2.91	3.88
7	1.79	2.37	2.88	3.40	4.52
8	2.05	2.71	3.29	3.86	5.17
9	2.31	3.05	3.71	4.37	5.82
10	2.57	3.39	4.12	4.86	6.47
12	3.08	4.07	4.95	5.83	7.75
14	3.59	4.74	5.77	6.81	9.05
16	4.11	5.42	6.60	7.78	10.34
18	4.63	6.10	7.43	8.76	11.64
20	5.15	6.78	8.25	9.73	12.93
22	5.65	7.46	9.08	10.71	14.93
24	6.16	8.14	9.91	11.68	15.52
26	6.67	8.82	10.73	12.66	16.82
28	7.18	9.50	11.56	13.63	18.11
30	7.71	10.18	12.38	14.60	19.41
32	8.22	10.84	13.21	15.58	20.71
34	—	11.52	14.05	16.55	22.00
36	—	12.20	14.90	17.53	23.29

Galvanized Steel Round Spiral Fittings

Weights of Galvanized Steel Round Spiral Fittings (pounds per piece)				
Diameter (inches)	**90° elbow**	**45° elbow**	**Coupling**	**Reducer**
3	1.3	1.0	0.5	1.0
4	2.2	1.3	0.6	1.2
5	3.3	1.9	0.7	1.4
6	4.3	2.5	0.9	1.8
7	5.8	3.3	1.0	2.0
8	7.3	4.3	1.2	2.4
9	8.8	5.3	2.6	3.3
10	11.8	7.5	2.9	4.4
12	16.3	10.0	3.5	5.3
14	22.0	13.0	4.1	6.2
16	28.3	15.8	4.6	6.9
18	34.5	19.0	5.2	7.8
20	41.5	23.5	5.8	8.7
22	48.3	27.5	6.4	9.6
24	57.5	32.0	6.9	10.3
26	68.8	37.5	7.5	11.2
28	76.8	42.8	8.1	12.1
30	87.0	48.0	8.7	13.0
32	99.5	55.0	9.2	13.8
34	112.0	61.0	9.8	14.7
36	161.0	89.5	10.4	15.6

Largest run diam. (inches)	Weights of Galvanized Steel Round Spiral Fittings (pounds per piece) Tee with reducing run and branch Branch diameter (inches)													
	3	**4**	**5**	**6**	**7**	**8**	**9**	**10**	**12**	**14**	**16**	**18**	**20**	**22**
4	2.3													
5	2.8	3.0												
6	3.2	3.5	3.7											
7	3.7	3.9	4.1	4.5										
8	4.1	4.4	4.7	5.0	5.3									
9	4.5	4.8	5.1	5.5	5.8	6.2								
10	6.4	6.7	7.1	7.5	7.8	8.2	8.6							
12	7.7	8.1	8.5	8.9	9.4	9.9	10.3	10.7						
14	8.9	9.4	10.0	10.5	11.0	11.5	12.1	12.6	13.6					
16	10.2	10.8	11.4	12.0	12.5	13.1	13.7	14.3	15.5	16.7				
18	11.5	12.1	12.7	13.5	14.1	14.8	15.5	16.1	17.5	18.8	19.2			
20	12.7	13.4	14.2	14.9	17.7	16.4	17.2	17.9	19.4	20.9	22.4	23.9		
22	14.0	14.8	15.6	16.4	17.2	18.1	18.9	19.7	21.3	23.0	24.6	26.2	27.9	
24	15.3	16.1	17.0	17.8	18.7	19.6	20.5	21.4	23.2	25.0	26.8	28.5	30.3	32.1
26	16.5	17.5	18.5	19.5	20.4	21.3	22.3	23.3	25.2	26.2	29.1	31.1	33.0	34.9
28	17.7	18.7	19.8	20.8	21.9	22.9	23.9	25.0	27.0	29.2	31.2	33.2	35.4	37.4
30	18.9	20.0	21.1	22.2	23.3	24.5	25.5	26.7	28.9	31.1	33.3	35.5	37.7	40.0
32	20.4	21.6	22.6	24.0	25.2	26.4	27.6	28.8	31.2	33.6	36.0	38.4	40.8	42.2
34	21.6	22.9	24.2	25.4	26.7	28.0	29.3	30.5	33.1	35.6	38.1	40.7	42.2	45.6
36	29.6	31.4	33.1	34.8	36.6	38.6	40.1	41.8	45.3	48.8	52.2	55.7	59.2	62.7

Run diam. (inches)	Weights of Galvanized Steel Round Spiral Fittings (pounds per piece) Cross with reducing run and branch Branch diameter (inches)													
	3	4	5	6	7	8	9	10	12	14	16	18	20	22
4	2.6													
5	3.2	3.6												
6	3.8	4.2	4.6											
7	4.4	4.9	5.4	5.8										
8	4.9	5.4	6.0	6.5	7.0									
9	5.4	6.0	6.6	7.2	7.7	8.3								
10	7.6	8.2	8.9	9.5	10.2	10.8	11.4							
12	8.8	9.6	10.3	11.0	11.7	12.5	13.2	13.8						
14	10.1	10.9	11.7	12.6	13.4	14.2	15.1	15.9	17.5					
16	11.3	12.2	13.1	14.0	15.0	15.9	16.8	17.7	19.5	21.4				
18	12.5	13.5	14.5	15.5	16.5	17.5	18.5	19.5	21.5	23.5	25.5			
20	13.8	15.0	16.1	17.2	18.2	19.4	20.6	21.7	23.9	26.2	28.4	30.6		
22	15.2	16.4	17.7	18.9	20.1	21.4	22.6	23.8	26.2	28.7	31.2	33.6	36.1	
24	16.6	17.9	19.3	20.6	22.0	23.3	24.6	26.0	28.7	31.5	34.1	36.7	39.4	42.1
26	18.0	19.4	20.9	22.3	23.8	25.2	26.7	28.2	31.1	33.9	36.9	39.8	42.7	45.6
28	19.3	20.8	22.4	24.0	25.5	27.1	28.6	30.2	32.3	36.4	39.6	42.7	45.8	48.9
30	20.6	22.3	23.9	25.6	27.3	28.9	30.6	32.3	35.6	38.9	42.3	45.6	49.9	52.2
32	22.2	24.0	25.8	27.6	29.4	31.2	33.0	34.8	38.4	42.0	45.6	49.1	52.8	56.5
34	23.5	25.4	27.3	29.2	31.1	33.1	34.9	36.9	40.7	44.5	48.3	52.1	55.9	59.8
36	32.2	34.8	37.4	40.0	42.8	45.3	48.0	50.5	55.7	61.0	66.2	71.5	76.7	82.0

Weights of 26 Gauge Galvanized Steel Rectangular Duct (pounds per LF)									
Size (inches)	4	5	6	7	8	9	10	11	12
4	1.21								
5	1.36	1.51							
6	1.51	1.66	1.81						
7	1.66	1.81	1.96	2.11					
8	1.81	1.96	2.11	2.27	2.42				
9	1.96	2.11	2.27	2.42	2.57	2.72			
10	2.11	2.27	2.42	2.57	2.72	2.87	3.02		
11	2.27	2.42	2.57	2.72	2.87	3.02	3.17	3.32	
12	2.42	2.57	2.72	2.87	3.02	3.17	3.32	3.47	3.62

Weights of 24 Gauge Galvanized Steel Rectangular Duct (pounds per LF)

Size (inches)	14	16	18	20	22	24	26	28	30
6	3.85	4.24	4.62	5.01	5.40	5.78	6.17	6.55	6.94
8	4.24	4.62	5.01	5.40	5.78	6.17	6.55	6.94	7.32
10	4.62	5.01	5.40	5.78	6.17	6.55	6.94	7.32	7.71
12	5.01	5.40	5.78	6.17	6.55	6.94	7.32	7.71	8.09
14	5.40	5.78	6.17	6.55	6.94	7.32	7.71	8.09	8.48
16	5.78	6.17	6.55	6.94	7.32	7.71	8.09	8.48	8.86
18	6.17	6.55	6.94	7.32	7.71	8.09	8.48	8.86	9.25
20	6.55	6.94	7.32	7.71	8.09	8.48	8.86	9.25	9.64
22	6.94	7.32	7.71	8.09	8.48	8.86	9.25	9.64	10.00
24	7.32	7.71	8.09	8.48	8.86	9.25	9.64	10.00	10.40
26	7.71	8.09	8.48	8.86	9.25	9.64	10.00	10.40	10.80
28	8.09	8.48	8.86	9.25	9.64	10.00	10.40	10.80	11.20
30	8.48	8.86	9.25	9.64	10.00	10.40	10.80	11.20	11.60

Weights of 22 Gauge Galvanized Steel Rectangular Duct (pounds per LF)

Size (inches)	32	34	36	38	40	42	44	46	48	50	52	54
8	9.37	9.84	10.3	10.8	11.3	11.7	12.2	12.7	13.1	13.6	14.1	14.5
10	9.84	10.3	10.8	11.3	11.7	12.2	12.7	13.1	13.6	14.1	14.5	15.0
12	10.3	10.8	11.3	11.7	12.2	12.7	13.1	13.6	14.1	14.5	15.0	15.5
14	10.8	11.3	11.7	12.2	12.7	13.1	13.6	14.1	14.5	15.0	15.5	15.9
16	11.3	11.7	12.2	12.7	13.1	13.6	14.1	14.5	15.0	15.5	15.9	16.4
18	11.7	12.2	12.7	13.1	13.6	14.1	14.5	15.0	15.5	15.9	16.4	16.9
20	12.2	12.7	13.1	13.6	14.1	14.5	15.0	15.5	15.9	16.4	16.9	17.3
22	12.7	13.1	13.6	14.1	14.5	15.0	15.5	15.9	16.4	16.9	17.3	17.8
24	13.1	13.6	14.1	14.5	15.0	15.5	15.9	16.4	16.9	17.3	17.8	18.3
26	13.6	14.1	14.5	15.0	15.5	15.9	16.4	16.9	17.3	17.8	18.3	18.7
28	14.1	14.5	15.0	15.5	15.9	16.4	16.9	17.3	17.8	18.3	18.7	19.2
30	14.5	15.0	15.5	15.9	16.4	16.9	17.3	17.8	18.3	18.7	19.2	19.7
32	15.0	15.5	15.9	16.4	16.9	17.3	17.8	18.3	18.7	19.2	19.7	20.2
34	15.5	15.9	16.4	16.9	17.3	17.8	18.3	18.7	19.2	19.7	20.2	20.6
36	15.9	16.4	16.9	17.3	17.8	18.3	18.7	19.2	19.7	20.2	20.6	21.1
38	16.4	16.9	17.3	17.8	18.3	18.7	19.2	19.7	20.2	20.6	21.1	21.6
40	16.9	17.3	17.8	18.3	18.7	19.2	19.7	20.2	20.6	21.1	21.6	22.0
42	17.3	17.8	18.3	18.7	19.2	19.7	20.2	20.6	21.1	21.6	22.0	22.5
44	17.8	18.3	18.7	19.2	19.7	20.2	20.6	21.1	21.6	22.0	22.5	23.0
46	18.3	18.7	19.2	19.7	20.2	20.6	21.1	21.6	22.0	22.5	23.0	23.4
48	18.7	19.2	19.7	20.2	20.6	21.1	21.6	22.0	22.5	23.0	23.4	23.9
50	19.2	19.7	20.2	20.6	21.1	21.6	22.0	22.5	23.0	23.4	23.9	24.4
52	19.7	20.2	20.6	21.1	21.6	22.0	22.5	23.0	23.4	23.9	24.4	24.8
54	20.2	20.6	21.1	21.6	22.0	22.5	23.0	23.4	23.9	24.4	24.8	25.3

Weights of 20 Gauge Galvanized Steel Rectangular Duct (pounds per LF)									
Size (inches)	**56**	**58**	**60**	**62**	**64**	**66**	**68**	**70**	**72**
18	20.4	21.0	21.5	22.1	22.6	23.2	23.7	24.3	24.8
20	21.0	21.5	22.1	22.6	23.2	23.7	24.3	24.8	25.4
22	21.5	22.1	22.6	23.2	23.7	24.3	24.8	25.4	25.9
24	22.1	22.6	23.2	23.7	24.3	24.8	25.4	25.9	26.5
26	22.6	23.2	23.7	24.3	24.8	25.4	25.9	26.5	27.0
28	23.2	23.7	24.3	24.8	25.4	25.9	26.5	27.0	27.6
30	23.7	24.3	24.8	25.4	25.9	26.5	27.0	27.6	28.2
32	24.3	24.8	25.4	25.9	26.5	27.0	27.6	28.2	28.7
34	24.8	25.4	25.9	26.5	27.0	27.6	28.2	28.7	29.3
36	25.4	25.9	26.5	27.0	27.6	28.2	28.7	29.3	29.8
38	25.9	26.5	27.0	27.6	28.2	28.7	29.3	29.8	30.4
40	26.5	27.0	27.6	28.2	28.7	29.3	29.8	30.4	30.9
42	27.0	27.6	28.2	28.7	29.3	29.8	30.4	30.9	31.5
44	27.6	28.2	28.7	29.3	29.8	30.4	30.9	31.5	32.0
46	28.2	28.7	29.3	29.8	30.4	30.9	31.5	32.0	32.6
48	28.7	29.3	29.8	30.4	30.9	31.5	32.0	32.6	33.1
50	29.3	29.8	30.4	30.9	31.5	32.0	32.6	33.1	33.7
52	29.8	30.4	30.9	31.5	32.0	32.6	33.1	33.7	34.2
54	30.4	30.9	31.5	32.0	32.6	33.1	33.7	34.2	34.8
56	30.9	31.5	32.0	32.6	33.1	33.7	34.2	34.8	35.3
58	31.5	32.0	32.6	33.1	33.7	34.2	34.8	35.3	35.9
60	32.0	32.6	33.1	33.7	34.2	34.8	35.3	35.9	36.4
62	32.6	33.1	33.7	34.2	34.8	35.3	35.9	36.4	37.0
64	33.1	33.7	34.2	34.8	35.3	35.9	36.4	37.0	37.5
66	33.7	34.2	34.8	35.3	35.9	36.4	37.0	37.5	38.1
68	34.2	34.8	35.3	35.9	36.4	37.0	37.5	38.1	38.6
70	34.8	35.3	35.9	36.4	37.0	37.5	38.1	38.6	39.2
72	35.3	35.9	36.4	37.0	37.5	38.1	38.6	39.2	39.7

Galvanized Steel Rectangular 90-Degree Elbows

Weights of Galvanized Steel Rectangular 90 Degree Elbow (pounds per piece)

Size (inches)	4	6	8	10	12	14	16	18	20
4	1.05	1.45	1.95	2.60	3.30	5.30	6.50	7.80	9.25
6	1.45	1.95	2.60	3.30	3.70	5.80	7.75	8.75	9.75
8	1.95	2.60	3.30	3.70	4.25	6.50	8.50	9.75	10.4
10	2.60	3.30	3.70	4.25	4.85	7.30	9.50	10.7	11.2
12	3.30	3.70	4.25	4.85	5.60	8.20	10.5	11.7	12.2
14	5.30	5.80	6.50	7.30	8.20	9.25	11.5	12.6	13.2
16	6.50	7.50	8.50	9.50	10.5	11.5	12.5	13.4	14.4
18	7.80	8.75	9.75	10.7	11.7	12.6	13.4	14.5	15.7
20	9.25	9.75	10.4	11.2	12.2	13.8	14.4	15.7	17.2
22	10.8	11.8	12.8	13.8	14.8	25.8	16.8	17.8	18.8
24	12.6	13.6	14.6	15.6	16.6	17.6	18.6	19.6	20.5
26	14.4	15.4	16.4	17.4	18.4	19.4	20.4	21.4	22.3
28	16.4	17.4	18.4	19.4	20.4	21.4	22.4	23.4	24.3
30	18.5	19.5	20.5	21.5	22.5	23.5	24.5	25.5	26.4
32	24.3	25.7	27.0	28.4	29.7	31.1	32.4	33.7	35.0
34	28.2	29.4	30.6	31.8	33.0	34.3	35.5	36.7	37.9
36	31.3	32.5	33.7	34.9	36.1	37.3	38.5	39.7	40.9
38	34.5	35.7	36.9	38.1	39.3	40.5	41.7	42.9	44.1
40	37.9	39.1	40.3	41.5	42.7	43.9	45.1	46.3	47.5
42	41.4	42.6	43.8	45.0	46.2	47.4	48.6	49.8	51.0
44	45.1	46.4	47.6	48.8	50.0	51.2	52.4	53.6	54.8
46	49.0	50.2	51.4	52.6	53.8	55.0	56.2	57.4	58.6
48	53.0	54.2	55.4	56.6	57.8	59.0	60.2	61.4	62.6
50	57.2	58.5	59.7	60.9	62.1	63.3	64.5	65.7	66.9

Note: Elbow weights do not include turning vanes.

Weights of Galvanized Steel Rectangular 90 Degree Elbows (pounds per piece)

Size (inches)	22	24	26	28	30	32	34	36	38
22	19.6	21.9	24.1	26.3	28.5	37.0	40.8	44.5	48.3
24	21.9	24.2	26.4	28.6	30.9	39.5	43.6	47.2	50.8
26	24.1	26.4	28.7	30.9	33.2	42.0	46.4	49.8	53.3
28	26.3	28.6	30.9	33.3	35.6	44.5	49.2	52.5	55.8
30	28.5	30.9	33.2	35.6	37.9	47.0	52.0	55.1	58.3
32	37.0	39.5	42.0	44.5	47.0	49.5	52.0	57.8	60.8
34	40.8	43.6	46.4	49.2	52.0	54.8	57.6	60.5	63.3
36	44.5	47.2	49.8	52.5	55.1	57.8	60.5	63.1	65.8
38	48.3	50.8	53.3	55.8	58.3	60.8	63.3	65.8	68.3
40	52.1	54.6	57.1	59.6	62.1	64.6	67.1	69.6	72.1
42	55.9	58.4	60.9	63.4	65.9	68.4	70.9	73.4	75.9
44	59.6	62.1	64.6	67.1	69.6	72.1	74.6	77.1	79.6
46	63.4	65.9	68.4	70.9	73.4	75.9	78.4	80.9	83.4
48	67.2	69.7	72.2	74.7	77.2	79.7	82.2	84.7	87.2
50	70.9	73.4	75.9	78.4	80.9	83.4	85.9	88.4	90.9
52	74.8	77.3	79.8	82.3	84.8	87.3	89.8	92.3	94.7
54	78.5	81.0	83.5	86.0	88.5	91.0	93.5	96.0	98.5
56	97.2	100	103	106	109	112	115	118	121
58	103	106	109	112	115	118	121	124	127
60	110	113	116	119	122	125	128	131	134
62	116	119	122	125	128	131	134	137	140
64	122	125	129	132	135	138	141	144	147
66	128	131	135	138	141	144	148	151	154
68	135	138	142	145	148	151	154	157	160
70	141	144	148	151	154	157	160	163	166
72	147	150	154	157	160	163	166	169	172

Note: Elbow weights do not include turning vanes.

Weights of Galvanized Steel Rectangular 90 Degree Elbows (pounds per piece)

Size (inches)	40	42	44	46	48	50	52	54	56
40	74.6	78.4	82.1	85.9	89.7	93.4	97.2	102	125
42	78.4	82.3	86.2	90.1	94.0	97.9	102	106	130
44	82.1	86.2	90.3	94.4	98.5	102	106	110	135
46	85.9	90.1	94.4	98.7	102	106	110	113	140
48	89.7	94.0	98.5	102	106	110	113	117	144
50	93.4	97.9	102	106	110	113	117	122	149
52	97.2	102	106	110	113	117	122	126	154
54	102	106	110	113	117	122	126	131	159
56	126	130	134	139	145	150	155	160	165
58	132	137	142	147	151	156	161	166	170
60	137	142	147	152	157	162	167	172	176
62	143	148	153	158	163	168	173	178	182
64	150	155	160	165	169	174	179	184	188
66	156	161	166	171	176	181	186	191	195
68	163	168	173	178	182	187	192	197	201
70	170	175	180	185	189	194	199	204	208
72	176	181	185	190	194	199	204	210	215

Note: Elbow weights do not include turning vanes.

Galvanized Steel Rectangular 90-Degree Elbows

Weights of Galvanized Steel Rectangular 90 Degree Elbows (pounds per piece)								
Size (inches)	**58**	**60**	**62**	**64**	**66**	**68**	**70**	**72**
58	176	182	187	194	200	207	214	221
60	182	187	194	200	207	214	221	227
62	187	194	200	207	214	221	227	233
64	194	200	207	214	221	227	233	239
66	200	207	214	221	227	233	239	246
68	207	214	221	227	233	239	246	252
70	214	221	227	233	239	246	252	259
72	221	227	233	239	246	252	259	266

Note: Elbow weights do not include turning vanes.

Weights of Galvanized Steel Rectangular Drops and Tap-In Tees (pounds per piece)									
Size (inches)	**6**	**8**	**10**	**12**	**16**	**20**	**24**	**28**	**32**
6	1.81	2.11	2.42	2.72	4.24	5.01	5.78	6.55	8.90
8	2.11	2.42	2.72	3.02	4.62	5.40	6.17	6.94	9.37
10	2.42	2.72	3.02	3.32	5.01	5.78	6.55	7.32	9.84
12	2.72	3.02	3.32	3.62	5.40	6.17	6.94	7.71	10.3
16	4.24	4.62	5.01	5.40	6.17	6.94	7.71	8.48	11.3
20	5.01	5.40	5.78	6.17	6.94	7.71	8.48	9.25	12.2
24	5.78	6.17	6.55	6.94	7.71	8.48	9.25	10.0	13.1
28	6.55	6.94	7.32	7.71	8.48	9.25	10.0	10.8	14.1
32	8.90	9.37	9.84	10.3	11.3	12.2	13.1	14.1	15.0
36	9.84	10.3	10.8	11.3	12.2	13.1	14.1	15.0	15.9
40	10.8	11.3	11.7	12.2	13.1	14.1	15.0	15.9	16.9
44	11.7	12.2	12.7	13.1	14.1	15.0	15.9	16.9	17.8
48	12.7	13.1	13.6	14.1	15.0	15.9	16.9	17.8	18.7
52	13.6	14.1	14.5	15.0	15.9	16.9	17.8	18.7	19.7
56	16.0	17.6	18.2	18.8	19.6	21.0	22.1	23.2	24.3
60	18.7	19.0	19.3	19.9	21.0	22.1	23.2	24.3	25.4

Note: Weights of drops and tap-in tees do not include splitter dampers.

Costs for Steel Spiral Duct. Costs per linear foot of duct, based on quantities less than 1,000 pounds. Add 50% to the material cost for lined duct. These costs include delivery, typical bracing, cleats, scrap, hangers, end closures, sealants and miscellaneous hardware.

Description	Craft@Hrs	Unit	Material $	Labor $	Equipment $	Total $

Galvanized steel spiral duct 26 gauge

Description	Craft@Hrs	Unit	Material $	Labor $	Equipment $	Total $
3"	S2@.030	LF	3.08	1.31	—	4.39
4"	S2@.041	LF	4.16	1.78	—	5.94
5"	S2@.051	LF	5.23	2.22	—	7.45
6"	S2@.061	LF	6.23	2.65	—	8.88
7"	S2@.072	LF	7.24	3.13	—	10.37
8"	S2@.082	LF	8.29	3.57	—	11.86
9"	S2@.092	LF	9.36	4.00	—	13.36
10"	S2@.103	LF	10.40	4.48	—	14.88
12"	S2@.123	LF	12.30	5.35	—	17.65
14" - 16"	S2@.174	LF	15.60	7.57	—	23.17
18" - 20"	S2@.195	LF	19.70	8.49	—	28.19
22" - 24"	S2@.236	LF	24.00	10.30	—	34.30
26" - 28"	S2@.277	LF	28.20	12.10	—	40.30
30" - 32"	S2@.318	LF	32.60	13.80	—	46.40

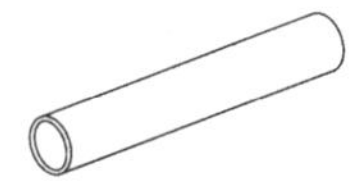

Galvanized steel spiral duct 24 gauge

Description	Craft@Hrs	Unit	Material $	Labor $	Equipment $	Total $
3"	S2@.041	LF	4.15	1.78	—	5.93
4"	S2@.054	LF	5.43	2.35	—	7.78
5"	S2@.067	LF	6.82	2.92	—	9.74
6"	S2@.081	LF	8.21	3.53	—	11.74
7"	S2@.095	LF	9.54	4.13	—	13.67
8"	S2@.108	LF	11.00	4.70	—	15.70
9"	S2@.122	LF	12.30	5.31	—	17.61
10"	S2@.135	LF	13.60	5.88	—	19.48
12"	S2@.163	LF	16.40	7.09	—	23.49
14" - 16"	S2@.203	LF	20.40	8.83	—	29.23
18" - 20"	S2@.257	LF	25.90	11.20	—	37.10
22" - 24"	S2@.312	LF	31.70	13.60	—	45.30
26" - 28"	S2@.366	LF	37.20	15.90	—	53.10
30" - 32"	S2@.419	LF	42.50	18.20	—	60.70
34" - 36"	S2@.474	LF	48.40	20.60	—	69.00

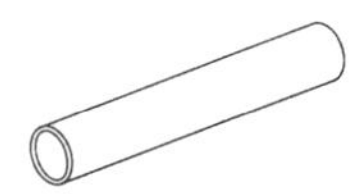

Galvanized Steel Spiral Duct

Galvanized steel spiral duct 22 gauge

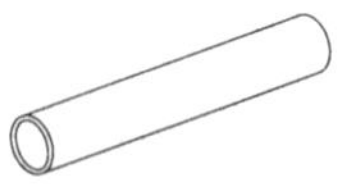

Description	Craft@Hrs	Unit	Material $	Labor $	Equipment $	Total $
3"	S2@.049	LF	5.09	2.13	—	7.22
4"	S2@.065	LF	6.72	2.83	—	9.55
5"	S2@.082	LF	8.46	3.57	—	12.03
6"	S2@.099	LF	10.10	4.31	—	14.41
7"	S2@.115	LF	11.80	5.00	—	16.80
8"	S2@.132	LF	13.20	5.74	—	18.94
9"	S2@.148	LF	15.10	6.44	—	21.54
10"	S2@.165	LF	17.10	7.18	—	24.28
12"	S2@.198	LF	20.30	8.62	—	28.92
14" - 16"	S2@.247	LF	25.10	10.70	—	35.80
18" - 20"	S2@.313	LF	32.40	13.60	—	46.00
22" - 24"	S2@.380	LF	38.80	16.50	—	55.30
26" - 28"	S2@.444	LF	45.90	19.30	—	65.20
30" - 32"	S2@.491	LF	52.80	21.40	—	74.20
34" - 36"	S2@.576	LF	59.70	25.10	—	84.80

Galvanized steel spiral duct 20 gauge

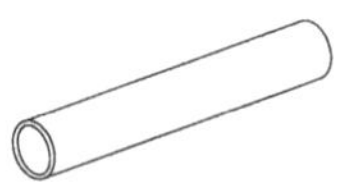

3"	S2@.058	LF	6.00	2.52	—	8.52
4"	S2@.078	LF	7.98	3.39	—	11.37
5"	S2@.097	LF	9.92	4.22	—	14.14
6"	S2@.116	LF	11.90	5.05	—	16.95
7"	S2@.136	LF	13.70	5.92	—	19.62
8"	S2@.154	LF	15.80	6.70	—	22.50
9"	S2@.175	LF	17.70	7.62	—	25.32
10"	S2@.194	LF	19.80	8.44	—	28.24
12"	S2@.233	LF	24.10	10.10	—	34.20
14" - 16"	S2@.291	LF	29.50	12.70	—	42.20
18" - 20"	S2@.369	LF	37.80	16.10	—	53.90
22" - 24"	S2@.446	LF	45.90	19.40	—	65.30
26" - 28"	S2@.525	LF	53.90	22.80	—	76.70
30" - 32"	S2@.604	LF	61.50	26.30	—	87.80
34" - 36"	S2@.681	LF	69.60	29.60	—	99.20

Galvanized steel spiral duct 18 gauge

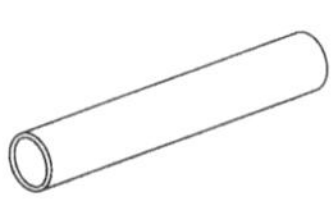

6"	S2@.155	LF	15.80	6.75	—	22.55
7"	S2@.181	LF	18.40	7.88	—	26.28
8"	S2@.207	LF	21.00	9.01	—	30.01
9"	S2@.233	LF	23.70	10.10	—	33.80
10"	S2@.259	LF	26.20	11.30	—	37.50
12"	S2@.310	LF	32.00	13.50	—	45.50
14" - 16"	S2@.387	LF	39.90	16.80	—	56.70
18" - 20"	S2@.491	LF	50.80	21.40	—	72.20
22" - 24"	S2@.608	LF	62.10	26.50	—	88.60
26" - 28"	S2@.696	LF	71.30	30.30	—	101.60
30" - 32"	S2@.803	LF	81.90	34.90	—	116.80
34" - 36"	S2@.905	LF	92.80	39.40	—	132.20

Costs for Steel Spiral Duct Fittings. Costs per fitting, based on quantities less than 1,000 pounds. For quantities over 1,000 pounds, deduct 15% from material costs. For lined duct add 50%. These costs include delivery, typical bracing, cleats, scrap, hangers, end closures, sealants and miscellaneous hardware.

Galvanized steel spiral duct 90-degree elbows

Description	Craft@Hrs	Unit	Material $	Labor $	Equipment $	Total $
3"	S2@.052	Ea	3.85	2.26	—	6.11
4"	S2@.088	Ea	6.49	3.83	—	10.32
5"	S2@.132	Ea	9.72	5.74	—	15.46
6"	S2@.172	Ea	12.70	7.49	—	20.19
7"	S2@.232	Ea	17.30	10.10	—	27.40
8"	S2@.292	Ea	21.80	12.70	—	34.50
9"	S2@.352	Ea	26.00	15.30	—	41.30
10"	S2@.470	Ea	34.80	20.50	—	55.30
12"	S2@.653	Ea	48.20	28.40	—	76.60
14" - 16"	S2@1.01	Ea	74.10	44.00	—	118.10
18" - 20"	S2@1.52	Ea	112.00	66.20	—	178.20
22" - 24"	S2@2.11	Ea	156.00	91.80	—	247.80
26" - 28"	S2@2.91	Ea	218.00	127.00	—	345.00
30" - 32"	S2@3.73	Ea	277.00	162.00	—	439.00
34" - 36"	S2@5.47	Ea	404.00	238.00	—	642.00

Galvanized steel spiral duct 45-degree elbows

Description	Craft@Hrs	Unit	Material $	Labor $	Equipment $	Total $
3"	S2@.040	Ea	2.95	1.74	—	4.69
4"	S2@.052	Ea	3.85	2.26	—	6.11
5"	S2@.076	Ea	5.60	3.31	—	8.91
6"	S2@.100	Ea	7.41	4.35	—	11.76
7"	S2@.132	Ea	9.72	5.74	—	15.46
8"	S2@.172	Ea	12.70	7.49	—	20.19
9"	S2@.212	Ea	15.60	9.23	—	24.83
10"	S2@.300	Ea	22.00	13.10	—	35.10
12"	S2@.400	Ea	29.50	17.40	—	46.90
14" - 16"	S2@.576	Ea	42.20	25.10	—	67.30
18" - 20"	S2@.850	Ea	62.90	37.00	—	99.90
22" - 24"	S2@1.19	Ea	87.60	51.80	—	139.40
26" - 28"	S2@1.61	Ea	119.00	70.10	—	189.10
30" - 32"	S2@2.06	Ea	153.00	89.70	—	242.70
34" - 36"	S2@3.01	Ea	221.00	131.00	—	352.00

Galvanized Steel Spiral Duct Fittings

Description	Craft@Hrs	Unit	Material $	Labor $	Equipment $	Total $

Galvanized steel spiral duct coupling

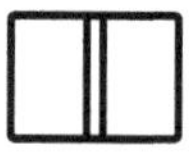

Description	Craft@Hrs	Unit	Material $	Labor $	Equipment $	Total $
3"	S2@.020	Ea	2.93	.87	—	3.80
4"	S2@.024	Ea	3.61	1.04	—	4.65
5"	S2@.028	Ea	4.10	1.22	—	5.32
6"	S2@.036	Ea	5.28	1.57	—	6.85
7"	S2@.040	Ea	5.91	1.74	—	7.65
8"	S2@.048	Ea	7.10	2.09	—	9.19
9"	S2@.104	Ea	15.30	4.53	—	19.83
10"	S2@.116	Ea	17.30	5.05	—	22.35
12"	S2@.140	Ea	20.70	6.09	—	26.79
14" - 16"	S2@.174	Ea	25.80	7.57	—	33.37
18" - 20"	S2@.220	Ea	32.10	9.57	—	41.67
22" - 24"	S2@.266	Ea	39.40	11.60	—	51.00
26" - 28"	S2@.312	Ea	46.00	13.60	—	59.60
30" - 32"	S2@.357	Ea	52.90	15.50	—	68.40
34" - 36"	S2@.403	Ea	59.60	17.50	—	77.10

Galvanized steel spiral duct with reducer

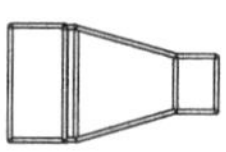

Description	Craft@Hrs	Unit	Material $	Labor $	Equipment $	Total $
3"	S2@.040	Ea	5.52	1.74	—	7.26
4"	S2@.048	Ea	6.63	2.09	—	8.72
5"	S2@.056	Ea	7.74	2.44	—	10.18
6"	S2@.072	Ea	9.98	3.13	—	13.11
7"	S2@.080	Ea	11.00	3.48	—	14.48
8"	S2@.096	Ea	13.00	4.18	—	17.18
9"	S2@.132	Ea	18.10	5.74	—	23.84
10"	S2@.176	Ea	24.10	7.66	—	31.76
12"	S2@.212	Ea	29.40	9.23	—	38.63
14" - 16"	S2@.262	Ea	36.30	11.40	—	47.70
18" - 20"	S2@.330	Ea	45.70	14.40	—	60.10
22" - 24"	S2@.398	Ea	54.80	17.30	—	72.10
26" - 28"	S2@.465	Ea	64.20	20.20	—	84.40
30" - 32"	S2@.536	Ea	73.40	23.30	—	96.70
34" - 36"	S2@.606	Ea	83.30	26.40	—	109.70

Costs for Steel Spiral Tees with Reducing Branch. Costs per tee, based on quantities less than 1,000 pounds. Deduct 15% from the material cost for quantities over 1,000 pounds. Add 50% to the material cost for lined duct. These costs include delivery, typical bracing, cleats, scrap, hangers, end closures, sealants and miscellaneous hardware.

Description	Craft@Hrs	Unit	Material $	Labor $	Equipment $	Total $
Galvanized steel spiral duct tees with 3" reducing branch						
4"	S2@.109	Ea	13.10	4.74	—	17.84
5"	S2@.112	Ea	15.40	4.87	—	20.27
6"	S2@.128	Ea	17.70	5.57	—	23.27
7"	S2@.148	Ea	19.90	6.44	—	26.34
8"	S2@.164	Ea	22.50	7.14	—	29.64
9"	S2@.180	Ea	24.10	7.83	—	31.93
10"	S2@.256	Ea	38.00	11.10	—	49.10
12"	S2@.308	Ea	45.10	13.40	—	58.50
14" - 16"	S2@.382	Ea	57.00	16.60	—	73.60
18" - 20"	S2@.485	Ea	71.60	21.10	—	92.70
22" - 24"	S2@.585	Ea	86.30	25.50	—	111.80
26" - 28"	S2@.683	Ea	101.00	29.70	—	130.70
30" - 32"	S2@.786	Ea	116.00	34.20	—	150.20
34" - 36"	S2@1.02	Ea	151.00	44.40	—	195.40

Description	Craft@Hrs	Unit	Material $	Labor $	Equipment $	Total $
Galvanized steel spiral duct tees with 4" reducing branch						
5"	S2@.120	Ea	17.20	5.22	—	22.42
6"	S2@.140	Ea	19.90	6.09	—	25.99
7"	S2@.156	Ea	21.80	6.79	—	28.59
8"	S2@.176	Ea	24.70	7.66	—	32.36
9"	S2@.192	Ea	26.20	8.36	—	34.56
10"	S2@.268	Ea	40.30	11.70	—	52.00
12"	S2@.324	Ea	48.90	14.10	—	63.00
14" - 16"	S2@.403	Ea	60.60	17.50	—	78.10
18" - 20"	S2@.508	Ea	76.60	22.10	—	98.70
22" - 24"	S2@.619	Ea	93.00	26.90	—	119.90
26" - 28"	S2@.724	Ea	109.00	31.50	—	140.50
30" - 32"	S2@.830	Ea	124.00	36.10	—	160.10
34" - 36"	S2@1.08	Ea	161.00	47.00	—	208.00

Description	Craft@Hrs	Unit	Material $	Labor $	Equipment $	Total $
Galvanized steel spiral duct tees with 5" reducing branch						
6"	S2@.148	Ea	21.80	6.44	—	28.24
7"	S2@.164	Ea	24.10	7.14	—	31.24
8"	S2@.188	Ea	27.00	8.18	—	35.18
9"	S2@.204	Ea	29.30	8.88	—	38.18
10"	S2@.284	Ea	43.60	12.40	—	56.00
12"	S2@.340	Ea	52.00	14.80	—	66.80
14" - 16"	S2@.428	Ea	65.40	18.60	—	84.00
18" - 20"	S2@.538	Ea	83.00	23.40	—	106.40
22" - 24"	S2@.651	Ea	99.20	28.30	—	127.50
26" - 28"	S2@.764	Ea	116.00	33.20	—	149.20
30" - 32"	S2@.873	Ea	134.00	38.00	—	172.00
34" - 36"	S2@1.15	Ea	175.00	50.00	—	225.00

Galvanized Steel Spiral Tees

Galvanized steel spiral duct tees with 6" reducing branch

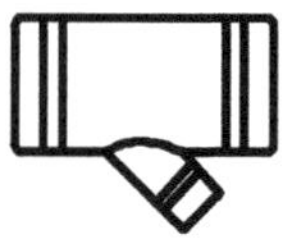

Description	Craft@Hrs	Unit	Material $	Labor $	Equipment $	Total $
7"	S2@.180	Ea	27.00	7.83	—	34.83
8"	S2@.200	Ea	29.30	8.70	—	38.00
9"	S2@.220	Ea	31.90	9.57	—	41.47
10"	S2@.300	Ea	46.70	13.10	—	59.80
12"	S2@.356	Ea	55.70	15.50	—	71.20
14" - 16"	S2@.449	Ea	69.70	19.50	—	89.20
18" - 20"	S2@.566	Ea	88.00	24.60	—	112.60
22" - 24"	S2@.683	Ea	106.00	29.70	—	135.70
26" - 28"	S2@.807	Ea	123.00	35.10	—	158.10
30" - 32"	S2@.924	Ea	142.00	40.20	—	182.20
34" - 36"	S2@1.20	Ea	186.00	52.20	—	238.20

Galvanized steel spiral duct tees with 7" reducing branch

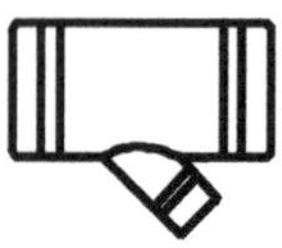

Description	Craft@Hrs	Unit	Material $	Labor $	Equipment $	Total $
8"	S2@.164	Ea	31.90	7.14	—	39.04
9"	S2@.180	Ea	34.60	7.83	—	42.43
10"	S2@.252	Ea	48.90	11.00	—	59.90
12"	S2@.304	Ea	59.10	13.20	—	72.30
14" - 16"	S2@.381	Ea	74.20	16.60	—	90.80
18" - 20"	S2@.482	Ea	93.00	21.00	—	114.00
22" - 24"	S2@.579	Ea	112.00	25.20	—	137.20
26" - 28"	S2@.683	Ea	133.00	29.70	—	162.70
30" - 32"	S2@.784	Ea	151.00	34.10	—	185.10
34" - 36"	S2@1.02	Ea	197.00	44.40	—	241.40

Galvanized steel spiral duct tees with 8" reducing branch

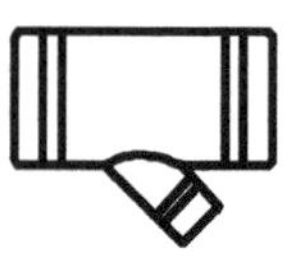

Description	Craft@Hrs	Unit	Material $	Labor $	Equipment $	Total $
9"	S2@.228	Ea	44.50	9.92	—	54.42
10"	S2@.268	Ea	52.00	11.70	—	63.70
12"	S2@.324	Ea	62.60	14.10	—	76.70
14" - 16"	S2@.401	Ea	77.90	17.50	—	95.40
18" - 20"	S2@.508	Ea	98.70	22.10	—	120.80
22" - 24"	S2@.617	Ea	119.00	26.90	—	145.90
26" - 28"	S2@.722	Ea	138.00	31.40	—	169.40
30" - 32"	S2@.830	Ea	159.00	36.10	—	195.10
34" - 36"	S2@1.08	Ea	209.00	47.00	—	256.00

Galvanized steel spiral duct tees with 9" reducing branch

Description	Craft@Hrs	Unit	Material $	Labor $	Equipment $	Total $
10"	S2@.284	Ea	55.30	12.40	—	67.70
12"	S2@.340	Ea	65.80	14.80	—	80.60
14" - 16"	S2@.426	Ea	83.00	18.50	—	101.50
18" - 20"	S2@.540	Ea	104.00	23.50	—	127.50
22" - 24"	S2@.649	Ea	124.00	28.20	—	152.20
26" - 28"	S2@.760	Ea	145.00	33.10	—	178.10
30" - 32"	S2@.875	Ea	172.00	38.10	—	210.10
34" - 36"	S2@1.15	Ea	220.00	50.00	—	270.00

Description	Craft@Hrs	Unit	Material $	Labor $	Equipment $	Total $

Galvanized steel spiral duct tees with 10" reducing branch

12"	S2@.360	Ea	69.70	15.70	—	85.40
14" - 16"	S2@.447	Ea	86.30	19.50	—	105.80
18" - 20"	S2@.564	Ea	110.00	24.50	—	134.50
22" - 24"	S2@.683	Ea	133.00	29.70	—	162.70
26" - 28"	S2@.805	Ea	156.00	35.00	—	191.00
30" - 32"	S2@.924	Ea	177.00	40.20	—	217.20
34" - 36"	S2@1.20	Ea	234.00	52.20	—	286.20

Galvanized steel spiral duct tees with 12" reducing branch

14" - 16"	S2@.572	Ea	95.30	24.90	—	120.20
18" - 20"	S2@.681	Ea	120.00	29.60	—	149.60
22" - 24"	S2@.820	Ea	143.00	35.70	—	178.70
26" - 28"	S2@.965	Ea	172.00	42.00	—	214.00
30" - 32"	S2@1.11	Ea	196.00	48.30	—	244.30
34" - 36"	S2@1.44	Ea	258.00	62.70	—	320.70

Galvanized steel spiral duct tees with 14" reducing branch

14" - 16"	S2@.572	Ea	110.00	24.90	—	134.90
18" - 20"	S2@.681	Ea	131.00	29.60	—	160.60
22" - 24"	S2@.820	Ea	158.00	35.70	—	193.70
26" - 28"	S2@.965	Ea	186.00	42.00	—	228.00
30" - 32"	S2@1.11	Ea	215.00	48.30	—	263.30
34" - 36"	S2@1.44	Ea	282.00	62.70	—	344.70

Galvanized steel spiral duct tees with 16" reducing branch

18" - 20"	S2@.739	Ea	142.00	32.20	—	174.20
22" - 24"	S2@.888	Ea	174.00	38.60	—	212.60
26" - 28"	S2@1.04	Ea	200.00	45.30	—	245.30
30" - 32"	S2@1.20	Ea	232.00	52.20	—	284.20
34" - 36"	S2@1.56	Ea	305.00	67.90	—	372.90

Galvanized steel spiral duct tees with 18" reducing branch

18" - 20"	S2@.837	Ea	160.00	36.40	—	196.40
22" - 24"	S2@.954	Ea	184.00	41.50	—	225.50
26" - 28"	S2@1.12	Ea	218.00	48.70	—	266.70
30" - 32"	S2@1.29	Ea	249.00	56.10	—	305.10
34" - 36"	S2@1.68	Ea	331.00	73.10	—	404.10

Galvanized steel spiral duct tees with 20" reducing branch

22" - 24"	S2@1.02	Ea	198.00	44.40	—	242.40
26" - 28"	S2@1.20	Ea	234.00	52.20	—	286.20
30" - 32"	S2@1.38	Ea	266.00	60.10	—	326.10
34" - 36"	S2@1.81	Ea	349.00	78.80	—	427.80

Galvanized steel spiral duct tees with 22" reducing branch

22" - 24"	S2@1.57	Ea	220.00	68.30	—	288.30
26" - 28"	S2@1.32	Ea	258.00	57.40	—	315.40
30" - 32"	S2@1.48	Ea	287.00	64.40	—	351.40
34" - 36"	S2@1.92	Ea	377.00	83.60	—	460.60

Galvanized Steel Spiral Tees

Costs for Steel Spiral Tees with Reducing Run and Branch. Costs per tee, based on quantities less than 1,000 pounds. Deduct 15% from the material cost for quantities over 1,000 pounds. Add 50% to the material cost for lined duct. These costs include delivery, typical bracing, cleats, scrap, hangers, end closures, sealants and miscellaneous hardware.

Description	Craft@Hrs	Unit	Material $	Labor $	Equipment $	Total $

Galvanized steel spiral duct tees with 3" run and branch

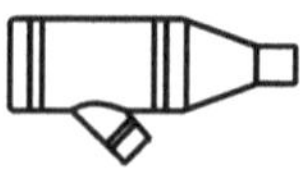

Description	Craft@Hrs	Unit	Material $	Labor $	Equipment $	Total $
4"	S2@.109	Ea	21.40	4.74	—	26.14
5"	S2@.112	Ea	21.80	4.87	—	26.67
6"	S2@.128	Ea	24.70	5.57	—	30.27
7"	S2@.148	Ea	28.70	6.44	—	35.14
8"	S2@.164	Ea	31.90	7.14	—	39.04
9"	S2@.180	Ea	34.60	7.83	—	42.43
10"	S2@.256	Ea	49.80	11.10	—	60.90
12"	S2@.308	Ea	59.30	13.40	—	72.70
14" - 16"	S2@.382	Ea	74.20	16.60	—	90.80
18" - 20"	S2@.485	Ea	93.60	21.10	—	114.70
22" - 24"	S2@.585	Ea	113.00	25.50	—	138.50
26" - 28"	S2@.683	Ea	133.00	29.70	—	162.70
30" - 32"	S2@.786	Ea	153.00	34.20	—	187.20
34" - 36"	S2@1.02	Ea	197.00	44.40	—	241.40

Galvanized steel spiral duct tees with 4" run and branch

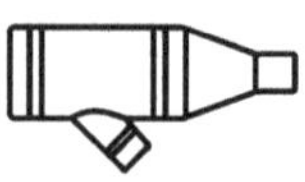

Description	Craft@Hrs	Unit	Material $	Labor $	Equipment $	Total $
5"	S2@.120	Ea	23.20	5.22	—	28.42
6"	S2@.140	Ea	27.00	6.09	—	33.09
7"	S2@.156	Ea	30.50	6.79	—	37.29
8"	S2@.176	Ea	33.90	7.66	—	41.56
9"	S2@.192	Ea	37.00	8.36	—	45.36
10"	S2@.268	Ea	52.00	11.70	—	63.70
12"	S2@.324	Ea	62.60	14.10	—	76.70
14" - 16"	S2@.403	Ea	78.20	17.50	—	95.70
18" - 20"	S2@.508	Ea	98.70	22.10	—	120.80
22" - 24"	S2@.619	Ea	119.00	26.90	—	145.90
26" - 28"	S2@.724	Ea	138.00	31.50	—	169.50
30" - 32"	S2@.830	Ea	159.00	36.10	—	195.10
34" - 36"	S2@1.08	Ea	210.00	47.00	—	257.00

Galvanized steel spiral duct tees with 5" run and branch

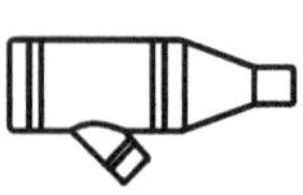

Description	Craft@Hrs	Unit	Material $	Labor $	Equipment $	Total $
6"	S2@.148	Ea	28.70	6.44	—	35.14
7"	S2@.164	Ea	31.90	7.14	—	39.04
8"	S2@.188	Ea	36.50	8.18	—	44.68
9"	S2@.204	Ea	39.60	8.88	—	48.48
10"	S2@.284	Ea	55.30	12.40	—	67.70
12"	S2@.340	Ea	65.80	14.80	—	80.60
14" - 16"	S2@.428	Ea	83.00	18.60	—	101.60
18" - 20"	S2@.538	Ea	104.00	23.40	—	127.40
22" - 24"	S2@.651	Ea	126.00	28.30	—	154.30
26" - 28"	S2@.764	Ea	147.00	33.20	—	180.20
30" - 32"	S2@.873	Ea	172.00	38.00	—	210.00
34" - 36"	S2@1.15	Ea	220.00	50.00	—	270.00

Galvanized steel spiral duct tees with 6" run and branch

Description	Craft@Hrs	Unit	Material $	Labor $	Equipment $	Total $
7"	S2@.180	Ea	34.60	7.83	—	42.43
8"	S2@.200	Ea	38.70	8.70	—	47.40
9"	S2@.220	Ea	42.10	9.57	—	51.67
10"	S2@.300	Ea	58.20	13.10	—	71.30
12"	S2@.356	Ea	69.20	15.50	—	84.70
14" - 16"	S2@.449	Ea	86.80	19.50	—	106.30
18" - 20"	S2@.566	Ea	110.00	24.60	—	134.60
22" - 24"	S2@.683	Ea	133.00	29.70	—	162.70
26" - 28"	S2@.807	Ea	156.00	35.10	—	191.10
30" - 32"	S2@.924	Ea	177.00	40.20	—	217.20
34" - 36"	S2@1.20	Ea	234.00	52.20	—	286.20

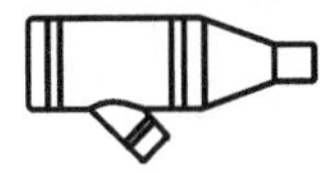

Galvanized steel spiral duct tees with 7" run and branch

Description	Craft@Hrs	Unit	Material $	Labor $	Equipment $	Total $
8"	S2@.212	Ea	41.00	9.23	—	50.23
9"	S2@.232	Ea	45.10	10.10	—	55.20
10"	S2@.312	Ea	60.00	13.60	—	73.60
12"	S2@.376	Ea	72.80	16.40	—	89.20
14" - 16"	S2@.470	Ea	91.40	20.50	—	111.90
18" - 20"	S2@.636	Ea	122.00	27.70	—	149.70
22" - 24"	S2@.717	Ea	138.00	31.20	—	169.20
26" - 28"	S2@.845	Ea	161.00	36.80	—	197.80
30" - 32"	S2@.969	Ea	190.00	42.20	—	232.20
34" - 36"	S2@1.27	Ea	245.00	55.30	—	300.30

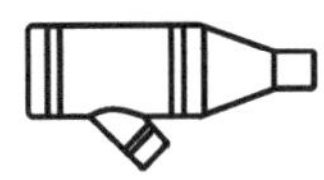

Galvanized steel spiral duct tees with 8" run and branch

Description	Craft@Hrs	Unit	Material $	Labor $	Equipment $	Total $
9"	S2@.248	Ea	48.20	10.80	—	59.00
10"	S2@.328	Ea	64.00	14.30	—	78.30
12"	S2@.396	Ea	76.50	17.20	—	93.70
14" - 16"	S2@.493	Ea	94.80	21.50	—	116.30
18" - 20"	S2@.623	Ea	120.00	27.10	—	147.10
22" - 24"	S2@.751	Ea	143.00	32.70	—	175.70
26" - 28"	S2@.882	Ea	173.00	38.40	—	211.40
30" - 32"	S2@1.02	Ea	196.00	44.40	—	240.40
34" - 36"	S2@1.33	Ea	258.00	57.90	—	315.90

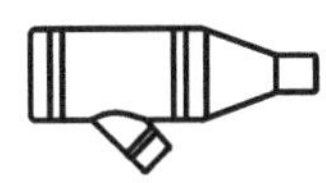

Galvanized steel spiral duct tees with 9" run and branch

Description	Craft@Hrs	Unit	Material $	Labor $	Equipment $	Total $
10"	S2@.344	Ea	66.40	15.00	—	81.40
12"	S2@.412	Ea	79.50	17.90	—	97.40
14" - 16"	S2@.515	Ea	99.90	22.40	—	122.30
18" - 20"	S2@.653	Ea	126.00	28.40	—	154.40
22" - 24"	S2@.779	Ea	153.00	33.90	—	186.90
26" - 28"	S2@.924	Ea	177.00	40.20	—	217.20
30" - 32"	S2@1.06	Ea	206.00	46.10	—	252.10
34" - 36"	S2@1.39	Ea	266.00	60.50	—	326.50

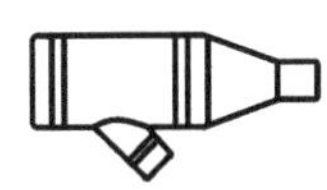

Galvanized Steel Spiral Tees

Description	Craft@Hrs	Unit	Material $	Labor $	Equipment $	Total $

Galvanized steel spiral duct tees with 10" run and branch

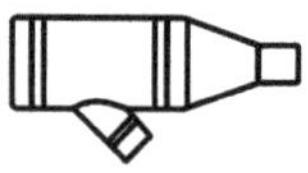

12"	S2@.427	Ea	41.60	18.60	—	60.20
14" - 16"	S2@.538	Ea	52.40	23.40	—	75.80
18" - 20"	S2@.681	Ea	130.00	29.60	—	159.60
22" - 24"	S2@.820	Ea	158.00	35.70	—	193.70
26" - 28"	S2@.965	Ea	186.00	42.00	—	228.00
30" - 32"	S2@1.11	Ea	215.00	48.30	—	263.30
34" - 36"	S2@1.44	Ea	280.00	62.70	—	342.70

Galvanized steel spiral duct tees with 12" run and branch

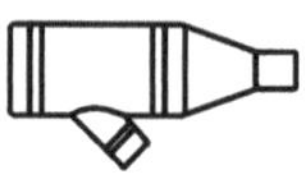

14" - 16"	S2@.581	Ea	112.00	25.30	—	137.30
18" - 20"	S2@.739	Ea	142.00	32.20	—	174.20
22" - 24"	S2@.888	Ea	174.00	38.60	—	212.60
26" - 28"	S2@1.04	Ea	200.00	45.30	—	245.30
30" - 32"	S2@1.20	Ea	232.00	52.20	—	284.20
34" - 36"	S2@1.57	Ea	305.00	68.30	—	373.30

Galvanized steel spiral duct tees with 14" run and branch

14" - 16"	S2@.666	Ea	128.00	29.00	—	157.00
18" - 20"	S2@.794	Ea	154.00	34.60	—	188.60
22" - 24"	S2@.959	Ea	184.00	41.70	—	225.70
26" - 28"	S2@1.11	Ea	215.00	48.30	—	263.30
30" - 32"	S2@1.29	Ea	249.00	56.10	—	305.10
34" - 36"	S2@1.69	Ea	331.00	73.50	—	404.50

Galvanized steel spiral duct tees with 16" run and branch

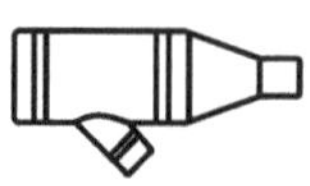

18" - 20"	S2@.833	Ea	159.00	36.30	—	195.30
22" - 24"	S2@1.03	Ea	198.00	44.80	—	242.80
26" - 28"	S2@1.20	Ea	234.00	52.20	—	286.20
30" - 32"	S2@1.39	Ea	266.00	60.50	—	326.50
34" - 36"	S2@1.81	Ea	349.00	78.80	—	427.80

Galvanized steel spiral duct tees with 18" run and branch

18" - 20"	S2@.956	Ea	184.00	41.60	—	225.60
22" - 24"	S2@1.09	Ea	210.00	47.40	—	257.40
26" - 28"	S2@1.29	Ea	248.00	56.10	—	304.10
30" - 32"	S2@1.48	Ea	287.00	64.40	—	351.40
34" - 36"	S2@1.93	Ea	377.00	84.00	—	461.00

Galvanized steel spiral duct tees with 20" run and branch

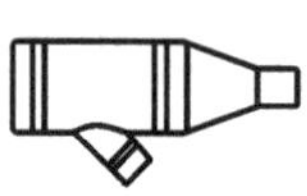

22" - 24"	S2@1.16	Ea	225.00	50.50	—	275.50
26" - 28"	S2@1.37	Ea	265.00	59.60	—	324.60
30" - 32"	S2@1.57	Ea	305.00	68.30	—	373.30
34" - 36"	S2@2.03	Ea	396.00	88.30	—	484.30

Galvanized steel spiral duct tees with 22" run and branch

22" - 24"	S2@1.28	Ea	248.00	55.70	—	303.70
26" - 28"	S2@1.45	Ea	282.00	63.10	—	345.10
30" - 32"	S2@1.64	Ea	320.00	71.40	—	391.40
34" - 36"	S2@2.16	Ea	417.00	94.00	—	511.00

Costs for Steel Spiral Crosses with Reducing Branches. Costs per cross, based on quantities less than 1,000 pounds. Deduct 15% from the material cost for quantities over 1,000 pounds. Add 50% to the material cost for lined duct. These costs include delivery, typical bracing, cleats, scrap, hangers, end closures, sealants and miscellaneous hardware.

Description	Craft@Hrs	Unit	Material $	Labor $	Equipment $	Total $
Galvanized steel spiral duct crosses with 3" branch						
4"	S2@.104	Ea	19.90	4.53	—	24.43
5"	S2@.128	Ea	24.70	5.57	—	30.27
6"	S2@.152	Ea	29.30	6.62	—	35.92
7"	S2@.176	Ea	33.90	7.66	—	41.56
8"	S2@.196	Ea	38.00	8.53	—	46.53
9"	S2@.216	Ea	41.60	9.40	—	51.00
10"	S2@.304	Ea	59.10	13.20	—	72.30
12"	S2@.352	Ea	68.70	15.30	—	84.00
14" - 16"	S2@.428	Ea	83.40	18.60	—	102.00
18" - 20"	S2@.525	Ea	102.00	22.80	—	124.80
22" - 24"	S2@.634	Ea	122.00	27.60	—	149.60
26" - 28"	S2@.745	Ea	142.00	32.40	—	174.40
30" - 32"	S2@.856	Ea	164.00	37.30	—	201.30
34" - 36"	S2@1.11	Ea	217.00	48.30	—	265.30

Description	Craft@Hrs	Unit	Material $	Labor $	Equipment $	Total $
Galvanized steel spiral duct crosses with 4" branch						
5"	S2@.144	Ea	28.00	6.27	—	34.27
6"	S2@.168	Ea	33.10	7.31	—	40.41
7"	S2@.196	Ea	38.00	8.53	—	46.53
8"	S2@.216	Ea	41.60	9.40	—	51.00
9"	S2@.240	Ea	46.70	10.40	—	57.10
10"	S2@.328	Ea	64.00	14.30	—	78.30
12"	S2@.384	Ea	75.00	16.70	—	91.70
14" - 16"	S2@.461	Ea	89.40	20.10	—	109.50
18" - 20"	S2@.568	Ea	110.00	24.70	—	134.70
22" - 24"	S2@.685	Ea	133.00	29.80	—	162.80
26" - 28"	S2@.805	Ea	156.00	35.00	—	191.00
30" - 32"	S2@.927	Ea	180.00	40.30	—	220.30
34" - 36"	S2@1.20	Ea	234.00	52.20	—	286.20

Description	Craft@Hrs	Unit	Material $	Labor $	Equipment $	Total $
Galvanized steel spiral duct crosses with 5" branch						
6"	S2@.184	Ea	35.90	8.01	—	43.91
7"	S2@.216	Ea	41.60	9.40	—	51.00
8"	S2@.240	Ea	46.70	10.40	—	57.10
9"	S2@.264	Ea	51.50	11.50	—	63.00
10"	S2@.356	Ea	69.20	15.50	—	84.70
12"	S2@.412	Ea	79.50	17.90	—	97.40
14" - 16"	S2@.497	Ea	95.60	21.60	—	117.20
18" - 20"	S2@.613	Ea	118.00	26.70	—	144.70
22" - 24"	S2@.741	Ea	142.00	32.20	—	174.20
26" - 28"	S2@.867	Ea	171.00	37.70	—	208.70
30" - 32"	S2@.993	Ea	192.00	43.20	—	235.20
34" - 36"	S2@1.29	Ea	249.00	56.10	—	305.10

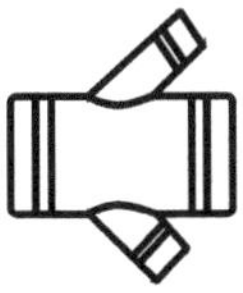

Galvanized steel spiral duct crosses with 6" branch

Description	Craft@Hrs	Unit	Material $	Labor $	Equipment $	Total $
7"	S2@.232	Ea	45.10	10.10	—	55.20
8"	S2@.260	Ea	50.70	11.30	—	62.00
9"	S2@.288	Ea	55.70	12.50	—	68.20
10"	S2@.380	Ea	74.10	16.50	—	90.60
12"	S2@.440	Ea	85.30	19.10	—	104.40
14" - 16"	S2@.532	Ea	103.00	23.20	—	126.20
18" - 20"	S2@.653	Ea	126.00	28.40	—	154.40
22" - 24"	S2@.790	Ea	154.00	34.40	—	188.40
26" - 28"	S2@.927	Ea	180.00	40.30	—	220.30
30" - 32"	S2@1.06	Ea	206.00	46.10	—	252.10
34" - 36"	S2@1.38	Ea	266.00	60.10	—	326.10

Galvanized steel spiral duct crosses with 7" branch

Description	Craft@Hrs	Unit	Material $	Labor $	Equipment $	Total $
8"	S2@.280	Ea	53.90	12.20	—	66.10
9"	S2@.308	Ea	59.30	13.40	—	72.70
10"	S2@.408	Ea	78.90	17.80	—	96.70
12"	S2@.470	Ea	90.50	20.50	—	111.00
14" - 16"	S2@.566	Ea	110.00	24.60	—	134.60
18" - 20"	S2@.692	Ea	134.00	30.10	—	164.10
22" - 24"	S2@.841	Ea	160.00	36.60	—	196.60
26" - 28"	S2@.986	Ea	191.00	42.90	—	233.90
30" - 32"	S2@1.13	Ea	219.00	49.20	—	268.20
34" - 36"	S2@1.48	Ea	287.00	64.40	—	351.40

Galvanized steel spiral duct crosses with 8" branch

Description	Craft@Hrs	Unit	Material $	Labor $	Equipment $	Total $
9"	S2@.310	Ea	64.30	13.50	—	77.80
10"	S2@.403	Ea	84.20	17.50	—	101.70
12"	S2@.470	Ea	96.20	20.50	—	116.70
14" - 16"	S2@.561	Ea	116.00	24.40	—	140.40
18" - 20"	S2@.690	Ea	142.00	30.00	—	172.00
22" - 24"	S2@.835	Ea	174.00	36.30	—	210.30
26" - 28"	S2@.980	Ea	200.00	42.60	—	242.60
30" - 32"	S2@1.12	Ea	234.00	48.70	—	282.70
34" - 36"	S2@1.47	Ea	305.00	64.00	—	369.00

Galvanized steel spiral duct crosses with 9" branch

Description	Craft@Hrs	Unit	Material $	Labor $	Equipment $	Total $
10"	S2@.427	Ea	88.00	18.60	—	106.60
12"	S2@.495	Ea	102.00	21.50	—	123.50
14" - 16"	S2@.596	Ea	123.00	25.90	—	148.90
18" - 20"	S2@.732	Ea	151.00	31.90	—	182.90
22" - 24"	S2@.882	Ea	183.00	38.40	—	221.40
26" - 28"	S2@1.04	Ea	215.00	45.30	—	260.30
30" - 32"	S2@1.19	Ea	245.00	51.80	—	296.80
34" - 36"	S2@1.55	Ea	320.00	67.50	—	387.50

Description	Craft@Hrs	Unit	Material $	Labor $	Equipment $	Total $
Galvanized steel spiral duct crosses with 12" branch						
14" - 16"	S2@.692	Ea	142.00	30.10	—	172.10
18" - 20"	S2@.850	Ea	176.00	37.00	—	213.00
22" - 24"	S2@1.03	Ea	214.00	44.80	—	258.80
26" - 28"	S2@1.18	Ea	245.00	51.40	—	296.40
30" - 32"	S2@1.38	Ea	287.00	60.10	—	347.10
34" - 36"	S2@1.80	Ea	377.00	78.30	—	455.30
Galvanized steel spiral duct crosses with 14" branch						
14" - 16"	S2@1.81	Ea	72.80	78.80	—	151.60
18" - 20"	S2@2.11	Ea	84.60	91.80	—	176.40
22" - 24"	S2@2.55	Ea	103.00	111.00	—	214.00
26" - 28"	S2@2.98	Ea	120.00	130.00	—	250.00
30" - 32"	S2@3.43	Ea	136.00	149.00	—	285.00
34" - 36"	S2@4.47	Ea	180.00	195.00	—	375.00
Galvanized steel spiral duct crosses with 16" branch						
18" - 20"	S2@1.08	Ea	209.00	47.00	—	256.00
22" - 24"	S2@1.30	Ea	251.00	56.60	—	307.60
26" - 28"	S2@1.53	Ea	296.00	66.60	—	362.60
30" - 32"	S2@1.76	Ea	339.00	76.60	—	415.60
34" - 36"	S2@2.31	Ea	448.00	101.00	—	549.00
Galvanized steel spiral duct crosses with 18" branch						
18" - 20"	S2@1.22	Ea	236.00	53.10	—	289.10
22" - 24"	S2@1.40	Ea	273.00	60.90	—	333.90
26" - 28"	S2@1.65	Ea	320.00	71.80	—	391.80
30" - 32"	S2@1.89	Ea	366.00	82.30	—	448.30
34" - 36"	S2@2.47	Ea	482.00	107.00	—	589.00
Galvanized steel spiral duct crosses with 20" branch						
22" - 24"	S2@1.51	Ea	292.00	65.70	—	357.70
26" - 28"	S2@1.77	Ea	343.00	77.00	—	420.00
30" - 32"	S2@2.05	Ea	402.00	89.20	—	491.20
34" - 36"	S2@2.65	Ea	515.00	115.00	—	630.00
Galvanized steel spiral duct crosses with 22" branch						
22" - 24"	S2@1.68	Ea	331.00	73.10	—	404.10
26" - 28"	S2@1.89	Ea	366.00	82.30	—	448.30
30" - 32"	S2@2.17	Ea	417.00	94.40	—	511.40
34" - 36"	S2@2.84	Ea	553.00	124.00	—	677.00

Galvanized Steel Rectangular Ductwork

Costs for Steel 26 Gauge Rectangular Duct. Costs per linear foot of duct, based on quantities less than 1,000 pounds of prefabricated duct. Deduct 15% from the material cost for quantities over 1,000 pounds. Add 50% to the material cost for lined duct. These costs include delivery, typical bracing, cleats, scrap, hangers, end closures, sealants and miscellaneous hardware.

Galvanized steel 4" rectangular duct 26 gauge

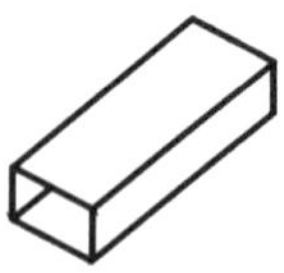

Description	Craft@Hrs	Unit	Material $	Labor $	Equipment $	Total $
4"	S2@.048	LF	4.78	2.09	—	6.87
5"	S2@.054	LF	5.28	2.35	—	7.63
6"	S2@.060	LF	5.91	2.61	—	8.52
7"	S2@.066	LF	6.53	2.87	—	9.40
8"	S2@.072	LF	7.12	3.13	—	10.25
9"	S2@.078	LF	7.65	3.39	—	11.04
10"	S2@.084	LF	8.26	3.66	—	11.92
11"	S2@.091	LF	8.87	3.96	—	12.83
12"	S2@.097	LF	9.48	4.22	—	13.70

Galvanized steel 5" rectangular duct 26 gauge

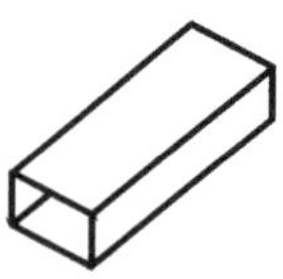

Description	Craft@Hrs	Unit	Material $	Labor $	Equipment $	Total $
5"	S2@.060	LF	5.91	2.61	—	8.52
6"	S2@.066	LF	6.53	2.87	—	9.40
7"	S2@.072	LF	7.12	3.13	—	10.25
8"	S2@.078	LF	7.65	3.39	—	11.04
9"	S2@.084	LF	8.26	3.66	—	11.92
10"	S2@.091	LF	8.87	3.96	—	12.83
11"	S2@.097	LF	9.48	4.22	—	13.70
12"	S2@.103	LF	10.00	4.48	—	14.48

Galvanized steel 6" rectangular duct 26 gauge

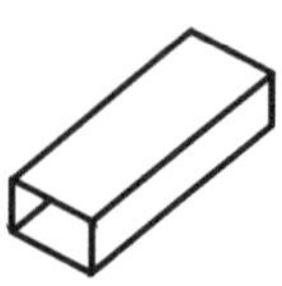

Description	Craft@Hrs	Unit	Material $	Labor $	Equipment $	Total $
6"	S2@.072	LF	7.12	3.13	—	10.25
7"	S2@.078	LF	7.65	3.39	—	11.04
8"	S2@.084	LF	8.26	3.66	—	11.92
9"	S2@.091	LF	8.87	3.96	—	12.83
10"	S2@.097	LF	9.48	4.22	—	13.70
11"	S2@.103	LF	10.00	4.48	—	14.48
12"	S2@.109	LF	10.60	4.74	—	15.34

Galvanized steel 7" rectangular duct 26 gauge

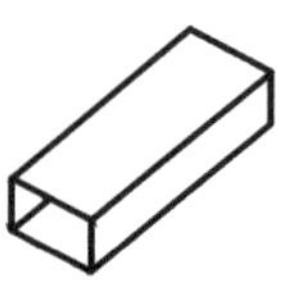

Description	Craft@Hrs	Unit	Material $	Labor $	Equipment $	Total $
7"	S2@.084	LF	8.26	3.66	—	11.92
8"	S2@.091	LF	8.87	3.96	—	12.83
9"	S2@.097	LF	9.48	4.22	—	13.70
10"	S2@.103	LF	10.00	4.48	—	14.48
11"	S2@.109	LF	10.60	4.74	—	15.34
12"	S2@.115	LF	11.20	5.00	—	16.20

Galvanized steel 8" rectangular duct 26 gauge

Description	Craft@Hrs	Unit	Material $	Labor $	Equipment $	Total $
8"	S2@.097	LF	9.48	4.22	—	13.70
9"	S2@.103	LF	10.00	4.48	—	14.48
10"	S2@.109	LF	10.60	4.74	—	15.34
11"	S2@.115	LF	11.20	5.00	—	16.20
12"	S2@.121	LF	11.80	5.27	—	17.07

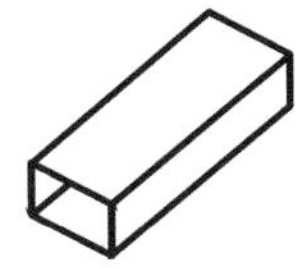

Galvanized steel 9" rectangular duct 26 gauge

9"	S2@.109	LF	10.60	4.74	—	15.34
10"	S2@.115	LF	11.20	5.00	—	16.20
11"	S2@.121	LF	11.80	5.27	—	17.07
12"	S2@.127	LF	12.30	5.53	—	17.83

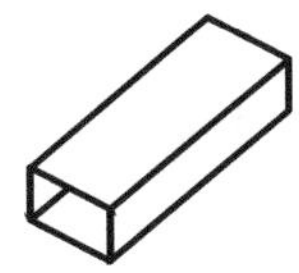

Galvanized steel 10" rectangular duct 26 gauge

10"	S2@.129	LF	11.80	5.61	—	17.41
11"	S2@.127	LF	12.30	5.53	—	17.83
12"	S2@.133	LF	12.90	5.79	—	18.69

Galvanized steel 11" rectangular duct 26 gauge

11"	S2@.133	LF	12.90	5.79	—	18.69
12"	S2@.139	LF	13.40	6.05	—	19.45

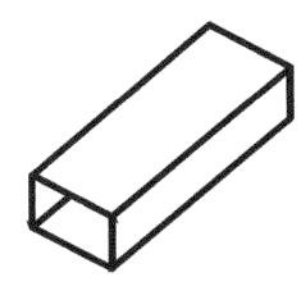

Galvanized steel 12" rectangular duct 26 gauge

12"	S2@.145	LF	13.90	6.31	—	20.21

Galvanized Steel Rectangular Ductwork

Costs for Galvanized 24 Gauge Rectangular Duct. Costs per linear foot of duct, based on quantities less than 1,000 pounds of prefabricated duct. Deduct 15% from the material cost for over 1,000 pounds. Add 50% to the material cost for lined duct. These costs include delivery, typical bracing, cleats, scrap, hangers, end closures.

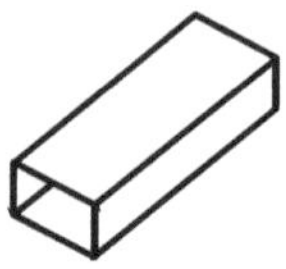
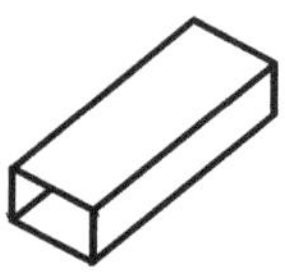
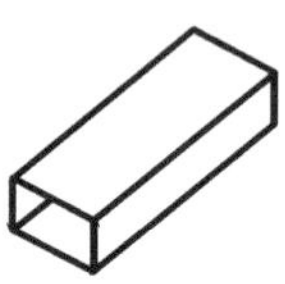

Description	Craft@Hrs	Unit	Material $	Labor $	Equipment $	Total $
Galvanized steel 14" rectangular duct 24 gauge						
6"	S2@.154	LF	15.10	6.70	—	21.80
8"	S2@.170	LF	16.60	7.40	—	24.00
10"	S2@.184	LF	18.20	8.01	—	26.21
12"	S2@.200	LF	19.70	8.70	—	28.40
14" - 16"	S2@.223	LF	21.90	9.70	—	31.60
18" - 20"	S2@.254	LF	24.90	11.10	—	36.00
22" - 24"	S2@.285	LF	28.20	12.40	—	40.60
26" - 28"	S2@.316	LF	31.50	13.80	—	45.30
30"	S2@.339	LF	33.50	14.80	—	48.30
Galvanized steel 16" rectangular duct 24 gauge						
6"	S2@.170	LF	16.60	7.40	—	24.00
8"	S2@.184	LF	18.20	8.01	—	26.21
10"	S2@.200	LF	19.70	8.70	—	28.40
12"	S2@.216	LF	21.40	9.40	—	30.80
14" - 16"	S2@.239	LF	23.50	10.40	—	33.90
18" - 20"	S2@.269	LF	26.50	11.70	—	38.20
22" - 24"	S2@.300	LF	29.30	13.10	—	42.40
26" - 28"	S2@.331	LF	33.10	14.40	—	47.50
30"	S2@.354	LF	35.40	15.40	—	50.80
Galvanized steel 18" rectangular duct 24 gauge						
6"	S2@.184	LF	18.20	8.01	—	26.21
8"	S2@.200	LF	19.70	8.70	—	28.40
10"	S2@.216	LF	21.40	9.40	—	30.80
12"	S2@.231	LF	22.70	10.10	—	32.80
14" - 16"	S2@.254	LF	24.90	11.10	—	36.00
18" - 20"	S2@.285	LF	28.20	12.40	—	40.60
22" - 24"	S2@.316	LF	31.50	13.80	—	45.30
26" - 28"	S2@.346	LF	33.90	15.10	—	49.00
30"	S2@.370	LF	36.60	16.10	—	52.70

Description	Craft@Hrs	Unit	Material $	Labor $	Equipment $	Total $

Galvanized steel 20" rectangular duct 24 gauge

Description	Craft@Hrs	Unit	Material $	Labor $	Equipment $	Total $
6"	S2@.200	LF	19.70	8.70	—	28.40
8"	S2@.216	LF	21.40	9.40	—	30.80
10"	S2@.231	LF	22.70	10.10	—	32.80
12"	S2@.247	LF	24.30	10.70	—	35.00
14" - 16"	S2@.269	LF	26.50	11.70	—	38.20
18" - 20"	S2@.300	LF	29.30	13.10	—	42.40
22" - 24"	S2@.331	LF	33.10	14.40	—	47.50
26" - 28"	S2@.362	LF	35.90	15.80	—	51.70
30"	S2@.385	LF	38.00	16.80	—	54.80

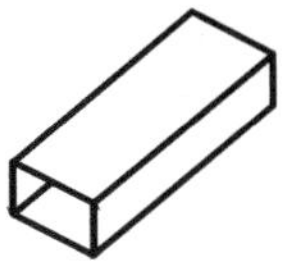

Galvanized steel 22" rectangular duct 24 gauge

Description	Craft@Hrs	Unit	Material $	Labor $	Equipment $	Total $
6"	S2@.216	LF	21.40	9.40	—	30.80
8"	S2@.231	LF	22.70	10.10	—	32.80
10"	S2@.247	LF	24.30	10.70	—	35.00
12"	S2@.262	LF	25.80	11.40	—	37.20
14" - 16"	S2@.285	LF	28.20	12.40	—	40.60
18" - 20"	S2@.316	LF	31.50	13.80	—	45.30
22" - 24"	S2@.346	LF	33.90	15.10	—	49.00
26" - 28"	S2@.377	LF	37.70	16.40	—	54.10
30"	S2@.400	LF	39.60	17.40	—	57.00

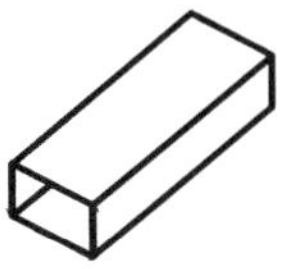

Galvanized steel 24" rectangular duct 24 gauge

Description	Craft@Hrs	Unit	Material $	Labor $	Equipment $	Total $
6"	S2@.231	LF	22.70	10.10	—	32.80
8"	S2@.247	LF	24.30	10.70	—	35.00
10"	S2@.262	LF	25.80	11.40	—	37.20
12"	S2@.278	LF	27.30	12.10	—	39.40
14" - 16"	S2@.300	LF	29.30	13.10	—	42.40
18" - 20"	S2@.331	LF	33.10	14.40	—	47.50
22" - 24"	S2@.362	LF	35.90	15.80	—	51.70
26" - 28"	S2@.392	LF	38.70	17.10	—	55.80
30"	S2@.415	LF	41.00	18.10	—	59.10

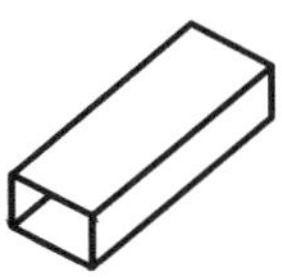

Galvanized steel 26" rectangular duct 24 gauge

Description	Craft@Hrs	Unit	Material $	Labor $	Equipment $	Total $
6"	S2@.247	LF	24.30	10.70	—	35.00
8"	S2@.262	LF	25.80	11.40	—	37.20
10"	S2@.278	LF	27.30	12.10	—	39.40
12"	S2@.292	LF	28.80	12.70	—	41.50
14" - 16"	S2@.316	LF	31.50	13.80	—	45.30
18" - 20"	S2@.346	LF	33.90	15.10	—	49.00
22" - 24"	S2@.377	LF	37.70	16.40	—	54.10
26" - 28"	S2@.407	LF	40.30	17.70	—	58.00
30"	S2@.431	LF	42.10	18.80	—	60.90

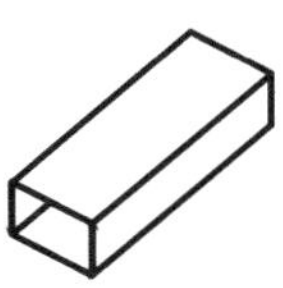

Galvanized Steel Rectangular Ductwork

Description	Craft@Hrs	Unit	Material $	Labor $	Equipment $	Total $
Galvanized steel 28" rectangular duct 24 gauge						
6"	S2@.262	LF	25.80	11.40	—	37.20
8"	S2@.278	LF	27.30	12.10	—	39.40
10"	S2@.292	LF	28.80	12.70	—	41.50
12"	S2@.308	LF	30.50	13.40	—	43.90
14" - 16"	S2@.331	LF	33.10	14.40	—	47.50
18" - 20"	S2@.362	LF	35.90	15.80	—	51.70
22" - 24"	S2@.392	LF	38.70	17.10	—	55.80
26" - 28"	S2@.423	LF	41.70	18.40	—	60.10
30"	S2@.448	LF	44.50	19.50	—	64.00

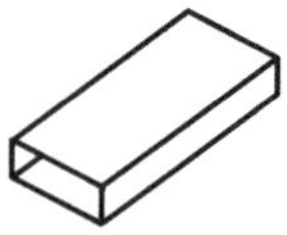

Description	Craft@Hrs	Unit	Material $	Labor $	Equipment $	Total $
Galvanized steel 30" rectangular duct 24 gauge						
6"	S2@.278	LF	27.30	12.10	—	39.40
8"	S2@.292	LF	28.80	12.70	—	41.50
10"	S2@.308	LF	30.50	13.40	—	43.90
12"	S2@.323	LF	32.00	14.10	—	46.10
14" - 16"	S2@.346	LF	33.90	15.10	—	49.00
18" - 20"	S2@.377	LF	37.70	16.40	—	54.10
22" - 24"	S2@.407	LF	40.30	17.70	—	58.00
26" - 28"	S2@.440	LF	43.20	19.10	—	62.30
30"	S2@.465	LF	45.60	20.20	—	65.80

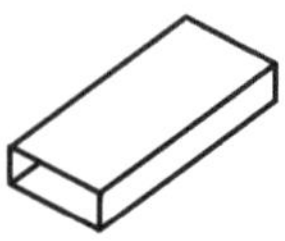

Costs for Galvanized 22 Gauge Rectangular Duct. Costs per linear foot of duct, based on quantities less than 1,000 pounds of prefabricated duct. Deduct 15% from the material cost for over 1,000 pounds. Add 50% to the material cost for lined duct. These costs include delivery, typical bracing, cleats, scrap, hangers, end closures.

Description	Craft@Hrs	Unit	Material $	Labor $	Equipment $	Total $
Galvanized steel 32" rectangular duct 22 gauge						
8"	S2@.374	LF	37.00	16.30	—	53.30
10"	S2@.393	LF	38.70	17.10	—	55.80
12"	S2@.412	LF	40.60	17.90	—	58.50
14" - 16"	S2@.442	LF	43.60	19.20	—	62.80
18" - 20"	S2@.478	LF	47.40	20.80	—	68.20
22" - 24"	S2@.517	LF	51.00	22.50	—	73.50
26" - 28"	S2@.553	LF	54.50	24.10	—	78.60
30" - 32"	S2@.589	LF	58.40	25.60	—	84.00
34" - 36"	S2@.628	LF	62.00	27.30	—	89.30
38" - 40"	S2@.664	LF	65.80	28.90	—	94.70
42" - 44"	S2@.702	LF	68.90	30.60	—	99.50
46" - 48"	S2@.739	LF	72.80	32.20	—	105.00
50" - 52"	S2@.777	LF	76.50	33.80	—	110.30
54"	S2@.807	LF	79.20	35.10	—	114.30

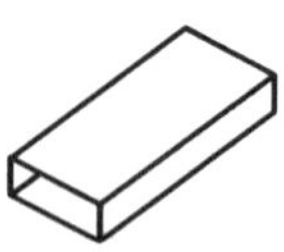

Description	Craft@Hrs	Unit	Material $	Labor $	Equipment $	Total $

Galvanized steel 34" rectangular duct 22 gauge

Description	Craft@Hrs	Unit	Material $	Labor $	Equipment $	Total $
8"	S2@.393	LF	38.70	17.10	—	55.80
10"	S2@.412	LF	40.60	17.90	—	58.50
12"	S2@.431	LF	42.10	18.80	—	60.90
14" - 16"	S2@.461	LF	45.30	20.10	—	65.40
18" - 20"	S2@.497	LF	49.70	21.60	—	71.30
22" - 24"	S2@.534	LF	52.40	23.20	—	75.60
26" - 28"	S2@.572	LF	56.70	24.90	—	81.60
30" - 32"	S2@.608	LF	59.90	26.50	—	86.40
34" - 36"	S2@.645	LF	64.00	28.10	—	92.10
38" - 40"	S2@.683	LF	67.30	29.70	—	97.00
42" - 44"	S2@.722	LF	71.40	31.40	—	102.80
46" - 48"	S2@.758	LF	75.00	33.00	—	108.00
50" - 52"	S2@.796	LF	78.20	34.60	—	112.80
54"	S2@.824	LF	81.50	35.90	—	117.40

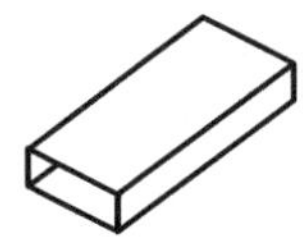

Galvanized steel 36" rectangular duct 22 gauge

Description	Craft@Hrs	Unit	Material $	Labor $	Equipment $	Total $
8"	S2@.412	LF	40.60	17.90	—	58.50
10"	S2@.431	LF	42.10	18.80	—	60.90
12"	S2@.453	LF	44.80	19.70	—	64.50
14" - 16"	S2@.478	LF	47.40	20.80	—	68.20
18" - 20"	S2@.517	LF	51.00	22.50	—	73.50
22" - 24"	S2@.553	LF	54.50	24.10	—	78.60
26" - 28"	S2@.589	LF	58.40	25.60	—	84.00
30" - 32"	S2@.628	LF	62.00	27.30	—	89.30
34" - 36"	S2@.664	LF	65.80	28.90	—	94.70
38" - 40"	S2@.702	LF	68.90	30.60	—	99.50
42" - 44"	S2@.739	LF	72.80	32.20	—	105.00
46" - 48"	S2@.777	LF	76.50	33.80	—	110.30
50" - 52"	S2@.816	LF	79.70	35.50	—	115.20
54"	S2@.841	LF	83.40	36.60	—	120.00

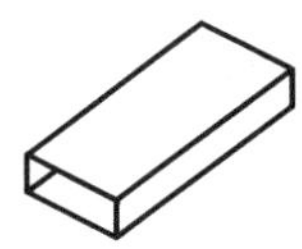

Galvanized steel 38" rectangular duct 22 gauge

Description	Craft@Hrs	Unit	Material $	Labor $	Equipment $	Total $
8"	S2@.431	LF	42.10	18.80	—	60.90
10"	S2@.453	LF	44.80	19.70	—	64.50
12"	S2@.470	LF	46.10	20.50	—	66.60
14" - 16"	S2@.497	LF	49.70	21.60	—	71.30
18" - 20"	S2@.534	LF	52.40	23.20	—	75.60
22" - 24"	S2@.572	LF	56.70	24.90	—	81.60
26" - 28"	S2@.608	LF	59.90	26.50	—	86.40
30" - 32"	S2@.645	LF	64.00	28.10	—	92.10
34" - 36"	S2@.683	LF	67.30	29.70	—	97.00
38" - 40"	S2@.722	LF	71.40	31.40	—	102.80
42" - 44"	S2@.758	LF	75.00	33.00	—	108.00
46" - 48"	S2@.796	LF	78.20	34.60	—	112.80
50" - 52"	S2@.833	LF	82.30	36.30	—	118.60
54"	S2@.863	LF	84.70	37.60	—	122.30

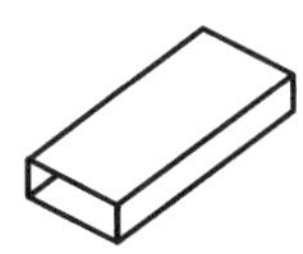

Galvanized Steel Rectangular Ductwork

Galvanized steel 40" rectangular duct 22 gauge

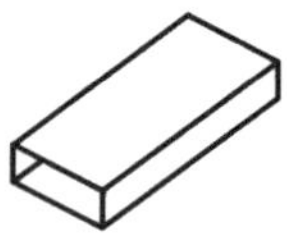

Description	Craft@Hrs	Unit	Material $	Labor $	Equipment $	Total $
8"	S2@.453	LF	44.80	19.70	—	64.50
10"	S2@.470	LF	46.10	20.50	—	66.60
12"	S2@.487	LF	48.30	21.20	—	69.50
14" - 16"	S2@.517	LF	51.00	22.50	—	73.50
18" - 20"	S2@.553	LF	54.50	24.10	—	78.60
22" - 24"	S2@.589	LF	58.40	25.60	—	84.00
26" - 28"	S2@.628	LF	62.00	27.30	—	89.30
30" - 32"	S2@.664	LF	65.80	28.90	—	94.70
34" - 36"	S2@.702	LF	68.90	30.60	—	99.50
38" - 40"	S2@.739	LF	72.80	32.20	—	105.00
42" - 44"	S2@.777	LF	76.50	33.80	—	110.30
46" - 48"	S2@.816	LF	79.70	35.50	—	115.20
50" - 52"	S2@.852	LF	84.20	37.10	—	121.30
54"	S2@.880	LF	86.30	38.30	—	124.60

Galvanized steel 42" rectangular duct 22 gauge

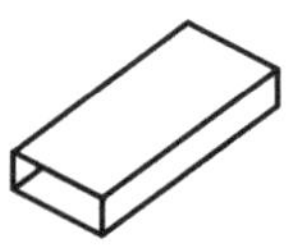

8"	S2@.470	LF	46.10	20.50	—	66.60
10"	S2@.487	LF	48.30	21.20	—	69.50
12"	S2@.508	LF	50.50	22.10	—	72.60
14" - 16"	S2@.534	LF	52.40	23.20	—	75.60
18" - 20"	S2@.572	LF	56.70	24.90	—	81.60
22" - 24"	S2@.611	LF	59.90	26.60	—	86.50
26" - 28"	S2@.645	LF	64.00	28.10	—	92.10
30" - 32"	S2@.683	LF	67.30	29.70	—	97.00
34" - 36"	S2@.722	LF	71.40	31.40	—	102.80
38" - 40"	S2@.758	LF	75.00	33.00	—	108.00
42" - 44"	S2@.798	LF	78.20	34.70	—	112.90
46" - 48"	S2@.833	LF	82.30	36.30	—	118.60
50" - 52"	S2@.871	LF	86.00	37.90	—	123.90
54"	S2@.901	LF	88.40	39.20	—	127.60

Galvanized steel 44" rectangular duct 22 gauge

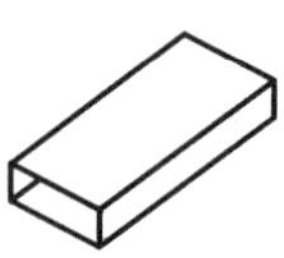

8"	S2@.487	LF	48.20	21.20	—	69.40
10"	S2@.508	LF	49.80	22.10	—	71.90
12"	S2@.525	LF	51.50	22.80	—	74.30
14" - 16"	S2@.555	LF	53.90	24.20	—	78.10
18" - 20"	S2@.589	LF	58.20	25.60	—	83.80
22" - 24"	S2@.628	LF	61.30	27.30	—	88.60
26" - 28"	S2@.666	LF	65.40	29.00	—	94.40
30" - 32"	S2@.705	LF	68.70	30.70	—	99.40
34" - 36"	S2@.739	LF	72.60	32.20	—	104.80
38" - 40"	S2@.777	LF	76.20	33.80	—	110.00
42" - 44"	S2@.816	LF	79.50	35.50	—	115.00
46" - 48"	S2@.852	LF	83.80	37.10	—	120.90
50" - 52"	S2@.890	LF	86.90	38.70	—	125.60
54"	S2@.918	LF	89.90	40.00	—	129.90

Galvanized steel 46" rectangular duct 22 gauge

Description	Craft@Hrs	Unit	Material $	Labor $	Equipment $	Total $
8"	S2@.508	LF	50.50	22.10	—	72.60
10"	S2@.525	LF	51.50	22.80	—	74.30
12"	S2@.542	LF	53.50	23.60	—	77.10
14" - 16"	S2@.572	LF	56.70	24.90	—	81.60
18" - 20"	S2@.611	LF	59.90	26.60	—	86.50
22" - 24"	S2@.649	LF	64.00	28.20	—	92.20
26" - 28"	S2@.683	LF	67.30	29.70	—	97.00
30" - 32"	S2@.722	LF	71.40	31.40	—	102.80
34" - 36"	S2@.758	LF	75.00	33.00	—	108.00
38" - 40"	S2@.796	LF	78.20	34.60	—	112.80
42" - 44"	S2@.833	LF	82.30	36.30	—	118.60
46" - 48"	S2@.871	LF	86.00	37.90	—	123.90
50" - 52"	S2@.909	LF	89.40	39.60	—	129.00
54"	S2@.935	LF	92.20	40.70	—	132.90

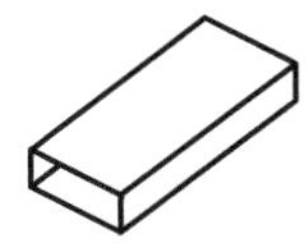

Galvanized steel 48" rectangular duct 22 gauge

Description	Craft@Hrs	Unit	Material $	Labor $	Equipment $	Total $
8"	S2@.525	LF	51.50	22.80	—	74.30
10"	S2@.542	LF	53.50	23.60	—	77.10
12"	S2@.564	LF	55.40	24.50	—	79.90
14" - 16"	S2@.589	LF	58.40	25.60	—	84.00
18" - 20"	S2@.628	LF	62.00	27.30	—	89.30
22" - 24"	S2@.664	LF	65.80	28.90	—	94.70
26" - 28"	S2@.702	LF	68.90	30.60	—	99.50
30" - 32"	S2@.739	LF	72.80	32.20	—	105.00
34" - 36"	S2@.777	LF	76.50	33.80	—	110.30
38" - 40"	S2@.816	LF	79.70	35.50	—	115.20
42" - 44"	S2@.852	LF	84.20	37.10	—	121.30
46" - 48"	S2@.890	LF	87.50	38.70	—	126.20
50" - 52"	S2@.927	LF	91.70	40.30	—	132.00
54"	S2@.956	LF	93.60	41.60	—	135.20

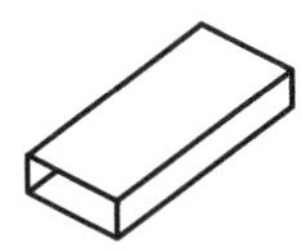

Galvanized steel 50" rectangular duct 22 gauge

Description	Craft@Hrs	Unit	Material $	Labor $	Equipment $	Total $
8"	S2@.542	LF	53.50	23.60	—	77.10
10"	S2@.564	LF	55.40	24.50	—	79.90
12"	S2@.581	LF	57.30	25.30	—	82.60
14" - 16"	S2@.611	LF	59.90	26.60	—	86.50
18" - 20"	S2@.645	LF	64.00	28.10	—	92.10
22" - 24"	S2@.683	LF	67.30	29.70	—	97.00
26" - 28"	S2@.722	LF	71.40	31.40	—	102.80
30" - 32"	S2@.758	LF	75.00	33.00	—	108.00
34" - 36"	S2@.798	LF	78.20	34.70	—	112.90
38" - 40"	S2@.833	LF	82.30	36.30	—	118.60
42" - 44"	S2@.871	LF	86.00	37.90	—	123.90
46" - 48"	S2@.909	LF	89.40	39.60	—	129.00
50" - 52"	S2@.946	LF	93.00	41.20	—	134.20
54"	S2@.974	LF	95.60	42.40	—	138.00

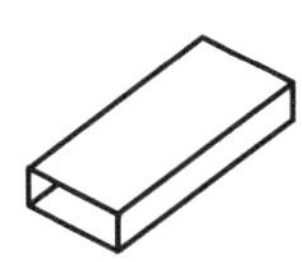

Description	Craft@Hrs	Unit	Material $	Labor $	Equipment $	Total $

Galvanized steel 52" rectangular duct 22 gauge

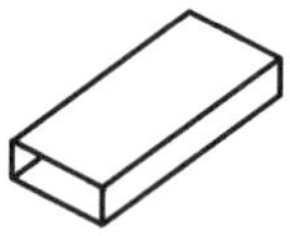

Description	Craft@Hrs	Unit	Material $	Labor $	Equipment $	Total $
8"	S2@.564	LF	55.40	24.50	—	79.90
10"	S2@.581	LF	57.30	25.30	—	82.60
12"	S2@.598	LF	59.10	26.00	—	85.10
14" - 16"	S2@.628	LF	62.00	27.30	—	89.30
18" - 20"	S2@.662	LF	65.80	28.80	—	94.60
22" - 24"	S2@.700	LF	68.90	30.50	—	99.40
26" - 28"	S2@.739	LF	72.80	32.20	—	105.00
30" - 32"	S2@.777	LF	76.50	33.80	—	110.30
34" - 36"	S2@.816	LF	79.70	35.50	—	115.20
38" - 40"	S2@.852	LF	84.20	37.10	—	121.30
42" - 44"	S2@.890	LF	87.50	38.70	—	126.20
46" - 48"	S2@.927	LF	91.70	40.30	—	132.00
50" - 52"	S2@.976	LF	94.80	42.50	—	137.30
54"	S2@.991	LF	97.30	43.10	—	140.40

Galvanized steel 54" rectangular duct 22 gauge

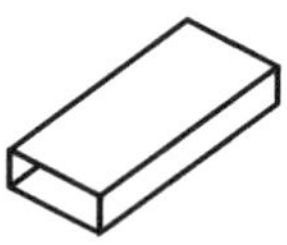

Description	Craft@Hrs	Unit	Material $	Labor $	Equipment $	Total $
8"	S2@.581	LF	57.30	25.30	—	82.60
10"	S2@.598	LF	59.10	26.00	—	85.10
12"	S2@.619	LF	60.80	26.90	—	87.70
14" - 16"	S2@.645	LF	64.00	28.10	—	92.10
18" - 20"	S2@.683	LF	67.30	29.70	—	97.00
22" - 24"	S2@.722	LF	71.40	31.40	—	102.80
26" - 28"	S2@.758	LF	75.00	33.00	—	108.00
30" - 32"	S2@.796	LF	78.20	34.60	—	112.80
34" - 36"	S2@.833	LF	82.30	36.30	—	118.60
38" - 40"	S2@.871	LF	86.00	37.90	—	123.90
42" - 44"	S2@.909	LF	89.40	39.60	—	129.00
46" - 48"	S2@.946	LF	93.00	41.20	—	134.20
50" - 52"	S2@.982	LF	96.20	42.70	—	138.90
54"	S2@1.01	LF	99.90	44.00	—	143.90

Costs for Galvanized 20 Gauge Rectangular Duct. Costs per linear foot of duct, based on quantities less than 1,000 pounds of prefabricated duct. Deduct 15% from the material cost for over 1,000 pounds. Add 50% to the material cost for lined duct. These costs include delivery, typical bracing, cleats, scrap, hangers, end closures, sealants and miscellaneous hardware.

Galvanized steel 56" rectangular duct 20 gauge

Description	Craft@Hrs	Unit	Material $	Labor $	Equipment $	Total $
18" - 20"	S2@.828	LF	82.30	36.00	—	118.30
22" - 24"	S2@.871	LF	86.30	37.90	—	124.20
26" - 28"	S2@.916	LF	90.50	39.90	—	130.40
30" - 32"	S2@.959	LF	95.30	41.70	—	137.00
34" - 36"	S2@1.00	LF	99.90	43.50	—	143.40
38" - 40"	S2@1.05	LF	104.00	45.70	—	149.70
42" - 44"	S2@1.09	LF	109.00	47.40	—	156.40
46" - 48"	S2@1.14	LF	113.00	49.60	—	162.60
50" - 52"	S2@1.18	LF	118.00	51.40	—	169.40
54" - 56"	S2@1.23	LF	121.00	53.50	—	174.50
58" - 60"	S2@1.27	LF	126.00	55.30	—	181.30
62" - 64"	S2@1.31	LF	130.00	57.00	—	187.00
66" - 68"	S2@1.36	LF	134.00	59.20	—	193.20
70" - 72"	S2@1.40	LF	138.00	60.90	—	198.90

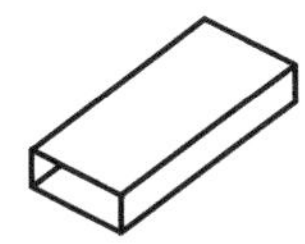

Galvanized steel 58" rectangular duct 20 gauge

Description	Craft@Hrs	Unit	Material $	Labor $	Equipment $	Total $
18" - 20"	S2@.850	LF	84.30	37.00	—	121.30
22" - 24"	S2@.895	LF	88.70	39.00	—	127.70
26" - 28"	S2@.937	LF	93.00	40.80	—	133.80
30" - 32"	S2@.980	LF	97.60	42.60	—	140.20
34" - 36"	S2@1.02	LF	102.00	44.40	—	146.40
38" - 40"	S2@1.07	LF	106.00	46.60	—	152.60
42" - 44"	S2@1.11	LF	111.00	48.30	—	159.30
46" - 48"	S2@1.16	LF	115.00	50.50	—	165.50
50" - 52"	S2@1.02	LF	120.00	44.40	—	164.40
54" - 56"	S2@1.25	LF	123.00	54.40	—	177.40
58" - 60"	S2@1.29	LF	127.00	56.10	—	183.10
62" - 64"	S2@1.33	LF	133.00	57.90	—	190.90
66" - 68"	S2@1.38	LF	136.00	60.10	—	196.10
70" - 72"	S2@1.42	LF	139.00	61.80	—	200.80

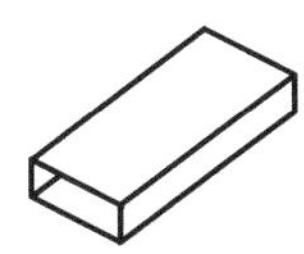

Galvanized steel 60" rectangular duct 20 gauge

Description	Craft@Hrs	Unit	Material $	Labor $	Equipment $	Total $
18" - 20"	S2@.871	LF	86.30	37.90	—	124.20
22" - 24"	S2@.916	LF	90.50	39.90	—	130.40
26" - 28"	S2@.959	LF	95.30	41.70	—	137.00
30" - 32"	S2@1.00	LF	99.90	43.50	—	143.40
34" - 36"	S2@1.05	LF	104.00	45.70	—	149.70
38" - 40"	S2@1.09	LF	109.00	47.40	—	156.40
42" - 44"	S2@1.14	LF	113.00	49.60	—	162.60
46" - 48"	S2@1.18	LF	118.00	51.40	—	169.40
50" - 52"	S2@1.23	LF	121.00	53.50	—	174.50
54" - 56"	S2@1.27	LF	126.00	55.30	—	181.30
58" - 60"	S2@1.31	LF	130.00	57.00	—	187.00
62" - 64"	S2@1.36	LF	134.00	59.20	—	193.20
66" - 68"	S2@1.40	LF	138.00	60.90	—	198.90
70" - 72"	S2@1.45	LF	142.00	63.10	—	205.10

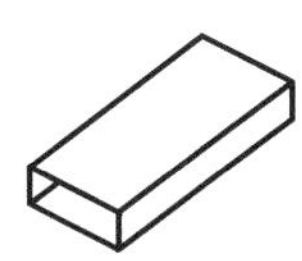

Description	Craft@Hrs	Unit	Material $	Labor $	Equipment $	Total $
Galvanized steel 62" rectangular duct 20 gauge						
18" - 20"	S2@.895	LF	88.70	39.00	—	127.70
22" - 24"	S2@.937	LF	93.00	40.80	—	133.80
26" - 28"	S2@.980	LF	97.60	42.60	—	140.20
30" - 32"	S2@1.02	LF	102.00	44.40	—	146.40
34" - 36"	S2@1.07	LF	106.00	46.60	—	152.60
38" - 40"	S2@1.11	LF	111.00	48.30	—	159.30
42" - 44"	S2@1.16	LF	115.00	50.50	—	165.50
46" - 48"	S2@1.20	LF	120.00	52.20	—	172.20
50" - 52"	S2@1.25	LF	123.00	54.40	—	177.40
54" - 56"	S2@1.29	LF	127.00	56.10	—	183.10
58" - 60"	S2@1.33	LF	133.00	57.90	—	190.90
62" - 64"	S2@1.38	LF	136.00	60.10	—	196.10
66" - 68"	S2@1.42	LF	139.00	61.80	—	200.80
70" - 72"	S2@1.47	LF	143.00	64.00	—	207.00
Galvanized steel 64" rectangular duct 20 gauge						
18" - 20"	S2@.916	LF	90.50	39.90	—	130.40
22" - 24"	S2@.959	LF	95.30	41.70	—	137.00
26" - 28"	S2@1.00	LF	99.90	43.50	—	143.40
30" - 32"	S2@1.05	LF	104.00	45.70	—	149.70
34" - 36"	S2@1.09	LF	109.00	47.40	—	156.40
38" - 40"	S2@1.14	LF	113.00	49.60	—	162.60
42" - 44"	S2@1.18	LF	118.00	51.40	—	169.40
46" - 48"	S2@1.23	LF	121.00	53.50	—	174.50
50" - 52"	S2@1.27	LF	126.00	55.30	—	181.30
54" - 56"	S2@1.31	LF	130.00	57.00	—	187.00
58" - 60"	S2@1.36	LF	134.00	59.20	—	193.20
62" - 64"	S2@1.40	LF	138.00	60.90	—	198.90
66" - 68"	S2@1.45	LF	142.00	63.10	—	205.10
70" - 72"	S2@1.49	LF	147.00	64.80	—	211.80
Galvanized steel 66" rectangular duct 20 gauge						
18" - 20"	S2@.937	LF	93.00	40.80	—	133.80
22" - 24"	S2@.980	LF	97.60	42.60	—	140.20
26" - 28"	S2@1.02	LF	102.00	44.40	—	146.40
30" - 32"	S2@1.07	LF	106.00	46.60	—	152.60
34" - 36"	S2@1.11	LF	111.00	48.30	—	159.30
38" - 40"	S2@1.16	LF	115.00	50.50	—	165.50
42" - 44"	S2@1.20	LF	120.00	52.20	—	172.20
46" - 48"	S2@1.25	LF	123.00	54.40	—	177.40
50" - 52"	S2@1.29	LF	127.00	56.10	—	183.10
54" - 56"	S2@1.33	LF	133.00	57.90	—	190.90
58" - 60"	S2@1.38	LF	136.00	60.10	—	196.10
62" - 64"	S2@1.42	LF	139.00	61.80	—	200.80
66" - 68"	S2@1.47	LF	143.00	64.00	—	207.00
70" - 72"	S2@1.51	LF	149.00	65.70	—	214.70

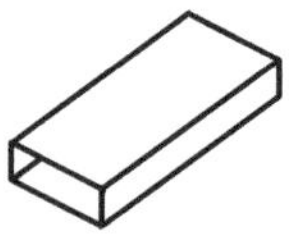

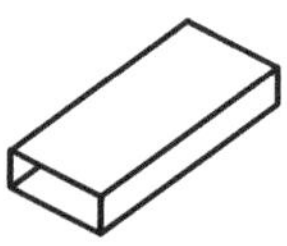

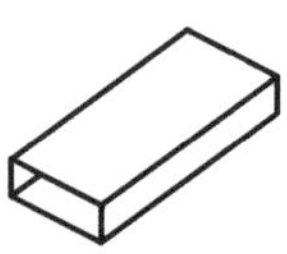

Galvanized steel 68" rectangular duct 20 gauge

Description	Craft@Hrs	Unit	Material $	Labor $	Equipment $	Total $
18" - 20"	S2@.959	LF	95.30	41.70	—	137.00
22" - 24"	S2@1.00	LF	99.90	43.50	—	143.40
26" - 28"	S2@1.05	LF	104.00	45.70	—	149.70
30" - 32"	S2@1.09	LF	109.00	47.40	—	156.40
34" - 36"	S2@1.14	LF	113.00	49.60	—	162.60
38" - 40"	S2@1.18	LF	118.00	51.40	—	169.40
42" - 44"	S2@1.23	LF	121.00	53.50	—	174.50
46" - 48"	S2@1.27	LF	126.00	55.30	—	181.30
50" - 52"	S2@1.31	LF	130.00	57.00	—	187.00
54" - 56"	S2@1.36	LF	134.00	59.20	—	193.20
58" - 60"	S2@1.40	LF	138.00	60.90	—	198.90
62" - 64"	S2@1.45	LF	142.00	63.10	—	205.10
66" - 68"	S2@1.49	LF	147.00	64.80	—	211.80
70" - 72"	S2@1.53	LF	153.00	66.60	—	219.60

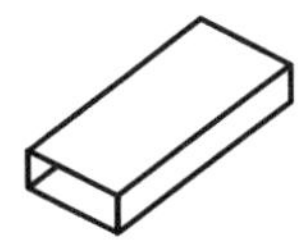

Galvanized steel 70" rectangular duct 20 gauge

Description	Craft@Hrs	Unit	Material $	Labor $	Equipment $	Total $
18" - 20"	S2@.980	LF	97.60	42.60	—	140.20
22" - 24"	S2@1.02	LF	102.00	44.40	—	146.40
26" - 28"	S2@1.07	LF	106.00	46.60	—	152.60
30" - 32"	S2@1.11	LF	111.00	48.30	—	159.30
34" - 36"	S2@1.16	LF	115.00	50.50	—	165.50
38" - 40"	S2@1.20	LF	120.00	52.20	—	172.20
42" - 44"	S2@1.25	LF	123.00	54.40	—	177.40
46" - 48"	S2@1.29	LF	127.00	56.10	—	183.10
50" - 52"	S2@1.33	LF	133.00	57.90	—	190.90
54" - 56"	S2@1.38	LF	136.00	60.10	—	196.10
58" - 60"	S2@1.42	LF	139.00	61.80	—	200.80
62" - 64"	S2@1.47	LF	143.00	64.00	—	207.00
66" - 68"	S2@1.51	LF	149.00	65.70	—	214.70
70" - 72"	S2@1.55	LF	154.00	67.50	—	221.50

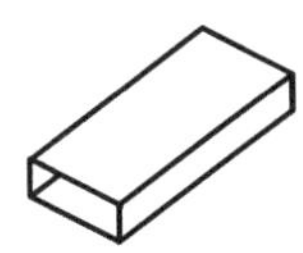

Galvanized steel 72" rectangular duct 20 gauge

Description	Craft@Hrs	Unit	Material $	Labor $	Equipment $	Total $
18" - 20"	S2@1.00	LF	99.90	43.50	—	143.40
22" - 24"	S2@1.05	LF	104.00	45.70	—	149.70
26" - 28"	S2@1.09	LF	109.00	47.40	—	156.40
30" - 32"	S2@1.14	LF	113.00	49.60	—	162.60
34" - 36"	S2@1.18	LF	118.00	51.40	—	169.40
38" - 40"	S2@1.23	LF	121.00	53.50	—	174.50
42" - 44"	S2@1.27	LF	126.00	55.30	—	181.30
46" - 48"	S2@1.31	LF	130.00	57.00	—	187.00
50" - 52"	S2@1.36	LF	134.00	59.20	—	193.20
54" - 56"	S2@1.40	LF	138.00	60.90	—	198.90
58" - 60"	S2@1.45	LF	142.00	63.10	—	205.10
62" - 64"	S2@1.49	LF	147.00	64.80	—	211.80
66" - 68"	S2@1.53	LF	153.00	66.60	—	219.60
70" - 72"	S2@1.58	LF	156.00	68.80	—	224.80

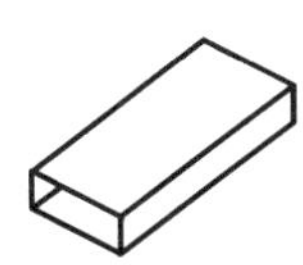

Galvanized Steel Rectangular Elbows

Costs for Galvanized Rectangular 90-Degree Elbows. Costs per elbow, based on quantities less than 1,000 pounds. Deduct 15% from the material cost for over 1,000 pounds. Add 50% to the material cost for lined duct. These costs include delivery, typical bracing, cleats, scrap, hangers, end closures, sealants and miscellaneous hardware. For turning vanes, please see page 342.

Galvanized steel 4" rectangular duct 90-degree elbow

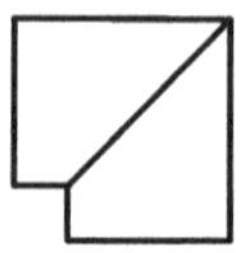

Description	Craft@Hrs	Unit	Material $	Labor $	Equipment $	Total $
4"	S2@.042	Ea	8.34	1.83	—	10.17
6"	S2@.058	Ea	11.50	2.52	—	14.02
8"	S2@.078	Ea	15.40	3.39	—	18.79
10"	S2@.104	Ea	20.60	4.53	—	25.13
12"	S2@.132	Ea	26.00	5.74	—	31.74
14" - 16"	S2@.236	Ea	46.70	10.30	—	57.00
18" - 20"	S2@.339	Ea	67.30	14.80	—	82.10
22" - 24"	S2@.468	Ea	92.20	20.40	—	112.60
26" - 28"	S2@.615	Ea	120.00	26.80	—	146.80
30" - 32"	S2@.854	Ea	172.00	37.20	—	209.20
34" - 36"	S2@1.19	Ea	235.00	51.80	—	286.80
38" - 40"	S2@1.45	Ea	287.00	63.10	—	350.10
42" - 44"	S2@1.73	Ea	339.00	75.30	—	414.30
46" - 48"	S2@2.04	Ea	403.00	88.80	—	491.80
50"	S2@2.28	Ea	453.00	99.20	—	552.20

Galvanized steel 6" rectangular duct 90-degree elbow

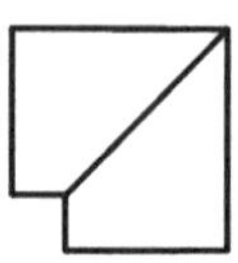

Description	Craft@Hrs	Unit	Material $	Labor $	Equipment $	Total $
4"	S2@.058	Ea	11.50	2.52	—	14.02
6"	S2@.078	Ea	15.40	3.39	—	18.79
8"	S2@.104	Ea	20.60	4.53	—	25.13
10"	S2@.132	Ea	26.00	5.74	—	31.74
12"	S2@.148	Ea	29.10	6.44	—	35.54
14" - 16"	S2@.266	Ea	52.90	11.60	—	64.50
18" - 20"	S2@.370	Ea	72.80	16.10	—	88.90
22" - 24"	S2@.506	Ea	100.00	22.00	—	122.00
26" - 28"	S2@.655	Ea	128.00	28.50	—	156.50
30" - 32"	S2@.905	Ea	177.00	39.40	—	216.40
34" - 36"	S2@1.24	Ea	245.00	54.00	—	299.00
38" - 40"	S2@1.49	Ea	293.00	64.80	—	357.80
42" - 44"	S2@1.78	Ea	349.00	77.50	—	426.50
46" - 48"	S2@2.09	Ea	413.00	91.00	—	504.00
50"	S2@2.34	Ea	464.00	102.00	—	566.00

Galvanized steel 8" rectangular duct 90-degree elbow

Description	Craft@Hrs	Unit	Material $	Labor $	Equipment $	Total $
4"	S2@.078	Ea	15.40	3.39	—	18.79
6"	S2@.104	Ea	20.60	4.53	—	25.13
8"	S2@.132	Ea	26.00	5.74	—	31.74
10"	S2@.148	Ea	29.10	6.44	—	35.54
12"	S2@.170	Ea	33.80	7.40	—	41.20
14" - 16"	S2@.300	Ea	59.10	13.10	—	72.20
18" - 20"	S2@.403	Ea	79.50	17.50	—	97.00
22" - 24"	S2@.549	Ea	108.00	23.90	—	131.90
26" - 28"	S2@.694	Ea	136.00	30.20	—	166.20
30" - 32"	S2@.950	Ea	190.00	41.30	—	231.30
34" - 36"	S2@1.28	Ea	252.00	55.70	—	307.70
38" - 40"	S2@1.54	Ea	308.00	67.00	—	375.00
42" - 44"	S2@1.83	Ea	362.00	79.60	—	441.60
46" - 48"	S2@2.13	Ea	421.00	92.70	—	513.70
50"	S2@2.39	Ea	478.00	104.00	—	582.00

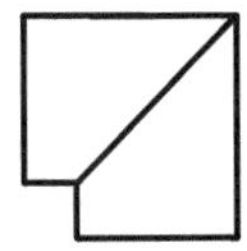

Galvanized steel 10" rectangular duct 90-degree elbow

Description	Craft@Hrs	Unit	Material $	Labor $	Equipment $	Total $
4"	S2@.104	Ea	20.60	4.53	—	25.13
6"	S2@.132	Ea	26.00	5.74	—	31.74
8"	S2@.148	Ea	29.10	6.44	—	35.54
10"	S2@.170	Ea	33.80	7.40	—	41.20
12"	S2@.194	Ea	38.30	8.44	—	46.74
14" - 16"	S2@.336	Ea	65.90	14.60	—	80.50
18" - 20"	S2@.438	Ea	86.30	19.10	—	105.40
22" - 24"	S2@.587	Ea	116.00	25.50	—	141.50
26" - 28"	S2@.737	Ea	143.00	32.10	—	175.10
30" - 32"	S2@.997	Ea	197.00	43.40	—	240.40
34" - 36"	S2@1.33	Ea	262.00	57.90	—	319.90
38" - 40"	S2@1.59	Ea	317.00	69.20	—	386.20
42" - 44"	S2@1.87	Ea	370.00	81.40	—	451.40
46" - 48"	S2@2.18	Ea	432.00	94.90	—	526.90
50"	S2@2.43	Ea	483.00	106.00	—	589.00

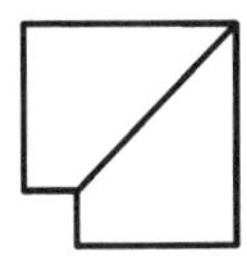

Galvanized steel 12" rectangular duct 90-degree elbow

Description	Craft@Hrs	Unit	Material $	Labor $	Equipment $	Total $
4"	S2@.132	Ea	26.00	5.74	—	31.74
6"	S2@.148	Ea	29.10	6.44	—	35.54
8"	S2@.170	Ea	33.80	7.40	—	41.20
10"	S2@.194	Ea	38.30	8.44	—	46.74
12"	S2@.224	Ea	44.50	9.75	—	54.25
18" - 20"	S2@.478	Ea	93.60	20.80	—	114.40
22" - 24"	S2@.628	Ea	123.00	27.30	—	150.30
26" - 28"	S2@.775	Ea	154.00	33.70	—	187.70
30" - 32"	S2@1.04	Ea	206.00	45.30	—	251.30
34" - 36"	S2@1.38	Ea	273.00	60.10	—	333.10
38" - 40"	S2@1.64	Ea	324.00	71.40	—	395.40
42" - 44"	S2@1.92	Ea	380.00	83.60	—	463.60
46" - 48"	S2@2.23	Ea	445.00	97.00	—	542.00
50"	S2@2.48	Ea	489.00	108.00	—	597.00

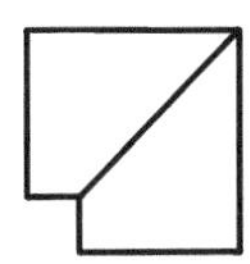

Galvanized Steel Rectangular Elbows

Description	Craft@Hrs	Unit	Material $	Labor $	Equipment $	Total $

Galvanized steel 14" rectangular duct 90-degree elbow

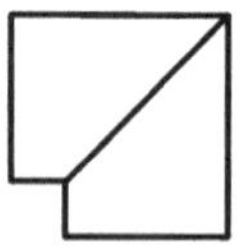

Description	Craft@Hrs	Unit	Material $	Labor $	Equipment $	Total $
4"	S2@.212	Ea	41.70	9.23	—	50.93
6"	S2@.232	Ea	46.10	10.10	—	56.20
8"	S2@.260	Ea	51.40	11.30	—	62.70
10"	S2@.292	Ea	57.70	12.70	—	70.40
12"	S2@.328	Ea	64.90	14.30	—	79.20
14" - 16"	S2@.415	Ea	82.30	18.10	—	100.40
18" - 20"	S2@.527	Ea	104.00	22.90	—	126.90
22" - 24"	S2@.867	Ea	131.00	37.70	—	168.70
26" - 28"	S2@.816	Ea	159.00	35.50	—	194.50
30" - 32"	S2@1.09	Ea	217.00	47.40	—	264.40
34" - 36"	S2@1.43	Ea	285.00	62.20	—	347.20
38" - 40"	S2@1.69	Ea	334.00	73.50	—	407.50
42" - 44"	S2@1.97	Ea	393.00	85.70	—	478.70
46" - 48"	S2@2.28	Ea	451.00	99.20	—	550.20
50"	S2@2.53	Ea	505.00	110.00	—	615.00

Galvanized steel 16" rectangular duct 90-degree elbow

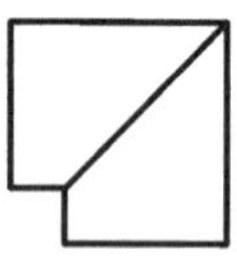

Description	Craft@Hrs	Unit	Material $	Labor $	Equipment $	Total $
4"	S2@.260	Ea	51.40	11.30	—	62.70
6"	S2@.310	Ea	61.30	13.50	—	74.80
8"	S2@.340	Ea	67.20	14.80	—	82.00
10"	S2@.380	Ea	75.50	16.50	—	92.00
12"	S2@.420	Ea	83.00	18.30	—	101.30
14" - 16"	S2@.480	Ea	94.70	20.90	—	115.60
18" - 20"	S2@.555	Ea	110.00	24.20	—	134.20
22" - 24"	S2@.707	Ea	138.00	30.80	—	168.80
26" - 28"	S2@.856	Ea	171.00	37.30	—	208.30
30" - 32"	S2@1.14	Ea	225.00	49.60	—	274.60
34" - 36"	S2@1.48	Ea	292.00	64.40	—	356.40
38" - 40"	S2@1.73	Ea	343.00	75.30	—	418.30
42" - 44"	S2@2.02	Ea	402.00	87.90	—	489.90
46" - 48"	S2@2.32	Ea	464.00	101.00	—	565.00
50"	S2@2.58	Ea	510.00	112.00	—	622.00

Galvanized steel 18" rectangular duct 90-degree elbow

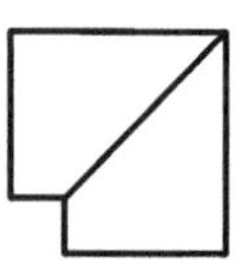

Description	Craft@Hrs	Unit	Material $	Labor $	Equipment $	Total $
4"	S2@.312	Ea	62.00	13.60	—	75.60
6"	S2@.350	Ea	69.50	15.20	—	84.70
8"	S2@.390	Ea	77.40	17.00	—	94.40
10"	S2@.427	Ea	84.60	18.60	—	103.20
12"	S2@.470	Ea	92.20	20.50	—	112.70
14" - 16"	S2@.519	Ea	103.00	22.60	—	125.60
18" - 20"	S2@.604	Ea	119.00	26.30	—	145.30
22" - 24"	S2@.747	Ea	147.00	32.50	—	179.50
26" - 28"	S2@.895	Ea	176.00	39.00	—	215.00
30" - 32"	S2@1.18	Ea	234.00	51.40	—	285.40
34" - 36"	S2@1.53	Ea	305.00	66.60	—	371.60
38" - 40"	S2@1.78	Ea	349.00	77.50	—	426.50
42" - 44"	S2@2.07	Ea	410.00	90.10	—	500.10
46" - 48"	S2@2.37	Ea	474.00	103.00	—	577.00
50"	S2@2.63	Ea	520.00	114.00	—	634.00

Description	Craft@Hrs	Unit	Material $	Labor $	Equipment $	Total $
Galvanized steel 20" rectangular duct 90-degree elbow						
4"	S2@.370	Ea	72.80	16.10	—	88.90
6"	S2@.390	Ea	77.40	17.00	—	94.40
8"	S2@.415	Ea	82.30	18.10	—	100.40
10"	S2@.448	Ea	88.50	19.50	—	108.00
12"	S2@.487	Ea	95.70	21.20	—	116.90
14" - 16"	S2@.553	Ea	109.00	24.10	—	133.10
18" - 20"	S2@.658	Ea	129.00	28.60	—	157.60
22" - 24"	S2@.786	Ea	156.00	34.20	—	190.20
26" - 28"	S2@.931	Ea	183.00	40.50	—	223.50
30" - 32"	S2@1.23	Ea	242.00	53.50	—	295.50
34" - 36"	S2@1.58	Ea	315.00	68.80	—	383.80
38" - 40"	S2@1.83	Ea	365.00	79.60	—	444.60
42" - 44"	S2@2.11	Ea	417.00	91.80	—	508.80
46" - 48"	S2@2.42	Ea	482.00	105.00	—	587.00
50"	S2@2.67	Ea	533.00	116.00	—	649.00
Galvanized steel rectangular 22" duct 90-degree elbow						
22" - 24"	S2@.828	Ea	161.00	36.00	—	197.00
26" - 28"	S2@1.01	Ea	198.00	44.00	—	242.00
30" - 32"	S2@1.31	Ea	258.00	57.00	—	315.00
34" - 36"	S2@1.71	Ea	338.00	74.40	—	412.40
38" - 40"	S2@2.01	Ea	402.00	87.50	—	489.50
42" - 44"	S2@2.31	Ea	453.00	101.00	—	554.00
46" - 48"	S2@2.61	Ea	515.00	114.00	—	629.00
50" - 52"	S2@2.91	Ea	577.00	127.00	—	704.00
54" - 56"	S2@3.51	Ea	695.00	153.00	—	848.00
58" - 60"	S2@4.26	Ea	843.00	185.00	—	1,028.00
62" - 64"	S2@4.76	Ea	934.00	207.00	—	1,141.00
66" - 68"	S2@5.25	Ea	1,040.00	228.00	—	1,268.00
70" - 72"	S2@5.76	Ea	1,130.00	251.00	—	1,381.00
Galvanized steel 24" rectangular duct 90-degree elbow						
22" - 24"	S2@.922	Ea	182.00	40.10	—	222.10
26" - 28"	S2@1.10	Ea	218.00	47.90	—	265.90
30" - 32"	S2@1.41	Ea	280.00	61.40	—	341.40
34" - 36"	S2@1.81	Ea	360.00	78.80	—	438.80
38" - 40"	S2@2.10	Ea	417.00	91.40	—	508.40
42" - 44"	S2@2.41	Ea	478.00	105.00	—	583.00
46" - 48"	S2@2.71	Ea	535.00	118.00	—	653.00
50" - 52"	S2@3.01	Ea	591.00	131.00	—	722.00
54" - 56"	S2@3.62	Ea	716.00	158.00	—	874.00
58" - 60"	S2@4.38	Ea	863.00	191.00	—	1,054.00
62" - 64"	S2@4.87	Ea	958.00	212.00	—	1,170.00
66" - 68"	S2@5.38	Ea	1,060.00	234.00	—	1,294.00
70" - 72"	S2@5.87	Ea	1,160.00	255.00	—	1,415.00

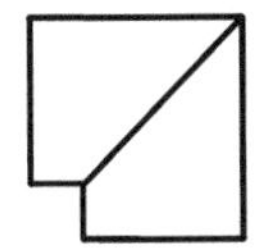

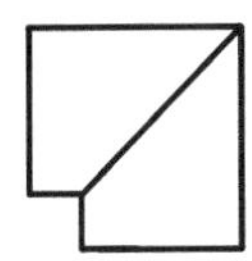

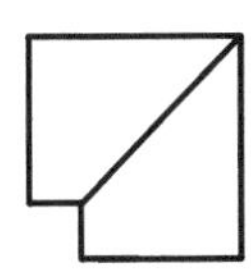

Galvanized Steel Rectangular Elbows

Description	Craft@Hrs	Unit	Material $	Labor $	Equipment $	Total $

Galvanized steel 26" rectangular duct 90-degree elbow

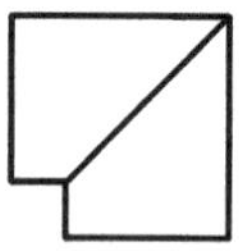

Description	Craft@Hrs	Unit	Material $	Labor $	Equipment $	Total $
22" - 24"	S2@1.01	Ea	198.00	44.00	—	242.00
26" - 28"	S2@1.19	Ea	235.00	51.80	—	286.80
30" - 32"	S2@1.50	Ea	296.00	65.30	—	361.30
34" - 36"	S2@1.92	Ea	380.00	83.60	—	463.60
38" - 40"	S2@2.21	Ea	436.00	96.20	—	532.20
42" - 44"	S2@2.51	Ea	498.00	109.00	—	607.00
46" - 48"	S2@2.81	Ea	553.00	122.00	—	675.00
50" - 52"	S2@3.11	Ea	620.00	135.00	—	755.00
54" - 56"	S2@3.73	Ea	742.00	162.00	—	904.00
58" - 60"	S2@4.50	Ea	887.00	196.00	—	1,083.00
62" - 64"	S2@5.02	Ea	987.00	218.00	—	1,205.00
66" - 68"	S2@5.53	Ea	1,100.00	241.00	—	1,341.00
70" - 72"	S2@6.04	Ea	1,190.00	263.00	—	1,453.00

Galvanized steel 28" rectangular duct 90-degree elbow

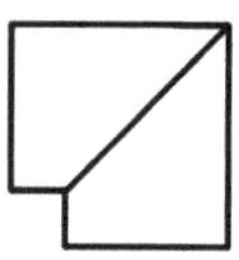

Description	Craft@Hrs	Unit	Material $	Labor $	Equipment $	Total $
22" - 24"	S2@1.10	Ea	217.00	47.90	—	264.90
26" - 28"	S2@1.28	Ea	252.00	55.70	—	307.70
30" - 32"	S2@1.60	Ea	319.00	69.60	—	388.60
34" - 36"	S2@2.03	Ea	403.00	88.30	—	491.30
38" - 40"	S2@2.31	Ea	453.00	101.00	—	554.00
42" - 44"	S2@2.61	Ea	515.00	114.00	—	629.00
46" - 48"	S2@2.91	Ea	577.00	127.00	—	704.00
50" - 52"	S2@3.21	Ea	639.00	140.00	—	779.00
54" - 56"	S2@3.84	Ea	762.00	167.00	—	929.00
58" - 60"	S2@4.61	Ea	916.00	201.00	—	1,117.00
62" - 64"	S2@5.15	Ea	1,010.00	224.00	—	1,234.00
66" - 68"	S2@5.66	Ea	1,110.00	246.00	—	1,356.00
70" - 72"	S2@6.15	Ea	1,200.00	268.00	—	1,468.00

Galvanized steel 30" rectangular duct 90-degree elbow

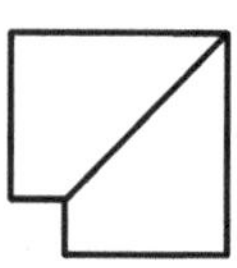

Description	Craft@Hrs	Unit	Material $	Labor $	Equipment $	Total $
22" - 24"	S2@1.19	Ea	234.00	51.80	—	285.80
26" - 28"	S2@1.37	Ea	273.00	59.60	—	332.60
30" - 32"	S2@1.70	Ea	335.00	74.00	—	409.00
34" - 36"	S2@2.14	Ea	421.00	93.10	—	514.10
38" - 40"	S2@2.41	Ea	478.00	105.00	—	583.00
42" - 44"	S2@2.71	Ea	535.00	118.00	—	653.00
46" - 48"	S2@3.01	Ea	591.00	131.00	—	722.00
50" - 52"	S2@3.31	Ea	654.00	144.00	—	798.00
54" - 56"	S2@3.95	Ea	779.00	172.00	—	951.00
58" - 60"	S2@4.74	Ea	932.00	206.00	—	1,138.00
62" - 64"	S2@5.25	Ea	1,040.00	228.00	—	1,268.00
66" - 68"	S2@5.79	Ea	1,140.00	252.00	—	1,392.00
70" - 72"	S2@6.28	Ea	1,230.00	273.00	—	1,503.00

Description	Craft@Hrs	Unit	Material $	Labor $	Equipment $	Total $

Galvanized steel 32" rectangular duct 90-degree elbow

22" - 24"	S2@1.53	Ea	305.00	66.60	—	371.60
26" - 28"	S2@1.73	Ea	339.00	75.30	—	414.30
30" - 32"	S2@1.93	Ea	383.00	84.00	—	467.00
34" - 36"	S2@2.25	Ea	448.00	97.90	—	545.90
38" - 40"	S2@2.51	Ea	498.00	109.00	—	607.00
42" - 44"	S2@2.81	Ea	553.00	122.00	—	675.00
46" - 48"	S2@3.11	Ea	613.00	135.00	—	748.00
50" - 52"	S2@3.41	Ea	673.00	148.00	—	821.00
54" - 56"	S2@4.06	Ea	797.00	177.00	—	974.00
58" - 60"	S2@4.85	Ea	956.00	211.00	—	1,167.00
62" - 64"	S2@5.38	Ea	1,060.00	234.00	—	1,294.00
66" - 68"	S2@5.89	Ea	1,160.00	256.00	—	1,416.00
70" - 72"	S2@6.40	Ea	1,260.00	279.00	—	1,539.00

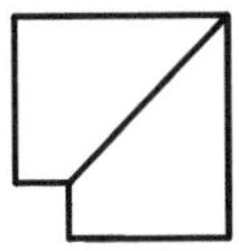

Galvanized steel 34" rectangular duct 90-degree elbow

22" - 24"	S2@1.67	Ea	332.00	72.70	—	404.70
26" - 28"	S2@1.91	Ea	380.00	83.10	—	463.10
30" - 32"	S2@2.12	Ea	417.00	92.30	—	509.30
34" - 36"	S2@2.36	Ea	467.00	103.00	—	570.00
38" - 40"	S2@2.60	Ea	515.00	113.00	—	628.00
42" - 44"	S2@2.91	Ea	577.00	127.00	—	704.00
46" - 48"	S2@3.21	Ea	627.00	140.00	—	767.00
50" - 52"	S2@3.51	Ea	695.00	153.00	—	848.00
54" - 56"	S2@4.17	Ea	826.00	181.00	—	1,007.00
58" - 60"	S2@4.97	Ea	985.00	216.00	—	1,201.00
62" - 64"	S2@5.49	Ea	1,090.00	239.00	—	1,329.00
66" - 68"	S2@6.04	Ea	1,190.00	263.00	—	1,453.00
70" - 72"	S2@6.51	Ea	1,270.00	283.00	—	1,553.00

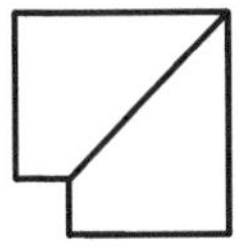

Galvanized steel 36" rectangular duct 90-degree elbow

22" - 24"	S2@1.83	Ea	365.00	79.60	—	444.60
26" - 28"	S2@2.04	Ea	406.00	88.80	—	494.80
30" - 32"	S2@2.26	Ea	448.00	98.40	—	546.40
34" - 36"	S2@2.47	Ea	489.00	107.00	—	596.00
38" - 40"	S2@2.70	Ea	535.00	118.00	—	653.00
42" - 44"	S2@3.01	Ea	591.00	131.00	—	722.00
46" - 48"	S2@3.31	Ea	654.00	144.00	—	798.00
50" - 52"	S2@3.61	Ea	714.00	157.00	—	871.00
54" - 56"	S2@4.27	Ea	843.00	186.00	—	1,029.00
58" - 60"	S2@5.10	Ea	1,000.00	222.00	—	1,222.00
62" - 64"	S2@5.61	Ea	1,110.00	244.00	—	1,354.00
66" - 68"	S2@6.15	Ea	1,200.00	268.00	—	1,468.00
70" - 72"	S2@6.64	Ea	1,300.00	289.00	—	1,589.00

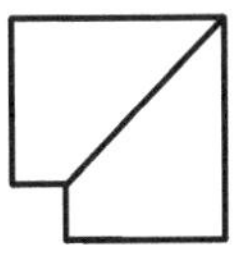

Galvanized steel 38" rectangular duct 90-degree elbow

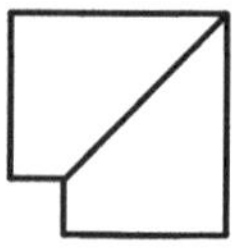

Description	Craft@Hrs	Unit	Material $	Labor $	Equipment $	Total $
22" - 24"	S2@1.98	Ea	393.00	86.20	—	479.20
26" - 28"	S2@2.18	Ea	432.00	94.90	—	526.90
30" - 32"	S2@2.38	Ea	474.00	104.00	—	578.00
34" - 36"	S2@2.58	Ea	514.00	112.00	—	626.00
38" - 40"	S2@2.81	Ea	553.00	122.00	—	675.00
42" - 44"	S2@3.11	Ea	613.00	135.00	—	748.00
46" - 48"	S2@3.41	Ea	673.00	148.00	—	821.00
50" - 52"	S2@3.71	Ea	729.00	161.00	—	890.00
54" - 56"	S2@4.38	Ea	863.00	191.00	—	1,054.00
58" - 60"	S2@5.21	Ea	1,030.00	227.00	—	1,257.00
62" - 64"	S2@5.74	Ea	1,130.00	250.00	—	1,380.00
66" - 68"	S2@6.28	Ea	1,230.00	273.00	—	1,503.00
70" - 72"	S2@6.75	Ea	1,330.00	294.00	—	1,624.00

Galvanized steel 40" rectangular duct 90-degree elbow

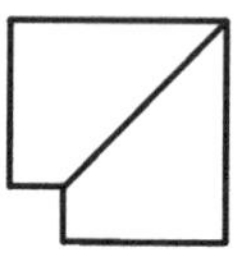

Description	Craft@Hrs	Unit	Material $	Labor $	Equipment $	Total $
40" - 42"	S2@3.06	Ea	606.00	133.00	—	739.00
44" - 46"	S2@3.36	Ea	659.00	146.00	—	805.00
48" - 50"	S2@3.66	Ea	725.00	159.00	—	884.00
52" - 54"	S2@3.98	Ea	785.00	173.00	—	958.00
56" - 58"	S2@5.17	Ea	1,020.00	225.00	—	1,245.00
60" - 62"	S2@5.59	Ea	1,110.00	243.00	—	1,353.00
64" - 66"	S2@6.11	Ea	1,200.00	266.00	—	1,466.00
68" - 70"	S2@6.66	Ea	1,310.00	290.00	—	1,600.00
72"	S2@7.05	Ea	1,370.00	307.00	—	1,677.00

Galvanized steel 42" rectangular duct 90-degree elbow

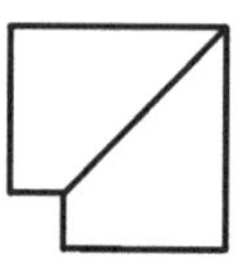

Description	Craft@Hrs	Unit	Material $	Labor $	Equipment $	Total $
40" - 42"	S2@3.21	Ea	639.00	140.00	—	779.00
44" - 46"	S2@3.52	Ea	697.00	153.00	—	850.00
48" - 50"	S2@3.83	Ea	762.00	167.00	—	929.00
52" - 54"	S2@4.16	Ea	823.00	181.00	—	1,004.00
56" - 58"	S2@5.34	Ea	1,050.00	232.00	—	1,282.00
60" - 62"	S2@5.81	Ea	1,140.00	253.00	—	1,393.00
64" - 66"	S2@6.32	Ea	1,240.00	275.00	—	1,515.00
68" - 70"	S2@6.85	Ea	1,340.00	298.00	—	1,638.00
72"	S2@7.22	Ea	1,420.00	314.00	—	1,734.00

Galvanized steel 44" rectangular duct 90-degree elbow

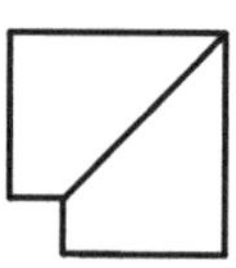

Description	Craft@Hrs	Unit	Material $	Labor $	Equipment $	Total $
40" - 42"	S2@3.36	Ea	659.00	146.00	—	805.00
44" - 46"	S2@3.69	Ea	728.00	161.00	—	889.00
48" - 50"	S2@4.01	Ea	789.00	175.00	—	964.00
52" - 54"	S2@4.32	Ea	853.00	188.00	—	1,041.00
56" - 58"	S2@5.51	Ea	1,090.00	240.00	—	1,330.00
60" - 62"	S2@6.00	Ea	1,180.00	261.00	—	1,441.00
64" - 66"	S2@6.51	Ea	1,270.00	283.00	—	1,553.00
68" - 70"	S2@7.05	Ea	1,380.00	307.00	—	1,687.00
72"	S2@7.39	Ea	1,430.00	322.00	—	1,752.00

Description	Craft@Hrs	Unit	Material $	Labor $	Equipment $	Total $

Galvanized steel 46" rectangular duct 90-degree elbow

40" - 42"	S2@3.52	Ea	697.00	153.00	—	850.00
44" - 46"	S2@3.86	Ea	765.00	168.00	—	933.00
48" - 50"	S2@4.16	Ea	823.00	181.00	—	1,004.00
52" - 54"	S2@4.32	Ea	884.00	188.00	—	1,072.00
56" - 58"	S2@5.72	Ea	1,120.00	249.00	—	1,369.00
60" - 62"	S2@6.19	Ea	1,210.00	269.00	—	1,479.00
64" - 66"	S2@6.70	Ea	1,330.00	292.00	—	1,622.00
68" - 70"	S2@7.26	Ea	1,420.00	316.00	—	1,736.00
72"	S2@7.60	Ea	1,490.00	331.00	—	1,821.00

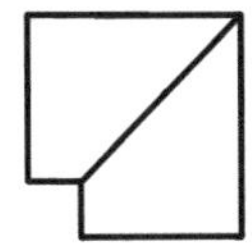

Galvanized steel 48" rectangular duct 90-degree elbow

40" - 42"	S2@3.67	Ea	726.00	160.00	—	886.00
44" - 46"	S2@4.01	Ea	789.00	175.00	—	964.00
48" - 50"	S2@4.32	Ea	853.00	188.00	—	1,041.00
52" - 54"	S2@4.61	Ea	905.00	201.00	—	1,106.00
56" - 58"	S2@5.91	Ea	1,180.00	257.00	—	1,437.00
60" - 62"	S2@6.40	Ea	1,260.00	279.00	—	1,539.00
64" - 66"	S2@6.90	Ea	1,350.00	300.00	—	1,650.00
68" - 70"	S2@7.41	Ea	1,450.00	322.00	—	1,772.00
72"	S2@7.77	Ea	1,540.00	338.00	—	1,878.00

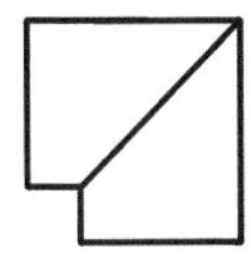

Galvanized steel 50" rectangular duct 90-degree elbow

40" - 42"	S2@3.82	Ea	762.00	166.00	—	928.00
44" - 46"	S2@4.16	Ea	823.00	181.00	—	1,004.00
48" - 50"	S2@4.46	Ea	884.00	194.00	—	1,078.00
52" - 54"	S2@4.78	Ea	936.00	208.00	—	1,144.00
56" - 58"	S2@6.11	Ea	1,200.00	266.00	—	1,466.00
60" - 62"	S2@6.60	Ea	1,300.00	287.00	—	1,587.00
64" - 66"	S2@7.09	Ea	1,380.00	309.00	—	1,689.00
68" - 70"	S2@7.62	Ea	1,510.00	332.00	—	1,842.00
72"	S2@7.94	Ea	1,570.00	346.00	—	1,916.00

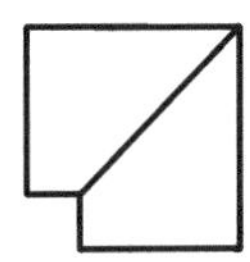

Galvanized steel 52" rectangular duct 90-degree elbow

40" - 42"	S2@3.98	Ea	785.00	173.00	—	958.00
44" - 46"	S2@4.32	Ea	853.00	188.00	—	1,041.00
48" - 50"	S2@4.61	Ea	905.00	201.00	—	1,106.00
52" - 54"	S2@4.95	Ea	976.00	215.00	—	1,191.00
56" - 58"	S2@6.32	Ea	1,240.00	275.00	—	1,515.00
60" - 62"	S2@6.79	Ea	1,340.00	296.00	—	1,636.00
64" - 66"	S2@7.30	Ea	1,420.00	318.00	—	1,738.00
68" - 70"	S2@7.81	Ea	1,540.00	340.00	—	1,880.00
72"	S2@8.16	Ea	1,590.00	355.00	—	1,945.00

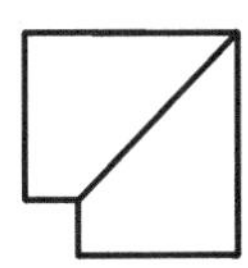

Galvanized Steel Rectangular Elbows

Galvanized steel 54" rectangular duct 90-degree elbow

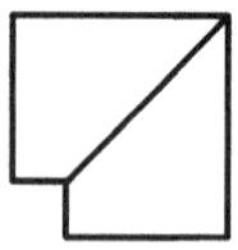

Description	Craft@Hrs	Unit	Material $	Labor $	Equipment $	Total $
40" - 42"	S2@4.16	Ea	823.00	181.00	—	1,004.00
44" - 46"	S2@4.46	Ea	884.00	194.00	—	1,078.00
48" - 50"	S2@4.78	Ea	936.00	208.00	—	1,144.00
52" - 54"	S2@5.15	Ea	1,010.00	224.00	—	1,234.00
56" - 58"	S2@6.51	Ea	1,270.00	283.00	—	1,553.00
60" - 62"	S2@7.00	Ea	1,370.00	305.00	—	1,675.00
64" - 66"	S2@7.49	Ea	1,470.00	326.00	—	1,796.00
68" - 70"	S2@8.01	Ea	1,580.00	349.00	—	1,929.00
72"	S2@8.41	Ea	1,640.00	366.00	—	2,006.00

Galvanized steel 56" rectangular duct 90-degree elbow

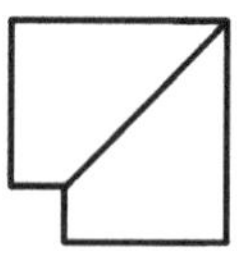

Description	Craft@Hrs	Unit	Material $	Labor $	Equipment $	Total $
40" - 42"	S2@5.10	Ea	1,000.00	222.00	—	1,222.00
44" - 46"	S2@5.49	Ea	1,090.00	239.00	—	1,329.00
48" - 50"	S2@5.85	Ea	1,160.00	255.00	—	1,415.00
52" - 54"	S2@6.26	Ea	1,230.00	272.00	—	1,502.00
56" - 58"	S2@6.68	Ea	1,310.00	291.00	—	1,601.00
60" - 62"	S2@7.15	Ea	1,390.00	311.00	—	1,701.00
64" - 66"	S2@7.66	Ea	1,510.00	333.00	—	1,843.00
68" - 70"	S2@8.18	Ea	1,590.00	356.00	—	1,946.00
72"	S2@8.58	Ea	1,720.00	373.00	—	2,093.00

Galvanized steel 58" rectangular duct 90-degree elbow

Description	Craft@Hrs	Unit	Material $	Labor $	Equipment $	Total $
58" - 60"	S2@7.15	Ea	1,420.00	311.00	—	1,731.00
62" - 64"	S2@7.62	Ea	1,510.00	332.00	—	1,842.00
66" - 68"	S2@8.13	Ea	1,600.00	354.00	—	1,954.00
70" - 72"	S2@8.69	Ea	1,740.00	378.00	—	2,118.00

Galvanized steel 60" rectangular duct 90-degree elbow

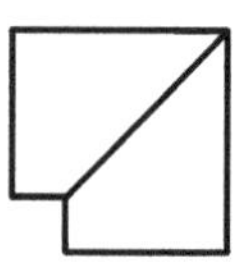

Description	Craft@Hrs	Unit	Material $	Labor $	Equipment $	Total $
58" - 60"	S2@7.37	Ea	1,450.00	321.00	—	1,771.00
62" - 64"	S2@7.88	Ea	1,570.00	343.00	—	1,913.00
66" - 68"	S2@8.41	Ea	1,710.00	366.00	—	2,076.00
70" - 72"	S2@8.95	Ea	1,770.00	390.00	—	2,160.00

Galvanized steel 62" rectangular duct 90-degree elbow

Description	Craft@Hrs	Unit	Material $	Labor $	Equipment $	Total $
58" - 60"	S2@7.62	Ea	1,510.00	332.00	—	1,842.00
62" - 64"	S2@8.13	Ea	1,590.00	354.00	—	1,944.00
66" - 68"	S2@8.69	Ea	1,730.00	378.00	—	2,108.00
70" - 72"	S2@9.18	Ea	1,820.00	400.00	—	2,220.00

Galvanized steel 64" rectangular duct 90-degree elbow

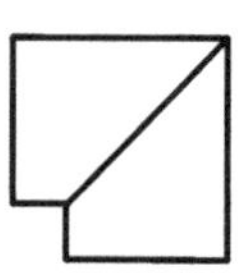

Description	Craft@Hrs	Unit	Material $	Labor $	Equipment $	Total $
58" - 60"	S2@7.88	Ea	1,560.00	343.00	—	1,903.00
62" - 64"	S2@8.41	Ea	1,660.00	366.00	—	2,026.00
66" - 68"	S2@8.95	Ea	1,760.00	390.00	—	2,150.00
70" - 72"	S2@9.44	Ea	1,860.00	411.00	—	2,271.00

Description	Craft@Hrs	Unit	Material $	Labor $	Equipment $	Total $

Galvanized steel 66" rectangular duct 90-degree elbow

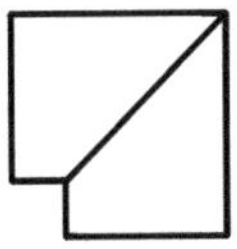

Description	Craft@Hrs	Unit	Material $	Labor $	Equipment $	Total $
58" - 60"	S2@8.13	Ea	1,590.00	354.00	—	1,944.00
62" - 64"	S2@8.69	Ea	1,730.00	378.00	—	2,108.00
66" - 68"	S2@9.18	Ea	1,820.00	400.00	—	2,220.00
70" - 72"	S2@9.69	Ea	1,920.00	422.00	—	2,342.00

Galvanized steel 68" rectangular duct 90-degree elbow

Description	Craft@Hrs	Unit	Material $	Labor $	Equipment $	Total $
58" - 60"	S2@8.41	Ea	1,660.00	366.00	—	2,026.00
62" - 64"	S2@8.95	Ea	1,760.00	390.00	—	2,150.00
66" - 68"	S2@9.44	Ea	1,860.00	411.00	—	2,271.00
70" - 72"	S2@9.95	Ea	1,970.00	433.00	—	2,403.00

Galvanized steel 70" rectangular duct 90-degree elbow

Description	Craft@Hrs	Unit	Material $	Labor $	Equipment $	Total $
58" - 60"	S2@8.69	Ea	1,730.00	378.00	—	2,108.00
62" - 64"	S2@9.18	Ea	1,820.00	400.00	—	2,220.00
66" - 68"	S2@9.69	Ea	1,920.00	422.00	—	2,342.00
70" - 72"	S2@10.2	Ea	2,000.00	444.00	—	2,444.00

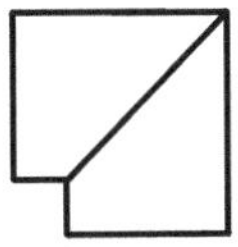

Galvanized steel 72" rectangular duct 90-degree elbow

Description	Craft@Hrs	Unit	Material $	Labor $	Equipment $	Total $
58" - 60"	S2@8.95	Ea	1,760.00	390.00	—	2,150.00
62" - 64"	S2@9.44	Ea	1,860.00	411.00	—	2,271.00
66" - 68"	S2@9.95	Ea	1,970.00	433.00	—	2,403.00
70" - 72"	S2@10.5	Ea	2,070.00	457.00	—	2,527.00

Galvanized Steel Rectangular Drops and Tees

Costs for Galvanized Rectangular Drops and Tap-in Tees. Costs per drop or tap-in tee, based on quantities less than 1,000 pounds. Deduct 15% from the material cost for over 1,000 pounds. Add 50% to the material cost for the lined duct. These costs include delivery, typical bracing, cleats, scrap, hangers, end closures, sealants and miscellaneous hardware. For turning vanes, please see page 342.

Description	Craft@Hrs	Unit	Material $	Labor $	Equipment $	Total $

Galvanized steel 6" rectangular duct drops and tap-in tees

6"	S2@.072	Ea	15.60	3.13	—	18.73
8"	S2@.084	Ea	18.20	3.66	—	21.86
10"	S2@.097	Ea	20.70	4.22	—	24.92
12" - 16"	S2@.139	Ea	30.00	6.05	—	36.05
20" - 24"	S2@.216	Ea	46.40	9.40	—	55.80
28" - 32"	S2@.309	Ea	65.90	13.40	—	79.30
36" - 40"	S2@.412	Ea	88.50	17.90	—	106.40
44" - 48"	S2@.489	Ea	105.00	21.30	—	126.30
52" - 56"	S2@.591	Ea	126.00	25.70	—	151.70
60"	S2@.747	Ea	159.00	32.50	—	191.50

Galvanized steel 8" rectangular duct drops and tap-in tees

6"	S2@.084	Ea	18.20	3.66	—	21.86
8"	S2@.097	Ea	20.70	4.22	—	24.92
10"	S2@.109	Ea	23.40	4.74	—	28.14
12" - 16"	S2@.152	Ea	33.20	6.62	—	39.82
20" - 24"	S2@.231	Ea	49.80	10.10	—	59.90
28" - 32"	S2@.326	Ea	70.60	14.20	—	84.80
36" - 40"	S2@.432	Ea	93.00	18.80	—	111.80
44" - 48"	S2@.506	Ea	109.00	22.00	—	131.00
52" - 56"	S2@.634	Ea	135.00	27.60	—	162.60
60"	S2@.760	Ea	161.00	33.10	—	194.10

Galvanized steel 10" rectangular duct drops and tap-in tees

6"	S2@.097	Ea	20.70	4.22	—	24.92
8"	S2@.109	Ea	23.40	4.74	—	28.14
10"	S2@.121	Ea	26.00	5.27	—	31.27
12" - 16"	S2@.167	Ea	35.90	7.27	—	43.17
20" - 24"	S2@.246	Ea	53.40	10.70	—	64.10
28" - 32"	S2@.343	Ea	74.20	14.90	—	89.10
36" - 40"	S2@.450	Ea	96.20	19.60	—	115.80
44" - 48"	S2@.525	Ea	112.00	22.80	—	134.80
52" - 56"	S2@.653	Ea	138.00	28.40	—	166.40
60"	S2@.773	Ea	164.00	33.60	—	197.60

Description	Craft@Hrs	Unit	Material $	Labor $	Equipment $	Total $

Galvanized steel 12" rectangular duct drops and tap-in tees

6"	S2@.109	Ea	23.40	4.74	—	28.14
8"	S2@.121	Ea	26.00	5.27	—	31.27
10"	S2@.133	Ea	28.70	5.79	—	34.49
12" - 16"	S2@.180	Ea	38.70	7.83	—	46.53
20" - 24"	S2@.262	Ea	56.70	11.40	—	68.10
28" - 32"	S2@.360	Ea	77.40	15.70	—	93.10
36" - 40"	S2@.470	Ea	100.00	20.50	—	120.50
44" - 48"	S2@.544	Ea	116.00	23.70	—	139.70
52" - 56"	S2@.675	Ea	143.00	29.40	—	172.40
60"	S2@.794	Ea	172.00	34.60	—	206.60

Galvanized steel 16" rectangular duct drops and tap-in tees

6"	S2@.170	Ea	36.50	7.40	—	43.90
8"	S2@.184	Ea	40.20	8.01	—	48.21
10"	S2@.200	Ea	43.20	8.70	—	51.90
12" - 16"	S2@.231	Ea	49.80	10.10	—	59.90
20" - 24"	S2@.293	Ea	62.60	12.80	—	75.40
28" - 32"	S2@.396	Ea	84.60	17.20	—	101.80
36" - 40"	S2@.506	Ea	109.00	22.00	—	131.00
44" - 48"	S2@.581	Ea	124.00	25.30	—	149.30
52" - 56"	S2@.709	Ea	153.00	30.90	—	183.90
60"	S2@.841	Ea	180.00	36.60	—	216.60

Galvanized steel 20" rectangular duct drops and tap-in tees

6"	S2@.200	Ea	43.20	8.70	—	51.90
8"	S2@.216	Ea	46.40	9.40	—	55.80
10"	S2@.231	Ea	49.80	10.10	—	59.90
12" - 16"	S2@.262	Ea	56.70	11.40	—	68.10
20" - 24"	S2@.324	Ea	69.70	14.10	—	83.80
28" - 32"	S2@.428	Ea	91.90	18.60	—	110.50
36" - 40"	S2@.544	Ea	116.00	23.70	—	139.70
44" - 48"	S2@.617	Ea	131.00	26.90	—	157.90
52" - 56"	S2@.758	Ea	160.00	33.00	—	193.00
60"	S2@.884	Ea	190.00	38.50	—	228.50

Galvanized Steel Rectangular Drops and Tees

Description	Craft@Hrs	Unit	Material $	Labor $	Equipment $	Total $

Galvanized steel 24" rectangular duct drops and tap-in tees

Description	Craft@Hrs	Unit	Material $	Labor $	Equipment $	Total $
6"	S2@.231	Ea	49.80	10.10	—	59.90
8"	S2@.247	Ea	53.40	10.70	—	64.10
10"	S2@.262	Ea	56.70	11.40	—	68.10
12" - 16"	S2@.293	Ea	62.60	12.80	—	75.40
20" - 24"	S2@.354	Ea	76.20	15.40	—	91.60
28" - 32"	S2@.462	Ea	98.50	20.10	—	118.60
36" - 40"	S2@.581	Ea	124.00	25.30	—	149.30
44" - 48"	S2@.655	Ea	138.00	28.50	—	166.50
52" - 56"	S2@.798	Ea	173.00	34.70	—	207.70
60"	S2@.927	Ea	198.00	40.30	—	238.30

Galvanized steel 28" rectangular duct drops and tap-in tees

Description	Craft@Hrs	Unit	Material $	Labor $	Equipment $	Total $
6"	S2@.262	Ea	56.70	11.40	—	68.10
8"	S2@.278	Ea	59.10	12.10	—	71.20
10"	S2@.292	Ea	62.60	12.70	—	75.30
12" - 16"	S2@.324	Ea	69.70	14.10	—	83.80
20" - 24"	S2@.385	Ea	82.60	16.80	—	99.40
28" - 32"	S2@.497	Ea	107.00	21.60	—	128.60
36" - 40"	S2@.617	Ea	131.00	26.90	—	157.90
44" - 48"	S2@.694	Ea	147.00	30.20	—	177.20
52" - 56"	S2@.837	Ea	180.00	36.40	—	216.40
60"	S2@.969	Ea	209.00	42.20	—	251.20

Galvanized steel 32" rectangular duct drops and tap-in tees

Description	Craft@Hrs	Unit	Material $	Labor $	Equipment $	Total $
6"	S2@.356	Ea	76.50	15.50	—	92.00
8"	S2@.374	Ea	81.20	16.30	—	97.50
10"	S2@.393	Ea	84.30	17.10	—	101.40
12" - 16"	S2@.432	Ea	93.00	18.80	—	111.80
20" - 24"	S2@.506	Ea	109.00	22.00	—	131.00
28" - 32"	S2@.581	Ea	124.00	25.30	—	149.30
36" - 40"	S2@.655	Ea	138.00	28.50	—	166.50
44" - 48"	S2@.730	Ea	156.00	31.80	—	187.80
52" - 56"	S2@.877	Ea	190.00	38.20	—	228.20
60"	S2@1.02	Ea	218.00	44.40	—	262.40

Description	Craft@Hrs	Unit	Material $	Labor $	Equipment $	Total $

Galvanized steel round duct snap-lock with slip joints

3"	S2@.080	LF	3.53	3.48	—	7.01
4"	S2@.090	LF	3.66	3.92	—	7.58
5"	S2@.095	LF	3.91	4.13	—	8.04
6"	S2@.100	LF	4.56	4.35	—	8.91
7"	S2@.110	LF	7.00	4.79	—	11.79
8"	S2@.120	LF	7.24	5.22	—	12.46
9"	S2@.140	LF	7.57	6.09	—	13.66
10"	S2@.160	LF	7.88	6.96	—	14.84

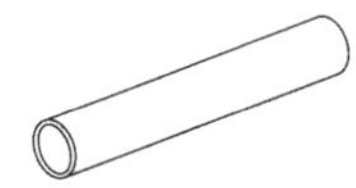

Galvanized steel round duct 90-degree adjustable elbow

3"	S2@.250	Ea	6.83	10.90	—	17.73
4"	S2@.300	Ea	9.42	13.10	—	22.52
5"	S2@.350	Ea	11.40	15.20	—	26.60
6"	S2@.400	Ea	14.70	17.40	—	32.10
7"	S2@.450	Ea	50.70	19.60	—	70.30
8"	S2@.500	Ea	56.70	21.80	—	78.50
9"	S2@.600	Ea	61.30	26.10	—	87.40
10"	S2@.700	Ea	67.00	30.50	—	97.50

Galvanized steel round duct 45-degree adjustable elbow

3"	S2@.250	Ea	6.83	10.90	—	17.73
4"	S2@.300	Ea	9.42	13.10	—	22.52
5"	S2@.350	Ea	11.40	15.20	—	26.60
6"	S2@.400	Ea	14.70	17.40	—	32.10
7"	S2@.450	Ea	50.70	19.60	—	70.30
8"	S2@.500	Ea	56.70	21.80	—	78.50
9"	S2@.600	Ea	61.30	26.10	—	87.40
10"	S2@.700	Ea	67.00	30.50	—	97.50

Fiberglass Ductwork

Standard fiberglass ductwork, or duct board, should not be used in systems where air velocities exceed 2,500 feet per minute, or where static pressures are higher than 2 inches of water gauge.

Additionally, it should not be used for the following applications:

1) Equipment rooms
2) Underground or outdoors
3) Final connections to air handling equipment
4) Plenums or casings
5) Fume, heat or moisture exhaust systems

The costs for manufacturing and installing fiberglass ductwork are based on the net areas of the duct and fittings.

Here are approximate current material costs, per square foot:

Fiberglass Type	Up to 2,500 SF Material $/SF	Over 2,500 SF Material $/SF
475	4.93	4.50
800	5.41	4.91
1400	6.62	6.36

Additional costs: Approximately 25 cents per square foot should be added for the cost of tie rods, hangers, staples, tape and waste.

Type 475 fiberglass board is used almost exclusively for residential and small commercial air systems, while Types 800 and 1400 are used in larger commercial and industrial systems.

Fiberglass Ductwork Fabrication Labor*						
Size ranges (inches)						
	Up through 16		17 through 32		33 through 48	
Activity	SF/Hour Output	Labor $/SF	SF/Hour Output	Labor $/SF	SF/Hour Output	Labor $/SF
Duct only	60	.73	58	.75	56	.78
Fittings only	40	1.09	38	1.15	35	1.24
Duct & 15% fittings	55	.79	53	.82	50	.87
Duct & 25% fittings	53	.82	51	.85	48	.91
Duct & 35% fittings	51	.85	49	.89	46	.95

**Labor rates are based on using a standard grooving machine and include all tie-rod reinforcing.*

Labor correction factors:

1) Hand grooving in lieu of machine grooving: 1.58
2) Use of an auto-closer: .82
3) Tie-rods enclosed in conduit: 1.36
4) T-bar or channel reinforcing in lieu of tie-rods: 1.25

In the following installation table the measurements refer to the "semi-perimeter", the width plus depth of the ductwork. Example – 12" duct (8" wide plus 4" high) has 2.00 SF of area for each linear foot.

Description	Craft@Hrs	Unit	Material $	Labor $	Equipment $	Total $

Fiberglass ductwork installation

Description	Craft@Hrs	Unit	Material $	Labor $	Equipment $	Total $
12" (2.00 SF/LF)	S2@.070	LF	—	3.05	—	3.05
14" (2.35 SF/LF)	S2@.082	LF	—	3.57	—	3.57
16" (2.65 SF/LF)	S2@.093	LF	—	4.05	—	4.05
18" (3.00 SF/LF)	S2@.105	LF	—	4.57	—	4.57
20" (3.35 SF/LF)	S2@.117	LF	—	5.09	—	5.09
22" (3.65 SF/LF)	S2@.128	LF	—	5.57	—	5.57
24" (4.00 SF/LF)	S2@.140	LF	—	6.09	—	6.09
26" (4.35 SF/LF)	S2@.152	LF	—	6.62	—	6.62
28" (4.65 SF/LF)	S2@.163	LF	—	7.09	—	7.09
30" (5.00 SF/LF)	S2@.175	LF	—	7.62	—	7.62
32" (5.35 SF/LF)	S2@.187	LF	—	8.14	—	8.14
34" (5.65 SF/LF)	S2@.198	LF	—	8.62	—	8.62
36" (6.00 SF/LF)	S2@.210	LF	—	9.14	—	9.14
38" (6.35 SF/LF)	S2@.222	LF	—	9.66	—	9.66
40" (6.65 SF/LF)	S2@.232	LF	—	10.10	—	10.10
42" (7.00 SF/LF)	S2@.245	LF	—	10.70	—	10.70
44" (7.35 SF/LF)	S2@.257	LF	—	11.20	—	11.20
46" (7.65 SF/LF)	S2@.268	LF	—	11.70	—	11.70
48" (8.00 SF/LF)	S2@.280	LF	—	12.20	—	12.20
50" (8.35 SF/LF)	S2@.292	LF	—	12.70	—	12.70
52" (8.65 SF/LF)	S2@.303	LF	—	13.20	—	13.20
54" (9.00 SF/LF)	S2@.315	LF	—	13.70	—	13.70
56" (9.35 SF/LF)	S2@.327	LF	—	14.20	—	14.20
58" (9.65 SF/LF)	S2@.338	LF	—	14.70	—	14.70
60" (10.0 SF/LF)	S2@.350	LF	—	15.20	—	15.20
64" (10.6 SF/LF)	S2@.373	LF	—	16.20	—	16.20
68" (11.4 SF/LF)	S2@.397	LF	—	17.30	—	17.30
72" (12.0 SF/LF)	S2@.420	LF	—	18.30	—	18.30
76" (12.6 SF/LF)	S2@.443	LF	—	19.30	—	19.30
80" (13.4 SF/LF)	S2@.467	LF	—	20.30	—	20.30
84" (14.0 SF/LF)	S2@.490	LF	—	21.30	—	21.30
88" (14.6 SF/LF)	S2@.513	LF	—	22.30	—	22.30
92" (15.4 SF/LF)	S2@.537	LF	—	23.40	—	23.40
96" (16.0 SF/LF)	S2@.560	LF	—	24.40	—	24.40

For estimating purposes, when measuring the length of ductwork, include the fitting. Then apply the labor per run foot to the total.

1¼" thick wire reinforced flexible fiberglass duct with vinyl cover

Description	Craft@Hrs	Unit	Material $	Labor $	Equipment $	Total $
4"	S2@.060	LF	9.51	2.61	—	12.12
5"	S2@.070	LF	9.83	3.05	—	12.88
6"	S2@.080	LF	10.80	3.48	—	14.28
8"	S2@.090	LF	13.70	3.92	—	17.62
9"	S2@.100	LF	14.60	4.35	—	18.95
10"	S2@.120	LF	15.40	5.22	—	20.62
12"	S2@.140	LF	19.70	6.09	—	25.79
14"	S2@.160	LF	23.60	6.96	—	30.56
16"	S2@.180	LF	27.40	7.83	—	35.23
18"	S2@.200	LF	31.20	8.70	—	39.90

Fiberglass Pipe Insulation

Fiberglass Pipe Insulation with AP-T Plus (all-purpose self-sealing) Jacket. By nominal pipe diameter. R factor equals 2.56 at 300 degrees F. These costs do not include scaffolding; add for same, if required. Also see fittings, flanges and valves at the end of this section.

Description	Craft@Hrs	Unit	Material $	Labor $	Equipment $	Total $
Jacketed fiberglass insulation for ½" pipe						
½" thick	P1@.038	LF	1.22	1.69	—	2.91
1" thick	P1@.038	LF	1.44	1.69	—	3.13
1½" thick	P1@.040	LF	2.94	1.78	—	4.72
Jacketed fiberglass insulation for ¾" pipe						
½" thick	P1@.038	LF	1.33	1.69	—	3.02
1" thick	P1@.038	LF	1.64	1.69	—	3.33
1½" thick	P1@.040	LF	3.05	1.78	—	4.83
Jacketed fiberglass insulation for 1" pipe						
½" thick	P1@.040	LF	1.42	1.78	—	3.20
1" thick	P1@.040	LF	1.88	1.78	—	3.66
1½" thick	P1@.042	LF	3.22	1.87	—	5.09
Jacketed fiberglass insulation for 1¼" pipe						
½" thick	P1@.040	LF	1.57	1.78	—	3.35
1" thick	P1@.040	LF	1.90	1.78	—	3.68
1½" thick	P1@.042	LF	3.44	1.87	—	5.31
Jacketed fiberglass insulation for 1½" pipe						
½" thick	P1@.042	LF	1.68	1.87	—	3.55
1" thick	P1@.042	LF	2.13	1.87	—	4.00
1½" thick	P1@.044	LF	3.69	1.96	—	5.65
Jacketed fiberglass insulation for 2" pipe						
½" thick	P1@.044	LF	1.75	1.96	—	3.71
1" thick	P1@.044	LF	2.45	1.96	—	4.41
1½" thick	P1@.046	LF	3.97	2.04	—	6.01
Jacketed fiberglass insulation for 2½" pipe						
1" thick	P1@.047	LF	2.53	2.09	—	4.62
1½" thick	P1@.048	LF	3.59	2.13	—	5.72
2" thick	P1@.050	LF	5.68	2.22	—	7.90
Jacketed fiberglass insulation for 3" pipe						
1" thick	P1@.051	LF	2.78	2.27	—	5.05
1½" thick	P1@.054	LF	3.92	2.40	—	6.32
2" thick	P1@.056	LF	6.30	2.49	—	8.79

Description	Craft@Hrs	Unit	Material $	Labor $	Equipment $	Total $
Jacketed fiberglass insulation for 4" pipe						
1" thick	P1@.063	LF	3.27	2.80	—	6.07
1½" thick	P1@.066	LF	4.37	2.93	—	7.30
2" thick	P1@.069	LF	7.02	3.07	—	10.09
Jacketed fiberglass insulation for 6" pipe						
1" thick	P1@.077	LF	3.81	3.42	—	7.23
1½" thick	P1@.080	LF	5.25	3.56	—	8.81
2" thick	P1@.085	LF	8.88	3.78	—	12.66
Jacketed fiberglass insulation for 8" pipe						
1" thick	P1@.117	LF	5.25	5.20	—	10.45
1½" thick	P1@.123	LF	6.40	5.47	—	11.87
2" thick	P1@.129	LF	10.50	5.73	—	16.23
Jacketed fiberglass insulation for 10" pipe						
1" thick	P1@.140	LF	6.25	6.22	—	12.47
1½" thick	P1@.147	LF	7.87	6.53	—	14.40
2" thick	P1@.154	LF	12.50	6.85	—	19.35
Jacketed fiberglass insulation for 12" pipe						
1" thick	P1@.155	LF	7.21	6.89	—	14.10
1½" thick	P1@.163	LF	8.88	7.25	—	16.13
2" thick	P1@.171	LF	13.60	7.60	—	21.20
Add for .016" aluminum jacket						
Per SF of surface	P1@.035	SF	1.03	1.56	—	2.59

Pipe fittings and flanges

For each fitting or flange use the cost for 3 LF of pipe of the same pipe size

Valves

Body only: for each valve body use the cost for 5 LF of pipe of the same pipe size

Body and bonnet or yoke: for each valve use the cost for 10 LF of pipe of the same pipe size

Flanged valves: add the cost for flanges per above

Calcium Silicate Pipe Insulation with Aluminum Jacket

Calcium Silicate Pipe Insulation with .016" Aluminum Jacket. By nominal pipe diameter. R factor equals 2.20 at 300 degrees F. Manufactured by Pabco. These costs do not include scaffolding; add for same, if required. Also see fittings, flanges and valves below.

Description	Craft@Hrs	Unit	Material $	Labor $	Equipment $	Total $
Jacketed calcium silicate insulation for 6" pipe						
2" thick	P1@.183	LF	22.60	8.13	—	30.73
4" thick	P1@.218	LF	49.00	9.69	—	58.69
6" thick	P1@.288	LF	68.80	12.80	—	81.60
Jacketed calcium silicate insulation for 8" pipe						
2" thick	P1@.194	LF	27.10	8.62	—	35.72
4" thick	P1@.228	LF	57.40	10.10	—	67.50
6" thick	P1@.308	LF	87.20	13.70	—	100.90
Jacketed calcium silicate insulation for 10" pipe						
2" thick	P1@.202	LF	31.60	8.98	—	40.58
4" thick	P1@.239	LF	66.90	10.60	—	77.50
6" thick	P1@.337	LF	105.00	15.00	—	120.00
Jacketed calcium silicate insulation for 12" pipe						
2" thick	P1@.208	LF	36.60	9.25	—	45.85
4" thick	P1@.261	LF	75.10	11.60	—	86.70
6" thick	P1@.388	LF	119.00	17.20	—	136.20

Pipe fittings and flanges

For each fitting or flange use the cost for 3 LF of pipe of the same pipe size

Valves

Body only: for each valve body use the cost for 5 LF of pipe of the same pipe size

Body and bonnet or yoke: for each valve use the cost for 10 LF of pipe of the same pipe size

Flanged valves: add the cost for flanges per above

Description	Craft@Hrs	Unit	Material $	Labor $	Equipment $	Total $

Closed Cell Elastomeric Pipe and Tubing Insulation. Semi split, by nominal pipe or tube diameter with insulation wall thickness as shown, no cover. R factor equals 3.58 at 220 degrees F. Manufactured by Rubatex. These costs do not include scaffolding; add for same, if required. Also see fittings, flanges and valves below.

¼", ½" thick	P1@.039	LF	1.18	1.73	—	2.91
3/8", ½" thick	P1@.039	LF	1.21	1.73	—	2.94
½", ½" thick	P1@.039	LF	1.35	1.73	—	3.08
¾", ½" thick	P1@.039	LF	1.54	1.73	—	3.27
1", ½" thick	P1@.042	LF	1.72	1.87	—	3.59
1¼", ½" thick	P1@.042	LF	1.92	1.87	—	3.79
1½", ½" thick	P1@.042	LF	2.25	1.87	—	4.12
2", ½" thick	P1@.046	LF	2.82	2.04	—	4.86
2½", 3/8" thick	P1@.056	LF	3.61	2.49	—	6.10
2½", ½" thick	P1@.056	LF	4.85	2.49	—	7.34
2½", ¾" thick	P1@.056	LF	6.24	2.49	—	8.73
3", 3/8" thick	P1@.062	LF	4.12	2.76	—	6.88
3", ½" thick	P1@.062	LF	5.44	2.76	—	8.20
3", ¾" thick	P1@.062	LF	7.09	2.76	—	9.85
4", 3/8" thick	P1@.068	LF	4.85	3.02	—	7.87
4", ½" thick	P1@.068	LF	6.38	3.02	—	9.40
4", ¾" thick	P1@.068	LF	8.31	3.02	—	11.33

Pipe fittings and flanges
For each fitting or flange use the cost for 3 LF of pipe of the same pipe size

Valves
Body only: for each valve body use the cost for 5 LF of pipe of the same pipe size

Body and bonnet or yoke: for each valve use the cost for 10 LF of pipe of the same pipe size

Flanged valves: add the cost for flanges per above

Description	Craft@Hrs	Unit	Material $	Labor $	Equipment $	Total $
Fiberglass flexible duct wrap insulation with foil facing vapor barrier to 250 degrees F						
1"	S2@.023	SF	.63	1.00	—	1.63
1½"	S2@.023	SF	.77	1.00	—	1.77
2"	S2@.025	SF	1.04	1.09	—	2.13
Deduct foil facing	—	%	-30.0	—	—	—
Fiberglass rigid board duct insulation plain, *without* foil facing vapor barrier to 450 degrees F						
1"	S2@.067	SF	1.21	2.92	—	4.13
1½"	S2@.067	SF	2.06	2.92	—	4.98
2"	S2@.075	SF	2.44	3.26	—	5.70
3"	S2@.085	SF	3.64	3.70	—	7.34
Fiberglass rigid board duct insulation with FSK (foil) facing vapor barrier to 250 degrees F						
1"	S2@.067	SF	1.62	2.92	—	4.54
1½"	S2@.067	SF	1.98	2.92	—	4.90
2"	S2@.075	SF	2.35	3.26	—	5.61
3"	S2@.085	SF	3.35	3.70	—	7.05
Fiberglass flexible blanket duct lining to 250 degrees F						
½"	S2@.032	SF	.74	1.39	—	2.13
1"	S2@.032	SF	.97	1.39	—	2.36
1½"	S2@.035	SF	1.38	1.52	—	2.90
2"	S2@.035	SF	1.80	1.52	—	3.32
Fiberglass rigid duct lining to 250 degrees F						
5/8"	S2@.032	SF	1.61	1.39	—	3.00
1"	S2@.032	SF	2.09	1.39	—	3.48
1½"	S2@.035	SF	2.74	1.52	—	4.26
2"	S2@.035	SF	3.49	1.52	—	5.01

Note: Material prices include all necessary tape, pins, mastic and 15% waste.

Description	Craft@Hrs	Unit	Material $	Labor $	Equipment $	Total $

Air balancing – air handling units

Description	Craft@Hrs	Unit	Material $	Labor $	Equipment $	Total $
Central station	S2@6.00	Ea	—	261.00	—	261.00
Multi-zone A/C	S2@4.00	Ea	—	174.00	—	174.00
Single-zone A/C	S2@3.00	Ea	—	131.00	—	131.00
Packaged A/C	S2@2.50	Ea	—	109.00	—	109.00
Built-up low pressure	S2@4.50	Ea	—	196.00	—	196.00
Built-up high pres.	S2@6.00	Ea	—	261.00	—	261.00
Built-up VAV	S2@6.00	Ea	—	261.00	—	261.00
Built-up dual duct	S2@6.00	Ea	—	261.00	—	261.00

Air balancing – terminal boxes

Description	Craft@Hrs	Unit	Material $	Labor $	Equipment $	Total $
Variable air volume	S2@0.40	Ea	—	17.40	—	17.40
Constant volume	S2@0.40	Ea	—	17.40	—	17.40
Dual duct box	S2@0.70	Ea	—	30.50	—	30.50
Terminal box and all downstream diffusers cost per zone	S2@1.75	Ea	—	76.20	—	76.20

Air balancing – diffusers

Description	Craft@Hrs	Unit	Material $	Labor $	Equipment $	Total $
Ceiling diffusers to 500 CFM	S2@.250	Ea	—	10.90	—	10.90
Ceiling diffusers over 500 CFM	S2@.350	Ea	—	15.20	—	15.20
Ceiling diffusers perforated	S2@.370	Ea	—	16.10	—	16.10
Sidewall diffusers double deflection	S2@.250	Ea	—	10.90	—	10.90
Linear slot diffusers 2 slot	S2@.060	LF	—	2.61	—	2.61
Linear slot diffusers 3 slot	S2@.070	LF	—	3.05	—	3.05
Linear slot diffusers 4 slot	S2@.080	LF	—	3.48	—	3.48
Plug-in diffuser	S2@.250	Ea	—	10.90	—	10.90
Light troffers 1 slot	S2@.195	Ea	—	8.49	—	8.49
Light troffers 2 slot	S2@.220	Ea	—	9.57	—	9.57
Laminar flow diff.	S2@.750	Ea	—	32.60	—	32.60
Laminar flow diffusers with HEPA filter	S2@3.00	Ea	—	131.00	—	131.00

Description	Craft@Hrs	Unit	Material $	Labor $	Equipment $	Total $

Air balancing – grilles

Return air grilles to 800 CFM	S2@.250	Ea	—	10.90	—	10.90
Return air grilles over 800 CFM	S2@.350	Ea	—	15.20	—	15.20
Bar type return air grilles 2 slot	S2@.060	LF	—	2.61	—	2.61
Bar type return air grilles 4 slot	S2@.070	LF	—	3.05	—	3.05
Bar type return air grilles 6 slot	S2@.080	LF	—	3.48	—	3.48

Air balancing – centrifugal fans

Supply and return air to 3,000 CFM	S2@1.00	Ea	—	43.50	—	43.50
Supply and return air to 5,000 CFM	S2@1.50	Ea	—	65.30	—	65.30
Supply and return air to 10,000 CFM	S2@2.00	Ea	—	87.00	—	87.00
Supply and return air to 20,000 CFM	S2@2.75	Ea	—	120.00	—	120.00
Roof exhaust to 6,000 CFM	S2@.725	Ea	—	31.60	—	31.60
Roof exhaust to 10,000 CFM	S2@.925	Ea	—	40.30	—	40.30
Roof exhaust to 20,000 CFM	S2@1.50	Ea	—	65.30	—	65.30
Range hood exhaust to 10,000 CFM	S2@1.75	Ea	—	76.20	—	76.20
Range hood exhaust to 20,000 CFM	S2@2.50	Ea	—	109.00	—	109.00

Air balancing – fan coil units

Vertical floor model	S2@.500	Ea	—	21.80	—	21.80
Horiz. ceiling mtd	S2@.650	Ea	—	28.30	—	28.30

Air balancing – fume hoods

Lab fume hood	S2@2.00	Ea	—	87.00	—	87.00
General purpose	S2@.450	Ea	—	19.60	—	19.60
Dust collectors	S2@1.50	Ea	—	65.30	—	65.30

Description	Craft@Hrs	Unit	Material $	Labor $	Equipment $	Total $
Wet balancing						
Boiler	S2@1.65	Ea	—	71.80	—	71.80
Chiller	S2@3.50	Ea	—	152.00	—	152.00
Cooling tower	S2@2.00	Ea	—	87.00	—	87.00
Circulating pump	S2@1.65	Ea	—	71.80	—	71.80
Heat exchanger	S2@1.75	Ea	—	76.20	—	76.20
AHU water coil	S2@2.50	Ea	—	109.00	—	109.00
Reheat coil	S2@.450	Ea	—	19.60	—	19.60
Wall fin radiation	S2@.350	Ea	—	15.20	—	15.20
Radiant heat panel	S2@.450	Ea	—	19.60	—	19.60
Fan coil unit, 2 pipe	S2@.350	Ea	—	15.20	—	15.20
Fan coil unit, 4 pipe	S2@.450	Ea	—	19.60	—	19.60
Cabinet heater	S2@.350	Ea	—	15.20	—	15.20
Unit ventilator	S2@.450	Ea	—	19.60	—	19.60
Unit heater	S2@.350	Ea	—	15.20	—	15.20

Commercial HVAC controls. Temperature, pressure and humidity control for any device. 0-10 Volts DC or 4-20 milliamp input. Honeywell T775U or equal. Complies with ANSI/ACCA standard 5QI-2007. With two modulating and two relay outputs, each with their own setpoint.

Description	Craft@Hrs	Unit	Material $	Labor $	Equipment $	Total $
Install control module enclosure	BE@1.00	Ea	—	48.40	—	48.40
Twisted pair Category 6 PVC cable	BE@.008	LF	.61	.39	—	1.00
Panel wiring	BE@.250	Ea	3.18	12.10	—	15.28
Relay and damper actuator wiring	BE@.250	Ea	2.86	12.10	—	14.96
Sensor wiring	BE@.300	Ea	2.86	14.50	—	17.36
Zone control module wiring	BE@.400	Ea	2.68	19.30	—	21.98
Control module and wiring	BE@2.00	Ea	509.00	96.70	—	605.70
Ground fault protection wiring	BE@1.25	Ea	159.00	60.40	—	219.40
Programming and system test	BE@3.00	Ea	—	145.00	—	145.00
Commissioning and final test	BE@8.00	Ea	—	387.00	—	387.00

Sensor and zone control modules. Honeywell Webstat TR23 or equal. 20K non-linear sensor with setpoint adjust 55-85, and override LED. With LonWorks interface card and power supply. Add the cost of control wiring.

Description	Craft@Hrs	Unit	Material $	Labor $	Equipment $	Total $
Room sensor	BE@.500	Ea	99.70	24.20	—	123.90

Damper actuators. Modutrol IV or equal. For rectangular volume control dampers. Add the cost of wiring and the damper from the section Ductwork Specialties.

Description	Craft@Hrs	Unit	Material $	Labor $	Equipment $	Total $
Lever arm actuator	BE@.500	Ea	620.00	24.20	—	644.20

Description	Craft@Hrs	Unit	Material $	Labor $	Equipment $	Total $

Remote CO_2 sensors. Honeywell C72321 or equal, with LonWorks interface card and power supply. Add the cost of wiring.

Panel mount	BE@.500	Ea	106.00	24.20	—	130.20

Monitoring and recording equipment. Honeywell DR4500A digital circular chart recorder or equal, with LonWorks card and power supply. Add the cost of wiring.

Panel mount	BE@.500	Ea	5,040.00	24.20	—	5,064.20

Thermostat, electric. Install and connect. Make additional allowances for sub-subcontractor markups and control wiring as required.

Heat/cool low voltage	BE@.575	Ea	53.00	27.80	—	80.80
Heat only low voltage	BE@.500	Ea	44.20	24.20	—	68.40
Heat /cool line voltage	BE@.650	Ea	160.00	31.40	—	191.40
Programmable – residential low voltage	BE@.850	LF	243.00	41.10	—	284.10
Programmable – commercial w/subbase low voltage	BE@1.10	LF	501.00	53.20	—	554.20

Thermostat, pneumatic. Install and connect. Make additional allowances for sub-subcontractor markups and pneumatic tubing as required.

Direct acting 55 to 85 Deg. F.	BE@.675	Ea	137.00	32.60	—	169.60
Reverse acting 55 to 85 Deg. F.	BE@.675	Ea	114.00	32.60	—	146.60
Floating	BE@.675	Ea	244.00	32.60	—	276.60

Control valve, 2-way, electric. 6.9 Volt Input. Proportional Spring Return. Connect only. N.O. means: "normally open." Make additional allowances for sub-subcontractor markups, control wiring and valve installation by piping trades as required. (See Plumbing and Piping Specialties)

½", 2-way, N.O.	BE@.400	Ea	195.00	19.30	—	214.30
¾", 2-way, N.O.	BE@.400	Ea	208.00	19.30	—	227.30
1", 2-way, N.O.	BE@.400	Ea	224.00	19.30	—	243.30
1¼" 2-way, N.O.	BE@.400	Ea	244.00	19.30	—	263.30
1½" 2-way, N.O.	BE@.400	Ea	525.00	19.30	—	544.30
2", 2-way, N.O.	BE@.400	Ea	574.00	19.30	—	593.30

Control valve, 3-way, electric. 6.9 Volt Input, Proportional Spring Return. Connect only. Make additional allowances for sub-subcontractor markups, control wiring and valve installation by piping trades as required. (See Plumbing and Piping Specialties)

½", 3-way	BE@.450	Ea	278.00	21.80	—	299.80
¾", 3-way	BE@.450	Ea	302.00	21.80	—	323.80
1", 3-way	BE@.450	Ea	362.00	21.80	—	383.80

Description	Craft@Hrs	Unit	Material $	Labor $	Equipment $	Total $

Control valve, 2-way, pneumatic. Proportional Spring Return, 3 to 7 psi Input, Connect only. N.O. means "normally open." Make additional allowances for sub-subcontractor markups, pneumatic tubing and valve installation by piping trades as required. (See Plumbing and Piping Specialties)

½", 2-way, N.O.	BE@.400	Ea	146.00	19.30	—	165.30
¾", 2-way, N.O.	BE@.400	Ea	171.00	19.30	—	190.30
1", 2-way, N.O.	BE@.400	Ea	201.00	19.30	—	220.30
1¼" 2-way, N.O.	BE@.400	Ea	238.00	19.30	—	257.30
1½" 2-way, N.O.	BE@.400	Ea	424.00	19.30	—	443.30
2", 2-way, N.O.	BE@.400	Ea	506.00	19.30	—	525.30

Control valve, 3-way, pneumatic. Proportional Spring Return, 3 to 7 psi Input, Connect only. Make additional allowances for sub-subcontractor markups, pneumatic tubing and valve installation by piping trades as required. (See Plumbing and Piping Specialties)

½", 3-way	BE@.450	Ea	177.00	21.80	—	198.80
¾", 3-way	BE@.450	Ea	213.00	21.80	—	234.80
1", 3-way	BE@.450	Ea	295.00	21.80	—	316.80
1¼", 3-way	BE@.450	Ea	342.00	21.80	—	363.80
1½", 3-way	BE@.450	Ea	491.00	21.80	—	512.80
2", 3-way	BE@.450	Ea	574.00	21.80	—	595.80

Air compressor. Reciprocating piston-type. Larger systems include HVAC actuator controls. Underwriters Laboratories and CSA rated. With ASME U-stamped pressure tank, pressure safety relief valve, silencer, lubrication system with lube cooling radiator and fan. Single-phase 110 volt electric motor, drive belt and OSHA-rated safety belt guard. Add the cost of compressor hose and fittings and hose reel, if required. Set in place only. Add installation costs from the table below.

½ hp	SL@3.00	Ea	481.00	132.00	—	613.00
¾ hp	SL@3.00	Ea	527.00	132.00	—	659.00
1 hp	SL@3.50	Ea	612.00	154.00	—	766.00
1½ hp	SL@3.50	Ea	678.00	154.00	—	832.00
2 hp	SL@3.50	Ea	721.00	154.00	—	875.00

Installation of air compressors.

Cut, frame support and gasket hole in sidewall	S2@1.00	Ea	35.30	43.50	—	78.80
Install piping for compressor line	P1@.100	LF	4.65	4.45	—	9.10
Install electrical wiring	BE@.150	LF	2.10	7.25	—	9.35
Install compressor controls	BE@.500	Ea	—	24.20	—	24.20
Commission and test	P1@2.00	Ea	—	88.90	—	88.90
Fine-tune system	P1@1.00	Ea	—	44.50	—	44.50

Ductile Iron, Class 153, Cement-Lined with Mechanical Joints

Cement-lined Class 153 ductile iron pipe with mechanical joints

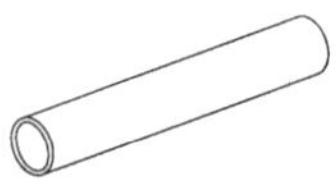

Description	Craft@Hrs	Unit	Material $	Labor $	Equipment $	Total $
4"	P1@.140	LF	15.40	6.22	—	21.62
6"	ER@.180	LF	22.30	8.53	3.07	33.90
8"	ER@.200	LF	30.50	9.48	3.41	43.39
10"	ER@.230	LF	45.20	10.90	3.93	60.03
12"	ER@.270	LF	61.60	12.80	4.61	79.01
14"	ER@.310	LF	93.60	14.70	5.29	113.59
16"	ER@.350	LF	120.00	16.60	5.98	142.58

Cement-lined Class 153 ductile iron 1/16 bend with mechanical joints

4"	P1@.900	Ea	61.10	40.00	—	101.10
6"	ER@1.30	Ea	94.20	61.60	22.20	178.00
8"	ER@1.70	Ea	148.00	80.60	29.00	257.60
10"	ER@2.15	Ea	202.00	102.00	36.70	340.70
12"	ER@2.40	Ea	280.00	114.00	41.00	435.00
14"	ER@2.70	Ea	614.00	128.00	46.10	788.10
16"	ER@3.00	Ea	693.00	142.00	51.20	886.20

Cement-lined Class 153 ductile iron 1/8 bend with mechanical joints

4"	P1@.900	Ea	63.00	40.00	—	103.00
6"	ER@1.30	Ea	93.60	61.60	22.20	177.40
8"	ER@1.70	Ea	151.00	80.60	29.00	260.60
10"	ER@2.15	Ea	205.00	102.00	36.70	343.70
12"	ER@2.40	Ea	285.00	114.00	41.00	440.00
14"	ER@2.70	Ea	611.00	128.00	46.10	785.10
16"	ER@3.00	Ea	714.00	142.00	51.20	907.20

Cement-lined Class 153 ductile iron 1/4 bend with mechanical joints

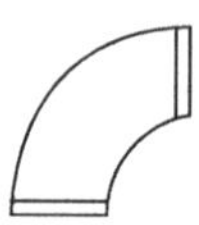

4"	P1@.900	Ea	67.50	40.00	—	107.50
6"	ER@1.30	Ea	110.00	61.60	22.20	193.80
8"	ER@1.70	Ea	230.00	80.60	29.00	339.60
10"	ER@2.15	Ea	346.00	102.00	36.70	484.70
12"	ER@2.40	Ea	473.00	114.00	41.00	628.00
14"	ER@2.70	Ea	1,210.00	128.00	46.10	1,384.10
16"	ER@3.00	Ea	1,270.00	142.00	51.20	1,463.20

Cement-lined Class 153 ductile iron tee with mechanical joints

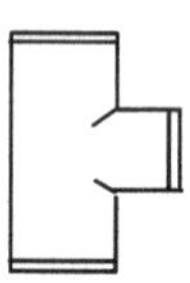

4"	P1@1.65	Ea	102.00	73.30	—	175.30
6"	ER@2.35	Ea	151.00	111.00	40.10	302.10
8"	ER@3.15	Ea	230.00	149.00	53.80	432.80
10"	ER@4.00	Ea	346.00	190.00	68.30	604.30
12"	ER@4.45	Ea	473.00	211.00	76.00	760.00
14"	ER@4.90	Ea	1,210.00	232.00	83.70	1,525.70
16"	ER@5.30	Ea	1,270.00	251.00	90.50	1,611.50

Description	Craft@Hrs	Unit	Material $	Labor $	Equipment $	Total $
Cement-lined Class 153 ductile iron wye with mechanical joints						
4"	P1@1.65	Ea	230.00	73.30	—	303.30
6"	ER@2.35	Ea	285.00	111.00	40.10	436.10
8"	ER@3.15	Ea	431.00	149.00	53.80	633.80
10"	ER@4.00	Ea	801.00	190.00	68.30	1,059.30
12"	ER@4.45	Ea	1,200.00	211.00	76.00	1,487.00
14"	ER@4.90	Ea	2,470.00	232.00	83.70	2,785.70
16"	ER@5.30	Ea	3,090.00	251.00	90.50	3,431.50
Cement-lined Class 153 ductile iron reducer with mechanical joints						
6"	ER@1.45	Ea	74.90	68.70	24.80	168.40
8"	ER@2.00	Ea	118.00	94.80	34.10	246.90
10"	ER@2.55	Ea	158.00	121.00	43.50	322.50
12"	ER@3.05	Ea	202.00	145.00	52.10	399.10
14"	ER@3.60	Ea	452.00	171.00	61.50	684.50
16"	ER@4.20	Ea	550.00	199.00	71.70	820.70

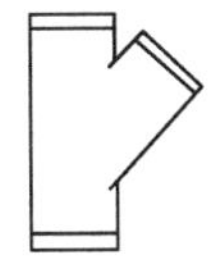

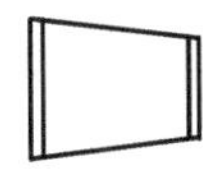

Ductile Iron, Class 153, Double Cement-Lined with Mechanical Joints

Double cement-lined Class 153 ductile iron pipe with mechanical joints

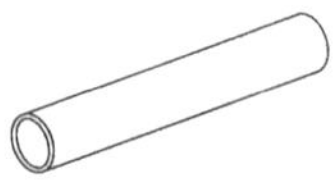

Description	Craft@Hrs	Unit	Material $	Labor $	Equipment $	Total $
4"	P1@.140	LF	16.30	6.22	—	22.52
6"	ER@.180	LF	23.60	8.53	3.76	35.89
8"	ER@.200	LF	32.20	9.48	3.41	45.09
10"	ER@.230	LF	48.10	10.90	3.93	62.93
12"	ER@.270	LF	65.60	12.80	4.61	83.01
14"	ER@.310	LF	99.70	14.70	5.29	119.69
16"	ER@.350	LF	128.00	16.60	5.98	150.58

Double cement-lined Class 153 ductile iron 1/16 bend with mechanical joints

Description	Craft@Hrs	Unit	Material $	Labor $	Equipment $	Total $
4"	P1@.900	Ea	71.90	40.00	—	111.90
6"	ER@1.30	Ea	111.00	61.60	22.20	194.80
8"	ER@1.70	Ea	177.00	80.60	29.00	286.60
10"	ER@2.15	Ea	238.00	102.00	36.70	376.70
12"	ER@2.40	Ea	326.00	114.00	41.00	481.00
14"	ER@2.70	Ea	730.00	128.00	46.10	904.10
16"	ER@3.00	Ea	827.00	142.00	51.20	1,020.20

Double cement-lined Class 153 ductile iron 1/8 bend with mechanical joints

Description	Craft@Hrs	Unit	Material $	Labor $	Equipment $	Total $
4"	P1@.900	Ea	71.90	40.00	—	111.90
6"	ER@1.30	Ea	108.00	61.60	22.20	191.80
8"	ER@1.70	Ea	173.00	80.60	29.00	282.60
10"	ER@2.15	Ea	237.00	102.00	36.70	375.70
12"	ER@2.40	Ea	322.00	114.00	41.00	477.00
14"	ER@2.70	Ea	710.00	128.00	46.10	884.10
16"	ER@3.00	Ea	827.00	142.00	51.20	1,020.20

Double cement-lined Class 153 ductile iron 1/4 bend with mechanical joints

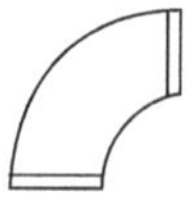

Description	Craft@Hrs	Unit	Material $	Labor $	Equipment $	Total $
4"	P1@.900	Ea	77.80	40.00	—	117.80
6"	ER@1.30	Ea	126.00	61.60	22.20	209.80
8"	ER@1.70	Ea	196.00	80.60	29.00	305.60
10"	ER@2.15	Ea	305.00	102.00	36.70	443.70
12"	ER@2.40	Ea	396.00	114.00	41.00	551.00
14"	ER@2.70	Ea	942.00	128.00	46.10	1,116.10
16"	ER@3.00	Ea	1,070.00	142.00	51.20	1,263.20

Description	Craft@Hrs	Unit	Material $	Labor $	Equipment $	Total $

Double cement-lined Class 153 ductile iron tee with mechanical joints

Description	Craft@Hrs	Unit	Material $	Labor $	Equipment $	Total $
4"	P1@1.65	Ea	119.00	73.30	—	192.30
6"	ER@2.35	Ea	173.00	111.00	40.10	324.10
8"	ER@3.15	Ea	264.00	149.00	53.80	466.80
10"	ER@4.00	Ea	396.00	190.00	68.30	654.30
12"	ER@4.45	Ea	542.00	211.00	76.00	829.00
14"	ER@4.90	Ea	1,400.00	232.00	83.70	1,715.70
16"	ER@5.30	Ea	1,460.00	251.00	90.50	1,801.50

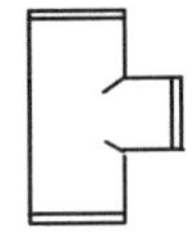

Double cement-lined Class 153 ductile iron wye with mechanical joints

Description	Craft@Hrs	Unit	Material $	Labor $	Equipment $	Total $
4"	P1@1.65	Ea	264.00	73.30	—	337.30
6"	ER@2.35	Ea	336.00	111.00	40.10	487.10
8"	ER@3.15	Ea	495.00	149.00	53.80	697.80
10"	ER@4.00	Ea	934.00	190.00	68.30	1,192.30
12"	ER@4.45	Ea	1,390.00	211.00	76.00	1,677.00
14"	ER@4.90	Ea	2,850.00	232.00	83.70	3,165.70
16"	ER@5.30	Ea	3,510.00	251.00	90.50	3,851.50

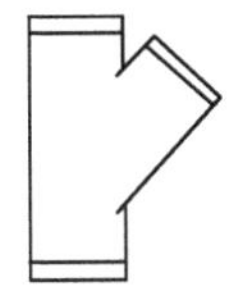

Double cement-lined Class 153 ductile iron reducer with mechanical joints

Description	Craft@Hrs	Unit	Material $	Labor $	Equipment $	Total $
6"	ER@1.45	Ea	86.40	68.70	24.80	179.90
8"	ER@2.00	Ea	123.00	94.80	34.10	251.90
10"	ER@2.55	Ea	160.00	121.00	43.50	324.50
12"	ER@3.05	Ea	235.00	145.00	52.10	432.10
14"	ER@3.60	Ea	502.00	171.00	61.50	734.50
16"	ER@4.20	Ea	632.00	199.00	71.70	902.70

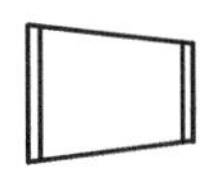

Ductile Iron, Class 110, Cement-Lined with Mechanical Joints

Cement-lined Class 110 ductile iron pipe with mechanical joints

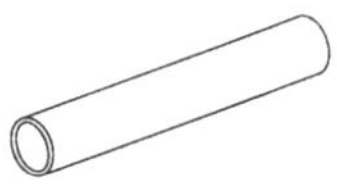

Description	Craft@Hrs	Unit	Material $	Labor $	Equipment $	Total $
4"	P1@.180	LF	17.90	8.00	—	25.90
6"	ER@.220	LF	25.30	10.40	3.76	39.46
8"	ER@.240	LF	35.00	11.40	4.10	50.50
10"	ER@.270	LF	52.30	12.80	4.61	69.71
12"	ER@.320	LF	71.10	15.20	5.46	91.76
16"	ER@.400	LF	139.00	19.00	6.83	164.83

Cement-lined Class 110 ductile iron 1/16 bend with mechanical joints

Description	Craft@Hrs	Unit	Material $	Labor $	Equipment $	Total $
4"	P1@1.40	Ea	179.00	62.20	—	241.20
6"	ER@2.00	Ea	238.00	94.80	34.10	366.90
8"	ER@2.50	Ea	327.00	119.00	42.70	488.70
10"	ER@3.20	Ea	498.00	152.00	54.60	704.60
12"	ER@3.60	Ea	634.00	171.00	61.50	866.50
16"	ER@4.50	Ea	1,650.00	213.00	76.80	1,939.80

Cement-lined Class 110 ductile iron 1/8 bend with mechanical joints

Description	Craft@Hrs	Unit	Material $	Labor $	Equipment $	Total $
4"	P1@1.40	Ea	176.00	62.20	—	238.20
6"	ER@2.00	Ea	236.00	94.80	34.10	364.90
8"	ER@2.50	Ea	327.00	119.00	42.70	488.70
10"	ER@3.20	Ea	480.00	152.00	54.60	686.60
12"	ER@3.60	Ea	633.00	171.00	61.50	865.50
16"	ER@4.50	Ea	1,620.00	213.00	76.80	1,909.80

Cement-lined Class 110 ductile iron 1/4 bend with mechanical joints

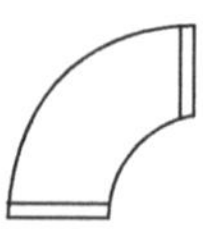

Description	Craft@Hrs	Unit	Material $	Labor $	Equipment $	Total $
4"	P1@1.40	Ea	194.00	62.20	—	256.20
6"	ER@2.00	Ea	260.00	94.80	34.10	388.90
8"	ER@2.50	Ea	374.00	119.00	42.70	535.70
10"	ER@3.20	Ea	579.00	152.00	54.60	785.60
12"	ER@3.60	Ea	867.00	171.00	61.50	1,099.50
16"	ER@4.50	Ea	2,110.00	213.00	76.80	2,399.80

Cement-lined Class 110 ductile iron tee with mechanical joints

Description	Craft@Hrs	Unit	Material $	Labor $	Equipment $	Total $
4"	P1@2.50	Ea	293.00	111.00	—	404.00
6"	ER@3.50	Ea	388.00	166.00	59.80	613.80
8"	ER@4.70	Ea	555.00	223.00	80.30	858.30
10"	ER@6.00	Ea	957.00	284.00	102.00	1,343.00
12"	ER@6.70	Ea	1,210.00	318.00	114.00	1,642.00
16"	ER@8.00	Ea	3,200.00	379.00	137.00	3,716.00

Cement-lined Class 110 ductile iron reducer with mechanical joints

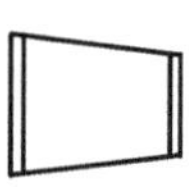

Description	Craft@Hrs	Unit	Material $	Labor $	Equipment $	Total $
6"	ER@2.20	Ea	194.00	104.00	37.60	335.60
8"	ER@3.00	Ea	267.00	142.00	51.20	460.20
10"	ER@3.80	Ea	438.00	180.00	64.90	682.90
12"	ER@4.60	Ea	517.00	218.00	78.50	813.50
16"	ER@6.30	Ea	1,410.00	299.00	108.00	1,817.00

Description	Craft@Hrs	Unit	Material $	Labor $	Equipment $	Total $
Class 150 cast iron pipe with mechanical joints						
1½"	P1@.080	LF	8.68	3.56	—	12.24
2"	P1@.100	LF	8.49	4.45	—	12.94
3"	P1@.120	LF	11.70	5.33	—	17.03
4"	P1@.140	LF	15.30	6.22	—	21.52
6"	ER@.180	LF	26.30	8.53	3.07	37.90
8"	ER@.200	LF	41.90	9.48	3.41	54.79

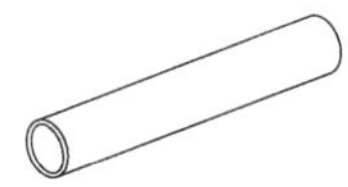

Description	Craft@Hrs	Unit	Material $	Labor $	Equipment $	Total $
Class 150 cast iron 1/8 bend with mechanical joints						
1½"	P1@.750	Ea	9.58	33.30	—	42.88
2"	P1@.800	Ea	10.70	35.60	—	46.30
3"	P1@1.00	Ea	14.40	44.50	—	58.90
4"	P1@1.20	Ea	18.20	53.30	—	71.50
6"	ER@1.70	Ea	42.80	80.60	29.00	152.40
8"	ER@2.25	Ea	123.00	107.00	38.40	268.40

Description	Craft@Hrs	Unit	Material $	Labor $	Equipment $	Total $
Class 150 cast iron 1/4 bend with mechanical joints						
1½"	P1@.750	Ea	13.30	33.30	—	46.60
2"	P1@.800	Ea	15.40	35.60	—	51.00
3"	P1@1.00	Ea	22.20	44.50	—	66.70
4"	P1@1.20	Ea	34.70	53.30	—	88.00
6"	ER@1.70	Ea	71.90	80.60	29.00	181.50
8"	ER@2.25	Ea	195.00	107.00	38.40	340.40

Description	Craft@Hrs	Unit	Material $	Labor $	Equipment $	Total $
Class 150 cast iron sanitary tee with mechanical joints						
1½"	P1@.750	Ea	16.00	33.30	—	49.30
2"	P1@1.00	Ea	17.60	44.50	—	62.10
3"	P1@1.20	Ea	21.20	53.30	—	74.50
4"	P1@1.65	Ea	32.60	73.30	—	105.90
6"	ER@2.35	Ea	93.00	111.00	40.10	244.10
8"	ER@3.15	Ea	298.00	149.00	53.80	500.80

Description	Craft@Hrs	Unit	Material $	Labor $	Equipment $	Total $
Class 150 cast iron wye with mechanical joints						
1½"	P1@.750	Ea	16.30	33.30	—	49.60
2"	P1@1.00	Ea	16.30	44.50	—	60.80
3"	P1@1.20	Ea	23.00	53.30	—	76.30
4"	P1@1.65	Ea	37.40	73.30	—	110.70
6"	ER@2.35	Ea	100.00	111.00	40.10	251.10
8"	ER@3.15	Ea	236.00	149.00	53.80	438.80

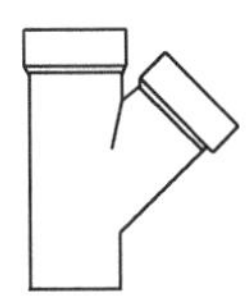

Description	Craft@Hrs	Unit	Material $	Labor $	Equipment $	Total $
Class 150 cast iron reducer						
2" x 1½"	P1@.600	Ea	8.75	26.70	—	35.45
3" x 2"	P1@.800	Ea	8.67	35.60	—	44.27
4" x 2"	P1@1.00	Ea	13.70	44.50	—	58.20
6" x 4"	ER@1.45	Ea	36.00	68.70	24.80	129.50
8" x 6"	ER@2.00	Ea	61.60	94.80	34.10	190.50

Asbestos-Cement, Class 2400 or 3000 with Mechanical Joints

Description	Craft@Hrs	Unit	Material $	Labor $	Equipment $	Total $

Asbestos-cement pipe with mechanical joints

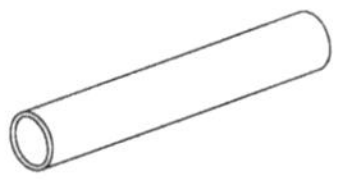

4"	ER@.080	LF	13.30	3.79	1.37	18.46
6"	ER@.100	LF	8.66	4.74	1.71	15.11
8"	ER@.140	LF	13.50	6.64	2.39	22.53
10"	ER@.200	LF	21.30	9.48	3.41	34.19
12"	ER@.260	LF	28.20	12.30	4.44	44.94
14"	ER@.300	LF	38.00	14.20	5.12	57.32
16"	ER@.350	LF	40.00	16.60	5.98	62.58

Asbestos-cement 1/16 bend with mechanical joints

4"	ER@1.00	Ea	195.00	47.40	17.10	259.50
6"	ER@1.25	Ea	256.00	59.30	21.30	336.60
8"	ER@1.70	Ea	351.00	80.60	29.00	460.60
10"	ER@2.20	Ea	539.00	104.00	37.60	680.60
12"	ER@2.70	Ea	617.00	128.00	46.10	791.10
14"	ER@3.40	Ea	1,280.00	161.00	58.10	1,499.10
16"	ER@3.90	Ea	1,780.00	185.00	66.60	2,031.60

Asbestos-cement 1/8 bend with mechanical joints

4"	ER@1.00	Ea	186.00	47.40	17.10	250.50
6"	ER@1.25	Ea	255.00	59.30	21.30	335.60
8"	ER@1.70	Ea	351.00	80.60	29.00	460.60
10"	ER@2.20	Ea	517.00	104.00	37.60	658.60
12"	ER@2.70	Ea	689.00	128.00	46.10	863.10
14"	ER@3.40	Ea	1,230.00	161.00	58.10	1,449.10
16"	ER@3.90	Ea	1,770.00	185.00	66.60	2,021.60

Asbestos-cement 1/4 bend with mechanical joints

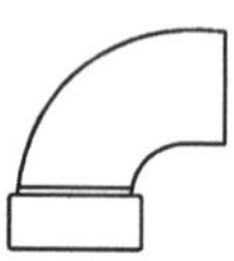

4"	ER@1.00	Ea	209.00	47.40	17.10	273.50
6"	ER@1.25	Ea	282.00	59.30	21.30	362.60
8"	ER@1.70	Ea	404.00	80.60	29.00	513.60
10"	ER@2.20	Ea	616.00	104.00	37.60	757.60
12"	ER@2.70	Ea	816.00	128.00	46.10	990.10
14"	ER@3.40	Ea	1,510.00	161.00	58.10	1,729.10
16"	ER@3.90	Ea	2,240.00	185.00	66.60	2,491.60

Asbestos-cement tee or wye with mechanical joints

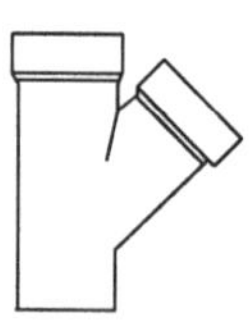

4"	ER@1.40	Ea	314.00	66.40	23.90	404.30
6"	ER@1.75	Ea	416.00	83.00	29.90	528.90
8"	ER@2.40	Ea	596.00	114.00	41.00	751.00
10"	ER@3.10	Ea	1,030.00	147.00	52.90	1,229.90
12"	ER@3.80	Ea	1,300.00	180.00	64.90	1,544.90
14"	ER@4.75	Ea	2,360.00	225.00	81.10	2,666.10
16"	ER@5.45	Ea	3,400.00	258.00	93.10	3,751.10

Asbestos-cement reducer with mechanical joints

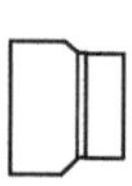

6" x 4"	ER@1.20	Ea	220.00	56.90	20.50	297.40
8" x 6"	ER@1.50	Ea	326.00	71.10	25.60	422.70
10" x 8"	ER@1.95	Ea	473.00	92.40	33.30	598.70
12" x 10"	ER@2.45	Ea	658.00	116.00	41.80	815.80

Description	Craft@Hrs	Unit	Material $	Labor $	Equipment $	Total $

Fiberglass septic tank. 2 compartments. Shallow burial (2 feet of earth cover). 32" manway entrance. One-piece construction. Including excavation, backfill and equipment costs.

750 gallon	P1@8.00	Ea	2,840.00	356.00	478.00	3,674.00
1,000 gallon	P1@9.00	Ea	3,770.00	400.00	538.00	4,708.00
1,250 gallon	P1@10.0	Ea	4,820.00	445.00	598.00	5,863.00
1,500 gallon	P1@11.5	Ea	5,460.00	511.00	687.00	6,658.00
1,750 gallon	P1@12.0	Ea	6,090.00	533.00	717.00	7,340.00
2,000 gallon	P1@12.5	Ea	6,710.00	556.00	747.00	8,013.00

Fiberglass septic tank. 2 compartments. Deep burial (7 feet of earth cover). 32" x 40" manway extension. One-piece construction. Including excavation, backfill and equipment costs.

750 gallon	P1@8.00	Ea	3,200.00	356.00	478.00	4,034.00
1,000 gallon	P1@9.00	Ea	4,200.00	400.00	538.00	5,138.00
1,250 gallon	P1@10.0	Ea	5,510.00	445.00	598.00	6,553.00
1,500 gallon	P1@11.5	Ea	6,190.00	511.00	687.00	7,388.00
1,750 gallon	P1@12.0	Ea	6,860.00	533.00	717.00	8,110.00
2,000 gallon	P1@12.5	Ea	7,550.00	556.00	747.00	8,853.00
Manway extension	P1@.350	LF	158.00	15.60	—	173.60

Fiberglass holding tank. Single compartment. Shallow burial (2 feet of earth cover). 32" manway entrance. One-piece construction. Including excavation, backfill and equipment costs. Can be used for drink water storage or sewage holding.

750 gallon	P1@8.00	Ea	2,470.00	356.00	478.00	3,304.00
1,000 gallon	P1@9.00	Ea	3,220.00	400.00	538.00	4,158.00
1,250 gallon	P1@10.0	Ea	4,200.00	445.00	598.00	5,243.00
1,500 gallon	P1@11.5	Ea	4,820.00	511.00	687.00	6,018.00
1,750 gallon	P1@12.0	Ea	5,460.00	533.00	717.00	6,710.00
2,000 gallon	P1@12.5	Ea	6,090.00	556.00	747.00	7,393.00

Fiberglass holding tank. Single compartment. Deep burial (7 feet of earth cover). 32" x 40" manway extension. One-piece construction. Including excavation, backfill and equipment costs. Can be used for drink water storage or sewage holding.

750 gallon	P1@8.00	Ea	2,890.00	356.00	478.00	3,724.00
1,000 gallon	P1@9.00	Ea	3,660.00	400.00	538.00	4,598.00
1,250 gallon	P1@10.0	Ea	4,890.00	445.00	598.00	5,933.00
1,500 gallon	P1@11.5	Ea	5,570.00	511.00	687.00	6,768.00
1,750 gallon	P1@12.0	Ea	6,220.00	533.00	717.00	7,470.00
2,000 gallon	P1@12.5	Ea	6,930.00	556.00	747.00	8,233.00
Manway extension	P1@.350	LF	158.00	15.60	—	173.60

Description	Craft@Hrs	Unit	Material $	Labor $	Equipment $	Total $

Plastic water tank. Suitable for potable water. Labor includes setting in place only. Freight, hoisting & rigging excluded.

Vertical tank, D31" x H55"						
165 gallon	P1@2.00	Ea	543.00	88.90	—	631.90
Vertical tank, D48" x H102"						
500 gallon	P1@4.00	Ea	1,660.00	178.00	—	1,838.00
Vertical tank, D64" x H79"						
1,000 gallon	P1@4.00	Ea	2,050.00	178.00	—	2,228.00
Vertical tank, D87" x H65"						
1,500 gallon	P1@5.00	Ea	3,100.00	222.00	—	3,322.00
Vertical tank, D95" x H89"						
2,500 gallon	P1@5.00	Ea	4,460.00	222.00	—	4,682.00
Vertical tank, D95" x H107"						
3,000 gallon	P1@6.00	Ea	6,370.00	267.00	—	6,637.00
Vertical tank, D102" x H125"						
4,000 gallon	P1@6.00	Ea	7,940.00	267.00	—	8,207.00
Vertical tank, D102" x H150"						
5,000 gallon	P1@8.00	Ea	14,900.00	356.00	—	15,256.00
Vertical tank, D141" x H192"						
12,000 gallon	P1@9.00	Ea	19,000.00	400.00	—	19,400.00

Plastic holding tank. Single compartment. One-piece polyethylene construction. Shallow burial (2 feet of earth cover). 28" manway entrance. Including excavation, backfill and equipment costs. Used for drink water storage.

30 gallon	P1@3.00	Ea	182.00	133.00	—	315.00
40 gallon	P1@3.00	Ea	207.00	133.00	—	340.00
75 gallon	P1@3.00	Ea	230.00	133.00	179.00	542.00
300 gallon	P1@5.00	Ea	1,020.00	222.00	299.00	1,541.00
525 gallon	P1@6.00	Ea	1,460.00	267.00	359.00	2,086.00
1,150 gallon	P1@8.00	Ea	2,560.00	356.00	478.00	3,394.00
1,700 gallon	P1@11.0	Ea	3,290.00	489.00	657.00	4,436.00
1,950 gallon	P1@12.0	Ea	3,510.00	533.00	717.00	4,760.00

Plastic holding tank. Single compartment. One-piece polyethylene construction. Above ground applications only. Used for drink water storage.

180 gallon	P1@3.00	Ea	609.00	133.00	179.00	921.00
200 gallon	P1@3.75	Ea	903.00	167.00	224.00	1,294.00
300 gallon	P1@4.25	Ea	1,170.00	189.00	254.00	1,613.00
525 gallon	P1@5.00	Ea	1,740.00	222.00	299.00	2,261.00
1,000 gallon	P1@5.75	Ea	2,340.00	256.00	344.00	2,940.00
1,150 gallon	P1@6.25	Ea	2,870.00	278.00	374.00	3,522.00
1,700 gallon	P1@7.00	Ea	3,800.00	311.00	418.00	4,529.00

Plastic septic tank. 750 gallon. Single compartment. One-piece polyethylene construction. Including excavation, backfill and equipment costs.

Trickle tank	P1@8.00	Ea	2,220.00	356.00	478.00	3,054.00
Shallow buried	P1@8.00	Ea	2,220.00	356.00	478.00	3,054.00
Deep buried	P1@8.00	Ea	2,340.00	356.00	478.00	3,174.00
Manway extension						
	P1@.350	LF	130.00	15.60	—	145.60

Description	Craft@Hrs	Unit	Material $	Labor $	Equipment $	Total $

Plastic sump pit. One-piece polyethylene construction. Complete with radon gasket electrical and discharge grommets. Including excavation, backfill and equipment costs.

Description	Craft@Hrs	Unit	Material $	Labor $	Equipment $	Total $
30 gallon	P1@3.00	Ea	194.00	133.00	—	327.00
40 gallon	P1@3.00	Ea	237.00	133.00	—	370.00
75 gallon	P1@3.00	Ea	279.00	133.00	179.00	591.00
110 gallon	P1@3.25	Ea	442.00	144.00	194.00	780.00
150 gallon	P1@3.75	Ea	597.00	167.00	224.00	988.00
250 gallon	P1@4.75	Ea	943.00	211.00	284.00	1,438.00
380 gallon	P1@5.50	Ea	1,470.00	244.00	329.00	2,043.00

Plastic sewage lift tank. One-piece polyethylene construction. Complete with radon gasket, inlet, vent, electrical and discharge grommets. Including excavation, backfill and equipment costs.

Description	Craft@Hrs	Unit	Material $	Labor $	Equipment $	Total $
30 gallon	P1@3.00	Ea	244.00	133.00	—	377.00
40 gallon	P1@3.00	Ea	269.00	133.00	—	402.00
75 gallon	P1@3.00	Ea	302.00	133.00	179.00	614.00
110 gallon	P1@3.25	Ea	512.00	144.00	179.00	835.00
150 gallon	P1@3.75	Ea	622.00	167.00	179.00	968.00
250 gallon	P1@4.75	Ea	1,020.00	211.00	194.00	1,425.00
380 gallon	P1@5.50	Ea	1,520.00	244.00	224.00	1,988.00

Plastic grey water treatment tank. 75 gallon. One-piece polyethylene construction. Including excavation, backfill and equipment costs.

Description	Craft@Hrs	Unit	Material $	Labor $	Equipment $	Total $
Buried	P1@3.00	Ea	364.00	133.00	179.00	676.00
Non-buried	P1@1.50	Ea	332.00	66.70	—	398.70

Trenching

Because of varying soil compositions and densities, trenching costs can sometimes be difficult to accurately estimate. The tables in this section are based on average soil conditions, which are loosely defined as dirt, soft clay or gravel mixed with rocks. The costs for trenching through limestone, sandstone or shale can increase by as much as 1,000 percent. For these conditions, it is best to consult with experts or to subcontract the work to specialists.

The primary consideration in excavations is to ensure the safety of the pipe layers. Current OSHA statutes require all excavations over 5 feet deep to be sloped, shored, sheeted or braced to prevent cave-ins. The use of shoring minimizes trenching costs by eliminating the need to slope the trench walls, but increases pipe-laying costs because of the difficulty in working between cross-bracing.

The estimator should carefully compare the costs for installing and removing shoring, and reduced pipe installation productivity, against the additional excavation costs for sloping the walls of the trench. For average soil conditions, not exceeding 6 to 8 feet in depth, it's usually more economical to slope the trench walls.

The current average cost for a backhoe with operator is $72.00 per hour. Some trenching contractors additionally charge $25.00 move-on and $50.00 move-off costs, particularly for small projects.

Total trenching, backfill and compaction costs are calculated by multiplying the backhoe hours from the following charts by $72.00 per hour, and adding move-on and move-off charges, if applicable.

Backhoe Trenching						
Trench walls at a 90-degree angle, based on average soil, including trenching, backfill and compaction. Hours per 100 linear feet of trench						
Trench depth (feet)	**Trench width (feet)**					
	1	**2**	**3**	**4**	**5**	**6**
2	2.6	5.1	7.7	10.2	12.8	15.3
3	3.8	7.7	11.5	15.3	19.2	23.0
4	5.1	10.2	15.3	20.4	25.6	30.7
5	6.4	12.8	19.2	25.6	32.0	38.3
6	7.7	15.3	23.0	30.7	38.3	46.0
7	8.9	17.9	26.8	35.8	44.7	53.7
8	10.2	20.4	30.7	40.9	51.1	61.3
9	11.5	23.0	34.5	46.0	57.5	69.0
10	12.8	25.5	38.3	51.1	63.9	76.6

Notes:

1) Compaction is based on the use of a compactor wheel.
2) Additional labor costs must be added if watering-in, stone removal or excess spoil removal is required.

Backhoe Trenching Trench walls at a 45-degree angle, based on average soil including trenching, backfill and compaction. Hours per 100 linear feet of trench						
	Trench width (feet)					
Trench depth (feet)	1	2	3	4	5	6
2	7.6	10.3	12.6	15.5	17.9	20.3
3	15.2	22.6	22.6	26.6	30.7	34.5
4	25.5	30.7	35.5	40.7	45.9	51.0
5	38.3	44.5	51.0	57.6	63.8	70.0
6	53.4	61.0	69.0	76.6	84.1	91.7
7	71.4	80.3	89.3	98.3	107.2	115.9
8	92.1	102.1	112.4	122.4	132.8	143.1
9	114.8	126.6	137.9	149.3	161.0	172.4
10	140.3	153.1	165.9	178.6	191.4	204.5

Notes:

1) Compaction is based on the use of a compactor wheel.
2) Additional labor costs must be added if watering-in, stone removal or excess spoil removal is required.

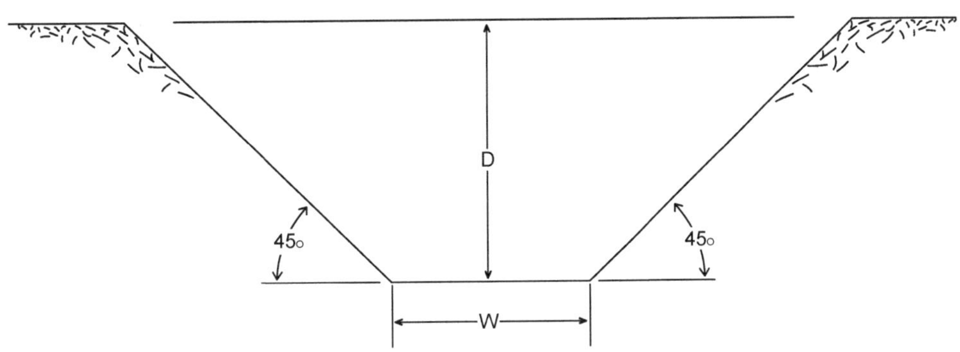

Equipment Rental			
Description	**Per day ($)**	**Per week ($)**	**Per month ($)**
Air compressor, gas or diesel, 100-125 CFM with 100' hose, air tool and 4 bits	301.00	891.00	2,230.00
Air compressor, gas or diesel, 165-185 CFM with 100' hose, air tool and 4 bits	348.00	1,040.00	2,650.00
Air compressor, gas or diesel, 100-125 CFM, no accessories	141.00	424.00	1,280.00
Air compressor, gas or diesel, 165-185 CFM, no accessories	188.00	570.00	1,700.00
Backhoe with bucket, 55 HP	403.00	1,460.00	3,390.00
Backhoe with bucket, 100 HP	447.00	1,410.00	4,200.00
Backhoe hydraulic compactor, 5-10 ton	211.00	542.00	1,350.00
Bender, hydraulic	98.60	364.00	1,030.00
Bevel machine with torch	32.20	106.00	260.00
Boom lift, 31' - 40'	532.00	1,320.00	2,710.00
Boom lift, 51' - 60'	641.00	1,380.00	2,860.00
Chain hoist, to 2 tons	33.20	115.00	349.00
Chain hoist, 10 tons	44.60	165.00	488.00
Come-along, ½ to 1 ton	30.20	41.20	234.00
Come-along, 1½ to 2 tons	52.30	135.00	410.00
Come-along, 4-ton	73.40	213.00	660.00
Compactor, soil, gas or diesel, 125 lb	129.00	393.00	835.00
Crane, truck-mounted, 13,000 lb capacity	457.00	1,330.00	4,150.00
Crane, truck-mounted, 30,000 lb capacity	571.00	1,760.00	4,990.00
Forklift, 2,000 pound capacity	234.00	623.00	1,680.00
Forklift, 4,000 pound capacity	225.00	612.00	1,850.00
Forklift, 6,000 pound capacity	295.00	1,310.00	2,090.00
Front-end loader, 1 yard capacity	402.00	1,160.00	3,270.00
Front-end loader, 2 yard capacity	654.00	1,850.00	4,890.00
Front-end loader, 3¼ yard capacity	971.00	1,650.00	7,570.00
Pipe machine, up to 2" capacity	73.40	220.00	614.00
Pipe machine, 2½" to 4" capacity	81.50	558.00	876.00
Ramset (nail gun)	30.10	120.00	361.00
Scissor-lift, to 20', diesel, self-propelled	123.00	307.00	799.00
Scissor-lift, 21' – 30'	341.00	729.00	1,480.00
Skid steer loader 1,750 lb capacity	328.00	955.00	2,160.00
Trailer, office, 25'	70.00	224.00	591.00
Truck, ¾-ton pickup	107.00	415.00	1,670.00
Truck, 1-ton pickup	161.00	508.00	1,610.00
Truck, 1-ton stake bed	218.00	657.00	1,930.00
Van, storage, 26'	—	—	406.00
Van, storage, 40'	—	—	505.00
Welding machine, gas, 200 amp, trailer-mounted with 50' bare lead	91.30	232.00	507.00
Welding machine, diesel, 400 amp, trailer-mounted with 50' bare lead	150.00	345.00	800.00

Description	Craft@Hrs	Unit	Material $	Labor $	Equipment $	Total $
1½" brass valve tags with black filled numbers						
Typical costs	P1@.150	Ea	3.94	6.67	—	10.61
1½" blank brass valve tags						
Typical costs	P1@.150	Ea	1.95	6.67	—	8.62
1½" plastic valve tags with engraved numbers						
Typical costs	P1@.150	Ea	3.27	6.67	—	9.94
Engraved plastic equipment nameplates						
Typical costs	P1@.400	Ea	8.47	17.80	—	26.27
Pipe and duct markers						
Typical costs	P1@.250	Ea	4.84	11.10	—	15.94
As-built drawings						
Typical costs	P1@.500	Ea	—	22.20	—	22.20
Instructing owner's operating personnel						
Typical costs	P1@8.00	Ea	—	356.00	—	356.00

Description	Craft@Hrs	Unit	Material $	Labor $	Equipment $	Total $

Remove rectangular galvanized ductwork. 22 to 30 gauge galvanized sheet metal (dimension/gauge per SMACNA standards). Per linear foot of sheet metal ductwork. The cost of removing ductwork includes the cost of removing the support hangers. Add the cost of disconnecting and removing heating coils, air volume zone boxes, air handing units, condensate drains, motorized dampers or other associated accessories, piping or wiring. These figures assume demolished materials are piled on site. Disposal costs or salvage values excluded. (SMACNA = Sheetmetal & Air Conditioning National Association.) Equipment may be needed for some sizes and locations. Refer to the equipment rental costs at the front of this section.

Description	Craft@Hrs	Unit	Material $	Labor $	Equipment $	Total $
4" to 12" rectangular/square duct, remove	S2@.015	LF	—	.65	—	.65
14" to 24" rectangular/square duct, remove	S2@.020	LF	—	.87	—	.87
26" to 36" rectangular/square duct, remove	S2@.065	LF	—	2.83	—	2.83
38" to 60" rectangular/square duct, remove	S2@.090	LF	—	3.92	—	3.92

Remove round galvanized ductwork. 22 to 30 gauge galvanized sheet metal (dimension/gauge per SMACNA standards). Per linear foot of sheet metal ductwork. The cost of removing ductwork includes the cost of removing the support hangers. Add the cost of disconnecting and removing heating coils, air volume zone boxes, air handing units, condensate drains, motorized dampers or other associated accessories, piping or wiring. These figures assume demolished materials are piled on site. Disposal costs or salvage values excluded. (SMACNA = Sheetmetal & Air Conditioning National Association.) Equipment may be needed for some sizes and locations. Refer to the equipment rental costs at the front of this section.

Description	Craft@Hrs	Unit	Material $	Labor $	Equipment $	Total $
4" to 12" round duct, remove	S2@.012	LF	—	.52	—	.52
14" to 24" round duct, remove	S2@.016	LF	—	.70	—	.70
26" to 36" round duct, remove	S2@.052	LF	—	2.26	—	2.26
38" to 60" round duct, remove	S2@.072	LF	—	3.13	—	3.13

Description	Craft@Hrs	Unit	Material $	Labor $	Equipment $	Total $

Remove welded steel ductwork. 10 to 16 gauge mild steel. Per linear foot of welded milled steel ductwork. The cost of removing ductwork includes the cost of removing the support hangers. Add the cost of disconnecting and removing exhaust fans, condensate drains, sprinkler heads or other associated accessories, piping or wiring. These figures assume demolished materials are piled on site. Disposal costs or salvage values excluded. Equipment may be needed for some sizes and locations. Refer to the equipment rental costs at the front of this section.

4" to 12", square or round duct, remove	S2@.045	LF	—	1.96	—	1.96
14" to 24", square or round duct, remove	S2@.060	LF	—	2.61	—	2.61
26" to 36", square or round duct, remove	S2@.095	LF	—	4.13	—	4.13
38" to 60", square or round duct, remove	S2@.150	LF	—	6.53	—	6.53

Remove diffuser or grille. These figures assume demolished materials are piled on site. Disposal costs or salvage values excluded. Equipment may be needed for some sizes and locations. Refer to the equipment rental costs at the front of this section.

6" to 12", square or round diffuser/grille, remove	S2@.150	Ea	—	6.53	—	6.53
14" to 24", square or round diffuser/grille, remove	S2@.180	Ea	—	7.83	—	7.83
26" to 36", square or round diffuser/grille, remove	S2@.400	Ea	—	17.40	—	17.40
38” to 60” square or round diffuser/grille, remove	S2@.800	Ea	—	34.80	—	34.80

Remove air mixing box. These figures include isolating and disconnecting hot water, chilled water, steam, refrigerant and/or pneumatic control tube or control wiring. These figures assume demolished materials are piled on site. Disposal costs or salvage values excluded. Equipment may be needed for some sizes and locations. Refer to the equipment rental costs at the front of this section.

Air box, 50 to 150 CFM, remove	S2@.500	Ea	—	21.80	—	21.80
Air box, 160 to 500 CFM, remove	S2@.650	Ea	—	28.30	—	28.30
Air box, 550 to 750 CFM, remove	S2@.900	Ea	—	39.20	—	39.20

Description	Craft@Hrs	Unit	Material $	Labor $	Equipment $	Total $

Remove air handling unit. These figures include isolating and disconnecting hot water, chilled water, steam or gas and/or ductwork, electrical connections and unit disassembly when required for removal. Disposal costs or salvage values excluded. Equipment may be needed for some sizes and locations. Refer to the equipment rental costs at the front of this section.

Description	Craft@Hrs	Unit	Material $	Labor $	Equipment $	Total $
AHU, 2,000-5,000 CFM, remove	P1@8.00	Ea	—	356.00	—	356.00
AHU, 5,000-10,000 CFM, remove	P1@12.0	Ea	—	533.00	—	533.00
AHU,10,000-20,000 CFM, remove	P1@18.0	Ea	—	800.00	—	800.00
AHU, over 20,000 CFM, remove	P1@28.0	Ea	—	1,240.00	—	1,240.00

Remove rooftop unit. These figures include isolating and disconnecting hot water, chilled water, steam or gas and/or ductwork, electrical connections and unit disassembly when required for removal. Disposal costs or salvage values excluded. Make additional hoisting allowances for roof heights above 40'. Equipment may be needed for some sizes and locations. Refer to the equipment rental costs at the front of this section.

Description	Craft@Hrs	Unit	Material $	Labor $	Equipment $	Total $
RTU, 2,000-5,000 CFM, remove	P1@6.00	Ea	—	267.00	—	267.00
RTU, 5,000-10,000 CFM, remove	P1@10.0	Ea	—	445.00	—	445.00
RTU. 10,000-20,000 CFM, remove	P1@16.0	Ea	—	711.00	—	711.00
RTU, over 20,000 CFM, remove	P1@26.0	Ea	—	1,160.00	—	1,160.00

Description	Craft@Hrs	Unit	Material $	Labor $	Equipment $	Total $

Remove boiler. These figures include isolating and disconnecting hot water, steam or gas and/or electrical connections and unit disassembly when required for removal. Disposal costs or salvage values excluded. Equipment may be needed for some sizes and locations. Refer to the equipment rental costs at the front of this section.

Description	Craft@Hrs	Unit	Material $	Labor $	Equipment $	Total $
10-50 KW electric boiler, remove	P1@8.00	Ea	—	356.00	—	356.00
50-100 KW electric boiler, remove	P1@12.0	Ea	—	533.00	—	533.00
100-200 KW electric boiler, remove	P1@26.0	Ea	—	1,160.00	—	1,160.00
200-500 KW electric boiler, remove	P1@36.0	Ea	—	1,600.00	—	1,600.00
over 500 KW electric boiler, remove	P1@44.0	Ea	—	1,960.00	—	1,960.00
to 150 Mbh gas-fired boiler, remove	P1@9.00	Ea	—	400.00	—	400.00
150-300 Mbh gas-fired boiler, remove	P1@14.0	Ea	—	622.00	—	622.00
300-750 Mbh gas-fired boiler, remove	P1@30.0	Ea	—	1,330.00	—	1,330.00
750-1,500 Mbh gas-fired boiler, remove	P1@46.0	Ea	—	2,040.00	—	2,040.00
1,500-2,000 Mbh gas-fired boiler, remove	P1@65.0	Ea	—	2,890.00	—	2,890.00
over 2,000 Mbh gas-fired boiler, remove	P1@80.0	Ea	—	3,560.00	—	3,560.00

Remove furnace. These figures include isolating and disconnecting plenums, fuel and/or electrical connections. Disposal costs or salvage values excluded. Equipment may be needed for some sizes and locations. Refer to the equipment rental costs at the front of this section.

Description	Craft@Hrs	Unit	Material $	Labor $	Equipment $	Total $
5-30 KW electric furnace, remove	P1@2.50	Ea	—	111.00	—	111.00
30-60 KW electric furnace, remove	P1@3.75	Ea	—	167.00	—	167.00
30-150 Mbh gas-fired furnace, remove	P1@5.00	Ea	—	222.00	—	222.00
150-225 Mbh gas-fired furnace, remove	P1@7.00	Ea	—	311.00	—	311.00

Description	Craft@Hrs	Unit	Material $	Labor $	Equipment $	Total $

Remove chiller. These figures include isolating and disconnecting chilled water and electrical connections, evacuating refrigerant to storage vessels and unit disassembly when required for removal. Disposal costs or salvage values excluded. Equipment may be needed for some sizes and locations. Refer to the equipment rental costs at the front of this section.

Description	Craft@Hrs	Unit	Material $	Labor $	Equipment $	Total $
Chiller, 20-50 tons, remove	P1@34.0	Ea	—	1,510.00	—	1,510.00
Chiller, 50-100 tons, remove	P1@48.0	Ea	—	2,130.00	—	2,130.00
Chiller, 100-150 tons, remove	P1@56.0	Ea	—	2,490.00	—	2,490.00
Chiller, 150-200 tons, remove	P1@68.0	Ea	—	3,020.00	—	3,020.00
Chiller, 200-250 tons, remove	P1@80.0	Ea	—	3,560.00	—	3,560.00
Chiller, over 250 tons, remove	P1@100	Ea	—	4,450.00	—	4,450.00

Remove air-cooled condenser. These figures include isolating and disconnecting refrigerant lines and electrical connections, evacuating refrigerant to storage vessels and unit disassembly when required for removal. Disposal costs or salvage values excluded. Make additional hoisting allowances for roof heights above 40'. Equipment may be needed for some sizes and locations. Refer to the equipment rental costs at the front of this section.

Description	Craft@Hrs	Unit	Material $	Labor $	Equipment $	Total $
Air-cooled condenser, 2-4 tons, remove	P1@4.00	Ea	—	178.00	—	178.00
Air-cooled condenser, 5-10 tons, remove	P1@8.00	Ea	—	356.00	—	356.00
Air-cooled condenser, 10-20 tons, remove	P1@20.0	Ea	—	889.00	—	889.00
Air-cooled condenser, 20-30 tons, remove	P1@28.0	Ea	—	1,240.00	—	1,240.00
Air-cooled condenser, 30-50 tons, remove	P1@36.0	Ea	—	1,600.00	—	1,600.00
Air-cooled condenser, 50-150 tons, remove	P1@44.0	Ea	—	1,960.00	—	1,960.00
Air-cooled condenser, 150-250 tons, remove	P1@52.0	Ea	—	2,310.00	—	2,310.00

Description	Craft@Hrs	Unit	Material $	Labor $	Equipment $	Total $

Remove cooling tower. These figures include isolating and disconnecting chilled water lines and electrical connections, and unit disassembly when required for removal. Disposal costs or salvage values excluded. Make additional hoisting allowances for roof heights above 40'. Equipment may be needed for some sizes and locations. Refer to the equipment rental costs at the front of this section.

Description	Craft@Hrs	Unit	Material $	Labor $	Equipment $	Total $
Cooling tower, 5-10 tons, remove	P1@8.00	Ea	—	356.00	—	356.00
Cooling tower, 10-20 tons, remove	P1@16.0	Ea	—	711.00	—	711.00
Cooling tower, 20-30 tons, remove	P1@24.0	Ea	—	1,070.00	—	1,070.00
Cooling tower, 30-50 tons, remove	P1@32.0	Ea	—	1,420.00	—	1,420.00
Cooling tower, 50-150 tons, remove	P1@40.0	Ea	—	1,780.00	—	1,780.00
Cooling tower, 150-250 tons, remove	P1@48.0	Ea	—	2,130.00	—	2,130.00
Cooling tower, 250-500 tons, remove	P1@70.0	Ea	—	3,110.00	—	3,110.00

Remove fan coil. These figures include isolating and disconnecting chilled water and/or hot water lines and/or duct and electrical connections, and unit disassembly when required for removal. Disposal costs or salvage values excluded. Equipment may be needed for some sizes and locations. Refer to the equipment rental costs at the front of this section.

Description	Craft@Hrs	Unit	Material $	Labor $	Equipment $	Total $
2-pipe fan coil, 300-500 CFM, remove	P1@4.00	Ea	—	178.00	—	178.00
2-pipe fan coil, 550-750 CFM, remove	P1@4.50	Ea	—	200.00	—	200.00
2-pipe fan coil, 800-1,200 CFM, remove	P1@6.00	Ea	—	267.00	—	267.00
4-pipe fan coil, 300-500 CFM, remove	P1@4.50	Ea	—	200.00	—	200.00
4-pipe fan coil, 550-750 CFM, remove	P1@5.00	Ea	—	222.00	—	222.00
4-pipe fan coil, 800-1,200 CFM, remove	P1@6.50	Ea	—	289.00	—	289.00

Description	Craft@Hrs	Unit	Material $	Labor $	Equipment $	Total $

Remove unit heater. These figures include isolating and disconnecting hot water, steam lines or gas and/or electrical connections. Disposal costs or salvage values excluded. Equipment may be needed for some sizes and locations. Refer to the equipment rental costs at the front of this section.

Description	Craft@Hrs	Unit	Material $	Labor $	Equipment $	Total $
Gas-fired unit heater, 50-150 Mbh, remove	P1@2.25	Ea	—	100.00	—	100.00
Gas-fired unit heater, 200-350 Mbh, remove	P1@3.50	Ea	—	156.00	—	156.00
Steam unit heater, 75-200 Mbh, remove	P1@3.25	Ea	—	144.00	—	144.00
Steam unit heater, 250-350 Mbh, remove	P1@4.75	Ea	—	211.00	—	211.00
Electric unit heater, 5-25 KW, remove	P1@2.00	Ea	—	88.90	—	88.90
Electric unit heater, 30-50 KW, remove	P1@2.75	Ea	—	122.00	—	122.00

Remove heat pump. These figures include isolating and disconnecting pipe, duct and electrical connections and evacuating refrigerant to storage vessels when required for removal. Disposal costs or salvage values excluded. Equipment may be needed for some sizes and locations. Refer to the equipment rental costs at the front of this section.

Description	Craft@Hrs	Unit	Material $	Labor $	Equipment $	Total $
Heat pump, split system, 2-5 ton, remove	P1@4.75	Ea	—	211.00	—	211.00
Heat pump, split system, 7.5-10 ton, remove	P1@8.50	Ea	—	378.00	—	378.00
Heat pump, packaged system, 2-5 ton, remove	P1@3.25	Ea	—	144.00	—	144.00
Heat pump, pkg. system, 7.5-10 ton, remove	P1@5.50	Ea	—	244.00	—	244.00

Description	Craft@Hrs	Unit	Material $	Labor $	Equipment $	Total $

Remove heat exchanger. These figures include isolating and disconnecting chilled water, hot water and/or steam line connections. Disposal costs or salvage values excluded. Equipment may be needed for some sizes and locations. Refer to the equipment rental costs at the front of this section.

Description	Craft@Hrs	Unit	Material $	Labor $	Equipment $	Total $
Heat exchanger, 6" x 30", remove	P1@2.75	Ea	—	122.00	—	122.00
Heat exchanger, 10" x 48", remove	P1@4.00	Ea	—	178.00	—	178.00
Heat exchanger, 12" x 72", remove	P1@5.25	Ea	—	233.00	—	233.00
Heat exchanger, 16" x 90", remove	P1@6.00	Ea	—	267.00	—	267.00

Remove duct-mounted coil. These figures include isolating and disconnecting steam, chilled water, hot water, condensate, refrigerant or electrical and duct connections. Disposal costs or salvage values excluded. Refrigerant coil removal costs include system refrigerant evacuation and storage. Equipment may be needed for some sizes and locations. Refer to the equipment rental costs at the front of this section.

Description	Craft@Hrs	Unit	Material $	Labor $	Equipment $	Total $
Single-row water coil, 12" to 18", remove	P1@2.00	Ea	—	88.90	—	88.90
Single-row water coil, 20" to 30", remove	P1@3.50	Ea	—	156.00	—	156.00
Single-row water coil, 32" to 60", remove	P1@8.00	Ea	—	356.00	—	356.00
Single-row steam coil, 12" to 18", remove	P1@2.25	Ea	—	100.00	—	100.00
Single-row steam coil, 20" to 30", remove	P1@3.75	Ea	—	167.00	—	167.00
Single-row steam coil, 32" to 60", remove	P1@9.00	Ea	—	400.00	—	400.00
Electric resistance, 12" to 18", remove	P1@1.25	Ea	—	55.60	—	55.60
Electric resistance, 20" to 30", remove	P1@1.75	Ea	—	77.80	—	77.80
Electric resistance, 32" to 60", remove	P1@4.00	Ea	—	178.00	—	178.00
Refrigerant coil, 2-3 ton, remove	P1@4.00	Ea	—	178.00	—	178.00
Refrigerant coil, 4-5 ton, remove	P1@6.50	Ea	—	289.00	—	289.00
Refrigerant coil, 6-10 ton, remove	P1@12.0	Ea	—	533.00	—	533.00
Refrigerant coil, 12-15 ton, remove	P1@16.0	Ea	—	711.00	—	711.00

Description	Craft@Hrs	Unit	Material $	Labor $	Equipment $	Total $

Remove steel piping. These figures include cutting out and removing threaded or welded steel piping. These figures assume demolished materials are piled on site. Equipment, fixture or apparatus disconnection costs and salvage values excluded. Equipment may be needed for some sizes and locations. Refer to the equipment rental costs at the front of this section.

Description	Craft@Hrs	Unit	Material $	Labor $	Equipment $	Total $
Steel pipe, Sch. 40, ½"-1"						
remove	P1@.038	LF	—	1.69	—	1.69
Steel pipe, Sch. 40, 1½"-2"						
remove	P1@.068	LF	—	3.02	—	3.02
Steel pipe, Sch. 40, 2½"-4"						
remove	P1@.108	LF	—	4.80	—	4.80
Steel pipe, Sch. 40, 6"-8"						
remove	P1@.230	LF	—	10.20	—	10.20
Steel pipe, Sch. 40, 10"-12"						
remove	P1@.330	LF	—	14.70	—	14.70
Steel pipe, Sch. 80, ½"-1"						
remove	P1@.042	LF	—	1.87	—	1.87
Steel pipe, Sch. 80, 1½"-2"						
remove	P1@.075	LF	—	3.33	—	3.33
Steel pipe, Sch. 80, 2½"-4"						
remove	P1@.118	LF	—	5.25	—	5.25
Steel pipe, Sch. 80, 6"-8"						
remove	P1@.290	LF	—	12.90	—	12.90
Steel pipe, Sch. 80, 10"-12"						
remove	P1@.360	LF	—	16.00	—	16.00
Remove steel pipe, Sch. 160,						
add	—	%	—	30.0	—	—

Remove copper piping. These figures include cutting out and removing copper piping. These figures assume demolished materials are piled on site. Equipment, fixture or apparatus disconnection costs and salvage values excluded.

Description	Craft@Hrs	Unit	Material $	Labor $	Equipment $	Total $
Copper pipe, ½"-1"						
remove	P1@.022	LF	—	.98	—	.98
Copper pipe, 1½"-2"						
remove	P1@.028	LF	—	1.24	—	1.24
Copper pipe, 2½"-4"						
remove	P1@.044	LF	—	1.96	—	1.96

Remove plastic piping. These figures include cutting out and removing plastic piping. These figures assume demolished materials are piled on site. Equipment, fixture or apparatus disconnection costs and salvage values excluded. Equipment may be needed for some sizes and locations. Refer to the equipment rental costs at the front of this section.

Description	Craft@Hrs	Unit	Material $	Labor $	Equipment $	Total $
Plastic pipe, ½"-1"						
remove	P1@.015	LF	—	.67	—	.67
Plastic pipe, 1½"-2"						
remove	P1@.030	LF	—	1.33	—	1.33
Plastic pipe, 2½"-4"						
remove	P1@.038	LF	—	1.69	—	1.69
Plastic pipe, 6"-8"						
remove	P1@.080	LF	—	3.56	—	3.56

Description	Craft@Hrs	Unit	Material $	Labor $	Equipment $	Total $

Remove pump. These figures include isolating and disconnecting pipe and electrical connections. Disposal costs or salvage values excluded. Equipment may be needed for some sizes and locations. Refer to the equipment rental costs at the front of this section.

Description	Craft@Hrs	Unit	Material $	Labor $	Equipment $	Total $
Base-mounted pump, ½-1 HP, remove	P1@1.00	Ea	—	44.50	—	44.50
Base-mounted pump, 1½-3 HP, remove	P1@1.75	Ea	—	77.80	—	77.80
Base-mounted pump, 5-10 HP, remove	P1@2.75	Ea	—	122.00	—	122.00
Base-mounted pump, 15-25 HP, remove	P1@3.75	Ea	—	167.00	—	167.00
Base-mounted pump, 30-50 HP, remove	P1@4.55	Ea	—	202.00	—	202.00
In-line mounted pump, 1/12-1/3 HP, remove	P1@.650	Ea	—	28.90	—	28.90
In-line mounted pump, ½-1 HP, remove	P1@.750	Ea	—	33.30	—	33.30
In-line mounted pump, 3-5 HP, remove	P1@1.85	Ea	—	82.20	—	82.20
In-line mounted pump, 7.5-10 HP, remove	P1@2.55	Ea	—	113.00	—	113.00
In-line mounted pump, 15-25 HP, remove	P1@3.90	Ea	—	173.00	—	173.00
In-line mounted pump, 30-50 HP, remove	P1@4.85	Ea	—	216.00	—	216.00

Remove valve. These figures include isolation and disconnection. Disposal costs or salvage values excluded. Equipment may be needed for some sizes and locations. Refer to the equipment rental costs at the front of this section.

Description	Craft@Hrs	Unit	Material $	Labor $	Equipment $	Total $
Soldered valve, ½"-1", remove	P1@.165	Ea	—	7.33	—	7.33
Soldered valve, 1½"-2", remove	P1@.225	Ea	—	10.00	—	10.00
Soldered valve, 2½"-4", remove	P1@.650	Ea	—	28.90	—	28.90
Threaded valve, ½"-1", remove	P1@.150	Ea	—	6.67	—	6.67
Threaded valve, 1½"-2", remove	P1@.235	Ea	—	10.40	—	10.40
Threaded valve, 2½"-4", remove	P1@.680	Ea	—	30.20	—	30.20
Flanged valve, ½"-1", remove	P1@.200	Ea	—	8.89	—	8.89
Flanged valve, 1½"-2", remove	P1@.280	Ea	—	12.40	—	12.40
Flanged valve, 2½"-4", remove	P1@.625	Ea	—	27.80	—	27.80
Flanged valve, 6"-8", remove	P1@1.70	Ea	—	75.60	—	75.60
Flanged valve, 10"-12", remove	P1@2.65	Ea	—	118.00	—	118.00

Description	Craft@Hrs	Unit	Material $	Labor $	Equipment $	Total $

Remove hot water tank. These figures include isolation and disconnection. Disposal costs or salvage values excluded. Equipment may be needed for some sizes and locations. Refer to the equipment rental costs at the front of this section.

Description	Craft@Hrs	Unit	Material $	Labor $	Equipment $	Total $
Gas-fired hot water tank, 40 gallon, remove	P1@2.00	Ea	—	88.90	—	88.90
Gas-fired hot water tank, 60 gallon, remove	P1@2.50	Ea	—	111.00	—	111.00
Gas-fired hot water tank, 100 gallon, remove	P1@3.00	Ea	—	133.00	—	133.00
Electric hot water tank, 40 gallon, remove	P1@1.60	Ea	—	71.10	—	71.10
Electric hot water tank, 60 gallon, remove	P1@2.00	Ea	—	88.90	—	88.90
Electric hot water tank, 100 gallon, remove	P1@2.60	Ea	—	116.00	—	116.00

Remove plumbing fixture. These figures include isolation and disconnection. Disposal costs or salvage values excluded. Equipment may be needed for some sizes and locations. Refer to the equipment rental costs at the front of this section.

Description	Craft@Hrs	Unit	Material $	Labor $	Equipment $	Total $
Floor-mount, tank-type water closet, remove	P1@.850	ea	—	37.80	—	37.80
Wall-mount, tank-type water closet, remove	P1@1.35	Ea	—	60.00	—	60.00
Floor-mount, flush valve water closet, remove	P1@.950	Ea	—	42.20	—	42.20
Wal- mount, flush valve water closet, remove	P1@1.55	Ea	—	68.90	—	68.90
Wall-mount, flush valve urinal, remove	P1@1.15	Ea	—	51.10	—	51.10
Wall-mount, tank-type urinal, remove	P1@1.65	Ea	—	73.30	—	73.30
Kitchen sink, remove	P1@.850	Ea	—	37.80	—	37.80
Bathroom sink, remove	P1@.850	Ea	—	37.80	—	37.80
Bathtub, remove	P1@2.25	Ea	—	100.00	—	100.00
Shower stall, remove	P1@2.00	Ea	—	88.90	—	88.90
Tub and shower combination, remove	P1@2.85	Ea	—	127.00	—	127.00
Laundry sink, remove	P1@.950	Ea	—	42.20	—	42.20
Mop sink, remove	P1@1.50	Ea	—	66.70	—	66.70
Slop sink, remove	P1@1.75	Ea	—	77.80	—	77.80
Drinking fountain, remove	P1@1.40	Ea	—	62.20	—	62.20

Description	Craft@Hrs	Unit	Material $	Labor $	Equipment $	Total $

Disconnect plumbing fixture. These figures include isolation and disconnection only. Fixtures removed by others. Disposal costs or salvage values excluded. Equipment may be needed for some sizes and locations. Refer to the equipment rental costs at the front of this section.

Description	Craft@Hrs	Unit	Material $	Labor $	Equipment $	Total $
Floor-mount, tank-type water closet, disconnect	P1@.450	Ea	—	20.00	—	20.00
Wall-mount, tank-type water closet, disconnect	P1@.650	Ea	—	28.90	—	28.90
Floor-mount, flush valve water closet, disconnect	P1@.550	Ea	—	24.40	—	24.40
Wall-mount, flush valve water closet, disconnect	P1@.650	Ea	—	28.90	—	28.90
Wall-mount, flush valve urinal, disconnect	P1@.650	Ea	—	28.90	—	28.90
Wall-mount, tank-type urinal, disconnect	P1@.950	Ea	—	42.20	—	42.20
Kitchen sink, disconnect	P1@.450	Ea	—	20.00	—	20.00
Bathroom sink, disconnect	P1@.450	Ea	—	20.00	—	20.00
Bathtub, disconnect	P1@.650	Ea	—	28.90	—	28.90
Shower stall, disconnect	P1@.650	Ea	—	28.90	—	28.90
Tub and shower combination, disconnect	P1@.650	Ea	—	28.90	—	28.90
Laundry sink, disconnect	P1@.450	Ea	—	20.00	—	20.00
Mop sink, disconnect	P1@.450	Ea	—	20.00	—	20.00
Slop sink, disconnect	P1@.750	Ea	—	33.30	—	33.30
Drinking fountain, disconnect	P1@.650	Ea	—	28.90	—	28.90
Gas-fired hot water tank, 40 gallon, disconnect	P1@.900	Ea	—	40.00	—	40.00
Gas-fired hot water tank, 60 gallon, disconnect	P1@1.15	Ea	—	51.10	—	51.10
Gas-fired hot water tank, 100 gallon, disconnect	P1@1.35	Ea	—	60.00	—	60.00
Electric hot water tank, 40 gallon, disconnect	P1@.600	Ea	—	26.70	—	26.70
Electric hot water tank, 60 gallon, disconnect	P1@.900	Ea	—	40.00	—	40.00
Electric hot water tank, 100 gallon, disconnect	P1@1.00	Ea	—	44.50	—	44.50

Description	Craft@Hrs	Unit	Material $	Labor $	Equipment $	Total $

Remove pipe insulation. These figures assume demolished materials are piled on site. Disposal costs or salvage values excluded.

Fiberglass pipe insulation, ½"-1", remove	P1@.020	LF	—	.89	—	.89
Fiberglass pipe insulation, 1½"-2", remove	P1@.025	LF	—	1.11	—	1.11
Fiberglass pipe insulation, 2½"-4", remove	P1@.035	LF	—	1.56	—	1.56
Fiberglass pipe insulation, 6"-8", remove	P1@.060	LF	—	2.67	—	2.67
Fiberglass pipe insulation, 10"-12", remove	P1@.090	LF	—	4.00	—	4.00
Pipe insul. with canvas jacket – add	—	%	—	10.0	—	—
Pipe insul. with alum. jacket – add	—	%	—	20.0	—	—
Calcium silicate pipe insul. – add	—	%	—	15.0	—	—

Remove duct insulation. These figures assume demolished materials are piled on site. Disposal costs or salvage values excluded.

Fiberglass blanket insulation, remove	S2@.020	SF	—	.87	—	.87
Fiberglass rigid board insulation, remove	S2@.025	SF	—	1.09	—	1.09
Insul. with canvas jacket – add	—	%	—	8.0	—	—
Insul. with alum jacket –s add	—	%	—	10.0	—	—

Budget Estimates for Plumbing and HVAC Work

General contractors and owners often ask mechanical subcontractors to prepare budget cost estimates for use in planning projects. A budget estimate isn't a bid for the contract and you probably won't be paid for preparing the estimate. That's why you won't want to spend too much time on these estimates. Accuracy within 8 to 10 percent of the final bid price will usually be acceptable. Normally you'll have only preliminary architectural drawings and possibly a guide specification to work from. The tables that follow will help you prepare budget estimates for nearly any plumbing or HVAC job.

The costs in these tables are averages and will vary considerably with design requirements and by geographic area. Make it clear to the owner or architect requesting the figures that these are budget estimates and will be revised when final plans and specifications are available.

Budget Estimates for Plumbing: First, count the number of plumbing fixtures. Multiply the number of fixtures of each type by the price per fixture shown in the table below. These prices include all the usual labor and material for a complete average plumbing installation. Prices are based on branch piping lengths averaging 35 feet between the fixture and the main. Typical main piping is also included in the prices.

Budget plumbing estimate, cast iron DWV, copper supply pipe

Description	Craft@Hrs	Unit	Material $	Labor $	Equipment $	Total $
Bathtubs	P1@15.0	Ea	2,730.00	667.00	—	3,397.00
Showers	P1@9.00	Ea	1,480.00	400.00	—	1,880.00
Lavatories	P1@11.0	Ea	2,550.00	489.00	—	3,039.00
Kitchen sinks	P1@12.0	Ea	2,670.00	533.00	—	3,203.00
Service sinks	P1@10.0	Ea	2,300.00	445.00	—	2,745.00
Bar sinks	P1@9.00	Ea	1,930.00	400.00	—	2,330.00
Floor sinks	P1@7.00	Ea	1,180.00	311.00	—	1,491.00
Wash fountains	P1@25.0	Ea	6,930.00	1,110.00	—	8,040.00
Can washers	P1@18.0	Ea	3,530.00	800.00	—	4,330.00
Floor drains	P1@8.00	Ea	1,090.00	356.00	—	1,446.00
Area drains	P1@6.00	Ea	937.00	267.00	—	1,204.00
Roof drains	P1@18.0	Ea	2,480.00	800.00	—	3,280.00
Overflow drains	P1@18.0	Ea	2,480.00	800.00	—	3,280.00
Deck drains	P1@4.00	Ea	772.00	178.00	—	950.00
Cleanouts	P1@5.00	Ea	834.00	222.00	—	1,056.00
Trap primers	P1@2.00	Ea	423.00	88.90	—	511.90
Sump pumps	P1@6.00	Ea	4,120.00	267.00	—	4,387.00
Urinals	P1@8.00	Ea	2,790.00	356.00	—	3,146.00
Water closets						
Tank-type	P1@11.0	Ea	1,930.00	489.00	—	2,419.00
Flush valve	P1@12.0	Ea	2,180.00	533.00	—	2,713.00
Drinking fountains						
(refrigerated)	P1@12.0	Ea	3,220.00	533.00	—	3,753.00
Water heaters						
Gas, 40 gal.	P1@16.0	Ea	4,760.00	711.00	—	5,471.00
Gas, 80 gal.	P1@17.0	Ea	5,280.00	756.00	—	6,036.00
Gas, 120 gal.	P1@18.0	Ea	6,180.00	800.00	—	6,980.00

Budget plumbing estimate, plastic DWV and plastic supply pipe

Description	Craft@Hrs	Unit	Material $	Labor $	Equipment $	Total $
Bathtubs	P1@12.0	Ea	2,130.00	533.00	—	2,663.00
Showers	P1@7.00	Ea	1,030.00	311.00	—	1,341.00
Lavatories	P1@9.00	Ea	1,970.00	400.00	—	2,370.00
Kitchen sinks	P1@10.0	Ea	2,200.00	445.00	—	2,645.00
Service sinks	P1@8.00	Ea	1,760.00	356.00	—	2,116.00
Bar sinks	P1@7.00	Ea	1,490.00	311.00	—	1,801.00
Floor sinks	P1@6.00	Ea	896.00	267.00	—	1,163.00
Urinals	P1@6.00	Ea	2,240.00	267.00	—	2,507.00
Drinking fountains						
(refrigerated)	P1@10.0	Ea	2,630.00	445.00	—	3,075.00
Wash fountains	P1@22.0	Ea	5,990.00	978.00	—	6,968.00
Can washers	P1@15.0	Ea	2,940.00	667.00	—	3,607.00
Floor drains	P1@7.00	Ea	749.00	311.00	—	1,060.00
Area drains	P1@5.00	Ea	613.00	222.00	—	835.00
Roof drains	P1@16.0	Ea	2,100.00	711.00	—	2,811.00
Overflow drains	P1@16.0	Ea	2,100.00	711.00	—	2,811.00
Deck drains	P1@3.00	Ea	520.00	133.00	—	653.00
Cleanouts	P1@4.00	Ea	636.00	178.00	—	814.00
Trap primers	P1@2.00	Ea	316.00	88.90	—	404.90
Sump pumps	P1@5.00	Ea	3,670.00	222.00	—	3,892.00
Water closets						
Tank-type	P1@8.00	Ea	1,390.00	356.00	—	1,746.00
Flush valve	P1@9.00	Ea	1,670.00	400.00	—	2,070.00
Water heaters						
Gas, 40 gal.	P1@15.0	Ea	4,390.00	667.00	—	5,057.00
Gas, 80 gal.	P1@16.0	Ea	4,810.00	711.00	—	5,521.00
Gas, 120 gal.	P1@17.0	Ea	5,590.00	756.00	—	6,346.00

Budget Estimates for HVAC Work: Budget estimates for heating, ventilating and air conditioning work are figured by the square foot of floor. Multiply the area of conditioned space by the cost per square foot in the table that follows. The result is the estimated cost including a complete heating, ventilating and air conditioning system and 20 percent overhead and profit. The table also shows average heating and cooling requirements for various types of buildings.

Description	Cooling (Btu/SF)	Heating (Btu/SF)	Total Selling Price ($) Per SF	Per Ton
Apartments and condominiums	25	25	16.00	5,900.00
Auditoriums	40	40	41.00	9,430.00
Banks	45	30	42.40	9,410.00
Barber shops	40	40	41.90	9,600.00
Beauty shops	60	30	49.70	7,600.00
Bowling alleys	40	40	32.70	7,500.00
Churches	35	35	35.50	9,250.00
Classrooms	45	40	64.60	13,100.00
Cocktail lounges	70	30	44.80	6,280.00
Computer rooms	145	20	91.40	6,160.00
Department stores	35	30	28.70	7,560.00
Laboratories	60	45	79.40	12,100.00
Libraries	45	35	47.40	9,680.00
Manufacturing plants	40	35	18.50	4,310.00
Medical buildings	35	35	31.30	8,260.00
Motels	30	30	22.00	6,710.00
Museums	35	40	35.00	9,180.00
Nursing homes	45	35	40.30	8,300.00
Office buildings, low-rise	35	35	31.90	8,390.00
Office buildings, high-rise	40	30	43.50	9,940.00
Residences	20	30	10.40	6,080.00
Retail shops	45	30	32.70	6,660.00
Supermarkets	30	30	17.10	5,780.00
Theaters	40	40	31.90	7,380.00

Note that these are budget estimates and can vary widely with conditions which may be unique to the job site. Many of these conditions are identified below. Avoid submitting even a budget estimate without either a site walk-through, a conference with the consulting engineer or architect or a careful review of the plans.

Be alert to any of the following conditions. Any of these can have a major impact on cost:

1. Do HVAC unit dimensions exceed the size of available openings?
2. If a crane is required, is there ample clearance for crane operations?
3. Will any doors, permanent ladders, scaffolding or load-bearing supports obstruct installation?
4. Do the plans show all vents, drains and cleanouts required by the code? Many items like these are not specified by the engineer but will be required by the inspector.
5. For retrofit work, consider carefully problems associated with demolition, including removal of the existing HVAC system, structural modifications and any encased asbestos that may be present.
6. In retrofit jobs, the type and size of HVAC equipment is limited by the window area and installed insulation.
7. How will available power, gas and water lines be extended to the new HVAC equipment?

The following pages contain business forms and letters most plumbing and HVAC contractors can use.

Change Orders
Change Estimate Takeoff
Change Estimate Worksheet
Change Estimate Summary
Change Order Log

Subcontract Forms
Standard Form Subcontract
Subcontract Change Order

Purchase Order

Construction Schedule

Letter of Intent

Submittal Data Cover Sheet

Submittal Index

Billing Breakdown Worksheet

An explanation is provided on how the forms function in the contractor's office, how to fill them out, and a sample provided. Where appropriate, a blank copy of the form is provided for your use.

Permission is hereby given by the publisher for the duplication of any of the blank forms on a copy machine or at a printer, provided they are for your own use and not to be sold at a cost exceeding the actual printing and paper cost.

Space has been provided at the top of the forms so you can add your own letterhead or logo.

Change Estimates

Very few projects are completed without some changes to the original plans. Most changes in the scope of work are a result of either errors by the designer or requests by the owner.

Here's the usual sequence in handling changes.

1) The project architect usually initiates a change by sending a Request for Quotation (R.F.Q.) to the general contractor with a detailed description of the work to be modified.

2) The general contractor prepares a Change Estimate (C.E.) if the work is to be done by the general.

3) The subcontractor prepares a change estimate which shows the work to be added to or deleted from the contract and the cost of changes, including markup. The general contractor reviews the change estimate and forwards a summary to the architect for decision.

4) If the decision is to go ahead with the change, the architect issues a Change Order (C.O.) to the general contractor. This change order is authorization to proceed with the change at the price quoted.

Your costs for making changes will normally be higher than your costs for similar items on the original estimate. Here's why:

Equipment and materials for changes are almost always purchased in smaller quantities. Volume discounts are seldom available.

Special handling and delivery of small orders will further increase equipment and material costs.

Scheduled work will be disrupted.

Overhead costs will be higher because special purchase orders or subcontracts will be required.

Vendor restocking charges for deleted items will range from 20 to 50 percent of the original cost.

Large change orders may require additional manpower, which usually results in decreased efficiency.

The cumulative effect of change orders will be to increase overhead costs if the project completion date is extended.

To compensate for these higher costs, most estimators use a higher markup on change estimates:

Add an extra 10 percent markup to equipment and material costs to cover increased overhead and handling costs, prior to adding final Overhead and Profit markup.

Increase competitive labor units (manhours required to perform a specific task) by 15 percent to compensate for inefficiencies such as disruptions in the work schedule, extra handling of equipment and material, and additional manpower.

Mark up subcontractor prices by 10 percent.

Figure overhead and profit markup at 15/10 or 10/10.

By following these suggestions, the mathematical gross profit will appear to be 40 to 45 percent, but the true gross will be closer to 25 percent because your costs for change order work are higher, as listed above.

Change Estimate Example

Let's assume that the general contractor on your job has sent a R.F.Q. to your attention. The extra work is a run of hot water piping that will connect to equipment being furnished and set by others. The first step is to list all material and labor required to do the work. Notice the items entered on the Change Estimate Take-off form on page 440.

When the take-off is complete, transfer the manhour total to the Change Estimate Worksheet on page 441. To complete this form:

1) Multiply manhours by the cost per hour.
2) Add the cost of supervision.
3) Total direct labor cost.
4) Add the cost of preparing the change order estimate (C.E.).
5) Add equipment and tool costs.
6) Add subcontract costs.

Then bring totals forward to the Change Estimate Summary on page 442:

1) Enter the material cost total from the Change Estimate Take-off form on page 444.
2) Enter direct and indirect labor cost totals from the Change Estimate Worksheet.
3) Enter equipment, tool and subcontract cost totals from the Change Estimate Worksheet.

Notice the additional bond charge on line M of the Change Estimate Summary. This is the extra premium charged by the bonding company to cover the increase in contract value. The next line, Service Reserve, is the allowance for service work (if any is required) during the warranty period. A sample Change Order Log, used to record the status of changes, is also included on page 443.

ACME

Mechanical Contractors

7600 Oak Avenue - Smallville, U.S.A. 12345-6789
Voice 1-234-5678 / FAX 1-234-5680

Change Estimate Take-off

Project CBC Bank Building **C. E. Number** One

Take-off by DCO **Job No** 7777

Date 2-2-2026

Equipment and Material			Material $		Labor Manhours	
Item	Description	Quantity	Unit	Total	Unit	Total
1	1" Type L hard copper pipe	40 LF	3.87	154.80	.038	1.52
2	1" 90 degree copper ell	6 Ea	4.21	25.26	.23	1.38
3	1" copper coupling	4 Ea	2.54	10.16	.24	.96
4	1" bronze ball valve	2 Ea	17.30	34.60	.36	.72
5	2" x 1" reducing copper tee	2 Ea	20.60	41.20	.40	.80
6	1" copper union	2 Ea	19.10	38.20	.26	.52
7	1" pipe hanger	8 Ea	4.33	34.64	.25	2.00
8	Miscellaneous material	1 Ea	LS	50.00	1.78	1.78
		Totals		388.86		9.68

ACME

Mechanical Contractors

7600 Oak Avenue - Smallville, U.S.A. 12345-6789
Voice 1-234-5678 / FAX 1-234-5680

Change Estimate Worksheet

Estimated by DCO **C. E. Number** One

Date 2-3-2026 **Job No** 7777

Direct Labor Expense (including benefits)

Plumber or pipefitter journeyman	9.68	manhours at $	41.85	per hour =	$	405.11
Plumber or pipefitter supervisor	1.00	manhours at $	45.00	per hour =	$	45.00
Sheet metal journeyman		manhours at $		per hour =	$	
Sheet metal supervisor		manhours at $		per hour =	$	
Laborer journeyman		manhours at $		per hour =	$	
		manhours at $		per hour =	$	
				Total Direct Labor Cost	$	450.11

Indirect Costs

Travel		$	
Subsistence		$	
Labor	Cost Estimate Preparation (1 Hour)	$	40.00
	Total Indirect Cost	$	40.00

Equipment and Tool Costs

Ramset rental	$	35.00
Propane	$	4.50
Total Equipment and Tool Cost	$	39.50

Subcontract Costs

Quikrap Insulation Co.	$	100.00
	$	
Total Subcontract Cost	$	100.00

ACME

Mechanical Contractors

7600 Oak Avenue - Smallville, U.S.A. 12345-6789
Voice 1-234-5678 / FAX 1-234-5680

Change Estimate Summary

Quotation to ABC General Contractors, Inc. **Date** 2-3-2026

Address 380 First Street **Change Order No.** One

Address ______ **Job Number** 7777

City / State / Zip Smallville, CA 63876

Attention: Mr. J.H. Smith

Job Name CBC Bank Building

Reference: Your R.F.O. No. 7 Labor and Material to furnish and install additional hot water piping.

A.	Materials and equipment:	$ 388.86		
B.	Sales tax: 7% .	$ 27.22		
C.	Direct labor:	$ 450.11		
D.	Indirect costs:	$ 40.00		
E.	Equipment and tools:	$ 39.50		
F.			**Subtotal**	$ 945.69
G.	Overhead at : 15% of line F:	$ 141.85		
H.	Subcontracts:	$ 100.00		
I.	Overhead at : 10% of line H:	$ 10.00		
J.			**Subtotal**	$ 1,197.54
K.	Profit at 10% of line J:	$ 119.75		
L.			**Subtotal**	$ 1,317.29
M.	Bond premium at 1% of line L:	$ 13.17		
N.	Service reserve at 0.5% of line L:	$ 6.59		
O.	**Total cost estimate, lines L thru N:**	☐ Add ☐ Deduct		**$** 1,337.05

P. Exclusions from this estimate: Per Base

Q. This quotation is valid for 30 days ☐

R. We require 1 day(s) extension of the contract time. ☐

S. We are proceeding with this work per your authorization. ☐

T. Please forward your confirming change order. ☐

Signed by: ______________________

Project Manager

Change Order Log

Job Name ______________________________ **Job Number** ____________

Change Number	Customer Number	Date	Amount	Date Approved	Amount Approved	Remarks

Use this space to insert your company logo and letterhead.

Change Estimate Take-off

Project ______________________________ C. E. Number ______________

Take-off by ___________________________ Job No ______________

Date ________________

Equipment and Material			Material $		Labor Manhours	
Item	Description	Quantity	Unit	Total	Unit	Total
		Totals				

Use this space to insert your company logo and letterhead.

Change Estimate Worksheet

Estimated by ______________________ **C. E. Number** ____________

Date ____________ **Job No** ____________

Direct Labor Expense (including benefits)

Plumber or pipefitter journeyman	______	manhours at $	______	per hour =	$ ______
Plumber or pipefitter supervisor	______	manhours at $	______	per hour =	$ ______
Sheet metal journeyman	______	manhours at $	______	per hour =	$ ______
Sheet metal supervisor	______	manhours at $	______	per hour =	$ ______
Laborer journeyman	______	manhours at $	______	per hour =	$ ______
______________	______	manhours at $	______	per hour =	$ ______
				Total Direct Labor Cost	$ ______

Indirect Costs

Travel	______________________	$ ______
Subsistence	______________________	$ ______
Labor	______________________	$ ______
	Total Indirect Cost	$ ______

Equipment and Tool Costs

______________________	$ ______
______________________	$ ______
Total Equipment and Tool Cost	$ ______

Subcontract Costs

______________________	$ ______
______________________	$ ______
Total Subcontract Cost	$ ______

Use this space to insert your company logo and letterhead.

Change Estimate Summary

Quotation to ______________________ **Date** ____________
Address ______________________ **Change Order No.** ________
Address ______________________ **Job Number** ________
City / State / Zip ______________________
Attention: ______________________
Job Name ______________________
Reference: ____

____

A. Materials and equipment: $ ________
B. Sales tax: ______ . $ ________
C. Direct labor: $ ________
D. Indirect costs: $ ________
E. Equipment and tools: $ ________
F. **Subtotal $** ________
G. Overhead at ______ of line F: $ ________
H. Subcontracts: $ ________
I. Overhead at ______ of line H: $ ________
J. **Subtotal $** ________
K. Profit at ______ of line J: $ ________
L. **Subtotal $** ________
M. Bond premium at 1% of line L: $ ________
N. Service reserve at 0.5% of line L: $ ________
O. **Total cost estimate, lines L thru N:** ☐ Add ☐ Deduct **$** ________
P. Exclusions from this estimate: ______________________
Q. This quotation is valid for ______ days
R. We require ______ day(s) extension of the contract time.
S. We are proceeding with this work per your authorization.
T. Please forward your confirming change order.

Signed by: ______________________________
Project Manager

Another term for the general contractor is *prime contractor*. Subcontractors working directly for the prime contractor are called *second-tier subcontractors*. Contractors working for these second-tier subs are usually known as *third-tier subcontractors*. On a large job, many second-tier subs may hire third-tier subs to perform specialized work such as insulation, fireproofing, test and balance, water treatment, instrumentation and trenching.

If a portion of your work will be subcontracted to others, I recommend drawing up a subcontract agreement for that work. Any subcontractors who furnish both material and labor for a job should have a written contract.

Most larger plumbing, heating and A/C subcontractors use a two-part *Standard Form Subcontract* published by the Associated General Contractors of America. This two-part form is used to write the base subcontract. The A.G.C. also publishes a *Subcontract Change Order* form which is used when adding or deleting items after the base subcontract is signed. These forms are sold at modest cost at most A.G.C. offices.

The first section of page one of the Standard Form Subcontract identifies the two contractors entering into the agreement. The party issuing the subcontract is the *contractor*. Recipient of the agreement is the *subcontractor*.

Under *Recitals,* the contractor must affirm the right to issue a subcontract by declaring that a legal contract exists with the *owner* (general contractor).

Section 1 - Entire Contract describes the contract documents: the project plans, specifications and addenda. This list of documents should always be the same as the documents identified in the second-tier subcontractor's contract with the general contractor.

Section 2 - Scope describes the specific scope of work to be performed.

Section 3 - Contract Price. Complete this section by inserting the agreed-upon price for doing the work described.

Section 4 - Payment Schedule on page two should show the same payment terms as your contract with the general contractor.

Section 7 - Special Provisions is used to list any special agreements or exclusions.

When the contract is typed, do not sign it. Instead, forward it to the third-tier subcontractor with instructions to sign it and return it intact. When the signed contract is received, sign it and send a copy to the third-tier subcontractor.

For contracts that comply with your state's regulations, please visit www.constructioncontractwriter.com

Standard Form Subcontract

Subcontract No. 7777-1

THIS AGREEMENT, made and entered into at Smallville, CA. this 3rd day of May, 2026, by and between Acme Mechanical Contractors hereinafter called **CONTRACTOR,** with principal office at 7600 Oak Avenue, Smallville, California, and Quikrap Insulation Co., 320 Maple Blvd., Smallville, CA hereinafter called a **SUBCONTRACTOR**.

RECITALS

On or about the 2nd day of January, 20 26, CONTRACTOR entered into a prime contract with

ABC General Contractors, Inc. hereinafter called OWNER, whose address is 380 First Street, Smallville, California to perform the following construction work: The installation of all mechanical and plumbing systems for the proposed Mile-Hi office building to be located at 2600 Second, North, Smallville, California.

Said work is to be performed in accordance with the prime contract and the plans and specifications. Said plans and specifications have been prepared by or on behalf of Quik-Draw and Associates, Smallville, California, **ARCHITECT**

SECTION 1 - ENTIRE CONTRACT

SUBCONTRACTOR certifies and agrees that he is fully familiar with all of the terms, conditions and obligations of the Contract Documents, as hereinafter defined, the location of the job site, and the conditions under which the work is to be performed, and that he enters into this Agreement based upon his investigation of all of such matters and is in no way relying upon any opinions or representations of CONTRACTOR. It is agreed that this Agreement represents the entire agreement. It is further agreed that the Contract Documents are incorporated in this Agreement by this reference, with the same force and effect as if the same were set forth at length herein, and that SUBCONTRACTOR and his subcontractors will be and are bound by any and all of said Contract Documents insofar as they relate in any part or in any way, directly or indirectly to the work covered by this Agreement. SUBCONTRACTOR agrees to be bound to CONTRACTOR in the same manner and to the same extent as CONTRACTOR is bound to OWNER under the Contract Documents, to the extent of the work provided for in this Agreement, and that where, in the Contract Documents, reference is made to CONTRACTOR and the work or specification therein pertains to SUBCONTRACTOR'S trade, craft, or type of work, then such work or specification shall be interpreted to apply to SUBCONTRACTOR instead of CONTRACTOR. The phrase "Contract Documents" is defined to mean and include:

Drawing Nos. A-1 thru A-17, C-1 thru C-4, S-1 thru S-7, E-1 thru E-6 and M-1 thru M-7, all dated 10-4-2025, Specifications dated 10-2-2025 and Addendum No. 1 dated 10-18-2025.

SECTION 2 - SCOPE

SUBCONTRACTOR agrees to furnish all labor, services, materials, installation, cartage, hoisting, supplies, insurance, equipment, scaffolding, tools and other facilities of every kind and description required for the prompt and efficient execution of the work described herein and to perform the work necessary or incidental to complete thermal insulation for all plumbing, HHW, CHW and condensate piping systems, including pumps P-1, P-2 and P-3 for the project in strict accordance with the Contract Documents and as more particularly, though not exclusively, specified in:

Division 15, Section 400 entitled "Thermal Insulation systems" and General Conditions, as applicable.

SECTION 3 - CONTRACT PRICE

CONTRACTOR agrees to pay SUBCONTRACTOR for the strict performance of his work, the sum of:

Forty-two thousand, three hundred dollars ($42,300.00), subject to additions and deductions for changes in the work as may be agreed upon, and to make payment in accordance with the Payment Schedule, Section 4.

SECTION 4 - PAYMENT SCHEDULE

CONTRACTOR agrees to pay SUBCONTRACTOR in monthly payments of 90 % of labor and materials which have been placed in position and for which payment has been made by OWNER to CONTRACTOR. The remaining 10 % shall be retained by CONTRACTOR until he receives final payment from OWNER, but not less than thirty-five days after the entire work required by the prime contract has been fully completed in conformity with the Contract Documents and has been delivered and accepted by OWNER, ARCHITECT, and CONTRACTOR. Subject to the provisions of the next sentence, the retained percentage shall be paid SUBCONTRACTOR promptly after CONTRACTOR receives his final payment from OWNER. SUBCONTRACTOR agrees to furnish, if and when required by CONTRACTOR, payroll affidavits, receipts, vouchers, release of claims for labor, material and subcontractors performing work or furnishing materials under this Agreement, all in form satisfactory to CONTRACTOR, and it is agreed that no payment hereunder shall be made, except at CONTRACTOR'S option, until and unless such payroll affidavits, receipts, vouchers or release; or any or all of them, have been furnished. And payment made hereunder prior to completion and acceptance of the work, as referred to above, shall not be construed as evidence of acceptance of any part of SUBCONTRACTOR'S work.

SECTION 5 - GENERAL SUBCONTRACT PROVISIONS

General Subcontract Provisions on back of Pages 1 and 2 are an integral part of this Agreement.

SECTION 6 - GENERAL PROVISIONS

1. *SUBCONTRACTOR* agrees to begin work as soon as instructed by the CONTRACTOR, and shall carry on said work promptly, efficiently and at a speed that will not cause delay in the progress of the CONTRACTOR'S work or work of other subcontractors. If, in the opinion of the CONTRACTOR, the SUBCONTRACTOR falls behind in the progress of the work, the CONTRACTOR may direct the SUBCONTRACTOR to take such steps as the CONTRACTOR deems necessary to improve the rate of progress, including, without limitation, requiring the SUBCONTRACTOR to increase the number of shifts, personnel, overtime operations, days of work, equipment, amount of plant, or other remedies and to submit to CONTRACTOR for CONTRACTOR'S approval an outline schedule demonstrating the manner in which the required rate of progress will be regained, without additional cost to the CONTRACTOR. CONTRACTOR may require SUBCONTRACTOR to prosecute, in preference to other parts of the work, such part or parts of the work as CONTRACTOR may specify.

The SUBCONTRACTOR shall complete the work as required by the progress schedule prepared by the CONTRACTOR, which may be amended from time to time. The progress schedule may be reviewed in the office of the CONTRACTOR and sequence of construction will be as directed by the CONTRACTOR.

The SUBCONTRACTOR agrees to have an acceptable representative (an officer of SUBCONTRACTOR if requested by the CONTRACTOR) present at all job meetings and to submit weekly progress reports in writing if requested by the CONTRACTOR. Any job progress schedules are hereby made a part of and incorporated herein by reference.

2. ***Reserved Gate Usage***

SUBCONTRACTOR shall notify in writing, and assign its employees, material men and suppliers, to such gates or entrances as may be established for their use by CONTRACTOR and in accordance with such conditions and at such times as may be imposed by CONTRACTOR. Strict compliance with CONTRACTOR'S gate usage procedures shall be required by the SUBCONTRACTOR who shall be responsible for such gate usage by its employees, material men, suppliers, subcontractors, and their material men and suppliers.

3. ***Staggered Days and Hours of Work and for Deliveries***

SUBCONTRACTOR shall schedule the work and the presence of its employees at the jobsite and any deliveries of supplies or materials by its material men and suppliers to the jobsite on such days, and at such times and during such hours, as may be directed by CONTRACTOR. SUBCONTRACTOR shall assume responsibility for such schedule compliance not only for its employees but for all its material men, suppliers and subcontractors, and their material men and suppliers.

SECTION 7 - SPECIAL PROVISIONS

1. Pipe hanger inserts will be furnished and installed by Contractor.
2. All trash pickup by Subcontractor; haul-away will be by General Contractor.

Contractors are required by law to be licensed and regulated by the Contractors' State License Board. Any questions concerning a contractor may be referred to the registrar of the board whose address is:

Contractors' State License Board -- P. O. Box 26000, Sacramento, California 95826

IN WITNESS WHEREOF: The parties hereto have executed this Agreement for themselves, their heirs, executors, successors, administrators, and assignees on the day and year first above written.

SUBCONTRACTOR	**CONTRACTOR**
Quikrap Insulation Company ______________	Acme Mechanical Contractors ______________
By ______________________	By ______________________
Name Title	Name Title
▭ Corporation ▭ Partnership ▭ Proprietorship	
(Seal)	
Contractor's State License No. ______________	Contractor's State License No. ______________

Revised 5-09 Published by AGC San Diego
Page 2 of 2

For contracts that comply with your state's regulations, please visit www.constructioncontractwriter.com

Subcontract Change Order

Number One

Date	6-11-2026	**Subcontract Number**	7777-1
To	Quickrap Insulation Co.	**Our Job Number**	
		Our Proposal Number	
		Architect's C.O. Number	
		Effective Date of Charge	6-17-2026

Subject to all the provisions of this Change Order, you are hereby directed to make the following change(s)

Delete HHW and CHW piping insulation located above Room No. 322 per ABC General Contractor's Change Order No. 18, dated 6-5-26.

The following change(s) will alter the price provided in your subcontract by the deduction ***of $*** 465.00

*Surety Consent		
This Change is Approved Date ______	Adjusted Subcontract Price Through C.O. Number 0	$ 42,300.00
Name: ______	Amount this C.O. Number:	$ (465.00)
By: ______ Title:	Current Adjusted Subcontract Price:	$ 41,835.00

When this Change Order is signed by both parties (and by Subcontractor's surety is subcontract is bonded), it constitutes their agreement:

(A) That the subcontract price is adjusted as shown above and that no further adjustment in that price by reason of the change(s) provided herein shall be made; and (B) That all the terms and conditions of the subcontract, except as modified by this and any previous changes, shall remain in full force and effect and apply to the work as so changed.

Accepted and Agreed:	**Date:** ______
By ______ Authorized Signature – Subcontractor	**By** ______ Authorized Signature – Contractor
Title	**Title**

* Subcontractor: Sign pink copy and return immediately. If subcontract is bonded, obtain consent of surety endorsed thereon. No payments on account of this Change Order will be made until you have complied with the foregoing.

Published by San Diego Chapter A.G.C.

A purchase order is a legal document used to order equipment and materials from a vendor, with the implied promise to pay for these goods within a specified time after receipt of the items. Purchase orders are sometimes used to write subcontracts whose value does not exceed one or two thousand dollars. Refer to the sample purchase order on page 452.

While most of the purchase order is self-explanatory, certain items need to be emphasized:

"Description" should include all pertinent data such as model numbers, arrangements, performance requirements, applicable standards, electrical characteristics and so on.

"Tag" is used to identify a particular piece of equipment to facilitate unloading it at the proper location at the jobsite. Tag numbers for all major equipment can usually be found on the contract drawings.

"48 hour delivery notice": This can be very important if the material or equipment being delivered is too heavy to be unloaded by hand. Time will be required to arrange for a crane or forklift to unload the truck.

"Sales tax" is the current state sales tax which must be included unless the goods are to be delivered and installed at a project located out-of-state.

"Terms": The usual payment terms are "Net 30" which means the entire amount of the purchase order must be paid within thirty days after receipt of all goods, undamaged. If items are received damaged, payment should be withheld until satisfactory repair or replacement is made.

ACME

Mechanical Contractors

7600 Oak Avenue - Smallville, U.S.A. 12345-6789
Voice 1-234-5678 / FAX 1-234-5680

Purchase Order

P.O. Number: 7777-36 **Date:** 1-6-2026

Job: CBC Bank Building

Job No: 7777

Seller:	**Ship to:**
Smallville Industrial Sales Co. 3657 Third Avenue Northwest Smallville, CA 63876	Acme Mechanical Contractors c/o CBC Bank Building 465 Commercial Street Smallville, CA 63876

Attn: Mr. L.H. Seller By: RJS

Delivery date: 2-28-26 ☐ **24 Hr. Delivery Notice Required** **Payment terms:** Net 30

Quantity	Description	Unit	Total
Two	Big-Blo Model No. 1JN 86-13 centrifugal	$443.00	$886.00
	fans with 480/3/60 one-half H.P., O.D.P.		
	motor. 3250 CFM @ 0.50 T.S.P.		
	Tag one EF-13 and one EF-23		

Number of copies of operation and maintenance manuals required 2.

All shipments are to be F.O.B. jobsite or shop unless otherwise noted.

Total before tax	$886.00
Sales tax	$62.02
Freight	Included
Total Order	$948.02

Buyer: RJS

Acceptance by seller ________________ Date ________

The general contractor's construction schedule may range from a simple bar chart to a complicated computer-generated CPM (Critical Path Method) diagram. More complex jobs require more complete and detailed schedules. On larger jobs, lower-tier subcontractors may be required to prepare a bar chart showing their construction schedule.

If you have to prepare a construction schedule, make conservative estimates of the time required to complete each part of the job. A slower schedule makes it possible to use smaller crews which are usually more efficient.

You can't schedule your own work until the general contractor has supplied a schedule for the balance of the project. When you have that schedule, prepare a list of tasks your crews will perform. Show this list to the general contractor to be certain there are no conflicts with the master schedule.

When the list of tasks to be performed has been approved, begin recording the manhours required for each task. Take these manhours from your estimate. When the duration for each task has been determined, figure the crew sizes required.

For example, suppose the estimate shows that it will take 1,040 manhours to install the underground plumbing for a project. The work must be completed in two months. Assuming one worker averages 173 hours of work in one month, how many workers will be required to meet the schedule? Here's the solution:

$$\frac{\text{1,040 manhours}}{\text{173 MH / Mo. x 2 months}} = \text{3 workers}$$

The first of the two bar charts on the following pages shows the schedule a general contractor might expect to receive from a plumbing and HVAC subcontractor. The chart on page 454 is intended for use by the subcontractor. It includes information a mechanical subcontractor needs to determine proper crew sizes for each task. The *Cumulative Manhours* line shows the budgeted manhours for each month of the project. If labor costs are to be kept within budget, actual hours should not exceed estimated hours. Monitor labor costs each month by comparing the line *Actual MH Used* (actual manhours used) with the line *Cumulative MH* (estimated cumulative manhours).

Construction Schedule

Company: Acme Mechanical Contractors **Project:** Mile-Hi Office Bldg **Job Number:** 7777 **Date:** 6/8/2026

	2026						2027				
Activity	**Jul**	**Aug**	**Sep**	**Oct**	**Nov**	**Dec**	**Jan**	**Feb**	**Mar**	**Apr**	**May**	**Crew Sizes**
Submittals												
Purchasing												
U/G Plumbing												
A/G Plumbing												
Fin. Plumbing												
HVAC Piping												
HVAC Ducting												
Equipment												
Insulation												
Temp. Controls												
Start-up												

Construction Schedule

Company: Acme Mechanical Contractors **Project:** Mile-Hi Office Bldg **Job Number:** 7777 **Date:** 6/8/2026

	2026						2027				
Activity	**Jul**	**Aug**	**Sep**	**Oct**	**Nov**	**Dec**	**Jan**	**Feb**	**Mar**	**Apr**	**May**	**Crew Sizes**
Submittals												N/A
Purchasing												N/A
U/G Plumbing												1040 MH ÷(173 x 2) = 3
A/G Plumbing												2075 MH ÷ (173 x 4) = 3
Fin. Plumbing												130 MH ÷ (173 x .5) = 1.5
HVAC Piping												2760 MH ÷ (173 x 4) = 4
HVAC Ducting												2790 MH ÷ (173 x 4) = 4
Equipment												690 MH ÷ (173 x 2) = 2
Insulation												Subcontract
Temp. Controls												Subcontract
Start-up												85 MH ÷ (173 x .5) = 1
Crew size/month			3	6	11	11	11	11.5	2	1		
Cumulative MH*			520	1159	3466	5373	7280	9143	9448	9573		MH Estimated = 9573
Actual MH used												
*Per estimate												
*Rounded												

Letter of Intent

On most large projects the contractor is required to submit drawings and technical data to the owner's representative for approval before work actually begins. Most of these submittals will be prepared by your suppliers and third-tier subs. Vendors and subs may be reluctant to prepare these documents without some assurance that they've been selected to do the work. That's the purpose of a letter of intent. It notifies a proposed vendor or subcontractor that you plan to contract with them when the owner approves the information submitted.

A letter of intent isn't a contract. It's an expressed intention to make a contract, which isn't a contract at all. But a vendor who gets your letter of intent and supplies all the information requested should get the contract. If you place the order with a different vendor, obviously, your letter of intent was worthless. That first vendor may be very reluctant to cooperate in the future.

A sample letter of intent follows this page.

ACME
Mechanical Contractors

7600 Oak Avenue - Smallville, U.S.A. 12345-6789
Voice 1-234-5678 / FAX 1-234-5680

December 20, 2025

Quikrap Insulation Company
320 Maple Boulevard
Smallville, U.S.A. 12345-6789

Attn: Mr. R. S. Quikrap

Dear Mr. Quikrap:

Acme Mechanical Contractors has been issued a subcontract to furnish and install the mechanical and plumbing systems for the proposed Mile-Hi Office Building to be located at 2600 Second Street in Smallville.

It is my intention to award your firm a subcontract to provide the thermal insulation for these systems in accordance with your proposal dated 12-13-2025 in the amount of $42,300.00.

Please prepare ten copies of your submittal data and forward them to me no later than January 15, 2026. Upon approval of your submittal, I will mail you the subcontract for your review and signature.

Thank you,

Bobby Thompson
Project Manager

CC: Project file

Submittal Data

The construction specifications may require that you submit manufacturer's technical data on certain equipment you plan to install. You may have to submit six to ten copies of this technical data for approval before buying the equipment or contracting with third-tier subcontractors.

First, find out how many submittal copies will be required. Request that number of copies plus at least two additional sets for your use. A subcontractor who provides only services (no materials) may have to submit a detailed written explanation of the work to be done.

The second step is to label each submittal with the paragraph number of the specification where that equipment, material or service is described. Type or write this number in the upper right corner of each submittal sheet. From this set of numbered submittals, prepare a submittal index, a list of submittals identified by title and paragraph number. This index should be arranged in specification paragraph order. A blank submittal index form follows this section.

I prefer to use a three-ring binder to hold submittals. That makes it easier to add and delete sections as needed. Be sure to include a cover sheet, usually on company letterhead, which identifies the project, job number and scope of work being submitted. A sample submittal cover sheet follows this section.

Purchase orders and subcontracts should not be written until an approved copy of the submittal brochure has been received.

ACME

Mechanical Contractors

7600 Oak Avenue - Smallville, U.S.A. 12345-6789
Voice 1-234-5678 / FAX 1-234-5680

Division 15

Heating, Ventilating, Air Conditioning and Plumbing Systems

Submittal Data for

Mile-Hi Office Building
2600 Second Street
Smallville, U.S.A.

Submittal Index

Specification Section	Specification Paragraph	Item Description	Proposed Vendor or Sub	As Specified	See Submittal Data Page

The format to be used for monthly progress billings is seldom specified in the contract documents. In any case, your bill has to include enough detail to satisfy the owner and the lender.

Before preparing your bill, ask the general contractor how much detail is required. Some general contractors will want detailed cost breakdowns showing costs for each system on each building floor. Others will require much less detail.

Find out if billings can include the cost of equipment and materials delivered and stored either on site or off site but not yet installed. If billings can include materials stored off site, do these materials have to be stored in a bonded warehouse? Do vendor's invoices have to accompany progress billings?

When you understand what's required for monthly progress billings, prepare a billing breakdown worksheet. A sample worksheet follows this section.

Column one, Activity Lists each work item in the order work will be performed. The first item listed will usually be mobilization. This includes the cost of the job site trailer, electrical and telephone hookups, office furniture and supplies and initial labor costs. Billings for mobilization costs are sometimes denied by the lender. Many contractors minimize these costs to increase the chance of approval. It's better to receive partial payment for these costs than none at all.

Column two, Cost ($) Lists the actual contractor's costs for each activity, including labor, material, equipment and sales taxes. Markup for overhead and profit are not included.

Column three, Factor This column shows the markup assigned to each activity. Activities scheduled to be completed first are usually assigned the highest markups. This is called front loading and is common in the construction industry. Most bills won't be paid for 30 to 60 days. When the bill is paid, the amount received probably won't include the percentage allowed for retention. Generally, retention is 10 percent and isn't released until the project is complete and accepted by the owner.

Front loading helps contractors carry the financial burden of work in progress. Few subcontractors have enough cash to pay all their bills when due and still wait months to collect for work that's been completed. Employees have to be paid in full and on time. Assigning a higher markup to work completed first accelerates the payment schedule and helps spread payments more evenly over the entire project. Front loading is routine with subcontractors and unpopular with owners and lenders. In practice, subcontractors have no choice but to place a higher price or higher markup on work completed first.

Note that figures in the Factor column are less than 1.0 for work done near the end of the project. That's because front loading doesn't change the contract price. It changes only when that money is received.

Column four, Sell ($) Values for this column are the product of the cost and factor columns. Use the figures in this column when submitting monthly progress invoices.

ACME

Mechanical Contractors

7600 Oak Avenue - Smallville, U.S.A. 12345-6789
Voice 1-234-5678 / FAX 1-234-5680

Billing Breakdown Worksheet

(CBC Bank Building - Job No. 7777)

Activity	Cost ($)	Factor	Sell ($)
Mobilization	2,000	1.50	3,000
Underground chilled and heated water piping	120,000	1.35	162,000
Underground plumbing & piping	87,000	1.35	117,450
Above ground chilled and heating water piping	115,000	1.20	138,000
Above ground plumbing & piping	87,000	1.20	104,400
HVAC ducting	96,000	1.20	115,200
Equipment	312,000	1.10	343,200
Duct and pipe insulation	119,000	1.00	119,000
Temperature controls	53,000	.90	47,700
Plumbing fixtures	43,000	.65	27,950
Water treatment	1,800	.55	990
Test and balance	13,000	.55	7,150
Validation	500	.50	250
	1,049,300		1,186,290

$$\text{Estimated gross profit} = \frac{\$1{,}186{,}290 - \$1{,}049{,}300}{\$1{,}049{,}300} = 13\%$$

Index

C

G

H

I

J

K

L

M

N

O

P

Q

R

S

T

U

V

W - X - Y - Z

Practical References for Builders

National Estimator Cloud

Generate professional construction estimates for all residential and commercial construction from your internet browser. Includes 10 Craftsman construction cost databases, over 40,000 labor and material costs for construction, in an easy-to-use format. Cost estimates are well-organized and thoroughly indexed to speed and simplify writing estimates for nearly any residential or light commercial construction project - new construction, improvement or repair. Convert the bid to an invoice - in either QuickBooks Desktop or QuickBooks Online. Access your estimates from anywhere and on any device with a Web browser. Monthly and one-time billing options available. Visit https://craftsman-book.com/national-estimator-cloud for more details.

Construction Contract Writer

Relying on a "one-size-fits-all" boilerplate construction contract to fit your jobs can be dangerous — almost as dangerous as a handshake agreement. *Construction Contract Writer* lets you draft a contract in minutes that precisely fits your needs and the particular job, and meets both state and federal requirements. You just answer a series of questions — like an interview — to construct a legal contract for each project you take on. Anticipate where disputes could arise and settle them in the contract before they happen. Include the warranty protection you intend, the payment schedule, and create subcontracts from the prime contract by just clicking a box. Includes a feedback button to an attorney on the Craftsman staff to help should you get stumped — *No extra charge.*
$199.95. *Download the Construction Contract Writer* at https://www.constructioncontractwriter.com

Plumber's Handbook Revised, 6th Edition

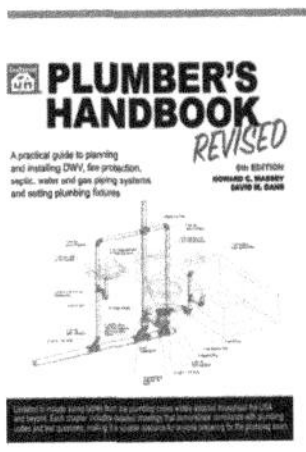

This new edition explains simply and clearly, in non-technical, everyday language, how to install all components of a plumbing system to comply not only with recent changes in the *International Plumbing Code* and the *Uniform Plumbing Code*, but with the requirements of the Americans with Disabilities Act. Originally written for working plumbers to assure safe, reliable, code-compliant plumbing installations that pass inspection the first time, Plumber's Handbook, because of its readability, accuracy and clear, simple diagrams, has become the textbook of choice for numerous schools preparing plumbing students for the plumber's exams. Now, with a set of questions for each chapter, full explanations for the answers, and with a 200-question sample exam in the back, this handbook is one of the best tools available for preparing for almost any plumbing journeyman, master or state-required plumbing contracting exam.
384 pages, 8½ x 11, $67.00
***eBook (PDF) also available;* $33.50** at https://craftsman-book.com

Paper Contracting: The How-To of Construction Management Contracting

Risk, and the headaches that go with it, have always been a major part of any construction project — risk of loss, negative cash flow, construction claims, regulations, excessive changes, disputes, slow pay — sometimes you'll make money, and often you won't. But many contractors today are avoiding almost all of that risk by working under a construction management contract, where they are simply a paid consultant to the owner, running the job, but leaving him the risk. This manual is the how-to of construction management contracting. You'll learn how the process works, how to get started as a CM contractor, what the job entails, how to deal with the issues that come up, when to step back, and how to get the job completed on time and on budget. Includes a link to free downloads of CM contracts legal in each state.
272 pages, 8½ x 11, $55.50
***eBook (PDF) also available;* $27.75** at https://craftsman-book.com

Building Code Compliance for Contractors & Inspectors

Have you ever failed a construction inspection? Have you ever dealt with an inspector who has his own interpretation of the Code and forces you to comply with it? This new book explains what it takes to pass inspections under the 2009 *International Residential Code*. It includes a Code checklist — with explanations and the Code section number — for every trade, covering some of the most common reasons why inspectors reject residential work. The author uses his 30 years' experience as a building code official to provide you with little-known information on what code officials look for during inspections. **232 pages, 8½ x 11, $32.50**
***eBook (PDF) also available;* $16.25** at https://craftsman-book.com

Craftsman eLibrary

Craftsman's eLibrary license gives you immediate access to 60+ PDF eBooks in our bookstore for 12 full months!
You pay only one low price. $149.99.
Visit https://craftsman-book.com **for more details.**

National Construction Estimator

Current building costs for residential, commercial, and industrial construction. Estimated prices for every common building material. Provides man-hours, recommended crew, and gives the labor cost for installation.
672 pages, 8½ x 11, $117.50. Revised annually
***eBook (PDF) also available;* $58.75** at https://craftsman-book.com

Builder's Guide to Accounting Revised

Step-by-step, easy-to-follow guidelines for setting up and maintaining records for your building business. This practical guide to all accounting methods shows how to meet state and federal accounting requirements, explains the new depreciation rules, and describes how the Tax Reform Act can affect the way you keep records. Full of charts, diagrams, simple directions and examples to help you keep track of where your money is going. Recommended reading for many state contractor's exams. Each chapter ends with a set of test questions, and a CD-ROM included FREE has all the questions in interactive self-test software. Use the Study Mode to make studying for the exam much easier, and Exam Mode to practice your skills. **360 pages, 8½ x 11, $61.50**
***Also available as an eBook (PDF),* $30.75 at** https://craftsman-book.com

Construction Forms for Contractors

This practical guide contains 78 practical forms, letters and checklists, guaranteed to help you streamline your office, organize your jobsites, gather and organize records and documents, keep a handle on your subs, reduce estimating errors, administer change orders and lien issues, monitor crew productivity, track your equipment use, and more. Includes accounting forms, change order forms, forms for customers, estimating forms, field work forms, HR forms, lien forms, office forms, bids and proposals, subcontracts, and more. All are also on the CD-ROM included, in Excel spreadsheets, as formatted Rich Text that you can fill out on your computer, and as PDFs. **360 pages, 8½ x 11, $48.50**
***eBook (PDF) also available;* $24.25** at https://craftsman-book.com

Roofing Construction & Estimating, Revised

Detailed, step-by-step instructions, with photographs and diagrams, for installing, repairing and estimating nearly every type of roof covering available today for residential and commercial structures: asphalt shingles, roll roofing, wood shingles and shakes, clay tile, slate, metal, built-up, elastomeric, TPO and more. Provides guidance on sheathing, synthetic and felt underlayment, as well as tips and tricks from an experienced pro for dealing with those difficult points on a roof that are prone to leaks, such as valleys and roof penetrations. For each roofing type, instructions are provided for estimating material quantities and labor costs, with formulas, easy-to-follow examples and sample estimates for you to test your skill. Use these methods to create reliable estimates that will help insure a profit on every job you take. **448 pages, 8½ x 11, $62.50**
***eBook (PDF) also available,* $31.25 at** https://craftsman-book.com